12th International Conference on Gas Discharge Plasmas and Their Applications (GDP 2015)

Journal of Physics: Conference Series Volume 652

Tomsk, Russia
6 – 11 September 2015

Editors:

Nikolai Koval
Nikolay Landl
Anna Bogdan
Artyom Yudin

ISBN: 978-1-5108-1726-5
ISSN: 1742-6588

Printed from e-media with permission by:

Curran Associates, Inc.
57 Morehouse Lane
Red Hook, NY 12571

Some format issues inherent in the e-media version may also appear in this print version.

Printed by Curran Associates, Inc. (2016)

For permission requests, please contact the Institute of Physics
at the address below.

Institute of Physics
Dirac House, Temple Back
Bristol BS1 6BE UK

Phone: 44 1 17 929 7481
Fax: 44 1 17 920 0979

techtracking@iop.org

Additional copies of this publication are available from:

Curran Associates, Inc.
57 Morehouse Lane
Red Hook, NY 12571 USA
Phone: 845-758-0400
Fax: 845-758-2634
Email: curran@proceedings.com
Web: www.proceedings.com

Table of contents

Volume 652

12th International Conference on Gas Discharge Plasmas and Their Applications
6–11 September 2015, Tomsk, Russia

Accepted papers received: 14 October 2015
Published online: 5 November 2015

Preface

Papers

Journal of Physics: Conference Series **652** (2015) 011001 doi:10.1088/1742-6596/652/1/011001

The **12th International Conference "Gas Discharge Plasmas and Their Applications" (GDP 2015)** was held in Tomsk, Russia, on September 6–11, 2015.

GDP 2015 represents a continuation of the conferences on physics of gas discharge held in Russia since 1984 and seminars and conferences on the technological applications of low temperature plasmas traditionally organized in Tomsk.

The six-day Conference brought together the specialists from different countries and organizations and provided an excellent opportunity to exchange knowledge, make oral contributions and poster presentations, and initiate discussions on the topics that are of interest to the Conference participants.

The selected papers of the Conference cover a wide range of technical areas and modern aspects of the physical processes in the generators of low-temperature plasma, the low and high-pressure discharges, the pulsed plasma sources, the surface modification, and other gas-discharge technologies.

The Conference was hosted by Institute of High Current Electronics SB RAS, Tomsk Polytechnic University, Tomsk Scientific Center, and Tomsk State University of Architecture and Building.

Organizing Committee of GDP 2015

Edited by:

Prof. Nikolai Koval
Institute of High Current Electronics SB RAS

Dr. Nikolay Landl,
Institute of High Current Electronics SB RAS

Anna Bogdan,
Tomsk Polytechnic University

Artyom Yudin,
Tomsk Polytechnic University

Journal of Physics: Conference Series **652** (2015) 011002 doi:10.1088/1742-6596/652/1/011002

Peer review statement

All papers published in this volume of *Journal of Physics: Conference Series* have been peer reviewed through processes administered by the proceedings Editors. Reviews were conducted by expert referees to the professional and scientific standards expected of a proceedings journal published by IOP Publishing.

Morphology of ceramic particles produced by plasma-chemical synthesis

V A Arkhipov[1], **S S Bondarchuk**[2], **V D Gol'din**[1], **I K Zharova**[1], **A S Zhukov**[1]

[1] Research Institute of Applied Mathematics and Mechanics of Tomsk State University, Lenin avenue, 36, Tomsk, 634050, Russia
[2] Tomsk State Pedagogical University, Kievskaya street, 60, Tomsk, 634041, Russia

Zharova@niipmm.tsu.ru

Abstract. Thermal and diffusion processes in the droplet weakly concentrated metal salt solution during its heating in the plasma chemical reactor, in order to synthesize metal oxide powders are considered. Numerical study is based on previously proposed physical-mathematical model. The results of numerical calculations are analyzed to assess the possible influence of the operation parameters of the reactor and the characteristics of the initial solution (precursor) on the morphology of the particles formed.

1. Introduction

The use of fine powders of metal oxides in the preparation of ceramic composite materials enables to synthesize products for various purposes with unique mechanical properties, create metal-substituting construction materials and protective coatings, as well as filters, membranes, and thermoinsulating and catalyst structures with a specified pore size. One of the most promising technologies for producing ceramic powders with controlled parameters is plasma-chemical synthesis, including the thermochemical decomposition of liquid-drop reagents (precursors) in a flow of high-temperature gaseous heat carrier [1, 2]. This technology is known as "spray pyrolysis" [3]. Depending on the chemical composition of the precursor, a wide range of metal oxide powders can be obtained: Al_2O_3 from an aluminum nitrate $Al(NO_3)_3$ solution, zirconium oxide ZrO_2 from a zirconium nitrate $Zr(NO_3)_3$ (or zirconium chloride $ZrOCl_2$) solution, as well as mixed oxides, such as $ZrO_2 + MgO$, $ZrO_2 + Y_2O_3$, $ZrO_2 + CaO$, $ZrO_2 + Y_2O_3 + Al_2O_3$, etc. [1–3].

Electron microscopic studies of powder particles prepared by the plasma□chemical method revealed that their morphology may vary, featuring solid spherical single-crystal formations, hollow spheres and their fragments, films, and porous or foamed particles [2, 3]. The internal structure of ceramic particles is the most important parameter that determines the quality of the product. While the production of construction ceramics requires fine spherical solid particles, the manufacture of heat-shielding materials and catalysts preferably uses porous and hollow particles.

Clearly, the particle morphology is defined by both the precursor characteristics (dispersion of drops, solution concentration) and thermochemical process parameters (speed of drops, time of their residence in the reactor, temperature regime, aerodynamic pattern of the two□phase flow in the reactor, thermophysical properties of the reagents, etc.). To ensure a controllable mode of operation of the plasma-chemical reactor in order to produce particles with desired characteristics, it is necessary to develop physical-mathematical models of the processes occurring in it. The development of such a mathematical apparatus involves multivariate analysis of the problem of calculating the parameters of a multicomponent two-phase chemically reacting medium with a detailed description of heat and mass transfer in the individual spray drops and between the drops and the gaseous heat carrier. However, the available studies are based on physical-mathematical models involving a number of serious simplifications in describing the processes within the solution drops [1–4], an approach that significantly reduces the possibility of predicting the characteristics of the synthesized powders of metal oxides, in particular, their morphology(internal structure).

The physical-mathematical model of heat and diffusion processes in a drop of slightly concentrated solution of a metal salt during is heating in a plasma-chemical reactor is presented in [5]. This model is designed to predict the morphology of the particles of ceramic powders depending on the characteristics of the solution and gaseous heat carrier.

2. Problem statement

The plasma-chemical synthesis of ceramic powders is a multistage process that includes the atomization of liquid precursor to form polydisperse drops in the reactor, the movement of the drops in the gaseous heat carrier flow, heating and evaporation of the drops (accompanied by the diffusion and crystallization of the metal salt in the drop bulk), the thermolysis of the salt, leading to the formation of a microporous metal oxide structure at high temperature, and finally, its sintering to form solid particles with a specific morphology (Figure 1).

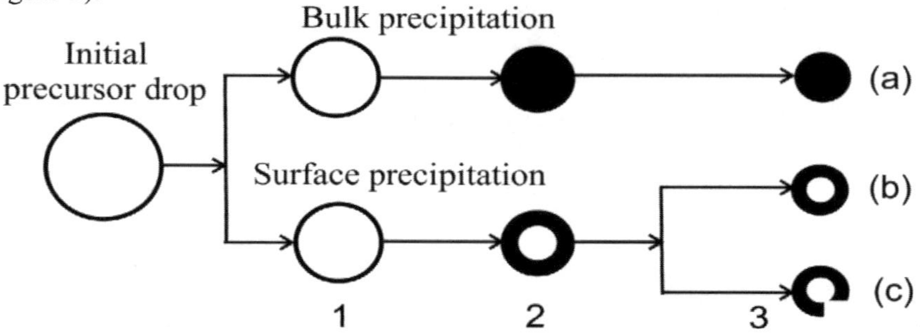

Figure 1. Effect of the conditions of salt precipitation on the morphology of the particle formed: (1) heating and evaporation, (2) precipitation and dehydration of the salt, and (3) thermolysis and sintering.

Figure 1 shows that plasma-chemical synthesis produces particles of different structures, depending on the conditions of salt deposition and the characteristics of the precursor salt. Solid particulates are produced by bulk deposition (Figure 1a). During surface deposition, either hollow particles (Figure 1b), if the gas permeability of the deposited salt layer is high enough, or fragments of hollow spheres (Figure 1c) are formed. In the latter case, because of a low permeability of the salt layer, the pressure of solvent vapor generated inside the particle destroys it, whereby producing polydisperse shell fragments.

When an aqueous solution of aluminum nitrate nonahydrate $Al(NO_3)_3 \cdot 9H_2O$ is used, the thermolysis of the salts crystallized in the drop bulk begins above 135°C. In this, the $Al(OH)_2NO_3 \cdot 1.5H_2O$ basic salt is first formed, while at higher temperatures (above 200°C), it decomposes to yield an amorphous alumina:

$$4Al(NO_3)_3 = 2Al_2O_3 + 12NO_2 \uparrow + 3O_2 \uparrow .$$

Analysis of the individual steps of the plasma-chemical synthesis revealed that the morphology of the powder particles arises at the stage of evaporation of the drop. Consider the main processes occurring simultaneously in the solution drop at the stage of evaporation, such as the temperature evolution of the drop during its heating, solvent evaporation from the drop surface, salt diffusion towards the drop center, salt crystallization, and removal of the remaining solvent (drop dehydration).

When a solution drop moves in the gaseous neat carrier flow, it is heated so that water evaporates from its surface. As a result, the dissolved salt concentration increases near the drop surface, thereby causing the salt to diffuse to the center of the drop. When the concentration of the salt reaches a certain value, it begins to crystallize to form a condensed precipitate.

The formation of the precipitate, a process determining the morphology of the final product particle, proceeds in two stages. During the first stage, the primary structure of the particle is formed. Let us assume that, when the concentration of salt in solution at the drop surface reaches a certain critical supersaturation concentration C_{cr}, the salt crystallizes instantaneously in the part of the drop where $C \geq C_*$ (C is the salt concentration, and C_* is the equilibrium concentration at a given temperature). In this case, depending on the rate of solvent evaporation and salt diffusion rate, there are two possibilities: bulk and surface precipitation.

Bulk precipitation, producing a solid particle, occurs if by the time t_{cr} of reaching C_{cr} at the surface of the drop, the inequality $C > C_*$ holds throughout its volume. Surface precipitation, producing a hollow spherical particle occurs if $C > C_*$ only for the peripheral part of the drop, within the layer limited by the critical radius $R_{cr}(t_{cr})$ and the actual radius $R(t)$ of the drop. The thickness of the wall of the hollow particle is $\delta_1 = R(t) - R_{cr}(t_{cr})$.

During the second stage, the solvent vapor is removed from the resulting porous structure. Assume that the primary precipitate has a finely porous structure permeable to water vapor.

In the case of bulk precipitation, removing the water vapor does not alter the primary structure of the particle: it is only densified due to the salt remaining in the solution. In the case of surface precipitation, we assume that the salt remaining in the hollow particle, with a mass of

$$m = 4\pi \int_0^{R_{cr}} C(r) r^2 dr$$

is uniformly deposited on the inner surface of radius $R_{cr}(t_{cr})$, forming an additional precipitate layer of thickness $\delta_2 = R_{cr}(t_{cr}) - R_1$, where R_1 is the radius of the spherical cavity inside the particle formed. The radius R_1 is given by

$$R_1 = \left[R_{cr}^3(t_{cr}) - 3m/(4\pi\rho_s)\right]^{1/3},$$

where ρ_s is the density of the salt precipitate.

3. Evaluation of the characteristic times

To construct a mathematical model of the processes at the stage of evaporation, we estimated their characteristic times.

•The characteristic time of the dynamic relaxation of the drop, indicating the lag between its velocity and the velocity of gaseous heat carrier is given by [6]

$$t_1 = 2\rho R_0^2 / 9\mu,$$

where ρ is the fluid density, μ is the dynamic viscosity of the gas, and R_0 is the initial radius of the drop.

•The characteristic time of temperature equilibration throughout in the drop is [7]

$$t_2 = R_0^2 / \kappa,$$

where κ is thermal diffusivity of the solution.

•The characteristic time of diffusion of salt in the drop bulk reads as

$$t_3 = R_0^2 / D,$$

where D is the diffusion coefficient.

•The characteristic time of evaporation of liquid from the drop surface according to the stagnant film model for evaporation can be written as [8]

$$t_4 = \frac{2\rho R_0^2 c_V}{\lambda_g \mathrm{Nu} \ln(1+B)},$$

where $B = c_V(T_e - T_V)/L$ is the Spaulding number, Nu is the Nusselt number, λ_g is the thermal conductivity of the gas at the boiling point, c_V is the specific heat at constant volume of solvent vapor, L is the latent heat of vaporization, is the temperature of the gaseous heat carrier, and is the temperature of equilibrium vaporization (boiling).

Note that the ratio between the characteristic times is independent of the initial radius R_0 of the drop and that, with increasing Nu and temperature difference $T_e - T_V$, the evaporation time decreases. Therefore, we can assume that, under all conditions, the following relationships hold:

$$t_1 \sim t_2 << t_3 \sim t_4.$$

Let us consider, as an example, a drop of a low concentration aqueous metal salt solution with an initial radius of $R_0 = 25$ μm. At atmospheric pressure and boiling temperature, the following values of the parameters must be used: $D = 10^{-9}$ m²/s, $\rho = 10^3$ kg/m³, $\kappa = 1.58 \cdot 10^{-7}$ m²/s, $c_V = 2.0 \cdot 10^3$ J/(kg K), $L = 2.26 \cdot 10^6$ J/kg, Nu = 2, $T_e - T_V = 50$ K, $T_V = 373$K, $\lambda_g = 3.3 \cdot 10^{-2}$ W/(m K), and $\mu_g = 2.2 \cdot 10^{-5}$ Pa s.

In this case, the characteristic times of the problem were found to be

$$t_1 = 6.3 \cdot 10^{-3}\ \text{s},\quad t_2 = 4.0 \cdot 10^{-3}\ \text{s},\quad t_3 = 0.63\ \text{s},\quad \text{and}\quad t_4 = 0.88\ \text{s}.$$

As shown in [5], the results depend on the parameter K:

Journal of Physics: Conference Series **652** (2015) 012001 doi:10.1088/1742-6596/652/1/012001

$$K = \frac{t_3}{t_4} = \frac{\lambda_g \mathrm{Nu} \ln(1+B)}{2\rho D c_V}.$$

4. Analysis of the results

When considering the heating and evaporation of a drop and the diffusion of salt in the drop prior to its crystallization, we assumed that:

•the drop has a spherical shape during the entire process;

•the thermophysical and transport properties of the solution are invariable;

•the density and specific heat of the solution and solvent are the same;

•thermal diffusion is ignored.

Assuming that evaporation of the drop begins only after reaching the boiling point of the solution, we can specify two stages of the process.

(1) The stage of initial heating, when the drop is heated but no evaporation occurs. The concentration of salt is uniform and the diffusion does not occur.

(2) When the surface temperature reaches the equilibrium vaporization temperature (boiling point), solvent evaporation starts, being accompanied by the decrease of the drop size, increase of the salt concentration near the surface, and occurrence of the diffusion process. The surface temperature of the drop remains constant and equal to the equilibrium temperature of evaporation.

Note that the proposed model adequately describes the thermal and diffusion processes for dilute solutions (initial concentration of salt is less than 5 wt %). Under these conditions, the assumptions underlying the formulation of the problem are correct. In many practical applications, the initial concentration of the solution ranges within 5–10 wt % [1–3].

The dimensionless mathematical formulation and numerical method of its solving are presented in [5]. Results of calculations show that the radial distributions of the salt concentration in a drop vary depending on the parameter K. At small K values, when the critical concentration C_{cr} at the drop surface is reached, the salt concentration inside the drop is greater than the equilibrium concentration C_* so that solid particles are formed. At $K \sim 1$, there is a region inside the drop where the concentration is below the equilibrium concentration C_*, and therefore, the process of crystallization produces a hollow particle.

Figure 2 displays the crystallization layer thickness normalized by the drop radius at the time of reaching the critical concentration C_{cr} at the drop surface as a function of K. The calculation results showed that, for the assumed initial concentration C_0 of the solution, the boundary value that separates the modes of formation of hollow and solid particles is $K_* = 0.75$. At $K < K_*$, vaporization is slower than diffusion, so solid particles are formed. At $K > K_*$, vaporization process is faster than diffusion, and therefore, hollow particles are produced. At fixed values of the thermal properties of the salt and solvent, the value of K is mainly determined by the temperature difference between the drop and carrying medium. At a small temperature difference, solid particles are formed, whereas at a large difference, hollow particles. At a large temperature difference, evaporation is faster than diffusion (K increases), and crystallization has time to occur only in the surface layer, and vice versa, at a low temperature difference, the salt concentration reaches an equilibrium value throughout the drop. Note that the ratio of the thicknesses of the primary deposited layer δ_1 and the layer δ_2 formed from the residual salt is determined by the value of the parameter K.

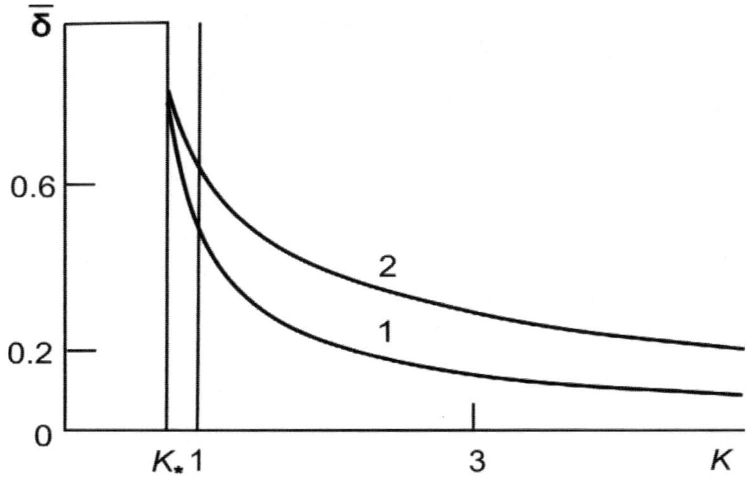

Figure 2. Dependence of the crystallization layer thickness on the parameter K: (1) primary crystallization layer and (2) overall layer after deposition of the residual salt inside the drop.

5. Conclusions

(1) Analysis of the results of numerical calculations within the framework of this model showed that the only dimensionless parameter that determines the structure (morphology) of the particle formed is the ratio of the characteristic times of diffusion and evaporation: $K = t_3/t_4$. At $K < K_*$, diffusion is faster than evaporation, and therefore, solid particles are formed; at $K > K_*$, hollow spherical particles are formed.

(2) In the case of the formation of hollow particles, the process of salt crystallization occurs in two stages: the primary precipitation to form a layer of thickness δ_1 followed by the deposition of the remaining salt onto the inner surface of the particles to form a layer of thickness δ_2. The ratio between δ_1 and δ_2 is determined by the value of K.

(3) In producing ceramic powders with a given morphology, the main parameters of the controlled plasma-chemical synthesis are the initial concentration of the precursor salt (salt solubility), temperature difference between the gaseous heat carrier and the precursor, as well as the difference between the critical supersaturation concentration C_{cr} and equilibrium saturation concentration C_*.

Acknowledgments

This work was supported by the Ministry of Education and Science of the Russian Federation within the framework of state job number 2014/223 (project No 1567) and of state job No 9.1024.2014/K.
The work was also supported by Ministry of Education and Science under the Federal Program «Research and development on priority directions of scientific-technological complex of Russia for 2014-2020», the agreement № 14.578.21.0034, unique identifier PNI RFMEFI57814X0034.

References

[1] Suris A L 1989 Plasma-Chemical Processes and Apparatus (Khimiya, Moscow) [in Russian]
[2] Larin V K, Kondakov V M, Malyi E N, et al. 2003 *Izv. Vyssh. Uchebn. Zaved. Tsvetn. Metallurg.*, **5** 59
[3] Messing G L, Shi-Chang Zhang, and Jayanthi G V 1993 *J. Am. Ceram. Soc.* **76** 2707
[4] Zhukov A S and Bondarchuk S S, 2010 *Izv. Vyssh. Uchebn. Zaved., Fiz.* **53**(12/2) 96
[5] Zhukov A S, Arkhipov V A, Bondarchuk S S, and Gol'din V D 2013 Evaluation of the Morphology of Particles Produced by Plasma-Chemical Synthesis of Ceramic Powders, *Russian Journal of Physical Chemistry B*, **7** 6 777-782
[6] Vasenin I M, Arkhipov V A, Butov V G, et al. 1986 Gas Dynamics of Two-Phase Nozzle Flows (Tomsk. Gos. Univ., Tomsk) [in Russian]
[7] Lykov A V 1967 The Theory of Heat Transfer (Vyssh. Shkola, Moscow) [in Russian].

[8] Pomerantsev V V, Aref'ev K M, Akhmedov D B, et al. 1986 Principles of the Practical Theory of Combustion (Energiya, Moscow) [in Russian]

Boriding of carbon steels by the electron beam treatment in vacuum

D E Dasheev, N N Smirnyagina, V M Khaltanova and A P Semenov
Institute of Physical Materials Science SB RAS, 670047, Ulan-Ude, Russia

E-mail: fokter@mail.ru

Abstract: Conditions of formation, structure and properties of boride iron layers on carbonaceous steels 3 and 45 at electron beam borating are investigated. New process to make layers of iron borides (Fe_2B, FeB) using electron beam is reported. The microstructure and microhardness of boride layers are investigated and are compared to layer properties obtained at solid phase borating. Formed layers were heterogeneous structure combining solid and weak components and resulting to fragility reduction of boride layer.

Keywords: electron beam, borides, micro hardness, structure, phases composition

1. Introduction

Surface modification of metals and alloys can increase the service life and reliability of the various machine parts and tools. Recently, along with the traditional in modern engineering methods for producing coatings that increase the hardness and wear parts, such as carburizing, nitriding, carbonitriding, borating, chromium etc., are increasingly using beam technologies, such as the processing of the laser beam and electron beam. The use of electronic heating with high ($>10^9$ W/cm^2) power density in a vacuum has the advantage due to the rapid inertia less achieve extremely high temperatures and ease of heating controls in a wide temperature range.

In this paper, the conditions for the formation, structure and properties of the layers based on iron borides on carbon steels 3 and 45 under the influence of the electron beam in a vacuum were investigated.

2. Experiment methods

The solid phase borating was carried out in a powder mixture in a container with a fusible mechanism on a following mode: powder composition – 97 wt. % B_4C and 3 wt. % KBF_4, temperature – 950°C, duration – 4 hours [1].

As a result on a surface it was formed a boride layer by thickness 70-100 μ. The subsequent electron beam processing carried out at following modes: an electron beam current 10 mA, an accelerating voltage 22 kV, time of treatment 10, 15 and 50 s.

Electron beam boriding. Samples were prepared by daub drawing on a previously prepared surface of steel. The daub composition was consisted of 1:1 the boriding compounds (boron carbide B_4C or amorphous boron) and the organic binding. The solution 1:10 glue BF-6 in acetone has been used in quality the binding. After drawing daubs by thickness 1 mm samples dried before complete removal acetone [2].

Self-propagating high-temperature synthesis (SHS) of boride layers. Layers Fe_2B and FeB were synthesized from reaction daub containing mixtures of boron carbide B_4C, oxide Fe_2O_3, carbon C (birch charcoal) and the organic binding. The electron beam treatment has been carried out in an electro-vacuum installation with a powerful industrial axial electron gun [3]. The pressure in chamber did not exceed 2×10^{-3} Pa.

The boride layers were analyzed by X-ray diffraction. An X-ray powder diffractometer Advance D8 Bruker using Cu Kα- radiation was employed for phase analysis and the determination of lattice parameters. Microhardness of prepared layers was measured by using PMT-3 hardness tester at a loading 0.5 and 1 H. Microstructure of the layers formed was determined by metallographic microscope "Neophot-21" or "Metam-21RB".

3. Results and Discussion

Figure 1 shows the microstructure of boride layers. The structures of surface layers after solid phase borating and electron beam boriding are distinct. The layer after solid phase borating showed a needle-like structure and the transition zone settles down under it's (Figure 1.a). Microhardness of boride needles come up to 1300-1350 MPa, cirrus selections is 300-330 MPa. Thickness – 70-90 μ.

Boride layers after electron beam boriding. It is established, that at electron beam boriding on a metal surface brightly expressed layers of the thickness about 350-360 μ (the daub from amorphous boron) and depths up 100-110 μ (the daub from B_4C) will be formed. In both cases, the fixed boundary between the layer and metal basic is founded. In comparison with the metal basic the layers have lower speed of the etching that testifies to them of considerably high corrosion stability.

The transition zone after electron beam boriding was not observed and the legible boundary between a layer and base metal was observable (Figure 1. b, c). The layer consists of rounded crystals, which are settling down on a surface and a eutectic.

The X-ray diffraction analysis is established that layers contain the iron borides Fe_2B and FeB. The relative maintenance of these borides has depended from daub composition. In case of an amorphous boron forest it FeB, and B_4C – Fe_2B. Besides, on X-ray diffraction patterns there are the lines of different intensity belonging to the cementite Fe_3C and ferrite α-Fe.

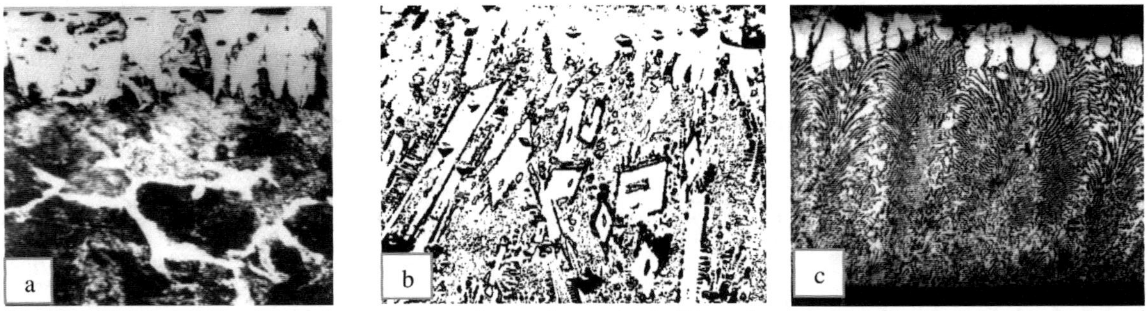

Figure 1. Layers boride microstructure formed on steel 45 surface: solid phase borating (a) and electron-beam boriding - daub from amorphous boron (b) and B_4C (c); a - ×250, b, c- ×500

The boride layer formed from daub B_4C (figure 1.c) consists from round off engagements, which locating on the layer surfaces and eutectic. The microhardness values 820-840 and 510-530 MPa for the layer surfaces and eutectic were obtained. The rounds off engagements are primary crystals of borides that answers entropic criterion of stability of the crystals limited form at the crystallization in conditions close to equilibrium. According to this criterion, if the value of the entropy of fusion (ΔS) is less than 2 kcal/mol×K to the crystals are rounded [4]. Obtained in work [5], the values of the entropy of melting for iron boride Fe_2B is ΔS=1.5 kcal/mol×K. In turn, the rounded forms borides determine the shape of the eutectic crystals.

Journal of Physics: Conference Series **652** (2015) 012002 doi:10.1088/1742-6596/652/1/012002

The boride layer formed from daub with amorphous boron has other structure (figure1.b). It consists of particles of the various forms: rhombic, prismatic, dendritically. On layer surface the continuous light film with needles, directed deep into of a sample is placed. Microhardness of film makes up 1200-1250 MPa. Inside this film the rare (1-2) large inclusions with microhardness 1750-1820 MPa is meet. Under the film there are the primary crystals and eutectic with microhardness 840-880 MPa and 500-540 MPa, respectively.

According to [6], the boride iron Fe_2B has tetragonal crystal lattice (Space. group I4/mcm with parameters of an elementary cell a=0.51087, c=0.42497 nm). At layer formation from daub with amorphous boron, the iron boride crystals have been inherited the form of an elementary cell. Therefore, the primary crystals of borides have the form of rhombuses, parallelograms etc., caused by different corner inclination of the crystal lattice (prism) to plane of polished specimen. It is necessary to note, that the similar forms of borides crystals are observed and at laser boriding [7].

Method of electron beam boriding [2], its mechanism is likely to be a modification of the method of solid-phase saturating boriding daub for thermochemical treatment. Its novelty is the impact of the electron beam on the boron containing daub in vacuum. The use of a highly concentrated source of energy allows you to quickly transfer the energy of the electron beam in its collision with the surface of the treated metal or alloy, heat the contact area to a very high temperature. This increases the diffusion of boron from saturating daub on the surface and its penetration into the bulk metal, the interaction and the formation of iron borides (figure 2). Boride layer thickness depends on the composition of boriding component. 300-360 μ thick layers obtained with boron, and thickness of 100-150 μ – boron carbide. Application of an electronic beam promotes increase in diffusion of boron in volume of metal, interaction and formation iron borides.

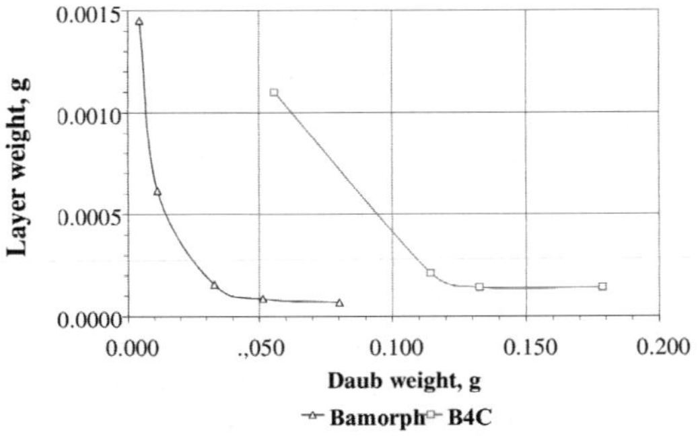

Figure 2. Dependence of mass boride layer on the number of layers boriding component (steel St3, W = 270 W, exposure time 5 min)

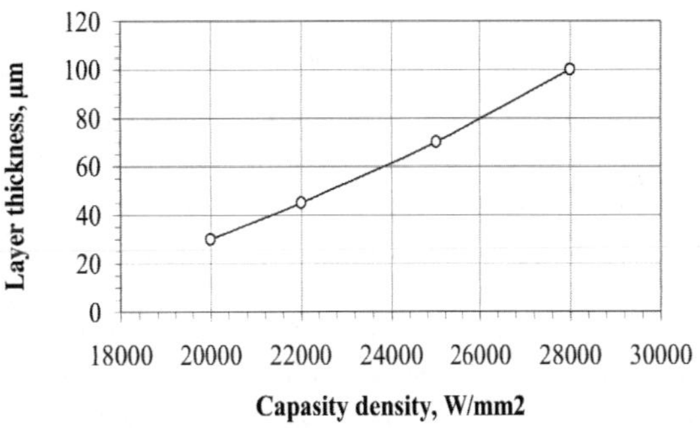

Figure 3. The influence of electron beam power to the thickness of boride layer (steel St3, coating based B amorphous, exposure time 1 min)

Figure 2 shows the dependence of the mass-forming layer of the total weight of the saturating coating. The total number of adhesive layers is ranged from 1 to 5 (amorphous boron) and 1 – 4 (B_4C). As follows from figure 2, for the formation of boride layer with the highest weight (thickness) is enough to put one boriding adhesive component, as in the case of boron, and the application of boron carbide. Initial samples steel 45 were pre-normalized at 930°C for 7 min, cooled in air.

Boride layer thickness depends on the power of the electron beam (figure 3). This relationship has the big practical importance, as allows control technological indicators (specific capacity and an electron beam current and accelerating voltage) depending on size of a preferred thickness of a boride layer.

Boride layers structure after the subsequent electron beam treatment. The boride layers microstructure is resulted on figure 1. Besides, in this figure the boride layers structure after the further electron beam processing is presented. It is known [1] that boride layers on low carbon steels have a needle structure at which borides, growing together in the bases, form a continuous layer. Directly borides needles adjoin allocation plumose carbo-boride phases. Microhardness of boride needles makes 1300-1350 MPa, plumose allocation 300-330 MPa. A layer thickness was 70-90 μ.

The mechanism of proceeding structural changes is studied by a method selective on depth Mossbauer spectroscopy [8].

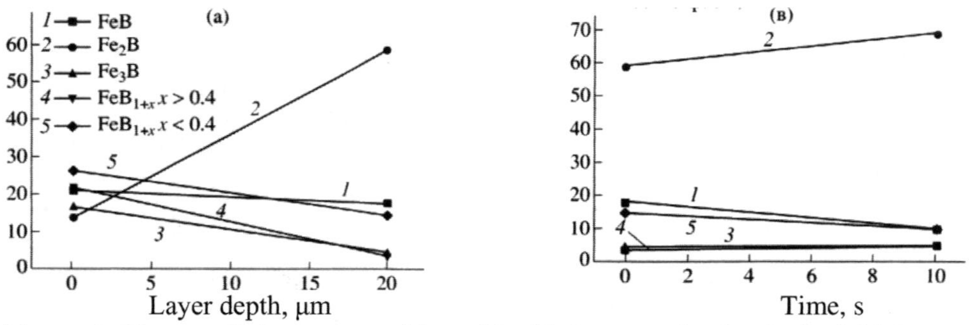

Figure 4. Change of phase composition of boride layer on depth: a)- the initial sample; b)- after electron beam processing (20 μ)

Microhardness measurement have shown, that on a surface of the sample it makes 2000 - 2050 MPa, and on depth 20-30 μ make 1600-1650 MPa, that will be corresponding with literary data of microhardness of iron borides FeB and Fe_2B.

For presentation the phase analysis of boride sample, received from Mossbauer spectra, is resulted on figure 4. Apparently from this figure, the surface layer of the sample on 70 % consists from enriched by boron of borides phases, while on depth 20 μ dominating (60 %) a crystal phase are Fe_2B. The quantity of phases rich with boron essentially decreases with depth. It is established, that after 19 seconds of electron beam processing the sharpest changes of phase composition occur in thin surface layer: very sharply increases (by 50 %) relative quantity of phase Fe_2B. With depth increase of surface layer to 20 μ of change in phase composition after an irradiation keep the same tendency, but changes become weaker: the maintenance of phase Fe_2B increases by 10 % and maintenance FeB decreases for 8 %. At comparison with metallographic analysis of these samples, it is visible, that this process speaks increase in density of needle layer Fe_2B and, thereby, reduction of the zone filled with phases rich with boron.

Because of a warming up surface boride layers under the influence of an electron beam (~ 1323-1373 K) dot defects begin diffused to borders of grains, forming a new defective zone that leads to essential increase in speed of diffusion of boron atoms in depth of the sample on cracks and between grains boundary owing to a gradient of temperatures. Besides, in parallel to it there is a sputtering of boron atoms from a layer under the influence of an electron beam [9].

Sputtering process, in our opinion, also leads to sharp reduction of quantity of the most sated disorder phase FeB_{1+x}located on a boron surface pine with concentration of boron 0.4. Thus in adjoining surface layer occur segregation and hashing of atoms. Because of all these processes, there is an impoverishment of boron phases in surface areas. Occurring destruction of disorder phases and boron diffusion in depth of the sample leads recrystallization of phases in favor of steadier at existing in surface layer to temperature ~ 1373 K phase Fe_2B. And speed of these transformations in thin surface layer depth ~ 0.3 μ above, than in thicker layer depth to 20 μ that is caused by a temperatures gradient and the limited layer process of sputtering of boron.

Electron beam fusing of SHS products. In this paper, we attempt to form the layers based on borides Fe_2B and FeB in the synthesis of their stoichiometric mixtures with Fe_2O_3, B and C on the surface of steel 45. To do this, take a mixture of the initial components in a ratio of Fe_2O_3: 3B: 3C

(Fe_2B) and Fe_2O_3: 2B: 3C (FeB), thoroughly ground in an agate mortar, mixed with an organic binder and the reaction surface of the sample steel 45. Electron beam processing was carried out in a vacuum at most 2×10^{-3} Pa at power electron beam W = 250-450 W for 1-3 min.

Figure 5. Microstructure of boride layer Fe_2B (a), FeB (b) on the surface of St45: ×400.

According to the X-ray phase analysis, boride layer is mainly composed of boride Fe_2B. Layer of the sample with a stoichiometric mixture of Fe_2O_3: 2B: 3C incorporates boride FeB.

The thickness of boride layers was 200-280 μ (Fe_2B) and 50-80 μ (FeB). Microstructure layer based Fe_2B presented at figure 5.a.

The structure of the complex includes primary boride crystals, including dendrites and eutectic. In figure 5.b shows the microstructure layer based boride FeB.

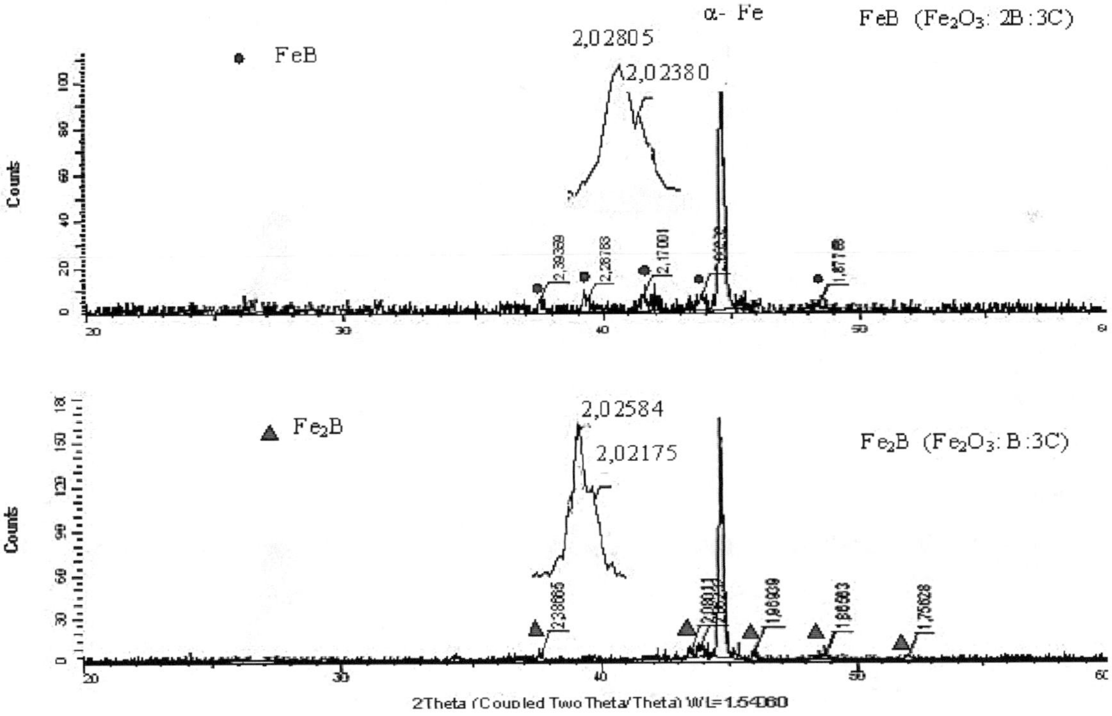

Figure 6. X-ray phase analysis of produced boride layers.

Figure 6 shows the X-ray phase analysis of produced boride layers. Dendrites are ferritic inclusions with parameters cells, a = 0.2821 nm). In the original steel 45 ferrite has a cubic volume-centered cell with a = 0.2869 nm. The use of high-resolution diffractometer D8 enabled to detect the x-ray lines reflexes plane (110) ferrite owned metal base. Also, found reflection from 20 % intensity, which, in our opinion, is the ferrite, which is formed during crystallization narrow melted near-surface zone.

Application of the protective layer of amorphous boron oxide contributes towards equilibrium boride layer (figure 7). Rounded and elongated inclusions were ordered arrangement in the layer in the layers of their microhardness were $Fe_2B + B_2O_3$ (1650 and 1850 MPa), $FeB + B_2O_3$ (1800 and 2350 MPa), respectively.

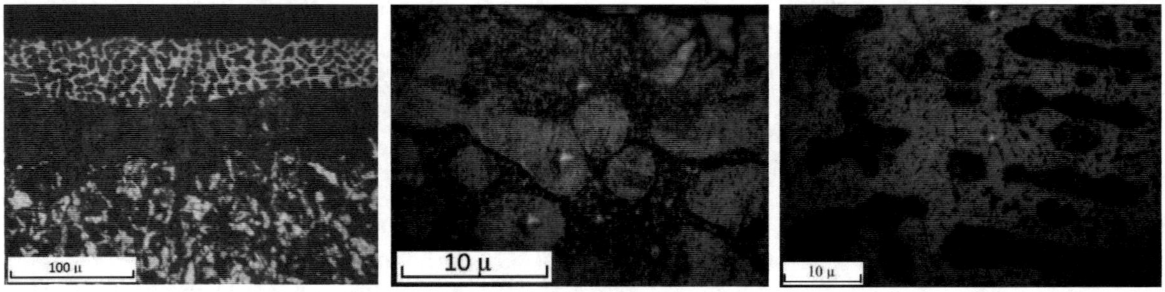

Figure 7. Layer structure $Fe_2B + B_2O_3$ at St45

For FeB: by reflex [110] cubic unit cell parameters: α-Fe (metal matrix) a=0.2868 nm, α-Fe (dendritic crystal), a=0.2821 nm. For Fe_2B: by reflex [110] cubic unit cell parameters: α-Fe (metal matrix) a=0.2865 nm, α-Fe (dendritic crystal), a=0.2859 nm.

4. Conclusions

Conditions of formation, structure and properties of boride iron layers on carbonaceous steels 3 and 45 at electron beam borating are investigated.

New process to make layers of iron borides (Fe_2B, FeB) using electron beam are reported. Carrying out boriding with participation of intensive electron beam has essentially changed layer structure, morphology of allocated phases, character and distribution formed in a pressure face zone of compounds. There was a basic possibility of a composite boride layer formation with heterogeneous disperse, instead of column (needle or tooth similar) morphology of boride crystals arrangement. Designing of the various composite layers combining another way "firm" and "soft" phases can give the chance to receive on a steels surface the boride layers possessing various plasticity.

Acknowledgments

This work was supported by the Russian Foundation for Basic Research, projects no. 14-14-08-31412_mol_a and 15-48-04217r_sibir_a.

References

[1] Directory ed. Liakhovich L S 1981 *Chemical and heat treatment of metals and alloys* (Moscow: Metallurgy)
[2] Semenov A P, Sizov I G, Smirnyagina N N etc. *Patent RU 2186872*, 2002. BI number 22.
[3] Semenov A P, Smirnyagina N N, Sizov I G 2001 *Metal technology* **4** 32-34
[4] Taran, V I 1978 *The structure of eutectic alloys* (Moscow: Metallurgy) p. 312
[5] Tavadze F N, Goribashvili V I, Nakaidze S H 1983 *Metallurgy and heat treatment of metals* **1** 2-3
[6] Kubaschewski O 1985 *Diagrams of binary systems based on iron* (Moscow: Metallurgy) p. 25-28
[7] Safonov A N 1998 *Metallurgy and heat treatment of metals* **1** 5-9

Journal of Physics: Conference Series **652** (2015) 012002 doi:10.1088/1742-6596/652/1/012002

[8] Novakova A A, Kuzmin R N 1989 *Mőssbauer conversion spectroscopy and their using* (Moscow: Lomonosov Moscow State University) p. 72
[9] Berish R 1986 *Sputtering of solid body by ion bombardment* (Moscow: Mir) p.448

Influence of plasma volume discharge in atmospheric-pressure air on the admittance of MIS structures based on MBE *p*-HgCdTe

A V Voitsekhovskii[1], S N Nesmelov[1], S M Dzyadukh[1], D V Grigoriev[1,2], V F Tarasenko[2], M A Shulepov[2]
1 National Research Tomsk State University, Lenin Avenue, 36, Tomsk, 634050, Russia, denn.grig@mail.tsu.ru,412772
2 Institute of High Current Electronics SB RAS, 2/3 Akademichesky Avenue, Tomsk, 634055, Russia

E-mail: vav43@mail.tsu.ru

Abstract. This article investigates the effect of a nanosecond plasma volume discharge, which is formed in an inhomogeneous electrical field at atmospheric pressure, on the electrical properties of MIS structures based on HgCdTe (MCT) epitaxial films. The MIS structure based on films exposed to the discharge significantly changed its main parameters. The most notable feature of the structure exposed to discharge is the significant increase in the positive fixed charge in the insulator. A possible reason for changes in the characteristics of MIS structure after exposure to discharge is the significant change in the impurity-defect system of the semiconductor near the interface. This is accompanied with a formation of an insulator film of nanometer thickness on the surface, which gives rise to positive fixed charge.

1. Introduction

One of the basic materials for the manufacture of IR photodetectors is the HgCdTe solid solution. This material has a low carrier concentration at the operating temperatures and a high quantum efficiency. The current task is to achieve controlled changing of material parameters for obtaining the desired semiconductor structures.

At the present moment, various types of discharge and electron beams are widely used for modifying the near-surface layers of materials [1]. Earlier studies investigated the influence of volume nanosecond pulse discharge in the air at atmospheric pressure in the epitaxial p-HgCdTe. The results showed that this effect leads to the formation of a surface layer with high electronic conductivity [2]. Further analysis [3] allowed us to suggest the formation of a layer of chemical MCT compounds with oxygen and/or nitrogen in the surface region. The formed layer has a built-in positive charge, which leads to the formation of an inversion layer of n-type conductivity at the interface of oxide / MCT. Additional studies of the surface properties of the irradiated material are required in order to confirm this hypothesis.

An MIS structure based on irradiated material was used to study the properties of the surface of semiconductor films. This paper presents the results of the studies of the effect of volume pulses nanosecond discharge in the air at atmospheric pressure on the electrical properties of MIS structures based on heteroepitaxial HgCdTe material.

2. Samples and experimental techniques

The samples of p-type conductivity ($p = 1.1 \div 2.5\times10^{16}$ cm^{-3}, $\mu_p = 300 \div 500$ cm^{2}V^{-1}s^{-1}) grown by molecular beam epitaxy in the Institute of Semiconductor Physics SB RAS (Novosibirsk) were prepared for experiment. The samples grown in this manner were put in a gas diode on a copper anode. The Radan-220 generator was used as a pulse-voltage source forming voltage pulses with the amplitude of ~230 kV (open-circuit voltage), the high-amplitude pulse duration of ~2 ns (at a matched load), and the rise time of ~ 0.5 ns. The samples were irradiated by 1,200 pulses in the periodic mode at the repetition rate of 1 Hz.

The MIS structures were formed on the basis of the initial and irradiated samples. The Al_2O_3 layer approximately 77 nm thick deposited using plasma atomic layer deposition was used as an insulator [4]. The metal electrodes are made of indium. Structures No.3 and No.4 were formed on the basis of the epitaxial film after the impact of the electromagnetic discharge. Structures No.1 and No.2 were used as control samples. The chemical mechanical surface treatment with HgCdTe was not conducted before the deposition of the insulator.

The measurements were performed using an automated setup of the nanoheterostructures admittance spectroscopy on the basis of a non-optical cryostat Janis and an immittance meter Agilent E4980A. When measuring the field dependences of the capacitance and conductance, the voltage change from negative values to positive ones was taken as the forward direction of sweep, and the voltage change from positive values to negative ones was taken as the reverse direction of sweep.

3. Experimental results and discussion

Figure 1 shows the normalized capacitance-voltage (CV) characteristics of structures No.1 and No.3. For structures No.1 and No.2, the capacitive characteristics at 77 K were similar. The CV characteristics of structure No.4 are qualitatively similar to the CV characteristics of structure No.3. The CV characteristics of structures No.3 and No.4 are shifted into the region of higher negative voltages. At the same frequency, the CV characteristics of structures No.3 and No.4 demonstrate a lower-frequency behavior than the CV characteristics of structures No.1 and No.2. For structures No.3 and No.4, the frequency dispersion of capacitance in the minimum of CV characteristics was observed, and the slope of the CV characteristics in the depletion at the forward voltage sweep is flatter than that at the reverse sweep. Also, for these structures, the capacitance modulation at the forward voltage sweep is higher (the value of the capacitance in the minimum of the low-frequency CV characteristic is lower), and there appears a much larger hysteresis of the CV characteristics.

Table 1 shows the main parameters of MIS structures No.1 - 4, which were obtained from the measurements of the CV characteristics at 77 K. The densities of fixed and mobility charges were determined from the shifts of the flat-band voltage in real and ideal structures. MIS structures based on HgCdTe exposed to the impact of discharge have a greater density of positive fixed charge. The mobile charge density in these samples is also higher.

An important characteristic of an MIS structure is the value of the differential resistance of the space-charge region (SCR) in the strong inversion mode (R_{SCR}) [5]. The research of the differential resistance of the SCR for MIS structures based on MBE *n (p)*-HgCdTe ($x = 0.21$-0.23) in a wide temperature range was carried out in other articles [6, 7]. The studies of the differential resistance at 200 kHz showed that, at 77 K, it was about 1,800 Ohm for structures No.1 and No.2. For structures No.3 and No.4, the values of the differential resistance of the SCR were so small that they did not exceed the error in determining R_{SCR} in a wide temperature range. The results of the measurements of the differential resistance of the SCR explain the low-frequency type of CV characteristics at sufficiently high frequencies for structures No.3 and No.4. The small values of the differential resistance of the SCR for structures No.3 and No.4 indicate a high rate of minority carriers (electrons) flow to the inversion layers. The most likely mechanism for the appearance of additional minority carriers in the inversion layer is the exchange of electrons with the inversion layer in the after-electrode region. This layer is induced by a significant positive fixed charge [5, 8].

One possible reason for the presence of a positive fixed charge can be the appearance of a thin insulator film on the surface of HgCdTe. This is due to the formation of oxides when exposed to the discharge. It is known that a positive fixed charge is typical for the anodic oxide film on HgCdTe [9, 10]. It is usually associated with the presence of oxygen vacancies in the anodic oxide [11]. In the system of MCT – anodic oxide at a distance of 10 nm from the interface – the slow surface states can exchange their charge with the semiconductor by means of tunneling. The fixed charge is located at the distance of more than 10 nm from the interface [10].

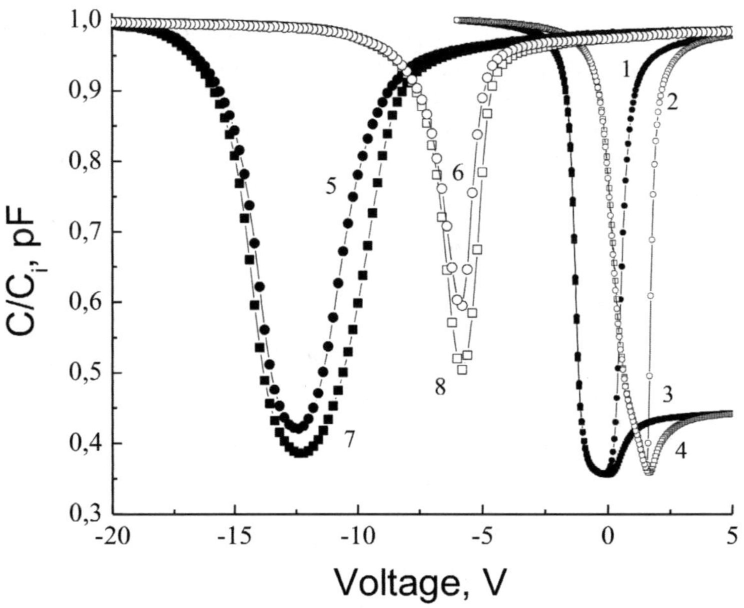

Figure 1. CV characteristics of MIS structures based on *p*-HgCdTe (control sample, structure No.1 (curves 1, 2, 3, 4) and No.3 (curves 5, 6, 7, 8)) measured at 77 K at the forward (curves 1, 3, 5, 7) and reverse (curves 2, 4, 6, 8) voltage sweeps at different frequencies. Curves 1, 2, 5, 6 - 10 kHz, the curves 3, 4, 7, 8 - 1 MHz

Table 1. Parameters of MIS-structures based on *p*-HgCdTe obtained from CV characteristics at 77 K

No.	Fixed charge density, cm^{-2}	Mobile charge density, cm^{-2}	The major carrier concentration obtained from minimum of the CV characteristic, cm^{-3}	The major carrier concentration obtained from $1/C^2(V)$ dependency in depletion mode, cm^{-3}
1	$1.15\cdot10^{11}$	$5.73\cdot10^{11}$	$4.05\cdot10^{15}$	$3.96\cdot10^{15}$
2	$9.49\cdot10^{10}$	$2.93\cdot10^{11}$	$4.60\cdot10^{15}$	$4.63\cdot10^{15}$
3	$3.28\cdot10^{12}$	$3.36\cdot10^{12}$	$6.30\cdot10^{15}$	$1.25\cdot10^{16}$
4	$6.77\cdot10^{12}$	$4.93\cdot10^{12}$	$9.22\cdot10^{15}$	$2.40\cdot10^{16}$

For the studied structures, the thickness of the insulator layer must be several nanometers. At a larger thickness, reduction in capacitance would be observed due to the insulator thickness increase. Therefore, an insulator layer with a small thickness has a higher density of the fixed positive charge than a typical anodic oxide. It is possible that, after the impact of the discharge in HgCdTe, a layer with a high density of build-in positive charges appears near the surface. These charges cannot change their state of charge due to the potential barriers.

4. Conclusions

The admittance of MIS structures based on MBE hetero-epitaxial p-$Hg_{0.78}Cd_{0.22}Te$ was investigated in a wide range of frequencies and temperatures. The investigated samples can be divided into two groups: MIS-structures that are based on the initial MCT films (control samples) and MIS-structures that are based on films exposed to the discharge. It is shown that the impact of the discharge substantially changes the characteristics of MIS structures.

The most notable feature of the structure exposed to discharge is a significant increase in the positive fixed charge in the insulator. This causes the appearance of an inversion layer in the after-electrode regions, which acts as the source of minority carriers and contributes to the formation of a sub-electrode inversion layer. Typical features of structures after exposure to discharge are extremely low values of the differential resistance of the SCR in the strong inversion mode and low-frequency behavior of the experimental CV characteristics at high frequencies.

The impact of the discharge increases the effective density of the mobile charge and the appearance of hysteresis of the electrical characteristics. A feature of the hysteresis is the difference of the capacitance in the minimum of the experimental low-frequency CV characteristics at the forward and reverse voltage sweeps and a change in the slope of the CV characteristics in the depletion mode. It is probable that the impact of the discharge increases the average density of the surface states in the depletion region.

MIS structure based on films exposed to the discharge significantly changed its main parameters. The heterogeneity of the changes in the properties over the structure area after the impact of the discharge points toward the heterogeneity of the discharge of its cross section. A possible reason for changes in the characteristics of MIS structures after exposure to discharge is the significant change in the impurity-defect system of the semiconductor near the interface. This is accompanied with a formation of an insulator film of nanometer thickness on the surface, which gives rise to positive fixed charge.

Acknowledgement

This work was supported by "The Tomsk State University Academic D.I. Mendeleev Fund Program" under grant (#8.2.10.2015).

References

[1] Shulepov M A et al. 2008 J. Technical Physics Letters **34** 296 - 299
[2] Voitsekhovskii A V et al. 2012 Russ. Phys. J. **10** 1152-1155.
[3] Voitsekhovskii A V et al. 2014 Izvestiya VUZov. Fizika **10/3** 126-130 (in Russ.)
[4] Fu R, Pattison J 2012 Optical Engineering **51** 104003
[5] Voitsekhovskii A V, Davyudov V N 1990 Fotoelektricheskie MDP-strukturyu iz uzkozonnyuh poluprovodnikov 327 (in Russ.)
[6] Voitsekhovskii A V et al. 2014 Russ. Phys. J. **4** 536-544
[7] Voitsekhovskii A V et al. 2014 Russ. Phys. J. **6** 707-716
[8] Sze S.M., Ng Kwok K. 2007 Physics of Semiconductor Devices 815
[9] Rogalski A2011 Infrared detectors: 2nd. ed. 876
[10] Kinch M.A. 1981 J. Semicond. Semimet. **18** 313-385
[11] Voitsekhovskii A V et al. 2008 Semiconductors **11** 1298-1303

Journal of Physics: Conference Series **652** (2015) 012004 doi:10.1088/1742-6596/652/1/012004

Development of dielectric barrier discharging power supply

Yinghui Gao[1,2], Kun Liu[1,2], Rongyao Fu[1,2], Yaohong Sun[1,2], Ping Yan[1,2],
[1]Institute of Electrical Engineering, Chinese Academy of Sciences
[2]Key Laboratory of Power Electronics and Electric Drive
No 6 zhongguancun beiertiao , Beijing, 100190

E-mail: gyh@mail.iee.ac.cn

Abstract. Due to the demand of a dielectric barrier discharge power supply, a high voltage and high frequency AC power supply was designed and implemented. Its output voltage is standard or approximate standard sine waveform with the frequency range of 1 kHz to 50 kHz. The output voltage and output frequency can be adjusted individually. The maximum output power of the power supply is 2 kW. It can be operated through local or remote control. The power supply has been used in the dielectric barrier discharging research under different conditions.

1. Introduction

Dielectric barrier discharging (DBD) is one of the important ways of obtaining low-temperature plasma. Many researches about DBD are carried out because of the wide application of low-temperature plasma in industrial and military fields. In order to study the influence factors of the low-temperature plasma produced by DBD, a flexible DBD power supply is needed.

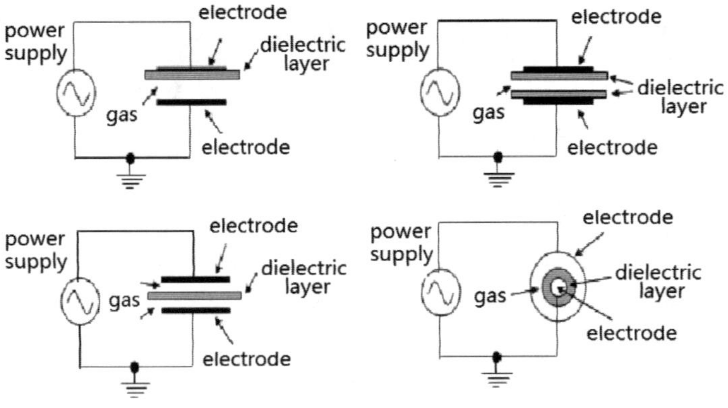

Figure 1. Four different DBD structures

The discharging structure of DBD is various. Figure 1 shows four different DBD discharging structures. With the same discharging structure the material and thickness of dielectric layer or discharging electrode and the working gas can be different. They all influence the production of low-temperature plasma. Researchers often hope to find the optimal combination condition of generating low-temperature. In the study process, a versatile power supply that can meet the various demands is essential. For an AC power supply, different material or structure DBD means different discharge voltage, different discharging output power. In addition, the output frequency is another influence factor of low-temperature generation. So the AC power supply should have adjustable output voltage and frequency, at the same time, its output power should be enough.

Considering of most DBD structure and control requirement, the specification of the AC power supply was designed and is showed in table 1.

Table 1. Specification of the AC power supply for DBD application.

Output voltage	0~30 kV peak
Maximum output power	2 kW
Output frequency	1 kHz~50kHz
Control mode	Local and remote(CAN bus, fiber, RS232 serial port)

2. Design of the AC power supply

In this section, the design of the high voltage high frequency AC power supply is introduced.

2.1. Choice of topological structure

The AC power supply is achieved through inverter technology usually. Multilevel inverter technology can output more standard sine wave. It only needs a high-frequency filter circuit which is easy to be achieved. While an H-bridge inverter which can be called two-level inverter can output a square wave only. The square wave has abundant low-frequency and high-frequency harmonic. In order to get a sine wave, a filter or frequency selected circuit with a large-scale of low-frequency and high-frequency is needed which is hard to be achieved. But the output frequency of the AC power supply is 50 kHz. If the multilevel inverter topology is selected, higher switching frequency than 50 kHz should be used in the multilevel inverter. For a 2kW output power the switching devices of the inverter is hard to be chose. So the H-bridge inverter topology was selected. Figure 2 shows the principle diagram of the AC power supply. The key designs in the power supply are the high voltage transformer and the filter.

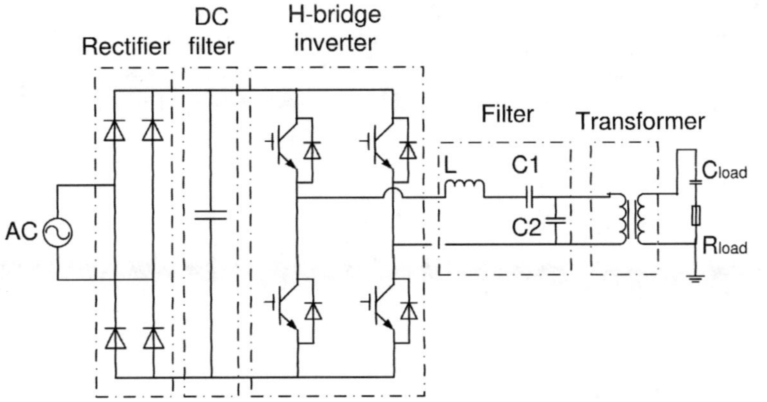

Figure 2. Circuit diagram of the AC power supply

In the topology shows in figure 2, the adjustment of output voltage of the power supply is realized through an electric controlled voltage regulator, which can adjust the input AC power. The adjustment of output frequency is realized through the inverter which is controlled by the control system.

2.2. Transformer design

Journal of Physics: Conference Series **652** (2015) 012004 doi:10.1088/1742-6596/652/1/012004

The high voltage transformer is a key component because its distributed parameters influence the filter parameter and lead to output voltage wave distortion especially in high frequency output state. Figure 3 shows the high frequency transformer equivalent circuit.

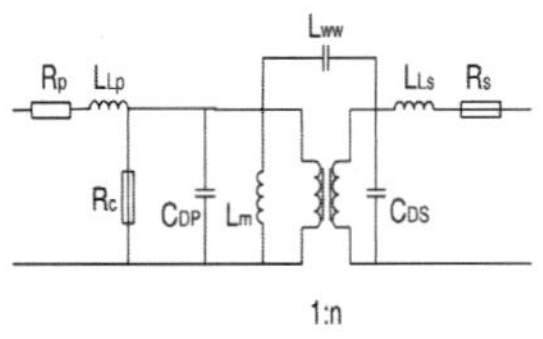

Figure 3. Equivalent circuit of transformer

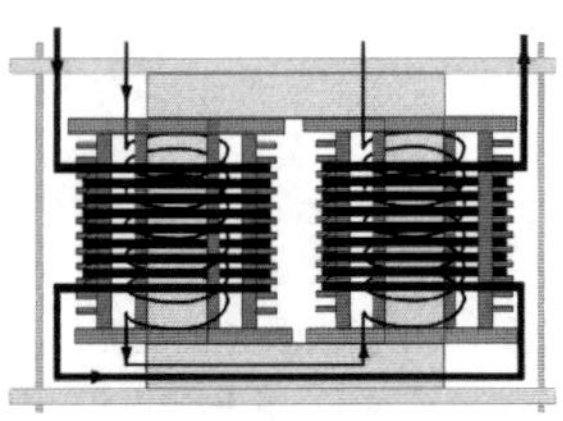

Figure 4. Transformer structure

For a high frequency and high voltage transformer, its secondary windings turns are more, so the distributed capacitance CDS and distributed inductance LLS are larger. When the CDS and LLS are converted to the primary of the transformer, the equivalent capacitance and inductance are larger because of the high ratio of the high voltage transformer. The distribute capacitance and inductance will influence the filter parameter especially at high frequency. When the distributed capacitor and inductance are larger than the filter parameter, none filter can realize the function and the frequency decided by the distributed capacitor and inductance will be the cut-off frequency of the sine wave outputted by the transformer. So the key of the high frequency transformer design is how to decrease the distribute capacitance and distribute inductance.

Ring transformer has least leakage inductance, but its insulation structure leads to the lower output voltage unless the diameter of the ring magnetic core is large enough. E-type or U-type transformer with coaxial structure has less leakage inductance, and its insulation voltage can be controlled through adjusting the distance of primary and secondary windings and the magnetic arm length of the E-type or U-type magnetic core. For a 30 kV output voltage, the U-type transformer with coaxial structure windings was selected. In order to increase the insulation voltage between the turns, the slot structure of high voltage winds is designed. In view of the limited magnetic arm length, the primary and secondary windings are divided into two equal pieces and fixed in the two magnetic arm of U-type magnetic core. The two primary and secondary windings are connected in series. Figure 4 shows the structure and winding connecting method of the transformer. The outside windings of the coaxial structure windings are the secondary windings. For a wide range output frequency, the ferrite core was selected.

2.3. Filter design

The output frequency of the power supply is from 1 kHz to 50 kHz, it is a wide frequency range. It is hard to design a filter without any gain or attenuation. So the whole output frequency is divided into eight subsection. Every subsection corresponds to a filter parameter or filter circuit. The correspondence of output frequency and filter circuit are showed in table 2.

Table 2. Correspondence of output frequency and filter parameter.

Output frequency	Filter circuit	Filter parameter
1~5 kHz	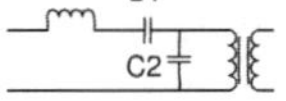	L1=5.35 mH C1=1.32 uF C2=0.5 uF
4.9~7 kHz		L1=5.35 mH C1=1.32 uF
6.1~7 kHz		L1=3.13 mH C1=1.32 uF
7~14 kHz		L1=2.28 mH C1=1.32 uF

Journal of Physics: Conference Series **652** (2015) 012004 doi:10.1088/1742-6596/652/1/012004

10.5~15 kHz	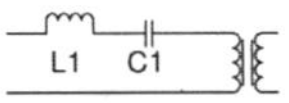	L1=925 uH	C1=1.32 uF
14~24.5 kHz		L1=413 uH	C1=1.32 uF
15~25 kHz		L1=104 uH	C1=1.32 uF
25~50 kHz	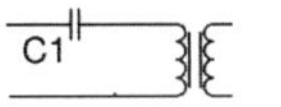	C1=1.32 uF	

The design rule of the filter circuit is to pass and gain the target frequency, so it can be design as a resonant circuit. The resonant frequency of filter is designed as the middle of the subsection frequency.

2.4. Control system design

The control system of the AC power supply fulfils the frequency adjustment and signal convert between the control computer and the AC power supply. So a digital signal processor (DSP) TMS320LF2407A was selected. Based on the function of TMS320LF2407A, a CAN bus interface and RS232 series port are designed. The RS232 series port has two output channels, one is the series port itself, and another is converted to a optical fiber interface through an electro-optic converter. In addition, based on a large number of I/O of TMS320LF2407A, a digital keyboard control channel is design which is installed on the front panel and fulfils the local control function. The remote control function is fulfiled through the CAN bus port and optical fiber interface. The RS232 series port was designed as a debug interface.

2.5. System assembly

The whole AC power has three parts of assembly. Figure 5 shows the photograph of the power supply cabinet. The top layer is the control system and inverter part; the middle layer is the band-pass filter part; the bottom layer is the high voltage transformer part.

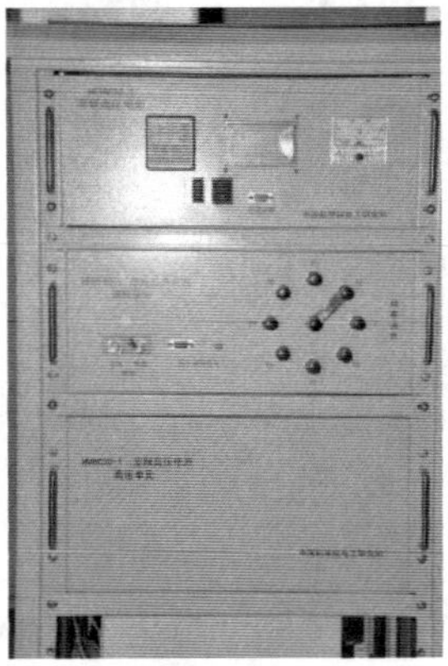

Figure 5. The AC power supply photograph

3. Experimental verification

In order to verify the performance, the experimental was carried out on an insulating material in case of load breakdown. Figure 6 and figure 7 show the output voltage waveform respectively at 1 kHz and 50 kHz. Between 5~50 kHz, the amplitude of output voltage can reach 30kV. But between 1~5 kHz, the amplitude of output voltage is lower and lower with the decrease of the output frequency. At 1 kHz, the maximum output voltage is 6.8 kV. The reason is that resonant circuit hasn't been design according to 1 kHz for the large volume of resonant induction. Figure 8 and figure 9 show the output voltage and discharge current waveforms at 5 mm air gap. Their output voltage is all 18 kV and output frequency is 7kHz and 14kHz respectively.

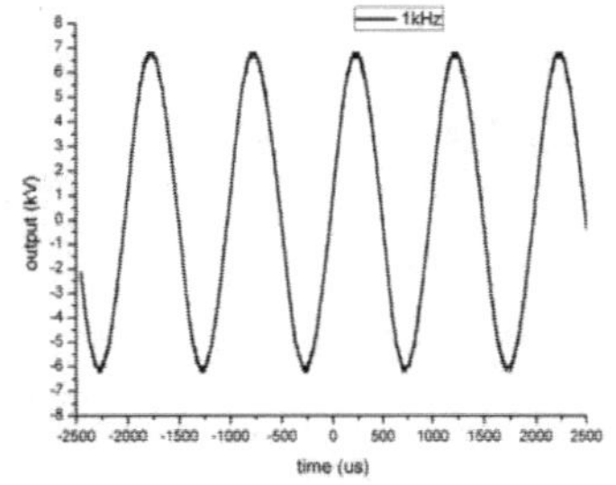

Figure 6. 1 KHz output voltage waveform

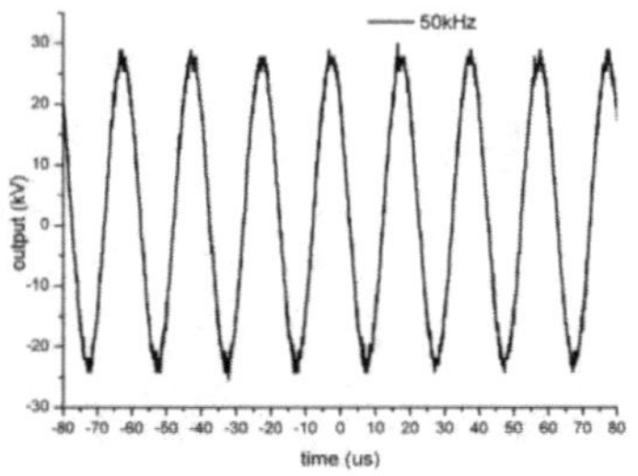

Figure 7. 50 kHz output voltage waveform

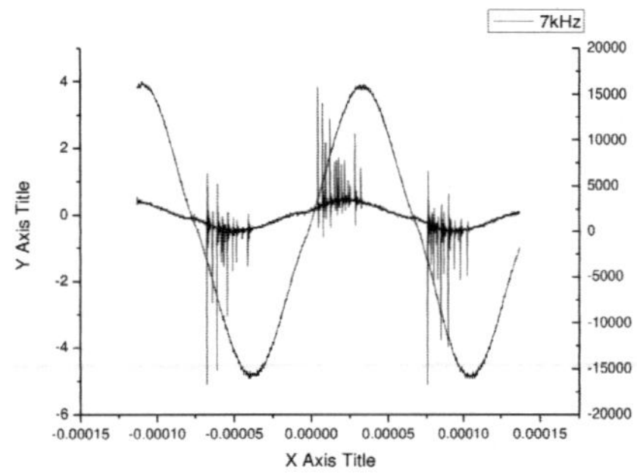

Figure 8. 1 KHz output voltage waveform

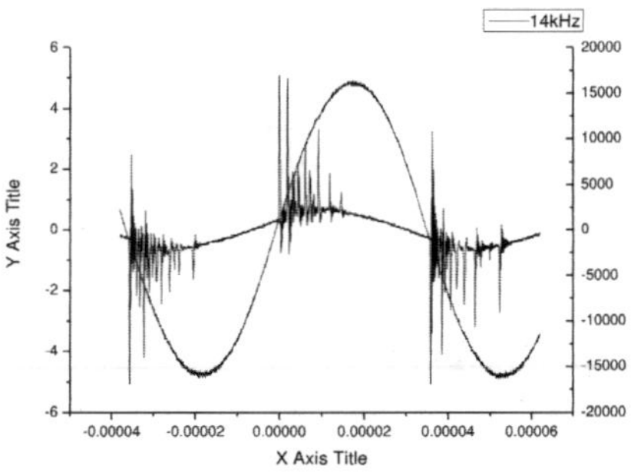

Figure 9. 50 kHz output voltage waveform

4. Conclusion

The AC power supply has used in the field of DBD and voltage withstand test of insulation materials at different frequency. The AC power supply is easy to operated and running well.

For the distributed capacitance and inductance the cut-off frequency of the transformer is 55 kHz only. Based on the test analysis the coupled inductor between the primary and secondary windings is the reason probably. It needs further validation. In addition, the filer parameter design is according to the resonant frequency, it lead to the more frequency subsection, it should be optimization.

Reference

[1] David F, Rafael D, Hubert P and Abdel K 2015 *IEEE Transactions on Industrial Electronics* Square-Shape Current-Mode Supply for Parametric Control of the DBD Excilamp Power, **62** 3 1451-60

[2] Young-chul S, Bongseong K, Kwang-cheol K *2010 IEEE IPMHVC* Considerations on the DBD Power Supply for Surface Change of Ozone Reactor 679-685

[3] Davari P, Zare F and Ghosh A 2013 *15th EPE* Analysing DBD Plasma Lamp Intensity Versus Power Consumption Using a Push-pull Pulsed Power Supply 1-8

[4] Bonnin X, Piquet H, Diez R, Florez D 2013 *15th EPE* Designing the High Voltage Transformer of Power Supplies for DBD: Wingdings Arrangment to Reduce the Parasitic Capacitive Effects 1-9

[5] Bhosle S, Zissis G, Damelincourt J.J, Dawson F.P 2004 39th Industry Applications Conference Calculation of the Impedance of an Axisymetric DBD Lamp for Power Supply Design Purposes **3** 1667-70

The change in the surface topography of magnesium under high-flux C ion irradiation

G V Potyomkin[1], A E Ligachev[2], M V Zhidkov[3,4], Y R Kolobov[3,4], G E Remnev[1], M Y Gazizova[3], S A Bozhko[3], S K Pavlov[1]

[1]Tomsk Polytechnic University, 30 Lenina ave., Tomsk 634050, Russia
[2]A.M. Prokhorov General Physics Institute, RAS, 119991 Moscow, Russia
[3] REC Nanostructured Materials and Nanotechnologies, Belgorod State National Research University, 308015 Belgorod, Russia
[4]Institute of Structural Macrokinetics and Materials Science, RAS, 142432 Chernogolovka, Russia

E-mail: carbin@yandex.ru

Abstract. The topography of the surface of the magnesium sample after irradiation by the high-intensity pulsed ion beam of a TEMP-4M accelerator was studied. The irradiation causes the formation of a regular comb structure and the creation of craters, their depth reaches 1-1.5 μm.

1. Introduction

The structural phase transformations which occur in metals and alloys under the action of high-intensity pulsed beams (HIPB) of the charged and neutral particles (electrons and ions), as well as the plasma and laser irradiation have been studied for over thirty years. Nevertheless, the mechanism of formation of topography on the surface and of the defects in the near-surface layer of fusible nonferrous materials after the HIPB action is still incomprehensible [1-5]. It especially concerns such metals as aluminum and magnesium. In this connection, the study of the high-intensity ion beam action on a beam surface layer is of a scientific interest.

2. Experimental setup and characterization techniques

The object of research was samples (30x15x3 mm) of technically pure magnesium which surface was mechanically burnished and polished using abrasive paper and diamond suspensions.
The samples were studied using a TEMP-4M accelerator at an accelerating pulse duration τ =100 ns and an energy density J (J/cm^2) on the surface of the target such as: 0.5 (Δ =100 pulses), 1.5 (Δ = 100 pulses), and 2.5 (Δ =10 pulses), J/cm^2 [6]. The particle flux from a self-magnetically insulated vacuum diode operating in a two-pulse mode consists of carbon ions C_n^{+1} admixed with protons and corresponding neutrals. The topography of the surface was studied using the Nova Nano SEM 650 ultrahigh-resolution field-emission scanning electron microscope (SEM) and the Quanta 200 3D thermal-emission SEM.

3. Results and discussion

The process of HIPB interaction with the matter includes the mode of short-pulse implantation of ions which are constituents of the beam, and high-speed heating (~100 ns) during the action of the beam current pulse. The density of the flux of carbon ions and protons for a pulse is within 10^{13}-10^{14} cm^{-2} depending on the beam current density. A high-rate heating of the target can cause the melting of the surface layer of the target. In the case of the uniform density of the particle flux density, the micro irregularities of the irradiated surface are primarily subjected to melting, because heat removal from the area of the micro irregularities is less than that from the flat surface. The melted parts can create a continuous field of liquid in the region of the beam action, when J is over 1.5 J/cm^2. The lifetime of melt is of an order of the duration of the beam current pulse [4].

When the HIPB action is terminated, the crystallization front in liquid [7] moves towards the surface with a rate determined by time of heat removal from the melting zone. In the case of high-speed cooling of the surface layer of magnesium, two-phase (porridge-like) region near the crystallization front is the place of gas bubble inception. The bubbles with a radius over critical can expand and rising to the surface to tear out the liquid shell forming the cavities in the shape of craters when hardening. When the state of magnesium is changed from liquid to solid [8], magnesium viscosity is increased by over 20 orders, therefore, some of these cavities do not have time to be covered in a quick-hardening matter, forming the cavities identified as craters. To our mind, it is one of the reasons of microcrater formation under the action of HIPB upon the metallic materials [9].

Thus, the action of HIPB on the surface of the sample causes the total or partial (depending on the average density of the power of an ion beam and the degree of homogeneity) melting of the near-surface layer of magnesium. The relief of the modified surface of the target, all other factors being equal, will depend on the density of the particle flux power acting on the surface of magnesium (J/cm^{2}s).

The crystallization of the melted layer on the surface of magnesium causes the formation of the wave (comb) structure (Figure 1, a). The alternation of the combs and cavities is most pronounced in Figures 1 (c, d). The reason for this alternation can be associated with the formation of shock waves in the layer of the liquid metal and the recoil momentum pressure emerging on the surface of the melting at the pulsed evaporation of the part of the layer of the metallic melting.

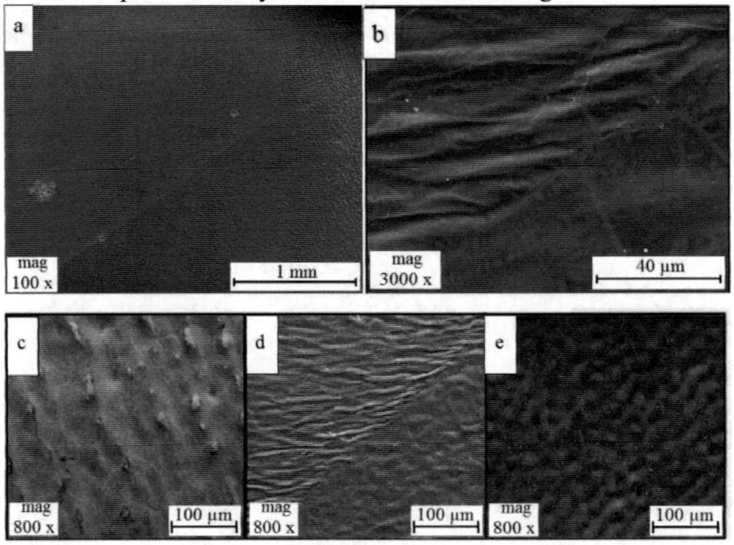

Figure 1. Structure of the surface of magnesium after irradiation (2.5 J/cm^2, 10 pulses) at a different enlargement of one and the same part of the surface under study.

As the structure of magnesium is coarse-grained (4-5 grains at the investigated surface with an area of 1 cm^2), the comb structure is not distorted with grain boundaries. Magnesium workpieces have been produced by a traditional casting technology with a minimum cooling rate. After solidification of liquid magnesium the large grains are formed, which have identical crystallographic structure, but located at various angles from each other. Consequently after the HPIB treatment of the surface of a sample, clipped from a workpiece, the partial ordering of surface occurs by means of the forming of periodic comb structure (Figure 1, b-e). It is especially seen in Figure 1, b, d, where the comb structure differs from grain to grain. In the top left corner of Figure 1, b, the number of combs per unit area of a grain is significantly higher than in the neighbour grain (bottom right corner of Figure 1, b). There are also craters (Figure 2) on the irradiated surface (mainly, on the tops of the combs, Figure 1, c). Two types of craters are formed during irradiation: craters with a cupola in the centre (Figure 2, a) and without a cupola (Figure 2, b).

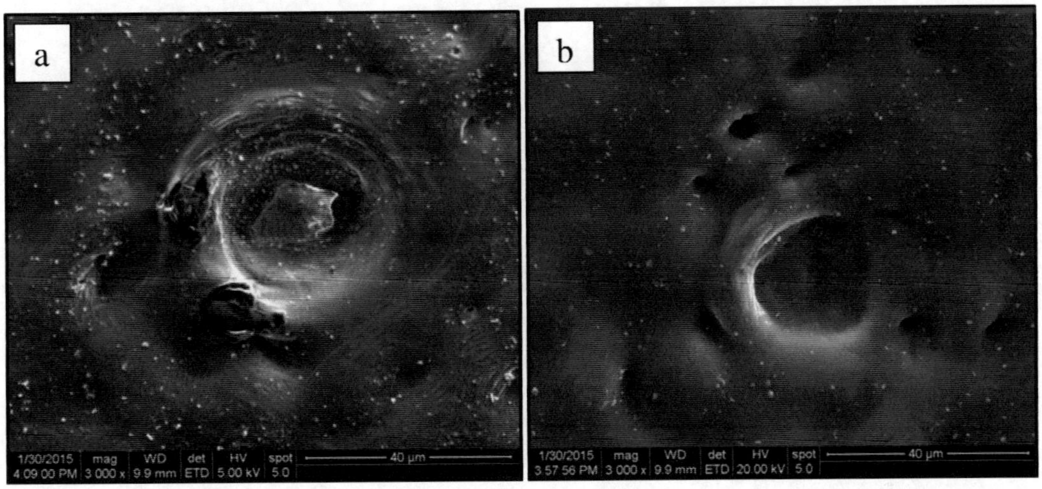

Figure 2. Craters on the surface of magnesium after irradiation (2.5 J/cm^2, 10 pulses).

The mechanism of crater formation under the action of HIPB differs fundamentally from blister formation under the action of the particle flux at a conventional ion implantation [10].

Conclusions

1. Melting the near-surface layer of magnesium with a high-intensity pulsed ion beam causes the formation of a regular wave (comb) structure, which is intermittent.
2. At the same time, round craters are formed on the tops; the formation mechanism requires additional investigation.

Acknowledgments

The work was supported by Russian Foundation for Basic Research (projects nos. 11-08-01165-a, 13-08-98061 and. 14-08-00632-a).

References

[1] Pogrebnjak A D, Remnev G E, Chistjakov S A, Ligachev A E 1987 *Izvestija Vuzov, Fizika* **1** 52
[2] Korotaev A D, Tjumencev A N, Pochivalov Ju I, Ovchinnikov S V, Remnev G E, Isakov I F 1996 *Fizika metallov i metallovedenie* **81** 112

[3] Shulov V A, Nochovnaya N A 1999 *Nucl. Instr. Meth. B* **148** 154
[4] Volkov N B, Maier A E, and Yalovets A P 2002 *Tech. Phys.* **47** 968
[5] Kovivchak V S, Panova T K, Michailov K A 2010 *Tech. Phys. Lett.* **36** 1092
[6] Remnev G E, Isakov I F, Opekounov M S, Kotlyarevsky G I, Lopatin V S, Matvienko V M, Potyomkin A V, Tarbokov V A, Kutuzov V L, Ovsyannikov M Yu 1997 *Surf. Coat. Technol.* **96** 103
[7] Anisimov S I, Luk'yanchuk B S 2002 *Phys. Usp.* **45** 293
[9] Kurz W, Fisher D J 1998 *Fundamental of Solidification* (Lausanne, Switzerland: Trans. Tech. Publications) p 300
[9] Shulov V A, Nochovnaia N A, Remnev G E 2002 *Surf. Coat. Technol.* **158-159** 488
[10] Komarov F F, 1990 *Ion Implantation in Metals* (Moscow: Metallurgiya, Moscow) p 216 [in Russian]

Investigation of properties of coal-water slurries produced by electric discharge methods

S Buyantuev[1], AKhmelev[1], A Kondratenko [2], FBaldynova[1],

[1]East Siberian State University of technology and management, 670013, Ulan-Ude, Republic of Buryatia, Russia
[2]Buryat State University, 670000, Ulan-Ude, Buryat Republic, Russia

E-mail: buyantuevsl@mail.ru

Abstract. The purpose of research is to obtain high quality coal-water slurry with minimum energy consumption. The paper presents the characteristics of coal-water slurries produced by electric discharge methods. The raw material is coal from different mines (Tugnuisky, Aduun-Chulunsky, etc.). Micrographs of the surface of the coal particles in the slurry and its chemical composition are obtained by scanning electron microscopy. Micrographs showed that the electric discharge treatment resulted in a significant dispersion of the coal particles. Elemental analysis showed a significant reduction of oxides of sulfur and nitrogen. Viscosity of slurries was determined by Brookfield rotational viscometer and corresponds to the standard of GB / T18856.4.

1. Introduction

Environmental problems associated with the use of coal fuel require the development of new effective coal technologies from economic and ecological points of views. So it is getting topical to use coal as a coal-water slurry fuel (CWSF). CWSF is a composite dispersion system and it consists of coal and a liquid phase –water [1].

Coal-water slurry (CWS) has technological properties of liquid fuels (transportation in road, rail tanks through pipelines, tankers and bulk vessels, storage in closed containers) and must retain its properties during prolonged storage and transportation, i.e. it must have sedimentation stability[2].

2. Experiment setup and methods

The solution is based on the use of plasma and electrolytic methods for the preparation of coal-water slurries without the use of plasticizers. The methods of the experiment and earlier received micrographs, and elemental analysis of coal before and after treatment were considered in this article in detail [3]. Determination of the viscosity using the rotary viscometer Brookfield was made. As a result the viscosity of the slurries corresponds to GB / T18856.4 (China). Coal from different mines (Tugnuisky, Aduun-Chulunsky etc.) was used as raw material.

This article shows the results of coal particles radius of CWS from Okinoklyuchevsky mine with sedimentation method and the speed of sedimentation.

Journal of Physics: Conference Series **652** (2015) 012006 doi:10.1088/1742-6596/652/1/012006

Sedimentation method is the determination of grain size composition based on the difference in the speed of sedimentation of particles in a viscous medium, depending on its size. Measuring the sedimentation speed, the radius of the settling particles by the Stokes' law can be determined [4]:

$$r = \sqrt{\frac{9 \cdot \eta \cdot U}{2(\rho_y \cdot \rho_B)g}}, \qquad (1)$$

where, r - particle radius, U -settling speed of particles in liquid medium, ρ_y – density of the coal powder, ρ_B – density of the liquid medium, g-acceleration of gravity,η_B-the viscosity of the liquid medium.

The sedimentation speed was measured on a special device for sediment analysis.

The liquid medium is water with a known value of viscosity. Coal powder is dispersed yielding 0.5% solution (by volume) of the suspension. After mixing, the suspension remains for sedimentation periodically weighing sediments fell on the weight pan. As a result of continuous determinations of the sediments' mass a sedimentation curve was plotted (Fig. 1).

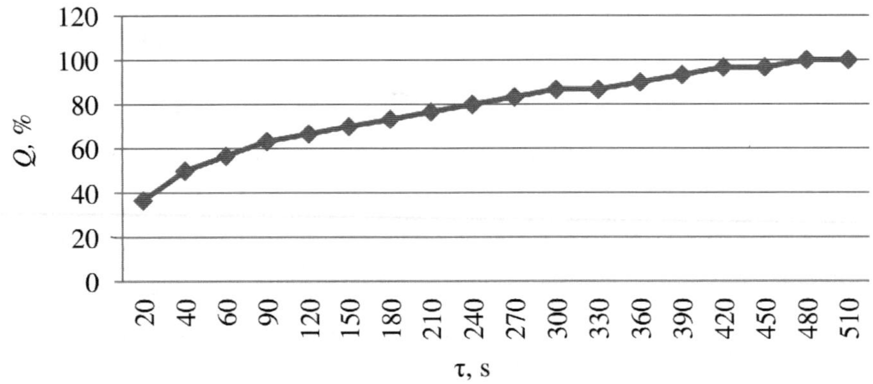

Figure 1. The sedimentation curve

Then sedimentation curve is graphically treated (by plotting tangents at the points of the curve, corresponding to different values of τ) and receive data to plot integral and differential distribution curves (Fig. 2, 3).

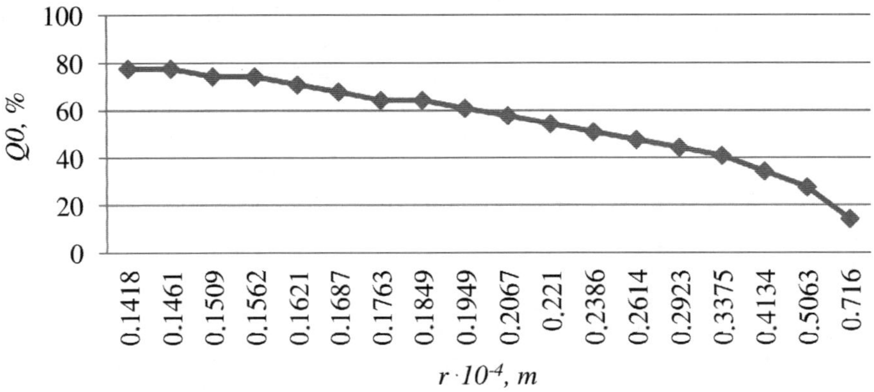

Figure 2. The integral distribution curve

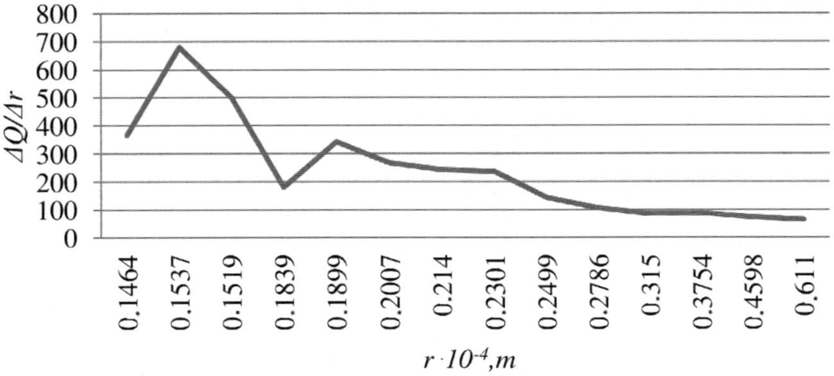

Figure 3. The differential distribution curve

The shape of the differential distribution curve shows the uniformity of coal particles size. However, when using a graphical method of calculating the distribution curves there may be errors related to insufficient accuracy in plotting the tangent to the curve. These disadvantages can be avoided with the analytical method of the construction of the distribution curves suggested by Tsyurupa N. (Fig. 4).

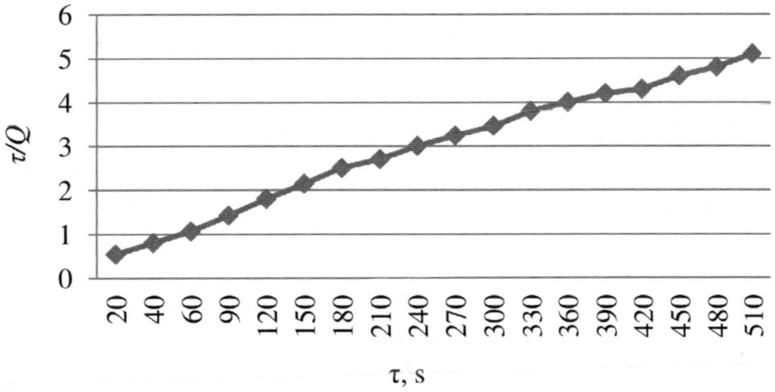

Figure 4. The sedimentation curve

In the analytical method, the process of sedimentation is described by equation:

$$Q = Q_m = \frac{\tau}{\tau + \tau_0} \tag{2}$$

Where Q –fraction weight, τ –time, $Q_m$and τ_0 - constants, and Q_m depends on the polydispersity and it is called the coefficient of polydispersity.

This equation is better to use in the linear form (Fig. 4):

$$\frac{\tau}{Q} = \frac{\tau_0}{Q_m} + \frac{\tau}{Q_m}, \tag{3}$$

After getting a line $\frac{\tau}{Q} = f(\tau)$, Q_m is calculated according to the values of cotangent angle to the x-axis:

$$Q_m = ctg\alpha = \frac{1}{tg\alpha} = \frac{1}{0.2671} = 4.6\ (\%), \tag{4}$$

$\frac{\tau_0}{Q_m}$ – – over the interval on the ordinate and $\tau_0 = \frac{\tau}{Q} Q_m$

$$\frac{\tau_0}{Q_m} = 0.4322, \ \tau_0 = 0.4322 \cdot 4.6 = 1.9881\ (s)$$

Using the equation $r_0 = \sqrt{\frac{K \cdot h}{\tau_0}}$, K is calculated:

$$r_0 = \sqrt{\frac{10.2521 \cdot 10^{-8} \cdot 16.5 \cdot 10^{-2}}{1.9881}} = 9.22 \cdot 10^{-5}(m), \quad (5)$$

$$K = \frac{9 \cdot \eta \cdot h}{2 \cdot (\rho_y \cdot \rho_{\text{в}}) \cdot g} = \frac{9 \cdot 0.897 \cdot 10^{-3} \cdot 16.5 \cdot 10^{-2}}{2(1.66 - 0.9971) \cdot 10^3 \cdot 9.8} = 10.2521 \cdot 10^{-8} \quad (6)$$

From the kinetic equation of sedimentation the following integral sedimentation distribution law can be derived [4]:

$$Q_0 = Q_m \left(\frac{r_0^2}{r^2 + r_0^2} \right)^2, (7)$$

where Q_0 - the proportion of fractions with a radius $\leq$ r; r_0 – index depending on the dispersion.The differential form of this equation is the following:

$$F = \frac{dQ_0}{dr} = 4Q_m \cdot r_0^4 \frac{r}{\left(r^2 + r_0^2\right)^3}. \quad (8)$$

Next, three major radius are determined characterizing the particles distribution curve.

3. Results

The minimum size of the particles:

$$r_{min} = r_0 \cdot \sqrt{0.1 \cdot (\sqrt{Q_m} - 1)} = 9.22 \cdot 10^{-5} \cdot \sqrt{0.1 \cdot (\sqrt{4.6} - 1)} = 3.11 \cdot 10^{-5}\ (m) \quad (9)$$

The most probable particles size:

$$r_{\text{н.в.}} = \frac{r_0}{2.24} = \frac{9.22 \cdot 10^{-5}}{2.24} = 4.12 \cdot 10^{-5}(m) \quad (10)$$

The maximum size of the particles:

$$r_{max} = 3 \cdot r_0 = 3 \cdot 9.22 \cdot 10^{-5} = 27.66 \cdot 10^{-5}(m). \quad (11)$$

The system is polydisperse, the sedimentation speed of the particles is the following:

$$U_{\text{сед}} = \frac{2gr^2(\rho - \rho_0)}{9\eta} \quad (12)$$

Where U- sedimentation speed, r – the radius of the particles of coal,g - acceleration of gravity.

The minimum size of the particles: r_{min}= $3.11 \cdot 10^{-5}(m)$, $U_{set.} = 15.58 \cdot 10^{-4}\ m/s$

The most probable particle size: $r_{m.p.} = 4.12 \cdot 10^{-5}(m)$, $U_{set.} = 27.35 \cdot 10^{-4}\ m/s$

The maximum size of the particles: $r_{max} = 27.66 \cdot 10^{-5}(m)$, $U_{set.} = 1232.59 \cdot 10^{-4}\ m/s$

4. Conclusions

Measuring the sedimentation speed the radius of the settling particles was determined by the Stokes' law. Basically predominate particle with a radius $r_{m.p.} = 4.12 \cdot 10^{-5}(m)$, the minimum radius of the particles $r_{min} = 3.11 \cdot 10^{-5}(m)$ and the maximum radius $r_{max} = 27.66 \cdot 10^{-5}(m)$.

All real disperse systems are polydisperse and therefore sedimentation speed is different: the large particles settle faster, the small ones more slowly.

References

[1] Senchurova J, Murko V,Fedyaev V,Dziuba D, Pusirev E 2008 *Research results spraying coal-water fuel injectors rotor* (Tomsk:Bulletin of the Tomsk Polytechnic University)

[2] Zhuravlev N, Murko V, Fedyaev VDziuba D, Senchurova Y, Zaostrovsky A 2009*The vortex combustion technology of coal-water slurry fuel: environmental aspects* (Kuzbass : Ecology and Industry of Russia)

[3] Buyantuev S, Kondratenko A, Khmelev A2014 A*nalysis of characteristics of coal-water slurries obtained by plasma and electric discharge methods* (Bali: ARPN Journal of Engineering and Applied Sciences)

[4] Baldynova F *Surface phenomena and disperse systems* (Ulan-Ude: ESSTU)

Industrial testing and analysis of activated carbons derived from electric arc plasma

S Buyantuev[1], A Kondratenko [2], I Starinsky[1], A Khmelev[1]

[1]East Siberian State University of technology and management, 670013, Ulan-Ude, Republic of Buryatia, Russia
[2]Buryat State University, 670000, Ulan-Ude, Republic of Buryatia, Russia

E-Mail: buyantuevsl@mail.ru

Abstract. The article presents the results of study of activated carbon produced by thermal destruction in electric arc plasma. There are measurement methods such as density measuring, measuring of total and summary porosity, porous surface microscopy, laboratory methods and industrial testing of sorption activity.

Key words. Sorbent, porosity, density (true, apparent, bulk), pore volume, specific surface, surface microscopy, electric arc plasma treatment.

1. Introduction

At present time, carbon absorbent (activated carbon) is used in many processes of chemical technologies. Gas and wastewater treatment is mainly based on activated carbon adsorption [1]. Characteristics of adsorption capacity are important for studying of carbon nature and for solving of many technical problems. Thermal destruction is the basic process of obtaining the target product [2]. Use of low temperature plasma is a new method of thermal activation of carbon [3]. Water purification enterprises and water treatment systems usually located at power stations, power plants are main consumers of activated carbon. Sewage treatment is complicated by the fact that this water is a complex heterogeneous system and consists of various substances in different states [4].In the article there are investigation results of structure and properties of coal after electric arc plasma treatment. The possibility of industrial use of these sorbents for sewage treatment is explored.

2. The experimental setup and methods of research

The use of electric arc plasma can significantly intensify heat treatment of coal, increase yield of desired synthesis product (sorbent) and thereby reduce production costs and considerably improve environmental performance. To investigate carbon activation process universal plasma modular installation of combined type has been created which opens possibility for integrated approach to the production of activated carbon and gas synthesis combined in unified process of carbon plasma treatment [3]. In the article sorption properties of coal mine “Okinoklyuchevsky” were investigated treated by means of electric arc plasma in the plasma reactor of combined type. Coal with the following characteristics was applied (see Table1).

In the experiment, coal passed through plasma chemical reactor with an arc rotating by magnetic field. In uniform rotation of anode arc spot between the annular cathodes in the reactor continuous plasma environment is formed with average temperature 2500-3000K. Due to this, coal particles supplied from the top into the reactor fully undergo plasma thermochemical treatment. The residence time of coal particles in plasma zone depends on the size of fraction and is regulated by installation of diaphragm in the bottom of the plasma reactor and ranges from 0.1 to 1 sec. The temperature and speed of rotation is adjusted by alteration of plasma current power source and coil. While passing through plasma zone all coal particles are exposed to short plasma treatment that leads to its thermal destruction. Treated solid particles fall into the bottom part of the plasma reactor - activation chamber. Generated synthesis gas is removed from the reaction zone by exhaust device. This principle of construction of plasma systems allows reaching industrial production level of carbon sorbents [3, 5].

In electric arc zone of the reactor of described type electric power loaded in the camera in the arc is converted into heat. Thus, electric arc zone is zone of generating thermal energy, its absorption, it covers entire section of the chamber, allows to adjust dwell time of coal dust, concentration power and hence temperature. This area is limited in height by planes perpendicular to the chamber axis and passing through the electrode end (top), and through the middle of the electromagnetic-solenoid coil (lower) [6]. Thermal energy is generated by the rotating arc in the zone and under influence of gas flow resulted from thermal effect on coal dust, is distributed throughout its entire volume.

Power of generating zone was determined by condition of its even distribution. In calculations the diameter of the reactor chamber was taken as key size and all basic regularities in the electric arc zone were expressed by means of it. It makes possible to extend the experiment results received for small reactors to geometrically similar reactors of larger power and hence larger diameter chambers [7].

Reactor current I_p is expressed by current density relevant to sectional area of the camera:

$$I_p = \Delta \mathrm{i}_\mathrm{p} \pi \frac{\mathrm{D_k}^2}{4}. \tag{1}$$

Accordingly, current density $I_p = \Delta \mathrm{i}_\mathrm{p} \pi \frac{\mathrm{D_k}^2}{4} = 1.132 \frac{A}{cm^2}$

Reactor voltage U_p is represented by means of voltage gradient of the electrode gap:

$$U_p = \mathrm{E_p} \frac{\mathrm{D_k}}{2} = 1.132. \tag{2}$$

Where $\frac{\mathrm{D_k}}{2}$ – interelectrode gap, which includes arc voltage for determining voltage gradient (in reactor with rod cathode).

Accordingly, voltage gradient $E_p = \frac{2\mathrm{U_p}}{D_k} = 26.7 \frac{V}{cm}$

As a result, taking into account expressions for I_p и U_p formula for reactor power is derived:

$$P_p = \mathrm{U_p} \cdot I_p = \frac{\pi}{4} \cdot \frac{1}{2E_p \Delta i_p Dk^3} \approx 0.39 E_p \Delta i_p Dk^3. \tag{3}$$

Which is equal to: $P_p \approx 0.39 E_p \Delta i_p Dk^3 = 0.39 \cdot 26.7 \frac{V}{cm} \cdot 1.132 \frac{A}{cm^2} \cdot (15)^3 cm^3 = 40 kW$.

As a result it can be concluded that generated in volume of zone heat power equal to loaded electrical power is proportional to the cube of the chamber diameter, it is also proportional to the current density Δi_p and voltage gradient E_p or volumetric power W_p, released in the zone. Value W_p, determining temperature level is, obviously, a constant for given production for geometrically similar reactors:

$$W_p = const. \tag{4}$$

Consequently, it is possible to adopt as constants Δi_p and E_p:

$$\Delta i_p \approx const. \tag{5}$$

$$E_p \approx const. \tag{6}$$

The value of $\Delta i_p \equiv I_д / Dk^2$ is determined by using models [8] provided high technological indicators of conditions in high-performance unit. Then, for given diameter of the reactor its value is provided by selection of current.

3. Results and discussion

At the initial stage of the study of sorption properties of coal, ablation of coal was determined and sorption activity was analyzed. Analysis results are presented in Table 1.

Table 1. The sorption properties of coal treated bylow-temperature plasma

№	Coal ablation, %	Sorption activity by iodine, cm^3/g	Sorption activity by benzene, cm^3/g	Sorption activity by iron, mg/g	Sorption activity by methylene blue, mg/g	Fractional composition		
						Up to 0.2 mm	From 0.2 to1 mm	More than 1mm
1		12.3	0.082		21.1	25	63	12
2		15.6			50.0			
3	72	17.8	0.085	3.10	123.85	8	78	16
4	72	20.3	0.133	2.60	123.89	12	74	16
5	72	19.4	0.157	2.24	123.9	13	81	6
6	72	19.1	0.192	2.54	133.5	11	57	32
7		15.6	0.046	14.52	58.3	38	61	1

As it can be seen from Table 1 ordinary low-grade coal (or a mixture of coal), treated by low temperature plasma has sorption properties. The quality of carbon sorbents in general was determined by nature of porous structure directly dependent on the following factors: true density of coal $\boldsymbol{\rho_a}$, apparent density of coal $\boldsymbol{\rho_{ap}}$, and bulk density of coal $\boldsymbol{\rho_b}$ [9]. Results ρ_a, ρ_{ap} *and* ρ_b are shown in Table 2.

Table 2. Determination of ρ_a, ρ_{ap} and ρ_b Okinoklyuchevsky brown coal properties before and after arc plasma treatment

№	ρ_a,(g/cm^3)	ρ_{ap}, (g/cm^3)	ρ_b, (g/cm^3)
1. Before treatment	1.45	1.24	0.67
2. After treatment	1.66	0.78	0.42

The received data allows us to estimate alteration of porous structure of sorbents. With this data important indicators of porous structure can be determined: effective porosity P_{ef}, total volume of open pores V_o, as well as one of the most important characteristics of sorbents - total porosity V_Σ. For layer of dispersed carbon materials interparticle porosity can be distinguished (Pd.i.) or interparticle pore volume (V_v) [2, 10]. Determination data Pef, V_o, $P_{d.i.}$,V_v and V_Σ shown in Table 3.

Table 3. Determination P_{ef}, V_o, $P_{d.i.}$,V_vи V_Σ of Okinoklyuchevsky brown coal properties before and after electric arc plasma treatment

№	$П_{ef}$, %	V_o, (cm^3/g)	$П_{d.i.}$, %	V_v, (cm^3/g)	V_Σ (cm^3/g)
1. Before treatment	14.5	0.12	46	0.7	-
2. After treatment	53.0	0.70	46	1.10	0.70

During testing it was observed that if to put coal treated by low-temperature plasma in water with steam supply (regular tap water was used), it separates into fractions: heavy fraction precipitated at the bottom (line 7, Table 2), as well as light fraction (line 6 Table 2) floated to the surface of water. Small part of coal particles were in water column. The analysis of sorption properties revealed that heavy (settling) fraction exhibits good chemisorption properties (sorption activity by iron 14.52 mg / g) and light (floating) fraction has good oil absorption properties.

Subsequently, possibility of industrial application of plasma treated sorbents at wastewater treatment plant LLC "Hydros" was studied.

It should be noted that all waste waters are complex heterogeneous systems typically contaminated by various substances in different states, namely: substances in undissolved, dissolved and colloidal conditions. Also all wastewaters have organic, inorganic and biological compounds.

To study sorption activity, chemical water analysis was conducted in compliance with procedures PNDF 14.1.2.3. and PNDF 14.1-95 ofState environmental control.

Initially, fraction of 1 mm was investigated, but due to its lightness the fraction floated to the surface so it was carried over with waste water making impossible to use this fraction.

At the next stage of investigation, fraction from 2 to 5 mm was taken. For testing a water sample prior to chlorination was taken, this sample was tested before and after passing the column with activated carbon. Investigation results of wastewater before and after sewage are shown in Table4.

Table4. The results of waste water investigationbefore and after sewage

Chem. substances, mg/dm^3	Without sorbent	After passing through sorbent	Purification, %
Iron	0.08	0.04	50
Nitrite	0.144	0.0395	72
Nitrates	21.90	11.48	48
Phosphates	2.20	1.73	21
Ammonia	0.34	0.105	69

At the final stage, investigation of pore surface of coal treated by plasma thermal activation was conducted by means of computer of 3D-modeling. When analyzing coal pore surface by 3D-modeling, samples of both initial and treated by thermal destruction in plasma were used. The analysis was carried out by means of optical microscope metallographic Altami MET 2 use. Micrographs of coal particles surface before and after heat treatment provides volumetric 3D interpretation of data surfaces of simulation using software environment analysis scans Image J.This 3D-simulation enabled to graphically evaluate nature of porosity of the surface, and to compare them before and after thermal activation in plasma [11, 12].

4. Conclusions

Thus, samples of carbon sorbents fraction of 2-5 mm, produced by electric arc plasma showed good chemisorption properties (Table. 2, 5), which allows to make conclusion about appropriateness of its use as sorbents for wastewater treatment and water treatment. Presented samples have high sorption properties: total porosity V_{Σ} = 0.70 cm^3/g, total volume of open pores V_o = 0.70 cm^3/g, interparticle porosity and interparticle pore volume $P_{d.i.}$= 46%, V_v = 1.10 cm^3/g, total porosity P_{ef}= 53%, making it correspond to parameters of industrial coal type BAU, KAD. The obtained data is in good agreement with the literature data [2, 4, 10].

Acknowledgement

The work was supported by the project of the Russian Federation Science order # 2014/23

References

[1] Greg S, Singh K 1984 *Adsorption, surface area, porosity* (Moscow: Mir)
[2] Kienle H, Bader E1984 *Active coals and their industrial applications* (Moscow: Chemistry)
[3] Kondratenko A 2013 *The sorption properties of coals after thermal destruction in the plasma arc* (Tomsk: News of higher educational institutions. Physics)
[4] Kolishkin D,Mikhailov K 1972 *Active coal properties and test methods* (Moscow: Chemistry)

Journal of Physics: Conference Series **652** (2015) 012007 doi:10.1088/1742-6596/652/1/012007

[5] Buyantuev S, Starinskiy I 2008 *RF patent №2314996*(Ulan-Ude: ESSTU)
[6] Bron O 1966 *The movement of the electric arc in the magnetic field* (Moscow: Electricity)
[7] Buyantuev S, Kondratenko A 2013 *The physical model of the electric zone in a plasma reactor of the combined type* (Ulan-Ude: BSU)
[8] Sergeev P1978 *Electric arc reactors* (Alma-Ata: Science)
[9] Buyantuev S, Kondratenko A 2012 *The structure and properties of carbon sorbents and carbon nanomaterials obtained by processing plasma arc*(Novosibirsk:ITP)
[10] Butyrin G 1976 *Highly porous carbon materials* (Moscow: Chemistry)
[11] Panteleyev V, Egorova O, Klykova E 2005 *Computer microscopy*(Moscow: Technosphere)
[12] Niederost M, Niederost J, Scucka J 2002 *Automatic 3D reconstruction and visualization of microscopic object from a monoscopic multifocus image sequence*(Czech Republic: CAS)

Journal of Physics: Conference Series **652** (2015) 012008 doi:10.1088/1742-6596/652/1/012008

Explosive-emission cathode with resistive decoupling in a high-current plasma-filled diode

P P Kiziridi, G E Ozur

Institute of High-Current Electronics, SB RAS, Tomsk, Russia

Email: kiziridi_pavel@mail.ru

Abstract. In the paper, the results of investigations of operation of multi-emitter explosive-emission resistively decoupled cathode in a plasma-filled diode are presented. The integral glow, current rise rate, and energy density distribution via cross section of a non-relativistic, high-current electron beam have been studied. It has been shown, that resistively decoupled cathodes look very perspective for improvement of stability and uniformity of high-current electron beams.

1. Introduction

Resistive decoupling is a well-known method stabilizing an operation of explosive-emission cathodes [1, 2]. This method allows one to suppress so-called emissive centers-leaders, consistently appearing while operation of wide-area explosive-emission cathodes. Availability of centers-leaders reduces a number of concurrent emission centers and therefore degrades the homogeneity and stability of the cathode operation. To suppress centers-leaders it is needed to limit the current through all emitters since the location of leaders may change from pulse to pulse. Current of each center may be limited by a space charge of electron flux emitted by respective center [3], or by connecting of active resistance in series to the circuit of each emitter. The first method may be realized at low current density (less than 30 A/cm^2) only, which is not suitable for us. Therefore, the second method was chosen. But the use of resistive decoupling in a plasma-filled diode requires a careful isolation of the cathode substrate because the emission can be initiated on a flat surface located under plasma. Therefore, the emission centers on the substrate may shunt the resistors placed in the emitter circuit which is unacceptable.

The cathode with resistive decoupling was firstly developed and applied for a high-current plasma-filled diode in [4]. This cathode represented a packet of resistors TVO-0.125 (made in Russia; 0.125 means 0.125 W of dissipated power) glued with epoxy compound. The diameter of emitting part of the cathode was 1.5 cm. The ends of the resistors were cut and polished to expose the carbon core, which served as emitter. However, carbon erosion products, which may deposit to the treated target, are usually undesirable. Therefore, we have developed an analogous cathode but with metal emitters. The present paper is devoted to investigations of high-current electron gun operation with this cathode.

2. Cathode design

Experiments were carried out on the "RITM-U" facility described in [5]. To create plasma anode, an usual high-current Penning discharge in argon [6] and hybrid discharge matching high-current Penning discharge with vacuum arcs [5] were used.

The cathode has been manufactured of TVO-1 resistors at face-value from 2.7 Ω to 5.6 Ω; their wire outputs of 6 mm in length serve as emitters (Figure 1). The opposite outputs of resistors were soldered with a brass disc. Total number of resistors was 156 and effective diameter of emitting area was about 58 mm. To avoid the contact of the plasma anode with the screen electrode, a diaphragm with the same diameter was installed on the anode edge addressed to the cathode. The packet of resistors was inserted into Teflon holder on which the screen electrode made of stainless steel was built-up. The gullets between Teflon holder and resistors were filled with sealant. Two versions of the cathode design are given in Figure 1. The second version (Figure 1, b) distinguishes from the first one (Figure 1, a) by the fact that the end surface of Teflon holder is protected by the ledge of the screen electrode.

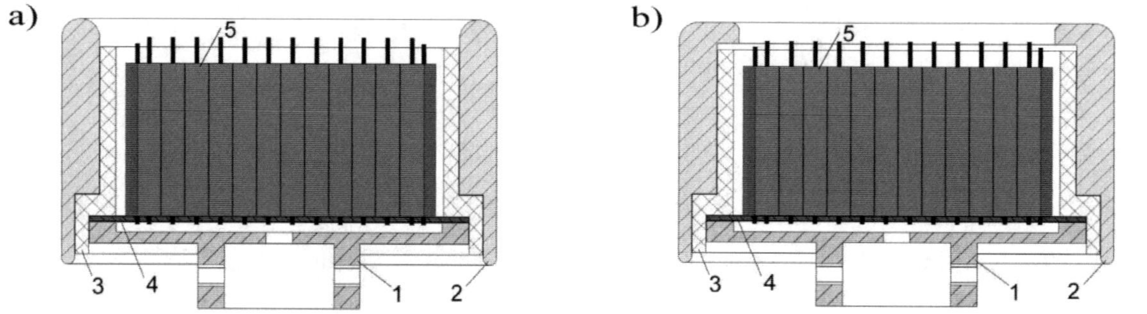

Figure 1. Explosive-emission cathode based on TVO-1 resistors. 1 - cathode holder; 2 - screen electrode; 3 - Teflon ferrule; 4 - brass disc; 5 - resistor TVO-1.

3. Results and discussion

3.1. Current rise rate

The current rise rate (*dI/dt*) is one of the key parameters characterizing the cathode operability. Evidently, the current rising is caused by the increasing of a number of emission centers and their emitting area. It was important to compare the *dI/dt* value of our new cathode with *dI/dt* provided by traditionally used cathode made of copper braid of radiofrequency cable [5, 6].

Typical waveforms of accelerating voltage and total cathode current are given in Figure 2. The studying of big number of waveforms has shown that in similar conditions, the current rise rate in the front of the pulse is higher by a factor of 20-30% than for the case of copper braid cathode.

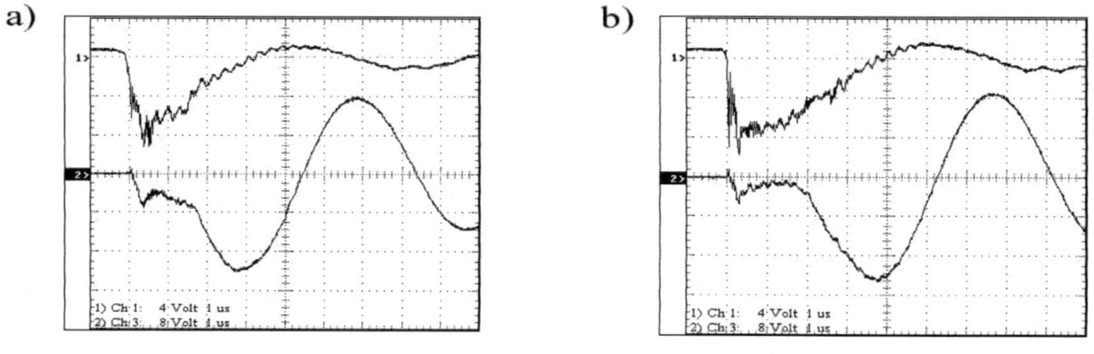

Figure 2. Typical waveforms of accelerating voltage (16 kV/div, upper traces) and cathode current (24 kA/div, lower traces). Plasma anode based on high-current Penning discharge (a) and plasma anode based on hybrid discharge (b). Argon pressure is 0.5 mTorr (a) and 0.4 mTorr (b). Guide magnetic field $B_z = 0.106$ T.

3.2. Cathode glow

Cathode glow integrated per pulse was registered with **CASIO QV-3000EX/Ir** digital camera in open shutter mode. Photographing of the glow was performed from the face of electron gun. At that, the trip-out of the beam electrons was performed onto tubular collector with magnetic cut-off analogously to the method described in [7]. To prevent camera from over-illumination, an interferometric light filter of type **HC-9** with thickness of 3 mm was installed in front of it.

Characteristic photos of the glow are given in Fig. 3. Rather bright spots referring to emitters can be seen on the diffused background. Note, these spots are observed practically on all emitters stably that confirms good operation of the cathode. Such behavior of the emitters is characteristic to both types of plasma anode.

Figure 3. Photos of glow. Argon pressure is 0.4 mTorr. Guide magnetic field induction $B_z = 0.106$ T.

It is logically to assume that bright spots localized on the emitters represent the glow of plasma of emissive centers or, in other words, cathode spots. Supposing that at the end of the current rise time all emission centers operate, it is easy to estimate the current per center as 70-150 A. At such a current, the lifetime of emission center is much more than pulse duration of the beam [8] which is positive factor for cathode operation. The voltage drop on resistors makes up 250-500 V and this is enough for "grab effect", i.e. appearance of a new emission center stimulating by plasma of the neighbor emission center.

Average distance between emission centers is defined by cross dimensions of the resistor and makes up about 3.6 mm, so the density of emission centers (number per area) is about 6 pcs/cm^2. In the case of the cathode made from copper braid, the density of emission centers is approximately 1.7-2 times lower [7]. If the cathode will be manufactured from resistors TVO-0.5 having lower cross sectional dimensions, so the density of emission centers may be increased twice - up to 11-12 pcs/cm^2. However, it increases expenses of the cathode manufacturing.

3.3. Energy density distribution

The uniformity of the beam energy density distribution, *W*(*r*), is a key (and difficulty achievable) parameter. In the present work, the energy density distributions were studied with thermal imaging diagnostics according to procedure described in [9, 10]. The block-scheme of measurements is given in Figure 4. 210-μm stainless steel foil served as beam collector (thermal imaging target). To increase the emissivity, the back side of the foil has been covered with black dull paint (20 μm).

Journal of Physics: Conference Series **652** (2015) 012008 doi:10.1088/1742-6596/652/1/012008

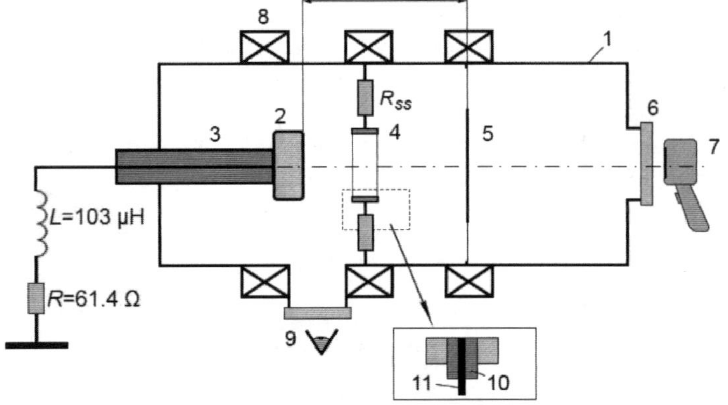

Figure 4. 1 – body of electron gun; 2 – cathode of electron gun; 3 – isolator; 4 – anode unit; 5 – collector (target); 6 – window made of CaF_2; 7 – thermal imager TESTO 875-1; 8 – solenoid; 9 – window for visual observation; 10 – ceramic tube; 11 – cathode of the arc source. R_{ss} = 110 Ω – resistor.

In the Figures 5 and 6, the thermal images (at the left) and corresponding energy density distributions (at the right) along dashed cross sections are given. The most uniform distributions have been obtained for the case of hybrid anode (Figure 5) at B_z = 0.106 T, which corresponds to the results obtained in [5]. The diameter of uniform part at the level of 0.9 from maximum value of energy density makes up 5.2 cm and this is the best our result achieved ever. In the case of Penning anode (without arc plasma sources in the anode), this diameter is lower - about 3.5 cm. If B_z value decreases to 0.06 T, so the beam uniformity becomes worse: the distribution $W(r)$ looks like bell (Figure 7).

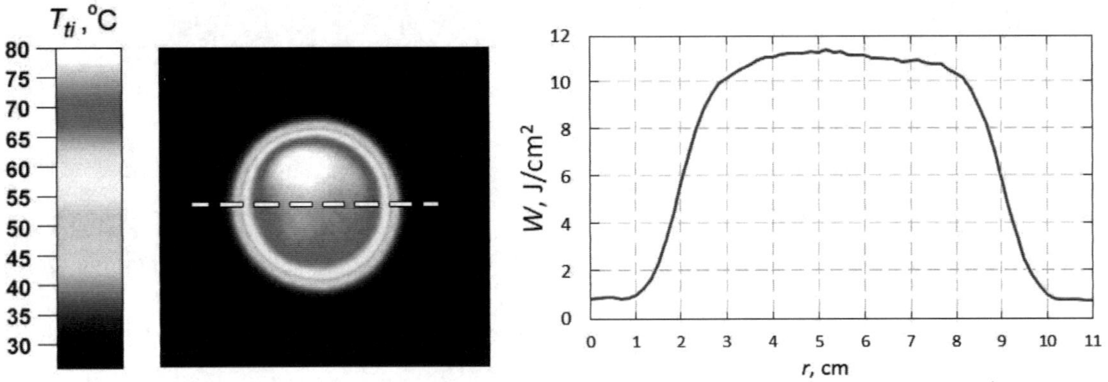

Figure 5. Thermal image and corresponding energy density distribution via horizontal cross section. Plasma anode based on hybrid discharge. Argon pressure is 0.4 mTorr. Guide magnetic field induction B_z = 0.106 T.

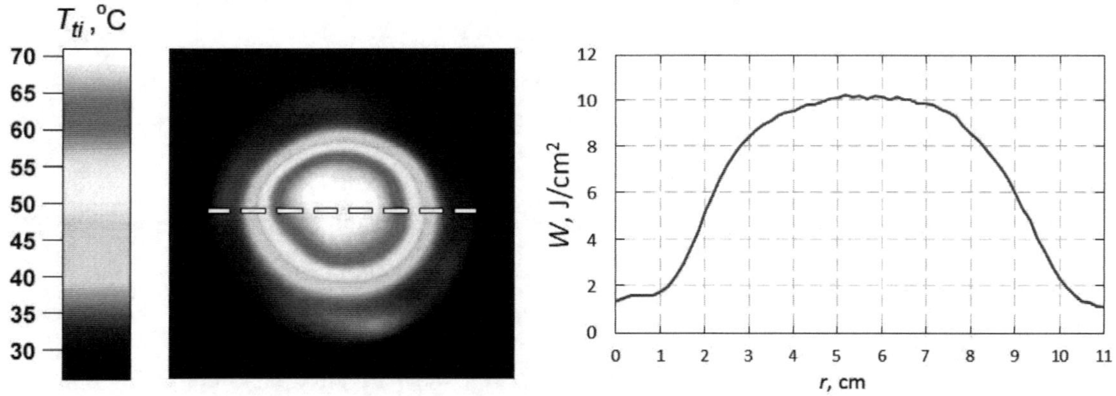

Figure 6. Thermal image and corresponding energy density distribution via horizontal cross section. Plasma anode based on high-current Penning discharge. Argon pressure is 0.5 mTorr. Guide magnetic field induction $B_z = 0.106$ T.

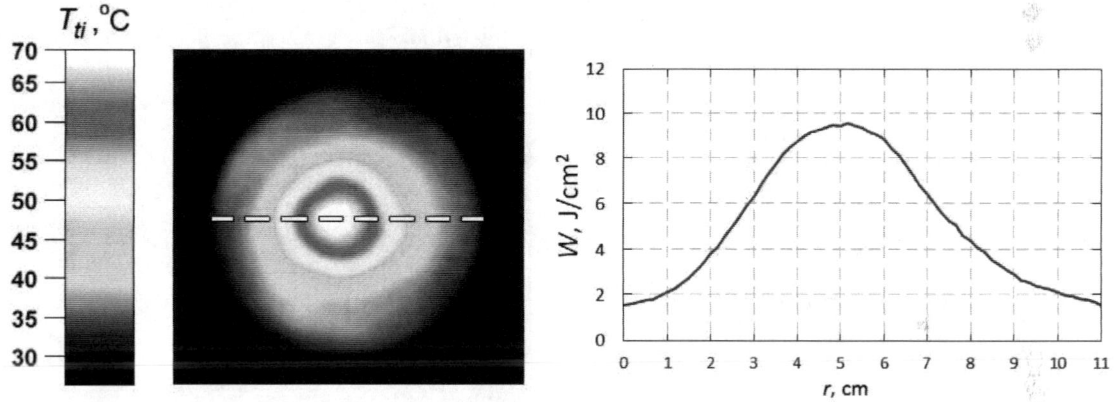

Figure 7. Thermal image and corresponding energy density distribution via horizontal cross section. Plasma anode based on hybrid discharge. Argon pressure is 0.4 mTorr. Guide magnetic field induction $B_z = 0.06$ T.

It should be noted that one can control the energy density distribution by varying the resistance in the circuit of each emitter. In our present experiments, the resistance of the resistors in the central part was twice higher than for peripheral part of the cathode. It has given some effect, and for further modification of the cathode, this difference in resistance will be increased for further improvement of the beam uniformity.

4. Conclusions

The main results of investigation of multi-point, explosive-emission cathode with resistive decoupling operation in a plasma-field diode are as follows.

- A stable operation of explosive-emission cathode with resistive decoupling of emitters based of resistors TVO-1 in a plasma-filled diode has been demonstrated.
- For this cathode, the current rise rate in the front of the pulse exceeds in average by a factor of 20-30% the corresponding value for multi-wire copper cathode.
- Satisfactory uniformity of energy density distribution in cross section of high-current electron beam has been obtained. The further improvement of the beam uniformity may be provide by

the increasing of the resistance in the circuits of emitters located in the central part of the cathode.

Acknowledgment
This work was supported by Russian Foundation for Basic Research (grant # 13-08-98066).

References

[1] Burtsev V A, Vasilevsky M A, Roife I M, Seredenko E V and Engelko V I 1978 *Technical Phys. Letters (Soviet)* **4** 1083.
[2] Vasilevsky M A, Roife I M and Engelko V I 1981 *Technical Physics (Soviet)* **51** 1183.
[3] Abdullin E N, Bazhenov G P, Erokhin G P and Ladyzhensky O B 1985 *In book: Pulsed electric discharge in dielectrics* (Novosibirsk: Nauka) 42-55.
[4] Ozur G E *Thesis on Degree of Cand.Phys. & Math.* 1991 (Tomsk: Institute of High Current Electronics).
[5] Kiziridi P P, Ozur G E, Zyulkova L A and Batrakov A V 2013 *Russian Phys. Journal* **55** Issue 7/2 42.
[6] Ozur G E, Proskurovsky D I and Karlik K V 2005 *Instruments & Experimental Technique* **48** 753
[7] Karlik K V, Ozur G E and Proskurovsky D I 2007 *Russian Phys. Journal* **50** Issue 9 264
[8] Mesyats G A and Proskurovsky D I 1989 *Pulsed electric discharge in vacuum* (Berlin: Springer, Chapter 10)
[9] Kiziridi P P, Markov A B, Ozur G E and Frolova V P 2014 *Russian Phys. Journal* **57** Issue 3/2 114.
[10] Kiziridi P P and Ozur G E 2015 *Technical Physics* **85** Issue 6 132.

Magnetron discharge sputtering for fabrication of nanogradient optical coatings

O D Volpian[1], A I Kuzmichev[2,4], G F Ermakov[1], A I Krikunov[1], Yu A Obod[1], N V Silin[3] and S V Shkatula[1]

[1] Scientific-Manufacturing Enterprise "Fotron-Auto Ltd.", 8 Novodanilovskaya naberezhnaya, Moscow 117105, Russia
[2] National Technical University of Ukraine "Kiev Polytechnical Institute", 37 prospect Pobedy, Kiev 03056, Ukraine
[3] Far Eastern Federal University, 8 Sukhanova street, Vladivostok 690950, Russia

E-mail: A.Kuzmichev@edd.ntu-kpi.kiev.ua

Abstract. The technology of the middle frequency pulse reactive magnetron sputtering for fabrication of nanogradient optical coatings with smooth variation of refractive index was developed and studied. The technology is based on programmable motion of a substrate over two magnetrons with targets of different materials. The feature of the deposition process is a constant composition of reactive gas medium and an invariable magnetron operation mode. To realize this technology, an automatic computer-controlled sputtering system additionally comprising a gas discharge activator of reactive gas (oxygen) and an *in situ* optical monitor-spectrovisor has been built. The dielectric oxide-based nanogradient coatings of photon-barrier type were successfully fabricated. The obtained results confirm the high potential of the middle frequency pulse reactive magnetron sputtering of silicon and metal targets for fabrication of nanogradient dielectric optical coatings with excellent properties.

1. Introduction

Coatings on various optical elements/substrates play an important role in optics and photonics. They are used for improving spectral, reflective or antireflective, angle, polarization, and phase properties; in particularly, very wide working spectrum band and wide operation angle may be provided. Typically, optical coatings are multilayer thin film interference structures with a stepwise periodic variation in refractive index n. In recent years in optics and laser technology, the increasing attention has been paid to coatings with a smooth, gradient (rather than abrupt) variation in index n, such as antireflective coatings and rugate rejection filters [1, 2]. The smooth variation in n values leads to the improvement of many characteristics of coatings including the excellent matching of wave impedance, thermomechanical strength and laser damage resistance thereof. When the refractive index $n = n(z)$ continuously, gradually distributed by a certain law within a medium, the latter is called as gradient one. If axis z is the direction of light propagation (it coincides typically with the normal to substrate surface), we have the so called longitudinal $1D_z$ gradient [2]. If change of n values occurs on distance less than 100 nm, such coating should be called as nanogradient one. Various refractive index profiles

[4] Corresponding author

Journal of Physics: Conference Series **652** (2015) 012009 doi:10.1088/1742-6596/652/1/012009

$n(z)$ – linear, quadratic, sinusoidal-like, polynomial of fifth degree, and others – may be used [1, 2]. Nowadays, the transformation optics arises as a new direction in optics where the effective control of radiation is conducted using a gradient medium of metamaterials (composites undergoing an electromagnetic reaction), which have been used to show the invisibility effect (the "invisibility cloak" effect) [3].

In order to fabricate a gradient coating, we need materials with variable n values. However, there is a limited range of natural optical materials with discrete n values; therefore, to obtain intermediate n values, it is necessary to prepare mixtures of available materials or derivatives thereof (nonstoichiometric oxides, oxynitrides, *etc.*) directly on the substrate by depositing the appropriate materials in atomized form. There are several ways to provide gradient variation in the refractive index n [2]. Among them, the common method is electron beam co-evaporation of two source materials but it is associated with the complexity and fundamental limitations of synchronized nonstationary evaporation of refractory materials from highly localized sources that constrains the industrial fabrication of precision nanogradient optical coatings. The other method is ion beam co-sputtering of several source materials but this method gives very low productivity. Meantime, the ion sputtering in magnetron discharge is able to provide very high sputtering rate and widely employed in industry. The magnetron discharge is a low pressure gas discharge in crossed electric and magnetic fields where the solid cathode serves as target sputtered by ions from discharge plasma. The deposition in magnetron discharge gives dense pore-free thin film coatings of excellent quality with good adhesion. However, the deposition of nanogradient optical coatings by the magnetron method is a relatively new field of technology [2] and needs intensive research and development for its implementation. The aim of the given work is examination and approbation of our recently developed magnetron sputtering method for deposition of dielectric layers with variable refractive index. As an example, we choose the process for fabrication of nanogradient optical coating of the "photon barrier" type.

2. Magnetron discharge sputtering apparatus and procedure for deposition of nanogradient optical coatings

The diagram of an automatic computer-controlled setup with a two-magnetron system for the ion sputtering of two materials in a reaction medium is shown in figure 1. In our experiments, targets of Si and metal (Nb, Ta or Ti) were sputtered in an argon atmosphere with an oxygen additive. The arrows above magnetrons M1 and M2 indicate the direction of propagation of the sputtered target materials; one can see some part of sputtered materials is captured by substrate S with coating formation on its lower surface.

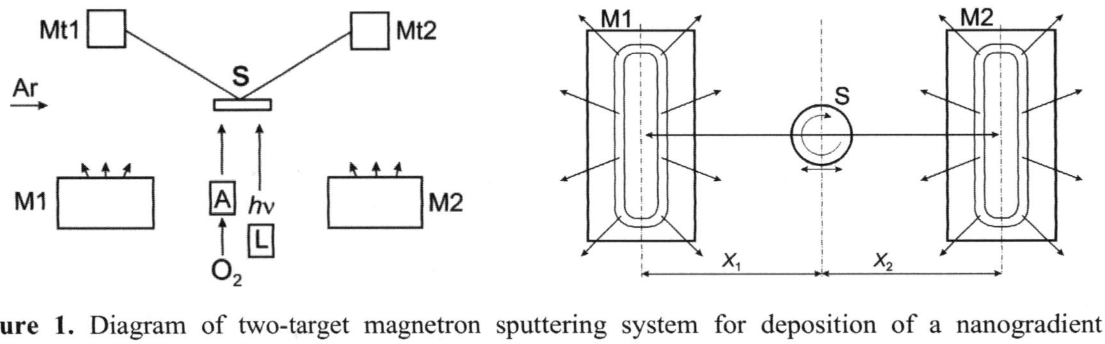

Figure 1. Diagram of two-target magnetron sputtering system for deposition of a nanogradient coatings of a mixtures of two oxides of two materials (the right picture shows top view): (Ar) flow of inert sputtering gas Ar; (A) activator of reactive oxygen; (L) laser; (M1, M2) sputtering magnetrons with Si and metal (Nb, Ta, Ti) targets; (Mt1, Mt2) optical parts of *in situ* monitoring system; (O_2) flow of reactive gas O_2; (S) rotating substrate, which can move along the main axis (bold line); the oval racetrack zones on the magnetron targets are the sputtered areas; the arrows above the magnetrons indicate the direction of propagation of the sputtered target material.

The composition of the coating material deposited on substrate S is determined by the substrate position (that is distances X_1 and X_2), the specified algorithm of the space–time motion of substrate S over magnetrons M1 and M2, and the electric power of the magnetrons. Depending on the composition, the refractive index n value of the coating varies. To prepare a coating with the longitudinal $1D_z$ gradient of n, it is necessary to provide a high uniformity of distribution of the n value along the radius of the substrate; therefore, the X_1 and X_2 distances are selected to meet the requirement for obtaining a maximum uniformity of the coating thickness along the radius of the rotating substrate with a minimum cross sputtering of the magnetron targets. To synthesize a stoichiometric dielectric coating with minimum optical losses, the reactants – the sputtered and deposited metals and the reactive gas (oxygen) – should be chemically activated. In the given setup, the activation occurs owing to the physical-chemical processes in the magnetron discharge plasma, the additional activation of oxygen in a special gas discharge device (activator A) through which the gas O_2 is fed onto the substrate, and the exposure of the substrate to ultraviolet laser L radiation.

The deposition and synthesis of the coatings were controlled by *in situ* optical monitoring of the coating properties using spectrovisors Mt1-Mt2. The optical thickness, refractive index, and spectral characteristics of the growing layer were determined.

The important issue is the choice of parameters of electrical power and gas supply of the magnetron system. Different approaches to power and gas regulation are known to be used for obtaining gradient coating [2]. We consider the most suitable way to ensure the stability and reproduction of the precision deposition process is keeping the invariable regime of power supply and constancy of gaseous parameters during the sputtering process. This provides constancy of temperature in the deposition chamber and rate of chemical reaction on the substrate surface. In the end, it means that the current composition of the synthesized coating material is controlled only by the law of substrate motion above the magnetrons.

The thickness of partial layer deposited at the given substrate position is determined by dose of captured atoms sputtered from magnetron targets. Each dose can be evaluated as the product of sputtered atom flow captured by the substrate and duration of substrate exposure by this flow. The exposure duration is the time period when the substrate is in the given position. The refractive index of the growing coating is determined by the ratio of partial doses of captures atoms from the two magnetron targets.

Thus, the deposition procedure is based on the variation of the substrate position X_1 (or X_2) and the exposure duration in the given position under computer control; herein the powers of magnetron discharges and values of gas parameters in the deposition chamber remain constant. Such approach ensures the stable sputtering rates of both the magnetrons in reactive gas medium.

Before the elaboration of computer control program, the deposition process has to be calibrated: the coating deposition rate and the refractive index n value are experimentally determined for each possible substrate positions at the given electrical powers of two magnetron discharges. Then, the computer control program describes step-by-step the trajectory and the time characteristic of substrate motion over the magnetron targets along the main axis shown in figure 1. To obtain the refractive index profile $n(\mathbf{z})$ in the coating with the minimal value, corresponding to the index of SiO_2, and the maximum value, corresponding to the index of other metal oxides, the computer provides movement of the substrate step-by-step from the position above 1st magnetron (M1) with Si target ($X_1 = 0$) to the position above 2nd magnetron (M2) with metal target ($X_2 = 0$), then to the position over 1st magnetron, and so on, if a periodical gradient structure is in need. The aforesaid calibration curve, presenting the calculated dependence of refractive index n upon the distance X_1 along the trajectory of silica substrate motion is shown in figure 2. The substrate diameter was 3 cm. The electrical and gas conditions during the calibration process are described below. Values of n from this curve were verified by means of ellipsometry of the single layer. Thus, the computer program contains information on the duration of each exposure in each position of the substrate accordingly to the prescribed refractive index profile.

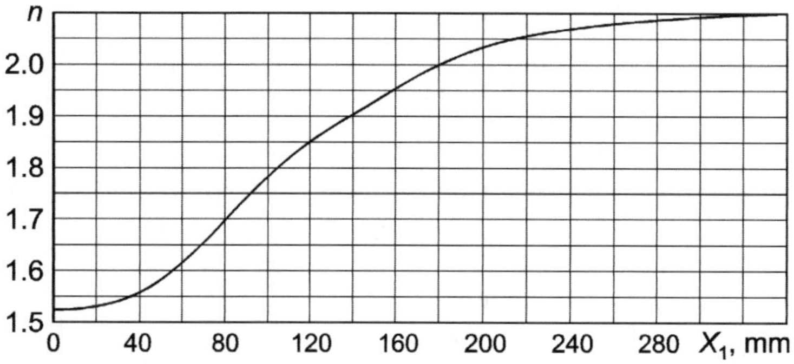

Figure 2. Calibration curve of the magnetron sputtering system, used in the given work (dependence of refractive index of deposited material of SiO_2 – Ta_2O_5 mixture against substrate position X_1).

For sputtering Si target and the second target of metal, the middle frequency (22 kHz) pulse mode of magnetron power supply has been used. Such mode excludes arcing on the magnetron targets in reactive gas medium. The arcing is caused by formation and electrical charging of dielectric films on the targets but they are discharged by plasma electrons during pause between magnetron current pulses. The calibration (see figure 2) of the deposition apparatus and fabrication of the nanogradient optical coating were fulfilled at the following magnetron operation conditions: the average current of both the magnetron discharges was 0.5 A (the discharge voltages were 380 V for Si target, M1, and 580 V for Ta target, M2), the total pressure of argon-oxygen gas mixture in the chamber was about 0.1 Pa. The substrate was prepared from quartz glass К8 with the refractive index $n_s = 1.51$. The calibration has indicated indeed that the minimal value of the refraction index ($n_m \approx 1.5$) was obtained when the substrate was above the magnetron M1 with Si target; the maximum value of the refraction index ($n \approx 2.1$) was obtained when the substrate was above the magnetron M2 with Ta target. The process duration for deposition of gradient layer with thickness $d = 140$ nm was about 1000 sec.

3. Results of nanogradient optical coatings deposition

Using the above described magnetron sputtering setup and the computer-controlled deposition process, the assigned periodical concave profile of variation in the refractive index $n(z)$ over the depth of the coating (figure 3) has been prepared by depositing a mixture of synthesized silicon and tantalum oxides on a quartz glass substrate via pulse sputtering of two targets (Si and Ta) in the oxygen medium. The shape of the given $n(z)$ profile has been proposed and theoretically grounded by Prof. A. B. Shvartsburg [4] as having heterogeneity-induced dispersion and photon barrier properties in the visible spectral range. The profile comprises five periods of variation in the n values.

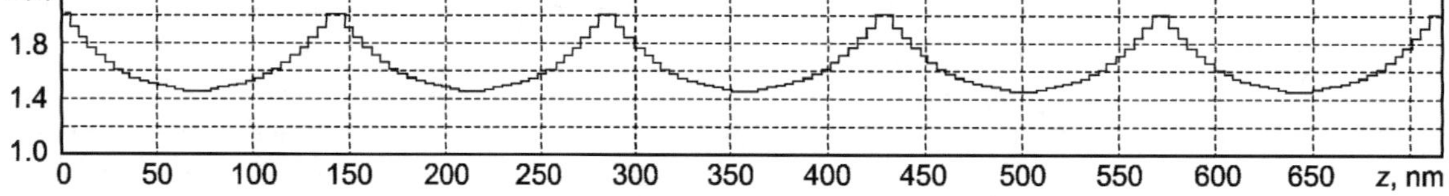

Figure 3. Nanogradient five-period profile of $n(z)$ for coating from SiO_2+Ta_2O_5 mixture.

Figure 4 shows the calculated and experimentally measured transmission spectra of the fabricated nanogradient coating; good matching of these spectra suggests that the real profile of variation in the refractive index over the depth of the coating is close to the assigned $n(z)$ profile. One can see such coating may be considered as a nanogradient rejection filter with band-stop, *i.e.* as a photon barrier, in the limited band in visible spectral range, and there is the wide band antireflection effect in the near infrared range.

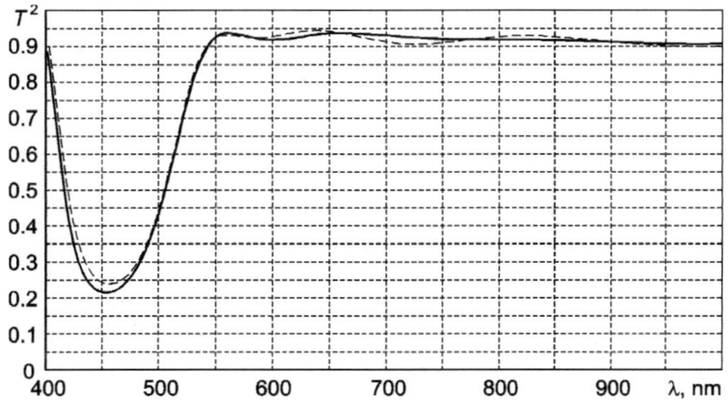

Figure 4. Calculated (dashed line) and experimental (solid line) transmission spectra of the nanogradient coating of SiO_2+Ta_2O_5 mixture with five-period $n(z)$ profile.

Some other properties of the fabricated coating may be mentioned. The coefficient of optical losses is less than $5 \cdot 10^{-5}$. The coating is resistant to laser radiation with power density W > 10^9 W/cm^2 for λ = 1064 nm, τ = 10 ns and the combination of thermal cycling ($\Delta t°$ = –60/+70°C, 100 cycles) with UV exposure at power density W > 100 W/cm^2 for λ = 300 nm, as well as to 3500 cycles of heating up to + 180°C with 5 minutes exposure and cooling down to – 190 °C (immersion into liquid nitrogen) with 1 minute exposure.

4. Conclusions

The obtained results confirm the middle frequency pulse reactive magnetron sputtering of silicon and metal targets is a perspective method for fabrication of precision nanogradient dielectric optical coatings with excellent spectral properties, in particular, with the photon barrier effect in visible spectral range and the wide band antireflection effect in the near infrared range. The main feature of the proposed technology is the programmable motion of the substrate over two magnetrons with targets of different materials at a constant composition of reactive gas medium and an invariable magnetron operation mode. Such technology is not very complex and may be easily adopted to industrial application. The promising application of this technology is fabrication of gradient optical metamaterials [4, 5], high performance solar cells and devices based on transformation optics. The dense pore-free microstructure and good adhesion of the coating, which are typical for magnetron sputtering, and absence of the internal sharp boundaries separating the layers with high and low n ensure the resistance to environmental factors, including thermomechanical ones. The nanogradient microstructure of the coating ensures the resistance to high power laser radiation damage.

Acknowledgments

The work was financially supported by the Ministry of Education and Science of the Russian Federation (Agreement with Fotron-Auto, Ltd. # 14.579.21.0066) and the Far Eastern Federal University (project # 14-08-2/3-20_u).

References

[1] Macleod H A 2010 *Thin-Film Optical Filters* (Fl., Boca Raton: CRC Press)
[2] Volpian O D and Kuzmichev A I 2013 *Russian Journal of General Chemistry* **83** 2182.
[3] Cai W and Shalaev V 2010 *Optical Metamaterials. Fundamental and Applications* (New York: Springer)

[4] Shvartsburg A B and Maradudin A A 2013 *Waves in Gradient Metamaterials* (WSPC)
[5] Volpian O D and Kuzmichev A I 2012 *Negative Refraction of Waves. Introduction into Physics and Technology of Electromagnetic Metamaterials* (Kiev: Avers) in Russian

Synthesis of transition metal borides layers under pulsed electron-beams treatment in a vacuum for surface hardening of instrumental steels

A S Milonov, B A Danzheev, N N Smirnyagina, D E Dasheev, T B Kim and A P Semenov
Institute of Physical Materials Science SB RAS, 670047, Ulan-Ude, Russia

E-mail: terwer81@mail.ru

Abstract: The saturation of the surface layers of metals and alloys with boron is conducted for increasing their surface hardness, wear resistance, etc. Multicomponent layers containing in its composition borides of refractory metals, as a rule, are formed by the methods of chemical-thermal processing in the interaction of boriding component with refractory one or by the method of saturation of refractory metal impurities or alloy with boron. In this work, we studied the features of vanadium and iron borides formation on the surface of instrumental steels U8A and R18 under the influence of intense electron beams in continuous and pulse modes.

Keywords: Electron beam, borides, alloys, microhardness, X-ray diffraction, structure, self-propagating high-temperature synthesis (SHS)

1. Introduction

The saturation of the surface layers of metals and alloys with boron is carried out to increase their surface hardness, wear resistance, etc. Application of electronic heating with high (> 10^9 W/m^2) power density in a vacuum owing to fast inertialess allows obtaining extremely high temperatures and facilitating the regulation of heating in a wide temperature range, as well as opens ample opportunities for creating protective layers on the basis of borides of refractory metals.

In [1] reported about formation of strengthening coatings based on TiB_2, CrB_2, VB_2, W_2B_5 on carbon steels under influence of an electron beam on boron-containing reactionary daubs in vacuum. The assumption of an active role of a surface metal has been made at electron-beam welding of self-heating synthesis products which was initiated by an electron beam and proceeding in reactionary daub of stoichiometric mixtures.

In this work, we studied the features of vanadium and iron borides formation on the surface of instrumental steels U8A and R18 under the influence of intense electron beams in continuous and pulse modes.

2. Layers of based on the boride VB_2 on steel U8A

The layers of vanadium boride VB_2 (later also borides of iron Fe_2B and FeB) are synthesized and formed on the surface of cutting plates in size 12×12×5 mm with roughness Rz = 4.41 ÷ 4.02 μm of instrumental steels U8A and R18 at the same time. Samples were prepared by applying the reactionary daubs on the pre-prepared (well-fat) surface of the steel. The composition of the daubs consisted of 1:1 by volume the stoichiometric mixture of oxide, boron-containing component and carbon, as well as organic binder - solution 1:10 of potassium BF-6 in acetone. Amorphous boron, charcoal (birch) and oxides V_2O_3, Fe_2O_3 were used as initial substances. Processing of samples was being conducted for 2-5 minutes with the power of electron beam of 150-300 watts. The pressure in the vacuum chamber did not exceed 2×10^{-3} Pa [2, 3].

X-ray diffraction revealed the prevalence of carbide phases (cementite) in the boride layer on steel U8A. This can be explained only by the deviation from stoichiometry at evaporation of the intermediate boron oxide. Application of the protective layer of the amorphous oxide B_2O_3 (1:1 by volume of reactionary daub: daub based on B_2O_3) allowed to form more uniform boride layers. It was found that the weight of the crystalline phases was 92.3%, and the amorphous phases' weight was 7.7%, while the crystallite size ranged from 15 to 70 nm.

The X-ray patterns of all the investigated boride layers of vanadium show the reflections of vanadium ferrite α-Fe-V, corresponding to phase α-Fe_9V of space group *Im3m*, with a cubic lattice parameter a = 0.2878 nm and cementite Fe_3C (PDF 00-003-0989) of space group *Pbnm* with elemental rhombic cell parameters a = 0.4518, b = 0.5069 and c = 0.6736 nm, z = 4 (Fig. 1).

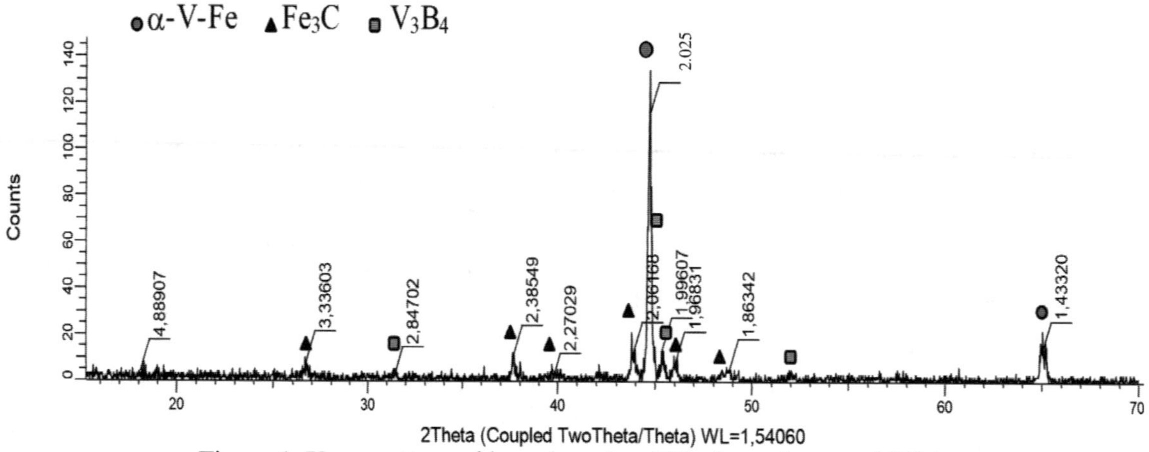

Figure 1. X-ray pattern of layer based on VB_2 formed on steel U8A

Boride V_3B_4 (PDF 03-065-2551, space group Immm, with elemental rhombic cell parameters a = 0.303, b = 1.318 and c = 0.2986 nm, z = 2) can be found in X-ray patterns of diboride vanadium layers. Formation of diboride vanadium VB_2 was not radiographically observed, although there were reflexes of light intensity related to the interplane distances of this phase. So, it should be investigated further.

Fig. 2 shows a general view (Figure 2.a) and the structure of the boride layer VB_2 on the surface of cutter plate steel U8A (Fig. 2.b). Layer thickness reaches up to 500 μm.

Microhardness testing showed its uneven distribution by thickness in the cross-section. Very rare separate inclusions have microhardness of HV≈ 24000 MPa and are located in the subsurface layers. Next, we observe two regions: in the first region the microhardness reaches 2500 MPa, and in the second one it is 1500 MPa. The metal base has microhardness of 200 MPa.

Journal of Physics: Conference Series **652** (2015) 012010 doi:10.1088/1742-6596/652/1/012010

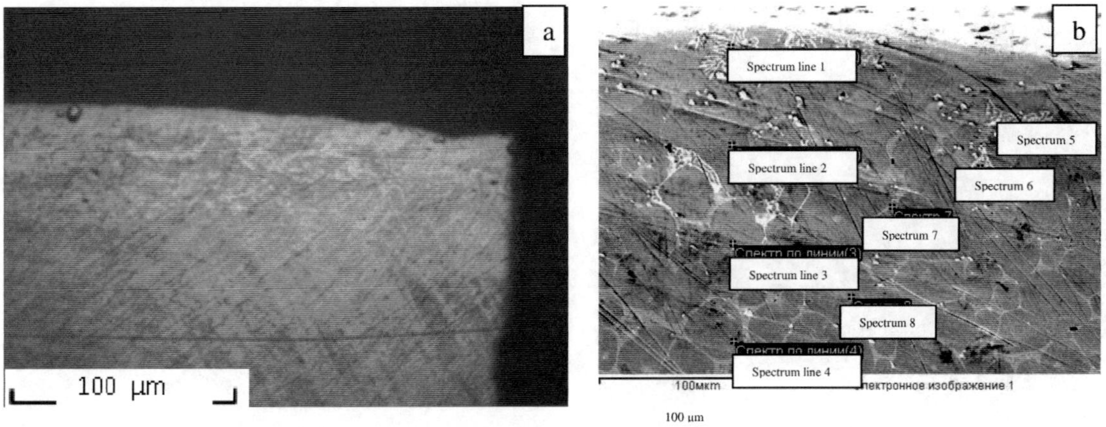

Figure 2. General view (etched layer) (a) and structure (b) layer VB_2 on steel U8A

3. Layers of iron borides Fe_2B and FeB on high-cutting steel R18.

X-ray diffraction (XRD) showed that after electron beam treatment in the samples of cutter plate steel R18 we observed the reflexes of the following phases: metal substrate - α-Fe and carbide W_3Fe_3C. It was found that the weight of the crystalline phase is 92.7% and the amorphous phase is 7.3%.

Boride FeB is located near the surface of the coating. This is evidenced by the results of the investigation of the end surface of the cutter plate steel R18 and X-ray analysis data (scanning electron microscope JSM-6510LV JEOL with microanalysis system INCA (Fig. 3).

Application of the protective layer of amorphous oxide B_2O_3 lead to obtaining equilibrium boride layers Fe_2B and FeB.

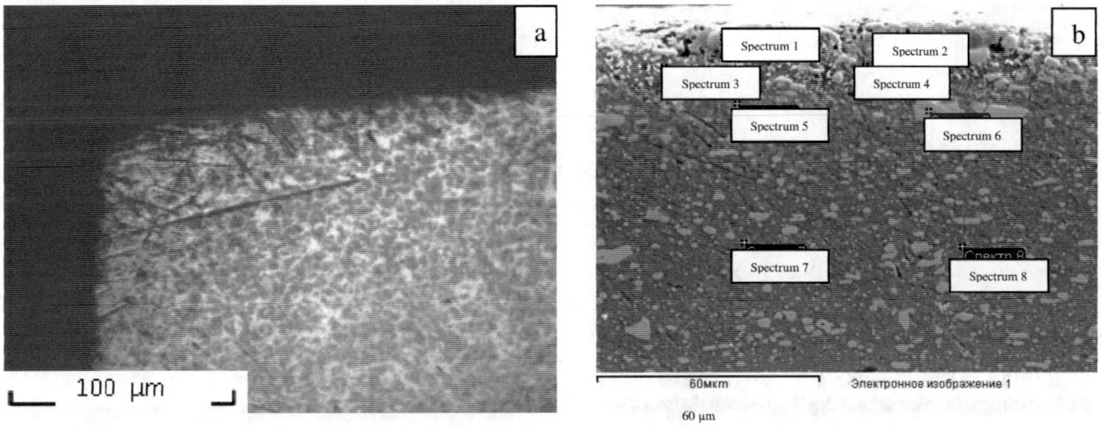

Figure 3. General view (a) and structure (b) of boride layer FeB on steel P18

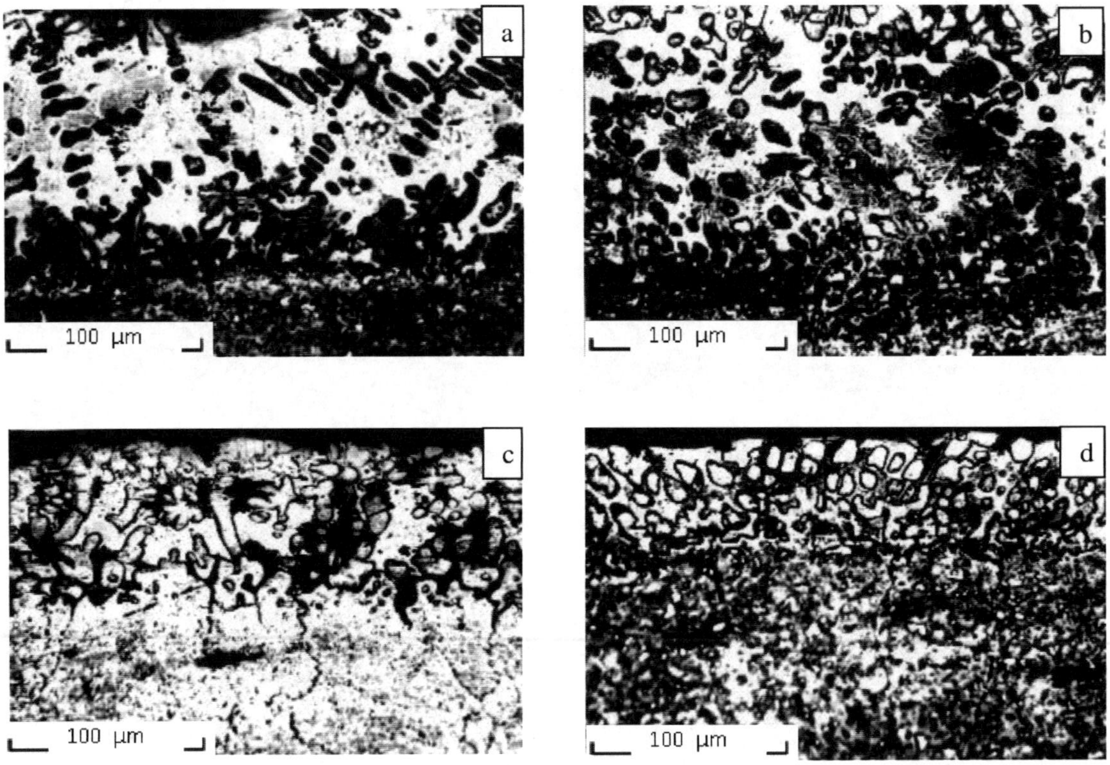

Figure 4. Structure of the surface layer on the steel R18 after electron beam boriding, a) at the power density of the electron beam J =2.8× 10^4 W/cm^2, b) at the power density of the electron beam J=2.5×10^4 W/cm^2, c) and d) at the power density of the electron beam up to J =2.2×10^4 W/cm^2

4. Wear resistance of boride layers on high-cutting steel R18

After the electron-beam boriding of steel R18 with the power density of the electron beam J =2.8× 10^4 W/cm^2 layer was formed as a result of deep weld penetration, which determined its structure. Fig. 4a shows that at directed crystallization the main axis of dendrites line is oriented to the direction of heat removal. The structure of dendrites (a chain of separate globules) indicates about the intermittent nature of their formation. When the the power density of the electron beam is J=2.5×10^4 W/cm^2 the layer mainly consists of stellate dendrites (Fig. 4b). When reducing the power density up to J =2.2×10^4 W/cm^2, surface layer also contains stellate dendrites, but their number is not dominant (Fig. 4 c, d). Microhardness of the layer is 1100-1860 kg/mm^2. The layer thickness is 25-100 μm.

After electron beam treatment of cutting plate with boron-containing daubs, a layer with thickness of 8-10 μm was formed on the surface (Fig. 5). The formed layer has a specified thickness that is practically along its entire length. The image at 500× magnification shows that the layer contains particles which are located not only on the layer, but also on a boundary of base layer. This allows concluding that these particles are carbides of alloying elements (tungsten, chromium, molybdenum and vanadium). The layer firmly holds on the metal base. The microhardness of the layer is slightly higher than that of the base, and it is 550 HV and 410 HV, respectively.

To evaluate the tribological properties of the resulting layer we conducted wear resistant tests of the cutters. Tests were performed on screw-cutting lathe 1A616 in the following modes: feed was 0.1 mm/rev, rate speed was 224 rev/min, depth of cut was 1.0 mm. Processed material was steel 45 and sus321. The cutters wear was determined by the clearance face using Brinell magnifying glass.

Geometric parameters of cutters are as follows: rake angle γ= 15°, main back angle α = 8°, main cutting edge angle φ = 45°. Criterion of cutters resistance is the time of reaching the limit flank wear of 0.6 mm.

The Fig. 6 shows that formation boride layers with thickness of 8-10 μm on the front surface, allows to increase the resistance of cutters almost in 1.5 times (for processed steel sus321).

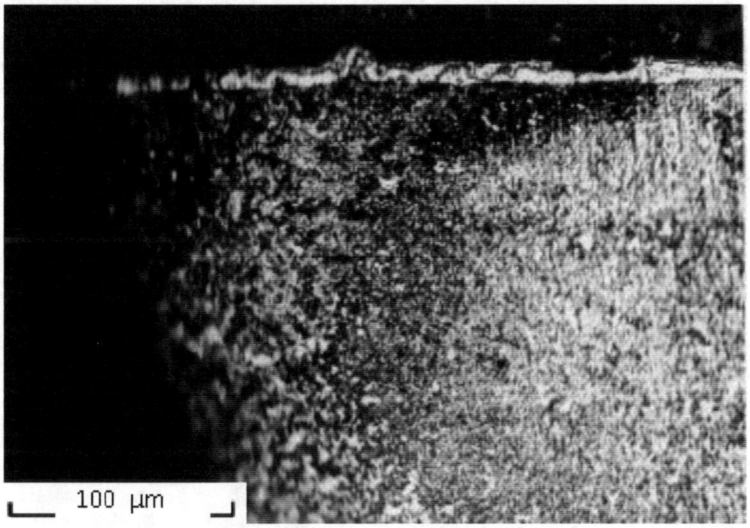

Figure 5. The microstructure of cutters from steel R18 after electron beam treatment

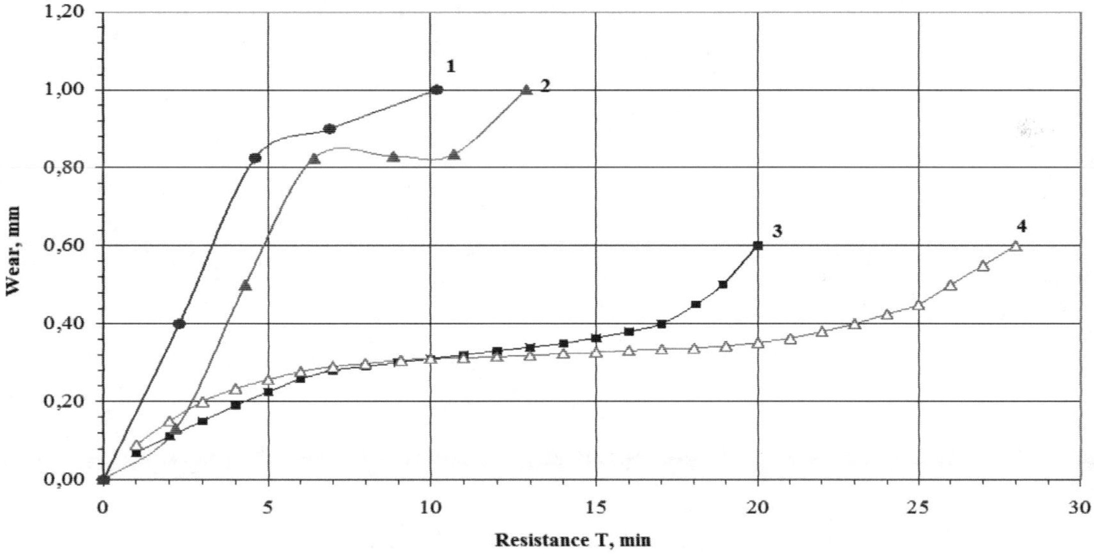

Figure 6. Wear kinetics of cutters from steel R18; 1-R18 hard (steel sus321), 2-R18 thermochemical treatment (steel 45), 3-R18 electron-beam boriding (steel sus321), 4-R18 hard (steel 45)

A more significant increase of cutters resistance is connected with increase of boride layer thickness. At electron beam boriding the thickness of boriding layer can reach 300 μm, but the processing temperature is 1100-1200°C. The processing time is 2-3 minutes. On that basis, the electron-beam boriding cannot be recommended as a final processing, but as an intermediate processing operation of the cutting tool, for example, before hardening. It should be noted that to achieve the same thickness of a layer when diffusion borating (for example, when boriding in hermetically sealed containers with safety valve) it take at least 3 hours.

5. Conclusion

The structure and properties of boride layers and their formation on the surface of cutting plates of carbon steel U8A formed during electron beam treatment in vacuum were investigated and discussed. It was found that at the directed crystallization the main axis of dendrites line is oriented to the direction of heat removal. The structure of dendrites (a chain of separate globuly) indicates about intermittent nature of their formation. The microhardness of layer is 1100-1860 HV; the thickness is 25-100 μm.

The formation of boride layers with the thickness of 8-10 μm on the front surface allows increasing the resistance of cutters almost in 1.5 times (for treated steel sus321).

Acknowledgments

This work was supported by the Russian Foundation for Basic Research, projects no. 14-08-31412_mol_a and 15-48-04217 r_sibir_a.

References

[1] Smirnyagina N N, Sizov I G, Semenov A P 2002 *Neorgan. Materials* **138 1** pp 48-54
[2] Dasheev D E, Smirnyagina N N 2012 *News of higher educational institutions. Physics* **55 12/2** pp 51-54
[3] Dasheev D E, Smirnyagina N N 2013 *J. Chem. Chem. Eng.* 7 **4** pp 319-325

Evaluation of Energy Characteristics of High Voltage Equipment for Electro-Blasting Destruction of Rocks and Concrete

N V Voitenko*, A S Yudin, N S Kuznetsova
Tomsk Polytechnic University, 30 Lenina ave., Tomsk, 634050, Russia

E-mail: tevn@hvd.tpu.ru

Abstract. The splitting off concrete monolith via electro-blast technology is described. High-voltage investigations were performed using the electric-discharge blasting unit with charging voltage up to 15 kV and maximum energy of 126 kJ. Several series of experiments of multi-borehole electro-blast with initiation of plasma channel by wire explosion have been carried out. The combined effect of pulse electric discharge and gas generating composition allows to increase the energy of electro-blasting. According to data obtained from experiments the average value of energy for concrete monolith splitting off of 1 m^3 is about 1554 kJ. The energy of electrical discharge pulse is about 95 kJ and gas-generating composition has energy contribution of 3.8 kJ/g. For carrying out of one electric-discharge blasting experiment 6-8 minutes has been spent. Therefore, the efficiency of pulse power system is about (0.85-0.9) m^3/h.

1. Introduction

Electro-blast method of solids destruction is based on the wave dynamics principles due to the rapid electrical energy release into the expanding plasma channel of the capillary electrical discharge in condensed media at impulse current flow through the channel [1].

A high voltage pulsed power electrical discharge through a solid can create a plasma channel with high energy density of about 25-30 kJ/cm^3. Inside the channel the electrical energy is transformed into internal thermodynamic energy which is subsequently used to deform and finally to destruct the surrounding solid. This process is basic for many technological applications where energy of 10-100 J/cm is deposited from a high voltage pulse generator in less than a few μs, leading to the increasing of plasma channel temperature and pressure of up to 10^4 K and 10^9 Pa respectfully [2-6].

The explosive expansion of plasma channel launches the shock-wave disturbances into surrounding solid and causes the stress-strain material state and formation of the mechanical compressive, tensile and shear stresses. Mechanical stresses are in the range of tens to hundreds of MPa and results in the plastoelastic deformations causing the solid destruction. The total effect of these factors leads to the radial crack initiation and propagation and, finally, to the solid media destruction. This method makes it possible to destruct the oversized rocks and concrete blocks, as well as carry out the directional concrete monolith splitting off.

The high energy efficiency of electro-blast method is caused by the impulse mode of loading of the material, providing the material brittle fracture with a predominance of tensile and shear stresses which required values are lower compared to the destruction by compression as in traditional mechanical method. The dynamic character of the loading provides the brittle fracture of solid with the

Journal of Physics: Conference Series **652** (2015) 012011 doi:10.1088/1742-6596/652/1/012011

minimal energy loss for the plastic deformation. The high degree of energy localization in the discharge channel ensure the crack initiation in solid at relatively low values of the single pulse energy, which is equivalent to grams of trinitrotoluene. The possibility of regulation over a wide range of the mode of plasma channel energy release allows to create the optimal conditions for solid loading and maintaining the propagation of cracks and destruction of the material depending on the nature and size of the destructible solid.

The efficiency of electro-blast method in comparison with the traditional mechanical methods strongly depends on the electrical properties of the destructible material and almost does not depend on its mechanical properties. Therefore, electro-blast method can be applied to the material of different mechanical strength, the greatest technical and economic effect of its application is reached with treatment of hard rocks and materials, and it makes a priority area of its use.

State-of-the-art of research in the field of the fundamentals of electro-blast technologies, despite the numerous and profound study of the discharge in condensed dielectrics [7-10], is characterized by a number of problems. There is still a matter of controversy, not only the role of parameters of electrical energy release in the plasma channel, but also the influence of wave processes on the formation of dynamically changing fields of mechanical stresses and deformations in solid. Electro-blast method of rocks and concrete fracture allows to control the plasma channel energy release and, thus, to affect the mechanical stress and deformation field around it, determining the material fracture pattern. The main goal in raising the efficiency of electro-discharge technologies is therefore matching of energy release rate conditions during the discharge and characteristics of the generated pressure wave causing the deformations in solid. At present the choice of regime parameters is carried out empirically and does not allow, not only to optimize them, but solve the fundamental task of technical and economic efficiency of blast-hole destruction.

The feature of electro-discharge installations is their impulse action on the destructible object. On the one hand, the radial cracks initiation requires the steep current impulse front, on the other hand - the main crack growth efficiency rises with increasing in pulse duration. To realize simultaneously both requirements in one installation is technically difficult. The efficiency of electro-discharge destruction of rocks and concrete depends on efficiency of conversion of stored electrical energy into a shock-wave disturbances and the duration of the energy release during the electro blast [11-12]. In order to increase the time of electro explosive effect on the destructible material walls, the researches have been carried out on the combined effects of the electrical discharge impulse and energy of the gas generating chemical composition.

2. Experimental Setup

An electric-discharge blasting unit was designed and fabricated at the High Voltage and Electro-physics Department of Tomsk Polytechnic University for carrying out the high voltage and high current investigations. The electric-discharge blasting unit is based on the custom high current pulse generator (HCPG) with operating voltage of up to 15 kV, peak current – up to 300 kA and stored energy in the battery of up to 126 kJ. The length of the bundle of 8 high voltage coaxial cables type RK-50-17-17 transmitting energy to borehole is 12 m. Pulse power system was mounted on the rigid metal frame and placed into a container with internal dimensions 275×175×180 cm.

3. Experimental Results of Reinforced Concrete Basement Destruction

The experiments of multi-borehole electric-discharge blasting on reinforced concrete basement were carried out. For splitting off concrete the electro-explosive cartridges (EC) as a working instrument were used. EC is a coaxial system with the thin copper wire which explodes at the moment of commutation of capacitor bank. Wire electro-explosion is the fast process of wire heating and evaporation by the high density current (up to 10^{12} A/m^2) which leads to plasma channel initiation accompanied with the further shock wave generation and its propagation into the surrounding material [1].

For increasing of the energy release duration with the aim of the borehole pressure growing as well as creation of additional mechanical tensile stress, the gas generating composition (GGC) was used. The GGC is a mixture of ammonium nitrate and diesel fuel. Each EC was filled with 18-20 grams of the mixture. The energy contribution of such chemical composition is 3.8 kJ/g. Under the certain conditions one can assume this composition as pyrotechnic mixture, not as explosives. The condition of assured detonation is provided with the minimum critical diameter of the cylindrical explosive when the detonation is possible. Since the critical diameter for such mixture is equal to 50 mm [13], and the diameter of the boreholes in the experiments was 26 mm, it is complies to conditions precluding the possibility of the transition of the combustion reaction to the detonation.

After the plasma channel initiation the GGC burns and generates a lot of gas, which creates additional pressure in the borehole. In order to increase the duration of heightened pressure in borehole, the borehole mouth was tamponed with construction gypsum. An arrangement of boreholes and the results of a series of experiments of multi-borehole electric-discharge blasting with GGC are shown in Fig. 1.

Fig.1. The experiments of multi-borehole electric-discharge blasting of concrete, (a) – an arrangement of boreholes, (b) – results of experiment series of electric-discharge blasting

The distance between the boreholes was 40-50 cm. The distance from the edge of the free surface was 40-45 cm, the depth of boreholes was 60-70 cm. Each electrode was sequentially connected with the coaxial cables and a high-voltage pulse with energy of about 95 kJ was applied. At the experiments four of the five previously prepared electrodes with EC were used for splitting off from the concrete monolith as it shown in Fig. 1 (a).

Table 1. The result of several series of the experiments

#	Quantity of boreholes	Consumed energy, kJ	Splitted off volume, m^3	Specific fracture energy, kJ/m^3
1	4	680	0.42	1619
2	4	680	0.38	1790
3	2	340	0.22	1545
4	2	340	0.25	1360
5	4	680	0.44	1545
6	5	850	0.58	1465

The results of calculations show that up to 30-35 % of stored energy dissipates as resistive losses in active resistance of the capacitors, coaxial cables and buses during the discharge. Only 9-10 % of released in EC energy was spent for material destruction. The rest part of energy was spent for discharge channel initiation, plasma formation and heating. The total energy consumed for splitting off concrete fragment with volume of about 0.42 m^3, considering the energy contribution from GGC, was equal to 680 kJ (Fig. 1 b). According to given technique a several series of experiments with different number of boreholes were carried out. The experiment results are given in Table 1.

4. Discussion

In [14], the specific fracture energy of various technologies of solid materials destruction is presented. The experimental results allow comparing the electric-discharge blasting technique with the other existing methods of materials destruction. Specific energy consumption of electric-discharge blasting technique is much lower while the total efficiency is considerably lower than traditionally used methods of rocks and concrete destruction. The possible reasons of that might be usage of one-time EC, imperfect of technological operations, small part of process automation et al.

As it was previously described, only a tenth of stored energy is allocated into the plasma channel. However, energy allocation for microsecond time period in the small volume leads to the increasing of plasma channel pressure up to GPa. At this moment the power of energy release reaches hundreds of MW, which results in the cracks initiation and propagation in the solid material. The usage of gas generating composition allows compensation of dissipated energy (resistive losses), and improving the electro-blasting method efficiency for solid destruction.

5. Summary

The experimental study of multi-borehole electric-discharge blasting accompanied with the gas generating composition was carried out to develop the approach of concrete splitting off. Combined effect of pulse electric discharge and gas generating composition allows compensation the much resistive energy losses with the small quantity of pyrotechnic mixture. Increasing in the energy release duration due to combustion reaction of GGC leads to the creation of additional mechanical tensile stresses, cracks formation enhancement and improving of the electro-blasting method efficiency.

The productivity of the electric-discharge blasting of solid materials destruction depends on the time of borehole drilling, installation of electrodes into the borehole, its connection to the electrode holder and finally charging of the capacitive energy storage. One can assume from the Table 1 that average energy required for 1 m^3 of lump splitting off from concrete monolith is about of 1554 kJ. It takes 6-8 minutes to complete one electric-discharge blasting. The time of removing split off pieces of concrete is not assumed. At the optimum mode of operation for distributing of such quantity of energy the 9-10 boreholes have been required. Therefore, the efficiency of present pulse power system is about (0.85-0.9) m^3/h.

The electric-discharge blasting unit can be applied for rock fragmentation, utilization of strong concrete constructions, dismantling of monolithic structures, tunnels construction or expansion and so on. The application of described technology in industrial scale requires the solution of some problems: the process automation, the usage of long life time operation instrument, the reducing of energy losses and improving of energy transfer efficiency into the discharge channel.

Acknowledgement

This work was supported by the Federal Target Program, project No. 14.575.21.0059 (RFMEFI57514X0059).

References

[1] Yudin A, Kuznetsova N, Lopatin V and Voitenko N 2014 *J. Phys.: Conf. Series*, **552** 012028
[2] Burkin V, Kuznetsova N and Lopatin V 2009 *J. Phys. D: Appl. Phys.* **43** 185204

[3] Inoue H, Lisitsyn I, Akiyama H, Nishizawa I, Drilling of hard rocks by pulsed power 2000 *IEEE ELECTR. INSUL. M.* **3** pp.19-25

[4] Silva C C M, Stellin A, Hennies W T, Costa E G Electrohydraulic Rock Blasting: An Alternative for Mining in Urban Areas 2002 *INT. J. MINING ENG.* pp.261-269

[5] Hofmann J, Th Weise H G G Pulsed Power Technologies for Commercial Material Reduction and Crushing Applications 1997 *Digest of 11th IEEE Pulsed Power Conference* (Baltimore) pp. 203-207

[6] Listisyn I, et al. Drilling and Demolition of Rocks by Pulsed Power 1999 *IEEE* pp. 169-172.

[7] Sjomkin B, Usov A and Kurets V 1995 *The principles of electric impulse destruction of materials* (St. Petersburg: Nauka) p 276

[8] Pronko S, Schofield G, Hamelin M and Kitzinger F Megajoule pulsed power experiments for plasma blasting mining applications 1993 *Proc. 9th IEEE Int. Pulsed Power Conf.* (Albuquerque) pp. 15-18

[9] Baltazar-Lopez M, Brandhorst H, Burell Z, Analysis and Simulations of Low Power Plasma Blasting for processing Lunar Materials 2009 *7th International Energy Conversion Engineering Conference* pp. 1-11.

[10] Chae J, Jeong Y, Shmelev V, Denicaev A, Poutchkov V, Ravi V Plasma Discharge Initiation of Explosives in Rock Blasting Application: A Case Study 2006 *PLASMA SOURCES SCI. T.* **8** pp. 165-168

[11] SikJin Y, Kim Y, Kim J Fabrication and testing of a 600-kJ pulsed power system 2013 *IEEE T. PLASMA SCI.* **41** pp. 2671-2674

[12] Bluhm H, Frey W, Giese H, Hoppe P, Schultheis C, Sträbner R Application of pulsed power HV discharges to material fragmentation and recycling 2000 *IEEE T DIELECT EL IN* **7** pp 625–636

[13] Shydlouski A 1973 *Fundamentals of pyrotechnics* (Moscow: Engineering) p 321

[14] Kurets V, Usov A, Zuckerman V 2002 *Electric pulse disintegration of materials* (Apatity: KSC RAS) p 324

Journal of Physics: Conference Series **652** (2015) 012012 doi:10.1088/1742-6596/652/1/012012

Blue and green jets in laboratory discharges initiated by runaway electrons

D.V. Beloplotov[1,2], M.I. Lomaev[1,2], D.A. Sorokin[1], V.F. Tarasenko[1,2]
[1]Institute of High Current Electronics, 2/3 Akademicheskii Ave., Tomsk, 634055, Russia.
[2]National Research Tomsk State University, 36 Lenin Ave., Tomsk, 634050, Russia.

Email: VFT@loi.hcei.tsc.ru

Abstract. Spectral and amplitude-temporal characteristics of plasma radiation of nanosecond pulse-periodic discharge in air, nitrogen and argon in pressure range of 30–760 Torr were investigated. Discharge gap geometry was a "point-to-plane". Voltage pulses of negative polarity (amplitude, FWHM and risetime was 13 kV, 10 ns and 4 ns, respectively) were applied to a pointed cathode made of different metals (stainless steel, aluminum and copper). Jets of different colour were observed near a cathode tip. They are formed due to explosive emission. Colour of jets depends on the cathode material. Intense lines of the atoms and ions of iron in the wavelength range of 200–600 nm, aluminum with $\lambda = 394.4$, 396.15 nm, including multiply charged ion Al VI with $\lambda = 360.39$ and 361.65 nm, copper with $\lambda = 324.8$, 327.3, 510.6, 515.3, 521.8, 522 nm were registered. The resonance energy transfer from metastable $A^3\Pi^+_u$ level of nitrogen molecule to $3d^{10}4p$ level of Cu I was found. As a result a luminescence duration of Cu I was about 1.5 ms at duration of discharge current of 1.5 μs. During constriction of the diffuse discharge the sputtering of material occurs in the direction perpendicular to the longitudinal axis of the discharge gap.

1. Introduction

Great attention has been given to study of pulse-periodic nanosecond discharges for the wide possibilities of their application in different fields of science and technologies [1, 2]. For example, a plasma of such discharge can be used for modification of different materials (metals, semiconductors, polymers) [1–3] as well as a point source of spontaneous radiation [4–7]. Depending on a discharge burning mode lines and bands of the different transitions can dominate in radiation spectrum. In diffuse discharge in nitrogen or air the radiation of the second positive (2^+) nitrogen system dominates [4, 5]. In a spark discharge a radiation of a cathode material vapour can become dominant in the emission spectrum. In paper [5] radiation of the atoms and ions of the iron was obtained at the interelectrode distance of 0.5 mm when the cathode made of a stainless steel was used. In paper [6] radiation of the atoms and ions of copper in discharge at atmospheric air pressure when the cathode made of a copper was obtained. In addition, bands of NO in a wavelength range of 200–300 nm were observed. Note that one of the features of the nanosecond discharge in "point-to-plane" gap is generation of the runaway electrons (REs) at the breakdown stage. REs are registered with collector behind the anode made of foil in both single and repetitive modes [8]. REs preionize gas in the gap and provide the diffuse discharge burning at high pressure.

The objective of the present work is study of spectral and amplitude-temporal characteristics of radiation of the pulse-periodic nanosecond discharge in air, nitrogen, argon at pressure range of 30-760 Torr with using different metallic cathodes (stainless steel, aluminum, copper).

2. Experimental setup

In experiments a repetitive pulser, a discharge chamber, spectrometers, a monochromator and a photomultiplier were used (figure 1). Voltage pulses of negative polarity with an amplitude of 13 kV, FWHM of 10 ns and risetime of 4 ns produced by the NPG-15/2000N pulser were applied to an electrode (cathode) made of different metals (stainless steel, aluminum, copper). Pulse repetition rate was 60 Hz. The cathode had a cone form. Diameter of cone base was 6 mm, and curvature radius of tip was about 0.2 mm. Grounded electrode (anode) was a disc with diameter of 38 mm. Interelectrode distance was 2 and 3 mm. Discharge current was registered with a shunt composed of low inductance chip-resistors. Spectra of the optical radiation of the discharge plasma from different regions of the discharge gap (shown in the red dashed frame in figure 1) were registered with the spectrometers HR4000 (Ocean Optics B.V., 1st: $\Delta\lambda = 330–425$ nm; 2nd: $\Delta\lambda = 200–305$ nm) and EPP-2000C (Stellar-Net Inc., $\Delta\lambda = 192–850$ nm). Waveforms of the discharge plasma radiation intensity from different regions of the discharge gap were registered with the MDR-23 monochromator and PMT-100 photomultiplier. Discharge chamber was filled with air, nitrogen and argon at pressure range of 30–760 Torr.

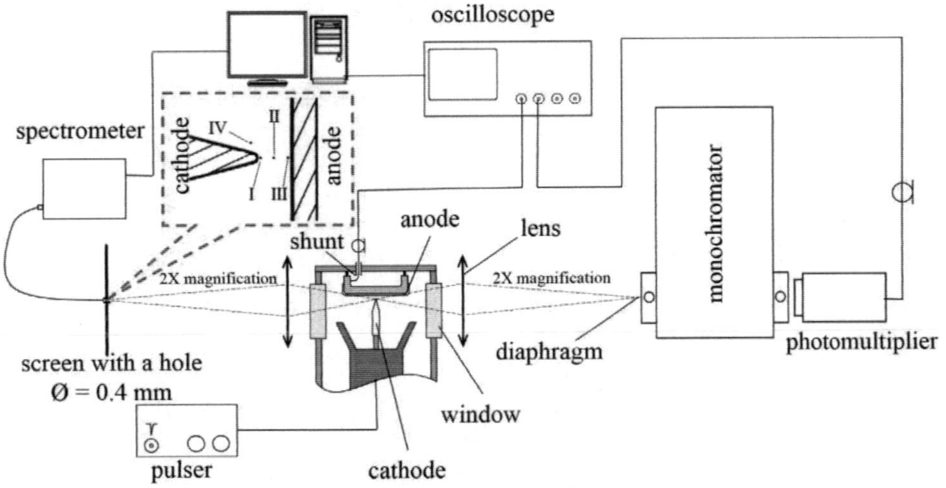

Figure 1. Scheme of experimental setup. Regions of spectrum registration are shown in red dashed frame.

3. Results

Let's consider results of experiment for each metal individually.

3.1. Stainless steel

The diffuse discharge was formed at pressure of gases up to 100 Torr. In this case radiation of (2^+) nitrogen system dominated in emission spectrum of discharge plasma. Subsequent pressure increasing led to formation of a spark discharge on the background of the diffuse discharge. Under these conditions, blue jets on the cathode surface around the one's tip were observed (figure 2). At gas pressure of 200 Torr and higher the blue jets were observed on the cathode tip only (figure 2b). The blue jets originate from bright spots (explosive emission centers) on the cathode surface. It indicates that the blue jets are a glow of metal vapor formed as a result of explosion of microinhomogeneities on the cathode surface and spark-erosion of the one.

Journal of Physics: Conference Series **652** (2015) 012012 doi:10.1088/1742-6596/652/1/012012

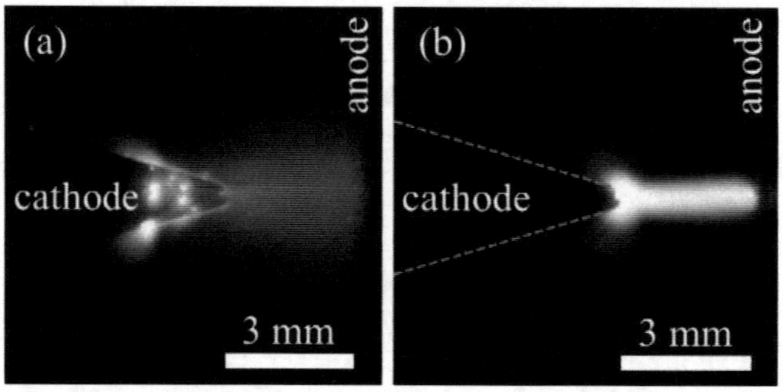

Figure 2. Image of the discharge plasma glow in air at pressure of 100 (a) and 200 Torr (b). Exposure time is 1 s. Pulse repetition rate is 60Hz. Cathode made of a stainless steel.

Increasing pressure leads to brightening of blue jets, but their size decreases. It's most likely that pressure increasing leads to reduction of injection distance of the metal vapor. The blue jets on the flat electrode surface were not observed. Spectrum of the discharge plasma radiation from regions IV and II at air pressure of 100 Torr and interelectrode distance of 2 mm are presented on figure 3. As is seen, radiation of (2^+) nitrogen system as well as the one of the nitrogen and oxygen atoms (N I, O I) and ions (N II, O II) dominates in the emission spectrum of the discharge plasma from the middle of the discharge gap, while near the cathode tip (region IV) a radiation of the iron atoms (Fe I) and ions (Fe II) dominates. There are about 12500 lines of Fe I and Fe II in the wavelength range of 200–600 nm [9].

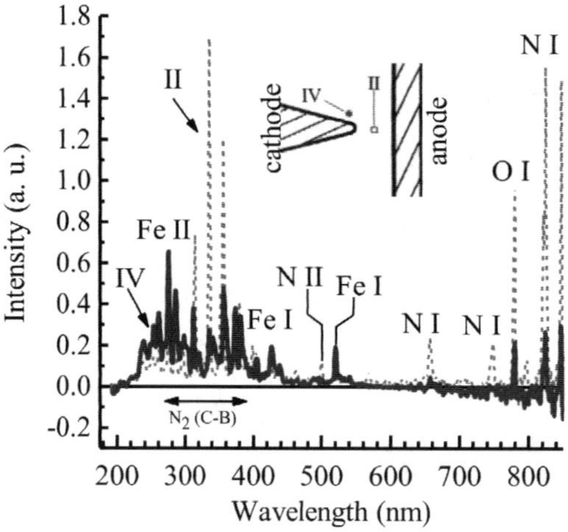

Figure 3. Spectrum of discharge plasma radiation in air at pressure of 100 Torr. Interelectrode distance is 2 mm. Cathode made of a stainless steel.

Dependence of discharge plasma radiation intensity from different regions of the discharge gap on gas pressure has a complex character. Spectra of the discharge plasma radiation from region I at argon pressure of 100 and 760 Torr and dependence of Fe I ($\lambda = 404.58$ nm) and Fe II ($\lambda_1 = 259.84$ nm, $\lambda_2 = 259.94$ nm) radiation intensity in different regions of the discharge gap on pressure of argon are shown in figure 4. It is seen that the radiation intensity of Fe II from regions I, II, III, IV (figure 4c) increases with pressure increasing up to about 400 Torr, but then it decreases. The radiation intensity of Fe I from region I (figure 4d) increases with pressure increasing up to 760 Torr, while one of regions II, III, IV reaches a maximum at pressure of about 400 Torr.

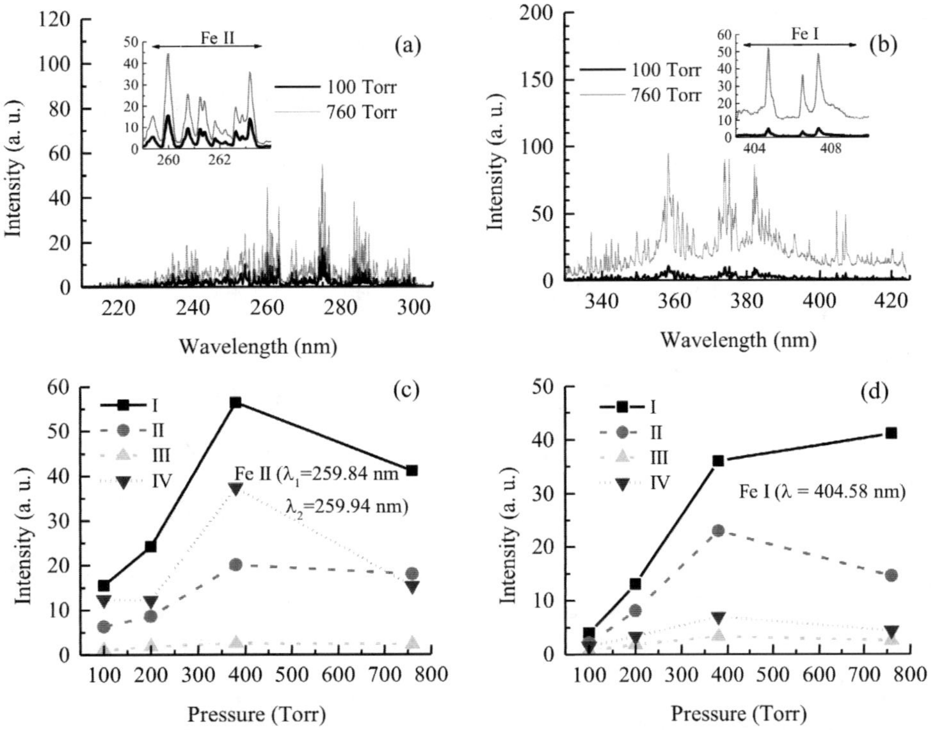

Figure 4. Spectra of the discharge plasma radiation from region I (a, b) and dependence of Fe I (λ= 404.58 nm) and Fe II (λ_1 = 259.84 nm, λ_2 = 259.94 nm) radiation intensity in different regions of the discharge gap on the argon pressure (c, d). Interelectrode distance is 2 mm.

As it was noted above, increasing of gas pressure leads to constriction of diffuse discharge. The discharge current density increases respectively. It contributes to erosion of the cathode and formation of the metal vapor. However, high pressure of gas reduces the injection distance of the metal vapor.

3.2. Aluminum

In the same way as for stainless steel the blue jets were observed on the cathode surface but not only near the cathode tip. At gas pressure of 30 and 50 Torr the discharge occupies whole volume of discharge chamber and the blue jets are observed on a cylindrical surface of the cathode. It is known that aluminum is more fusible as compared with stainless steel. Hence, the explosive emission centers on the aluminum cathode surface are formed at lower density of discharge current than on the stainless steel cathode surface. At gas pressure of 100 Torr and higher the blue jets were observed near the cathode tip only. In this pressure range the discharge formed between the cathode tip and the flat anode (like in figure 2b).

Spectral investigation with high-resolution spectrometers HR4000 shown that there are intensive radiation of the aluminum atoms (Al I) and ions (Al II) in emission spectrum of the discharge plasma. Radiation of Al I with λ = 394.4, 396.15 nm that impart blue colour of jets (upper level is $3s^24s$ ($^2S_{1/2}$) (3.14 eV), lower levels are $3s^23p$ ($^2P^0{}_{1/2}$) (0 eV) and $3s_23p$ ($^2P^0{}_{3/2}$) (0.014 eV) respectively [9]) as well as the one of Al II with λ = 280.12, 280.55, 281.62 nm are registered. Further still radiation of multiple charged ion Al VI with λ = 360.39 and 361.65 nm (upper levels are $2s^22p^3(^2D^0)3p$ ($^3D_2$) (127.84 eV) and $2s^22p^3(^2D^0)3p$ ($^3D_1$) (127.82 eV) respectively, lower level is $2s^22p^3(^2P^0)3s$ ($^1P^0{}_1$) (124.4 eV) [9]) were registered. Dependence of the radiation intensity of Al I on pressure of gases is similar to the one of Fe I (figure 4b). However maximum of the radiation intensity in region II, III, IV is observed at pressure of about 200 Torr. Waveforms of the radiation intensity of Al I with λ = 396.15 nm and (2^+)

nitrogen system with λ = 380.1 nm as well as waveform of the discharge current at the nitrogen pressure of 200 Torr are presented in figure 5. The waveform of the discharge current consists from some short spikes (about 40 ns) due to multiple reflections of the voltage pulse. It is caused by an impedance mismatch. Full duration of the discharge current is about 1.5 μs. Duration of one spike on the base is about 35 ns. Radiation of (2^+) nitrogen system with λ = 380.1 nm is registered during first spike (the diffuse stage of the discharge). Duration of the Al I radiation with λ= 396.15 nm is about 3 μs and exceeds the one of the discharge current. Likely, such character of the waveform of the Al I radiation is determined by a recombination afterglow.

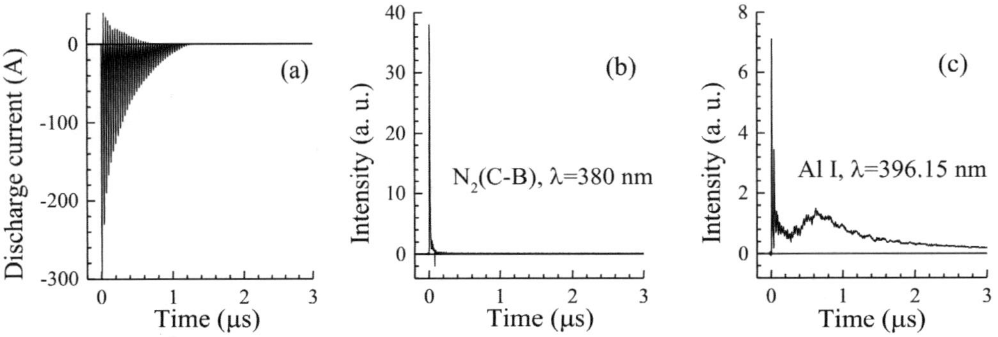

Figure 5. Waveforms of discharge current (a), radiation intensity of N_2(C–B) with λ = 380 nm (b) and radiation intensity of Al I with λ = 396.15 nm. Interelectrode distance is 2 mm. Nitrogen pressure is 200 Torr.

3.3. Copper

When the cathode was made of a copper green jets were observed. Like in case of the aluminum cathode, the green jets are observed on whole cathode surface at gas pressure of 30 and 50 Torr. At pressure above 100 Torr the green jets are observed near the cathode tip only. In emission spectrum of the discharge plasma radiation Cu I lines with λ = 324.8, 327.3, 510.6, 515.3, 521.8, 522 nm were registered. It is necessary to note that radiation intensity of Cu I with λ = 510.6 and 578.2 nm (laser lines) was a far fewer as compared with the one of Cu I with λ = 324.8, 327.3, 515.3, 521.8, 522 nm. Dependence of the radiation intensity of Cu I on gas pressure is similar to the one of Al I.

In air and nitrogen a glow of discharge plasma had an interest feature. Orange jets from discharge channel flowing along the anode surface were observed (figure 6a). However, the orange jets were not observed in argon and/or when the cathode was made of aluminum or stainless steel.

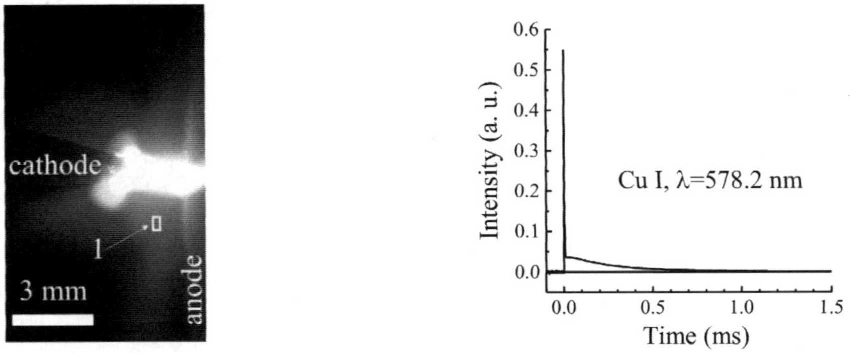

(a) (b)

Figure 6. Image of the discharge plasma glow in nitrogen at pressure of 200 Torr (a) (exposure time is 1 s, pulse repetition rate is 60 Hz, cathode made of a copper) and waveform of radiation intensity of Cu I with λ = 578.2 nm from region 1 (b). Interelectrode distance is 2 mm.

In the paper [10] it is noted that there is the resonance energy transfer from metastable $A^3\Pi^+_u$ level of nitrogen molecule (spontaneous lifetime is 13 s) to $3d^{10}4d$ level of Cu I in plasma. Moreover, high-energy vibrational levels of $X^1\Sigma^+_g$ state of nitrogen molecule can transmit energy to $3d^{10}4p$ level of Cu I. It leads to increasing of the luminescence duration of the Cu I with $\lambda = 578.2$ nm up to 1.5 ms (figure 5b).

Discharge plasma products sputter in radial direction at the discharge constriction. Copper atoms contained in the ones provide the orange glow of the jets. Thereby the sputtering of the discharge plasma products is visualized. Material deposition on the inner side surface of the discharge chamber in region that stood in the way of the orange jets was found. Deposited material consists from homogeneous particles with thickness about 0.5 μm and length about 3–5 μm.

4. Conclusion

We studies experimentally the spectral and amplitude-temporal characteristics of optical radiation of the nanosecond pulse-periodic discharge in air, nitrogen and argon at pressure range of 30–760 Torr. It was found that in diffuse discharge jets of different colour are formed on the cathode surface due to explosive emission. It wass shown, that on the one hand, pressure increasing leads to increasing intensity of the metal vapour formation due to constrictions of the diffuse discharge but in other hand, the one leads to reduction of the injection distance of the metal vapour. Colour of jets depends on the cathode material. Intense lines of the atoms and ions of iron, copper and aluminum, including the multiple charged ion Al VI with $\lambda = 360.39$ and 361.65 nm, were registered. It was found that due to the resonance energy transfer from metastable $A^3\Pi^+_u$ level of nitrogen molecule to $3d^{10}4d$ level of Cu the luminescence duration of Cu I with $\lambda = 578.2$ nm reaches about 1.5 ms. During constriction of the diffuse discharge the sputtering of material occurs in the direction perpendicular to the longitudinal axis of the discharge gap.

Acknowledgments

The work is performed in the framework of the Russian Science Foundation (the project #14-29-00052).

References

[1] Edited by Paul K. Chu, XinPei Lu 2014 *Low Temperature Plasma Technology* (CRC Press, Taylor & Francis Group)

[2] Hippler R, Kersten H, Schmidt M, Schoenbach K H (Eds.) 2014 *Low Temperature Plasma* (Fundamentals, Technologies)

[3] Baksht E H, Burachenko A G, Kostyrya I D, Lomaev M I, Rybka D V, Shulepov M A and Tarasenko V F 2009 *J. Phys. D: Appl. Phys.* **42** 185201

[4] Erofeev M V, Baksht E Kh, Tarasenko V F, Shut'ko Yu V 2010 *Quantum Electron* **40(6)** 561–564

[5] Baksht E Kh, Tarasenko V F, Shut'ko Yu V, Erofeev M V 2012 *Quantum Electron* **42(2)** 153–156

[6] Shuaibov A K, Laslov G E, Kozak Ya Yu 2014 *Optics and Spectroscopy* **116(4)** 552–556

[7] Shuaibov A K, Laslov G E, Minya A I, Gomoki Z T 2014 *Technical Physics Letters* **40(11)** 943–945

[8] Edited by Tarasenko V.F. 2014 *Runaway Electrons Preionized Diffuse Discharges*. (New York: Nova Science Publishers, Inc)

[9] Kramida A, Ralchenko Yu, Reader J and NIST ASD Team 2014 *NIST Atomic Spectra Database (ver. 5.2)* [Online] Available: http://physics.nist.gov/asd [2015, May 17]. National Institute of Standards and Technology, Gaithersburg, MD

[10] Mikheyev P A, Shepelenko A A, Kupryayev N V, Voronov A I 2002 *Proc. 3rd Int. Symp. On Theoretical and Applied Psasma Chemistry (Plios. Ivanovo: ISUCT)* vol 3 pp 138–141 [in Russian]

Structure and properties of commercially pure titanium nitrided in the plasma of a low-pressure gas discharge produced by a PINK plasma generator

Yu F Ivanov[1, 2, 3], Yu H Akhmadeev[1], I V Lopatin[1], E A Petrikova[1, 2], O V Krysina[1, 2], N N Koval[1, 2]

[1] Institute of High Current Electronics SB RAS, Tomsk, 634055, Russia
[2] National Research Tomsk State University, Tomsk, 634050, Russia
[3] National Research Tomsk Polytechnic University, Tomsk, 634050, Russia

E-mail: yufi55@mail.ru

Abstract. The paper analyzes the surface structure and properties of commercially pure VT1-0 titanium nitrided in the plasma of a low-pressure gas discharge produced by a PINK plasma generator. The analysis demonstrates that the friction coefficient of the nitrided material decreases more than four times and its wear resistance and microhardness increases more than eight and three times, respectively. The physical mechanisms responsible for the enhancement of strength and tribological properties of the material are discussed.

1. Introduction

Titanium and its alloys are widely used in industry and medicine due to their low specific weight, high corrosion resistance, and biological compatibility [1]. However, the materials display low hardness and low wear resistance, and this is a barrier to further extension of their applications. One of the promising ways to enhance the functional properties of materials, including titanium and its alloys, is their diffusion doping [2] which nitrogen is often used for. Nitriding considerably increases the surface strength and wear resistance of materials, make them less prone to tears, and provide their high cavitation and corrosion resistances in atmosphere, fresh water, and vapor. The deformation of materials during nitriding is minimal and their nitrided layers are easily grinded and polished.

The aim of the present study is to analyze the surface structure and properties of commercially pure titanium subjected to low-temperature nitriding.

2. Materials and methods of study

The material under study was commercially pure VT1-0 titanium [3], hereinafter also Ti, shaped as plates of thickness 4.5 mm and dimensions of 15 × 15 mm. The surface of the specimens was subjected to nitriding in the plasma of a low-pressure gas discharge produced by a PINK plasma generator which provides the possibility to increase the plasma density at comparatively low discharge operating voltages, vary the discharge current irrespective of the latter and of the pressure in the working chamber, and produce the plasma with no microdroplets [4]. The pressure at which a discharge operates in such systems can be varied from ~0.1 to ~5 Pa, making possible efficient ion cleaning of treated surfaces. The energy of bombarding ions is controlled using an additional bias voltage of 20–1000 V, and this allows one to control the rate of surface etching and vary the temperature of treated objects in the range 200–700°C. In our study, nitriding was realized at a temperature of 500 °C, 600 °C, and 650 °C for 1–5 h. Before nitriding, the specimens were mechanically polished to a roughness of Ra = 0.02 μm. One of the flat surfaces of each specimen (let term it the specimen back) was tightly pressed to the specimen holder and the other surface (its face) was free. The nitriding temperature was measured at the specimen holder with the use of a chromel-alumel thermocouple the thermal junction of which was insulated from the holder by a quartz cup. The structure of the modified material was examined by optical and scanning electron microscopy. The phase composition was determined by X-ray diffraction analysis. The surface properties were studied by estimating the microhardness, wear resistance, and friction coefficient.

3. Experimental results and their discussion

Tests of the specimens at an indenter load of 0.5 and 0.2 N show that the surface hardness of the treated material is more than three times higher than that of the untreated one (Fig. 1). The data in Fig, 1 suggest that the hardness of VT1-0 titanium depends on both the temperature and duration of nitriding. Nitriding at 500 °C fails to provide surface hardening and this is likely because of low nitrogen penetration into the surface layers. Conversely, nitriding at 600 °C and 650 °C does reveal surface hardening the degree of which is higher on the face of the specimens (Fig. 1, *a*) and lower on their back (Fig. 1, *b*). The surface hardening of both the face and back of the Ti specimens nitrided at 650 °C is damping (Fig. 1, curves *5* and *6*). The surface hardening of the specimen face nitrided at 600 °C is also damping (Fig. 1, *a*, curves *3* and *4*); on the back of the specimens, the surface hardness increases linearly with nitriding duration (Fig. 1, *b*, curves *3* and *4*).

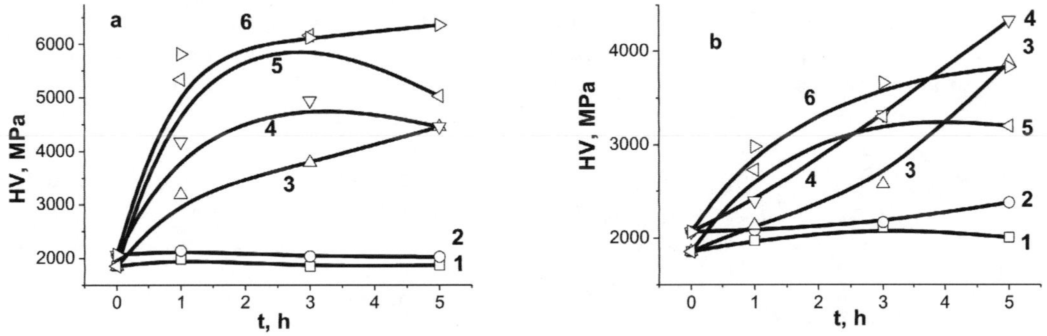

Figure 1. Surface microhardness HV of the specimen face (a) and back (b) vs the duration of nitriding **t** at 500 °C (1, 2), 600 °C (3, 4), and 650 °C (5, 6); P = 0.5 N (1, 3, 5); P = 0.2 N (2, 4, 6).

Nitriding in the low-pressure gas discharge plasma with ion bombardment of the Ti surface causes its etching. The roughness of both the specimen back and face (Fig. 2, curves *1*, *2* and *3*, *4*, respectively) increases whatever the nitriding temperature. At the same nitriding temperature and time, the roughness of the specimen face is 1.5-2.0 times higher than that of the specimen back.

Titanium and its alloys, despite their good strength and corrosion properties, are of limited use as structural materials because of their low wear resistance. The wear resistance of Ti-based alloys, where its high values are required, is increased by special surface treatment (nitriding, deposition of wear-resistant coatings, implantation). Figure 3 shows dependences of the friction coefficient on the number of loading cycles (friction track length, number of revolutions with respect to a counterbody made of hard VK8 (WC-Co) alloy) for the back and face of the nitrided Ti specimens. It is seen that the minimum friction coefficient of the specimen back is 0.13 (Fig. 3, *a*, stage *1*), and after ~50 cycles, the friction coefficient reaches ~0.46 and then, with regard to its fluctuation, varies in the range 0.38–0.54 (Fig. 3, *a*, stage *2*). Note that the friction coefficient of untreated VT1-0 titanium reaches ~0.46 even in the first cycles and then, with regard to its fluctuations, varies in the range 0.38–0.56. The face of the nitrided Ti specimen features a rather extended initial stage of wear with a low (0.09–0.23) and stable friction coefficient (Fig. 3, *b*, stage *1*), suggesting that the wear at this stage is mostly abrasive. Following the initial stage, the friction coefficient and its fluctuation amplitude increases, reaching values close to those of commercially pure VT1-0 titanium (Fig. 3, *b*, stage *2*).

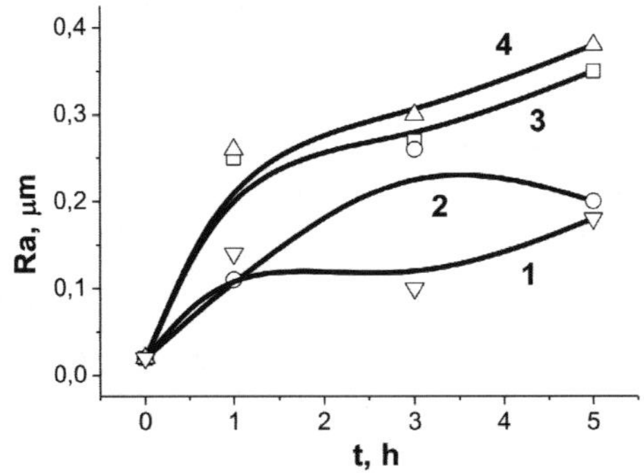

Figure 2. Surface roughness Ra of the specimen back (1, 2) and face (3, 4) vs the duration of nitriding **t** at 600 °C (2, 3) and 650 °C (1, 4).

Figure 4 shows 3D images of wear tracks for the initial and nitrided Ti specimens. It can be concluded from the images that the initial Ti specimens undergo mostly adhesive wear (Fig. 4, *a*) in which volumes of the material are torn from its depth, transferred from one friction surface to another, and the thus formed jogs affect the conjugate surface [5]. The wear track of the nitrided specimen is smooth (Fig. 4, *b*, *c*), suggesting that the wear is abrasive and is due to cutting or scratching. A comparative profilometric analysis shows that the surface hardness of the specimen nitrided at 650 °C for 3 h increases more than three times, its friction coefficient decreases more than four times, and this increases the wear resistance of the specimen more than eight times.

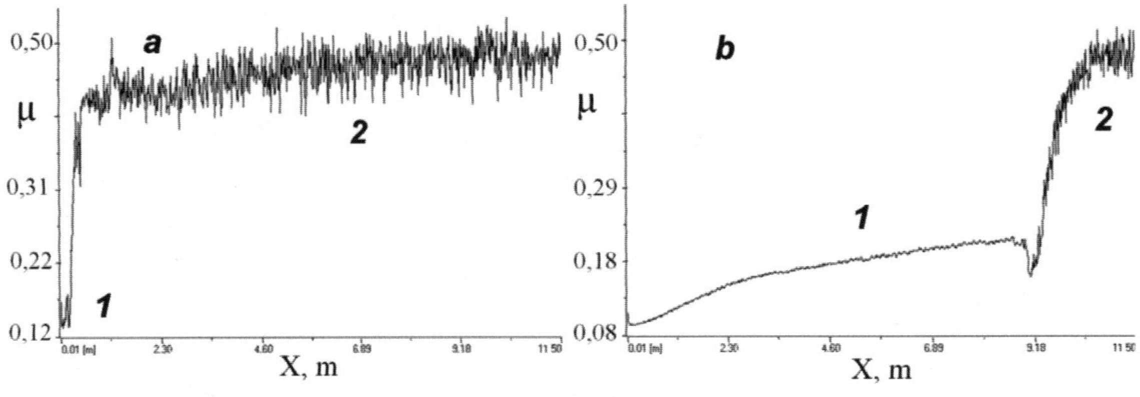

Figure 3. Friction coefficient **µ** vs the friction track length for the back (a) and face (b) of the Ti specimen nitrided at 650 °C for 3 h and tested with an indenter (VK8 ball) of diameter 3 mm at a track radius of 2 mm, applied load of 3 N, and indenter path of 11.5 m

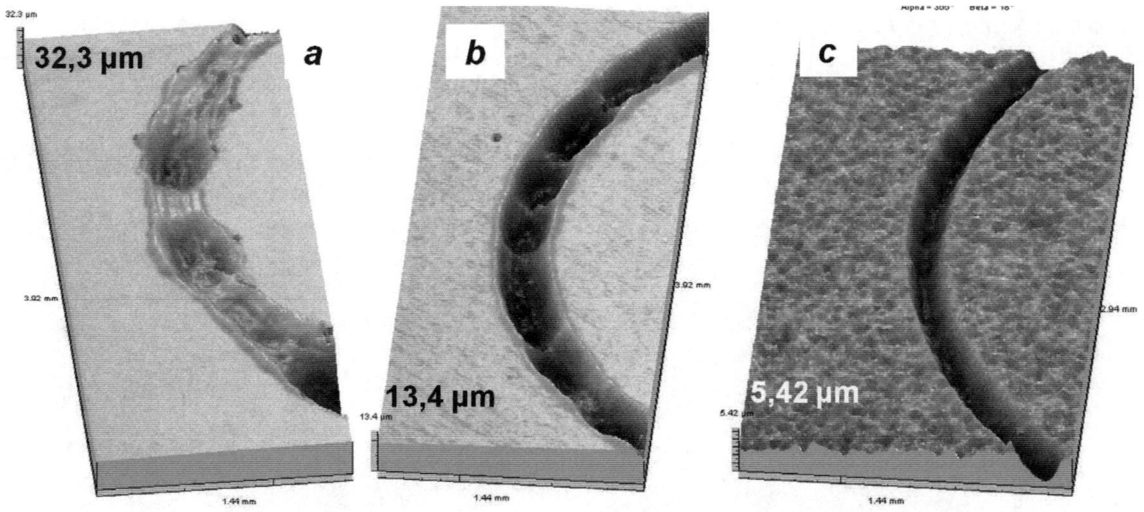

Figure 4. Images of wear tracks in the initial titanium (a) and titanium nitrided at 650 °C for 3 h (b) and tested with an indenter (VK8 ball) of diameter 3 mm at a track radius of 2 mm, applied load of 3 N, and indenter path of 11.5 m.

The many-fold improvement of the microhardness, wear resistance, and frictional properties of the Ti specimens nitrided in the low-pressure gas discharge plasma produced by the PINK generator is evidence for considerable changes in the material surface structure. The elemental composition and defect substructure of the Ti surface after nitriding were examined by optical and scanning electron microscopy for which the Ti specimens nitrided at 650 °C for 3 h were used.

The study shows that before nitriding the titanium represents a polycrystalline material with an average grain size of 25.8 μm. After nitriding at 650 °C for 3 h, the average grain size in its surface layers is 27.0 μm on the specimen back and 22.5 μm on the specimen face, which is close to the average grain size in the initial material (Fig. 5, *a*, *c*). In the grain volume of the nitrided specimen, a subgrain structure with an average subgrain size of 0.75 μm on the specimen back and 2.1 μm on the specimen face is formed (Fig. 5, *b*, *d*).

The grain and subgrain boundaries in the nitrided Ti specimen are decorated with second phase particles. The specimen back is characterized by an average particle size of 0.3 μm at the grain boundaries and 0.1 μm at the subgrain boundaries. For the specimen face, the average particle size at the grain and subgrain boundaries is 0.25 μm. The data of scanning electron microscopy shows that the nitrided specimen contains nitrogen atoms the concentration of which on the specimen back and face is 33 at. % and 36 at. %, respectively. Likely the second phase particles at the grain and subgrain boundaries are titanium nitride particles. And this is really so, as evidenced by X-ray studies which reveal the presence of an ε-phase (Ti_2N) with a content of 24.8 % on the specimen face. Assuming the formation of Ti_2N particles on the specimen back as well and taking into account the average particle size on both sides, the volume fraction of ε-phase particles is by estimate 1.8 %.

Nitriding, while involving the formation of nitride particles in the titanium, implies nitrogen saturation of the Ti lattice with its attendant distortion and change of its parameters. According to the X-ray data, the α-Ti lattice parameters in the nitrided titanium are a = 0.29454 nm, c = 0.47091 nm, c/a = 1.599, and $\Delta d/d = 2.7*10^{-3}$, whereas their table values for the α-Ti lattice are a = 0.29510 nm, c = 0.46970 nm, and c/a = 1.587. Thus, as evidenced by the change in the lattice parameters a and c, the Ti lattice is saturated with nitrogen.

The foregoing research data on the elemental composition, phase composition, and defect substructure in the nitrided surface layer of commercially pure VT1-0 titanium suggest that the strength and tribological properties of the material are increased through saturation of its lattice with nitrogen atoms, subgrain structure formation, and Ti_2N particle precipitation. When nitrided, the entire face of the Ti specimen is almost uniformly involved in Ti_2N precipitation, whereas this precipitation on the specimen back dominates at the grain and subgrain boundaries.

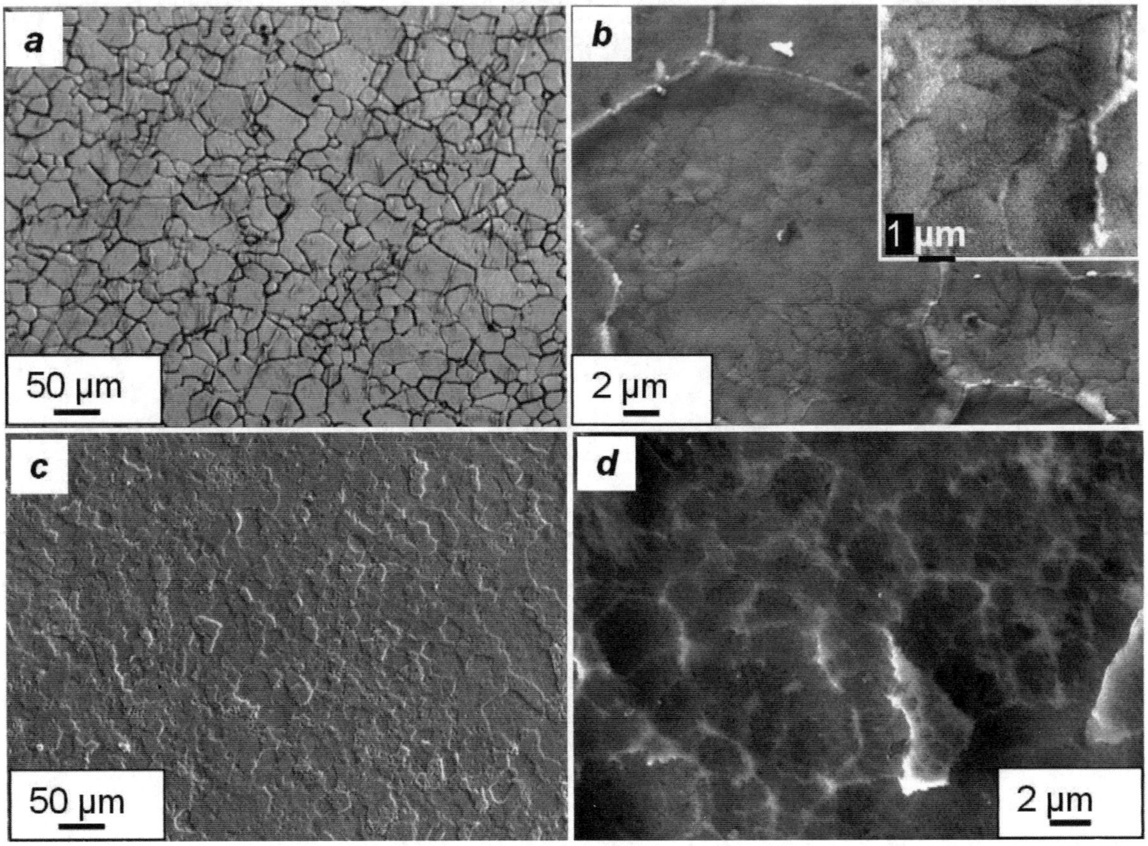

Figure 5. Surface structure on the back (*a*, *b*) and face (*c*, *d*) of commercially pure VT1-0 titanium nitrided at 650 °C for 3 h.

Formation of surface alloy is followed by change of strength characteristics of material. In Figure 1 the results of hardness change for the alloyed aluminum surface layer are presented. It is clear that the hardness and the Young's modulus of surface layer smoothly decreases from the values which is repeatedly exceeding strength characteristics of initial material to the values corresponding to A7 alloy. Therefore, the modified layer smoothly passes into the main volume of a sample. The boundary, which can contain concentrators of tension, is absent.

4. Conclusion

Thus, we analyzed the structure and properties of commercially pure VT1-0 titanium nitrided in the low-pressure gas discharge plasma produced by the PINK plasma generator. It is shown that the nitriding decreases the friction coefficient of the material more than four times and increases its hardness and wear resistance more than three and eight times, respectively. It is demonstrated that the strength and tribological properties of the material are enhanced due to nitrogen saturation of its lattice, subgrain structure formation, and Ti_2N particle precipitation. The research results allow us to suppose that the face of the Ti specimen is nitrided by intercrystalline mechanisms through diffusion at its internal interfaces and by intracrystalline mechanisms through diffusion at dislocation cores and interstitial sites of its crystal lattice. The back surface of the Ti specimen is nitrided mostly by intercrystalline mechanisms.

Acknowledgement

This work was supported by RSF grant (project No. 14-29-00091).

References

[1] Iliin A A, Kolachev B A, Polkin I S 2009 *Titanium alloys: composition, structure, properties*. – (M. MATI) 519
[2] Lakhtin Yu M, Kogan Ya D, Shpis G, Bemer Z 1991 *Theory and technology of nitriding*. – (M. Metallurgia) 320
[3] Sorokin V G 2001 *Grades of steels and alloys. Reference book*. – (M. Intermet Engineering) 608
[4] Schanin P M, Koval N N, Goncharenko I M, Grigorev S V 2001 Nitriding constructional steel in gas discharges of low pressure Fiz. Khim Obr. Mater. No. 3 16-19
[5] GOST 27674-88: *Friction, wear, and lubrication. Terms and definitions*

Journal of Physics: Conference Series **652** (2015) 012014 doi:10.1088/1742-6596/652/1/012014

Surface structure of commercially pure VT1-0 titanium irradiated by an intense pulsed electron beam

Yu F Ivanov[1, 2, 3], A D Teresov[1], E A Petrikova[1, 2], O V Ivanova[4], I A Ikonnikova[4]

[1] Institute of High Current Electronics SB RAS, Tomsk, 634055, Russia
[2] National Research Tomsk State University, Tomsk, 634050, Russia
[3] National Research Tomsk Polytechnic University, Tomsk, 634050, Russia
[4] Tomsk State University of Architecture and Building, Tomsk, 634002, Russia

E-mail: yufi55@mail.ru

Abstract. It is shown that pulsed electron beam irradiation of commercially pure titanium at a beam energy density of 10 J/cm^2, pulse duration of 150 μs, number of pulses of N = 5 pulses, and pulse repetition frequency of 0.3 Hz with attendant polymorphic $\alpha \rightarrow \beta \rightarrow \alpha$ transformations allows a more than five-fold decrease in the grain and subgrain sizes of the material structure.

1. Introduction

Electron beams are widely used for vacuum treatment of materials, including their welding, melting, and annealing, as well as for modification of polymers, curing of lacquer coatings, sterilization of foodstuff and medical tools, microwave generation, and excitation of active media in gas lasers [1–4]. The technology of surface modification by intense pulsed electron beams holds much promise due to the possibility of increasing the wear resistance of materials, their corrosion or oxidation resistance, fatigue life, etc. [5–7].

The aim of the present study is to analyze the structure and properties of surface layers in commercially pure titanium irradiated by a submillisecond intense pulsed electron beam.

2. Materials and methods of study

The material under study was commercially pure VT1-0 titanium [8], hereinafter also titanium or Ti, shaped as plates of dimensions 15 × 15 mm and thickness 4.5 mm. Before electron beam irradiation, the specimens were mechanically polished to Ra = 0.02 μm. The specimen surface was irradiated on a SOLO setup by a submillisecond intense pulsed electron beam at a beam energy density of (10–25) J/cm^2, pulse duration of 100 μs and 150 μs, and pulse repetition frequency of 0.3 Hz; the number of pulses was N = 5. The surface structure of the irradiated specimens was examined by optical, scanning, and transmission electron microscopy.

3. Model

In the model proposed, the problem of finding the temperature field in a certain range of electron energy density is reduced to solving the heat conduction equation for the one-dimensional case of heating and cooling of a plate of thickness *d*. The coordinate system is such that the axis x is directed deep into (inward) the specimen. At $x = 0$, a heat flux is specified, and the backside of the plate is free from heat exchange. In the coordinate form, the heat conduction equation is written as

$$c\rho\frac{\partial T}{\partial t} = \lambda\frac{\partial^2 T}{\partial x^2}. \quad (1)$$

Here c is the specific heat capacity, ρ is the density, and λ is the heat conductivity of a material.

The boundary conditions for pulsed electron beam treatment have the form

$$-\lambda\frac{\partial T}{\partial x} = q(t), \quad (2)$$

where the heat flux from the surface deep into the specimen is

$$q(t) = \begin{cases} q_0, & 0 < t \le t_0; \\ 0, & t > t_0. \end{cases} \quad (3)$$

Here q_0 is the average heat flux during the time of electron beam action t_0. The initial temperature is $T(0, x) = T_0$ throughout the specimen depth $0 < x < d$.

For numerical solution of the problem, we use a difference grid with time and space steps τ and h: $t_j = j\tau$ and $x_i = ih$, where j is the number of a time step and i is the number of a space step. The temperature is determined at the grid nodes. The differential heat conduction equation is approximated with the first order in time and second order in space by an explicit scheme:

$$T_{j+1,i} = T_{j,i} + \frac{\tau(T_{j,i+1} - 2T_{j,i} + T_{j,i-1})}{c\rho h^2}. \quad (4)$$

The numerical solution of the heat conduction equation was used to calculate the temperature fields in Ti specimens irradiated by an intense electron beam with an energy density of (10–40) J/cm^2 and pulse duration of (50, 100, 150, 200) μs. The surface layer thickness was $d = 10^{-3}$ m; the observation time was 300 μs.

For melting and evaporation of the material, we used the following model representation [5]. As the melting temperature T_{mel} is reached in a cell, the cell temperature is taken to be fixed and equal to this melting temperature, and the overall heat

$$q = \frac{\left(T_{j,i} - T_{пл}\right)c}{q_{пл}}, \quad (5)$$

where T_{ji} is the cell temperature, is expended in melting the specimen; $q = 1$ means that the material transform into a liquid state. The model representation of crystallization is the same, except that q is decreased from 1 to 0, and so is the model representation of evaporation and condensation. As the boiling temperature T_{boil} is reached in a cell, the cell temperature is taken to be fixed and equal to this boiling temperature, and the overall heat

$$q = \frac{\left(T_{j,i} - T_{кип}\right)c}{q_{кип}} \quad (6)$$

is expended in evaporating the matter; $q = 1$ means that the matter transforms into vapor.

The thermophysical properties of titanium in calculations were the following [9, 10]: heat conductivity $\lambda = 20*10^{-2}$ W/(cm·K) and specific heat capacity $c_p = 684*10^{-3}$ J/(g·K) both at T = 1000 K; density $\rho = 4.5$ g/cm^3; melting temperature $T_{mel} = 1881$ K; evaporation temperature $T_{ev} = 3560$ K; melting heat $q_{mel} = 358$ kJ/kg; and evaporation heat $q_{ev} = 9720$ kJ/kg.

Journal of Physics: Conference Series **652** (2015) 012014 doi:10.1088/1742-6596/652/1/012014

Table 1 presents calculated and experimental values of the melted layer thickness for VT1-0 titanium subjected to pulsed electron beam irradiation.

It is clearly seen that the calculation and experimental data on the melted Ti layer thickness are in good qualitative and quantitative agreement. This suggests that the model representation adequately reflects the processes occurring in the Ti surface layer subjected to intense pulsed electron beam irradiation and can be used for interpretation of experimental data.

Table 1. Calculated and experimental thickness of the melted Ti surface layer subjected to intense electron beam irradiation at a pulse duration of 50 μs

Calculation		Experiment	
Es, J/cm^2	d, μm	Es, J/cm^2	d, μm
10	0.0	12	0.5
15	7	18	6
20	13	20	8.5
25	17	25	14

Figure 1 shows calculated dependences of the melted Ti layer thickness on the beam energy density. Analysis of these data allows the conclusion that with a beam energy density of 10 J/cm^2, the surface layer of VT1-0 titanium is involved in solid phase transformations at both pulse durations (100 and 150 μs). The initial stage of surface melting (melted layer thickness ≈ 1.5 μm) is observed at a pulse duration of 100 μs with a beam energy density of 15 J/cm^2 and at a pulse duration of 150 μs with a beam energy density of 20 J/cm^2.

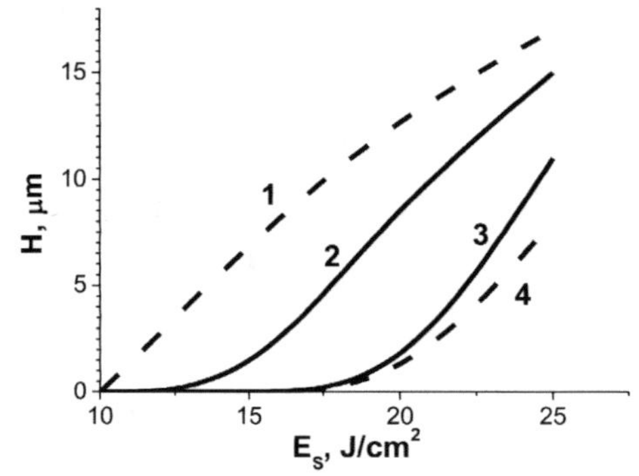

Figure 1. Melted Ti layer thickness vs the beam energy density at a pulse duration of 50 (*1*), 100 (*2*), 150 (*3*), and 200 μs (*4*).

Figure 2 shows calculated dependences of the melt lifetime vs the pulse duration and beam energy density. It is clearly seen that the melt lifetime reaches its maximum (slightly more than 96 μs) at a pulse duration of 100 μs and beam energy density of 25 J/cm^2.

According to the calculations performed, the maximum temperature on the surface of commercially pure VT1-0 titanium varies in the range 1760–4000 K at a pulse duration of 100 μs and in the range 1500–3300 K at a pulse duration of 150 μs. Remind that the beam energy density in both cases was varied from 10 to 25 J/cm^2 (Fig. 3).

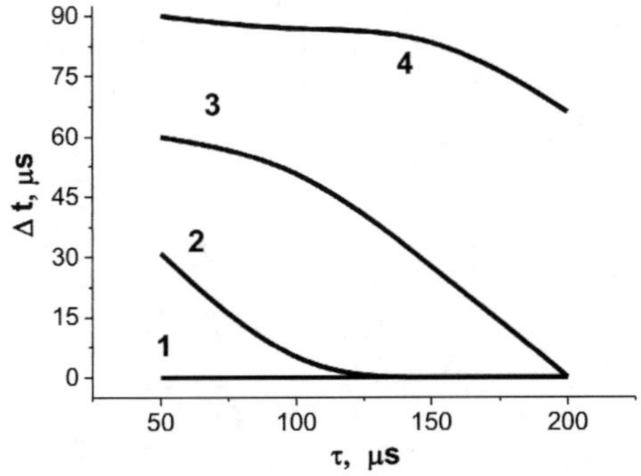

Figure 2. Melt lifetime on the Ti surface vs the pulse duration at a beam energy density of 10 (*1*), 15 (*2*), 20 (*3*), and 25 J/cm^2 (*4*).

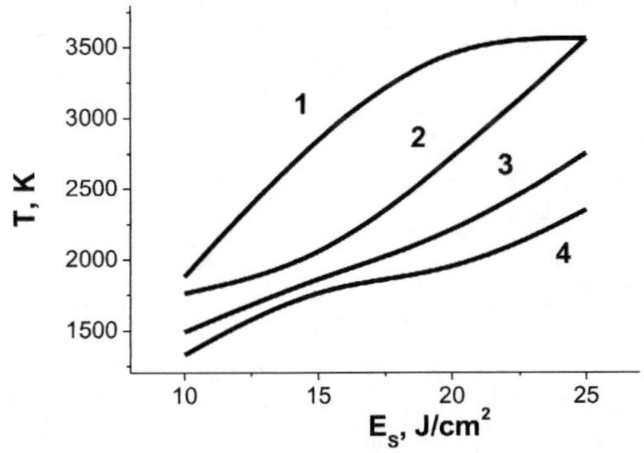

Figure 3. Maximum temperature on the Ti surface vs the beam energy density at a pulse duration of 50 (*1*), 100 (*2*), 150 (*3*), and 200 μs (*4*).

Thus, the calculations suggest that the model adequately describes the thermal processes occurring in the surface layer of commercially pure VT1-0 titanium irradiated by an intense pulsed electron beam and allows one to estimate different parameters the main of which are presented in Figs. 1–3.

4. Experimental results and their discussion

Figure 4 shows images of the surface structure in commercially pure VT1-0 titanium before and after intense electron beam irradiation.

It is readily seen that the initial titanium represents a polycrystalline material with an average grain size of 26 μm (Fig. 4, a). After intense electron beam irradiation, the structure of its surface layer is considerably refined (Fig. 4, b–d). The average grain size revealed on the irradiated surface by optical microscopy is 4–5 μm (Fig. 4, b), 8–10 μm (Fig. 4, c), and 12–13 μm (Fig. 4, d). Hence, the polymorphic $\alpha \rightarrow \beta \rightarrow \alpha$ transformations involved in pulsed electron beam irradiation at 10 J/cm^2, 150 μs, N = 5 pulses, and 0.3 Hz allows a many-fold decrease in the grain and subgrain sizes of the Ti structure. Electron beam irradiation at 15 J/cm^2, 100 μs, N = 5 pulses, 0.3 Hz with attendant melting of the surface layer and its high-rate crystallization provides the formation of a structure in which the minimum average grain size is 8–10 μm (Fig. 4, c).

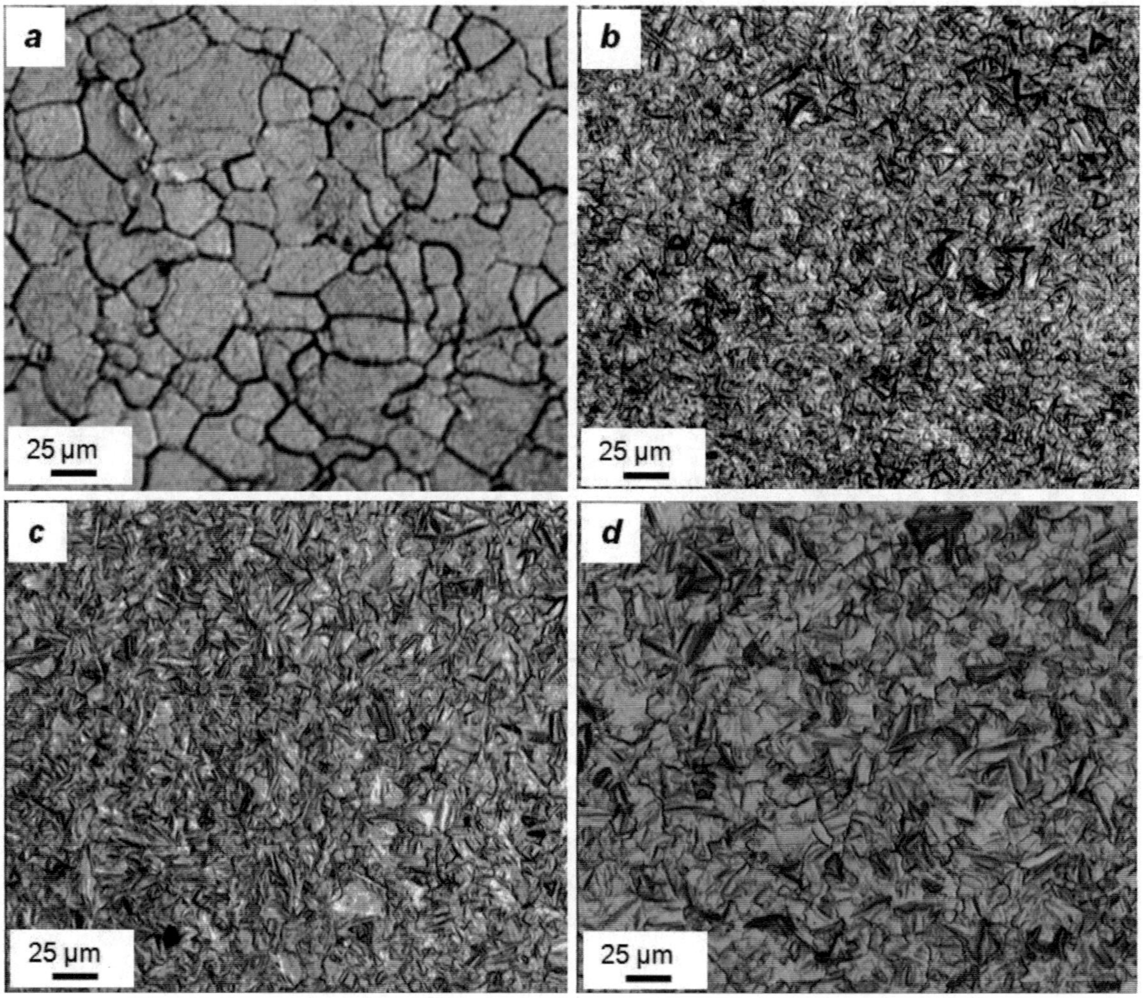

Figure 4. Etched surface of the initial titanium (a) and its surface after electron beam irradiation at 150 µs, 10 J/cm^2 (b); 100 µs, 15 J/cm^2 (c); and 150 µs, 20 J/cm^2 (d).

5. Conclusion

Thus, the one-dimensional model used in the study allowed us to calculate the temperature field arising in commercially pure VT1-0 titanium irradiated by an intense pulsed electron beam. The calculation results were compared with experimental data on the structural phase state of the material subjected to electron beam irradiation at different beam energy densities and different pulse durations. It is shown that pulsed electron beam irradiation at 10 J/cm^2, 150 µs, N = 5 pulses, and 0.3 Hz with attendant polymorphic $\alpha \rightarrow \beta \rightarrow \alpha$ transformations allows a many-fold decrease in the grain and subgrain sizes of the material structure.

Acknowledgement

This work was supported by RSF grant (project No. 14-29-00091).

References

[1] Bugaev S P, Kreindel Yu E, Schanin P M 1984 *Electron beams of large cross-section*. – (M.

Energoatomizdat) 113
[2] Nazarov D S, Ozur G E, Proskurovsky D I 1997 Production of low-energy, high-current electron beams in a reflected-discharge plasma-anode gun *Proc. 11th IEEE Int. Pulsed Power Conf.*, (Baltimore, USA) **II**. 1335–1340.
[3] Engelko V, Mueller G, Bluhm H 2001 Influence of particle fluxes from target on characteristics of intense electron beams *Vacuum* 62/2-3 97–103
[4] Koval N N, Oks E M, Protasov Yu S, Semashko N N 2009 *Emission electronics* (M. Bauman Moscow State Tech. Univ.) 596
[5] Rotshtein V, Ivanov Yu, Markov A 2006 Surface treatment of materials with low-energy, high-current electron beams *Materials Surface Processing by Directed Energy Techniques*, (P. Elsevier) **6** 205–240
[6] Koval N N, Ivanov Yu F 2008 Nanostructuring the surface of metal-ceramic and ceramic materials under pulsed electron-beam treatment *Izv. Vyssh. Uchebn. Zaved. Fiz.* No. 5 60–70
[7] Belov A B, Bytsenko O A, Krainikov A V 2012 *High-current pulsed electron beams for aircraft engines*. (M. Dipak) 292
[8] Borisov E A, Bochvar G A, Brun M Ya 1980 *Titanium alloys. Metallography of titanium alloys* (M. Metallurgia) 464
[9] Babichev A P, Babushkina N A, Bratkovsky A M 1991 *Physical quantities: Reference book* (M. Energoatomizdat) 1232
[10] *Physical Encyclopedia* 1998 (M. Great Russian Encyclopedia) **5** 687.

Features of formation of structural-phase states on the surface of titanium alloy VT1-0 after electron-ion-plasma treatment

Yu F Ivanov[1, 2, 3], A D Teresov [1], O V Ivanova[4], V E Gromov[5], E A Budovskikh[5], V A Vlasov[4], A A Klopotov[3, 4]

[1] Institute of High Current Electronics SB RAS, Tomsk, 634055, Russia
[2] National Research Tomsk Polytechnic University, Tomsk, 634050, Russia
[3] National Research Tomsk State University, Tomsk, 634050, Russia
[4] Tomsk State University of Architecture and Building, Tomsk, 634002, Russia
[5] Siberian State Industrial University, 654007, Novokuznetsk, Russia

E-mail: yufi55@mail.ru

Abstract. Complex modification of a surface of commercially pure titanium is realized. Firstly plasma is created by electrical explosion of a carbon-graphite fiber, of which surface was placed nanosized TiB_2 powder. Then the surface of technically pure titanium is processed with this plasma. Finally, the modified surface was irradiated by an electron beam. Formation of multi-layer multiphase nanosized structure is revealed. It is shown that the maximum microhardness reached in a near-surface layer exceeds microhardness of a initial material more than by 10 times. Wear resistance of a blanket increases in 7.5; the friction coefficient decreases by 1.15 times.

1. Introduction

Currently, one of the most promising approaches to creation of wear-resistant coatings is a method of formation of composite materials with titanium diboride. Such a composite material has a high load-bearing capacity combined with high physical and mechanical properties (high hardness, low values of the thermal expansion coefficient, low sensitivity to heat shock) and is not subject to plastic deformation which contributes to accumulation of defects in a crystalline lattice and the subsequent damage during a tribological contact. Such a surface is possible to create using pulsed melting with simultaneous saturation of the material surface layers with doping elements followed by crystallization and formation of strengthening phases, carried out by plasma formed during an electric explosion of a conductive material (electroexplosive doping), is one of the promising methods for modification of the structure and properties of metals and alloys.

This paper presents the results of the study on the combined treatment effect, which includes electroexplosive doping and the subsequent exposure to a high-intensity pulsed electron beam, on the structural-and-phase state of the surface layer of the alloy based on titanium VT1-0.

2. Materials and methods of study

Doping of the surface layer of the alloy based on titanium VT1-0 was carried out by exposure to plasma formed during an electric explosion of a graphitized carbon fiber, on the surface of which in the explosion area a sample of the powder of titanium diboride TiB_2 was placed. Additionally, pulsed melting of the modified layer was carried out with a high-intensity electron beam on the installation

"Solo", under the mode: electron energy 18 keV, energy density of the electron beam (45-60) J/cm^2, duration and the number of exposure pulses 100 ms, 10 pulses; 200 ms, 20 pulses, the pulse repetition rate 0.3 s^{-1}. Structural studies were carried out using methods of scanning and electron diffraction microscopy, X-ray analyses; physical-and-mechanical properties of the surface layer were studied by measuring the microhardness, the wear resistance, and the friction coefficient.

3. Experimental results and their discussion

According to the literature data on the enthalpy, formation of compounds in systems Ti-B [1] and Ti-C [2] is significant and it indicates a high stability of compounds TiB_2 and TiC_X (Fig. 1).

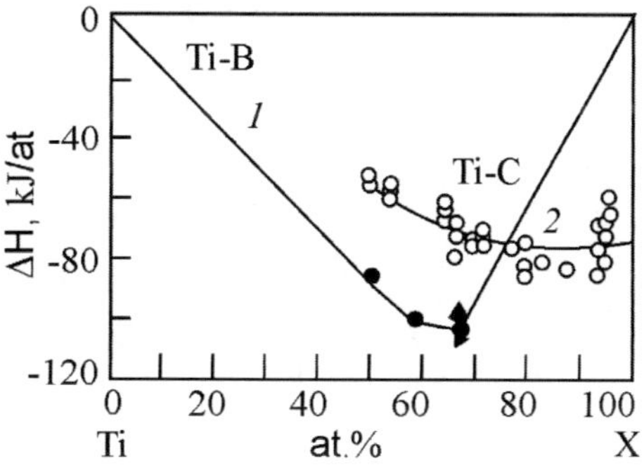

Figure 1. The enthalpy of compound formation in systems Ti-B (curve 1) [1] and Ti-C (curve 2) [2] at 298 K

In the system Ti-B during the crystallization three intermediate phases from the side B are formed: TiB, Ti_3B_4, TiB_2 [3] (Fig. 2). Mutual solubility of components in the solid state is virtually absent. At a temperature of 1540 ± 10 °C the eutectic (βTi) + TiB is crystallized. The compound TiB_2 melts congruently at 3225 ± 25 °C, it has a small homogeneity range. Compounds Ti_3B_4 and TiB are formed by peritectic reactions at temperatures of 2200 °C and 2180 °C, respectively [4]. Titanium is a transition metal and has an unfilled 3*d*-band [5]. Among elements of the group IVA it is the most electronegative metal and has an atomic radius (R_{Ti} = 0.1462 nm). Formation of compounds with a variety of elements of the periodic system is characteristic of titanium; as well as formation of limited solid solutions.

There are several known crystalline modifications of boron (Fig. 2) [6]. For example, in the structure of silver boron atoms form a three-dimensional framework, whose basic structural unit is an icosahedron with 12 boron atoms at its vertices. Icosahedra are interconnected by means of intermediate boron atoms, which are not part of any of the icosahedron, and boron atoms that are part of icosahedra. This leads to the fact that boron atoms in crystals have different coordination numbers: 4, 5, 6, and 5 + 2 (5 near "neighboring" and 2 more distant). According to the electronic configuration $1s^22s^22p^1$ there are three electron atoms of boron on the outer shell, and this leads to the fact that each bond in the crystalline boron has less than 2 electrons. Thus, in the crystalline state the main bond type for boron is covalent with electron deficiency [6].

In the system Ti-C there is one carbide compound TiC(δ) with a structure B1 (prototype NaCl, Pearson symbol sF8, sp. gr. *Fm3m*) [4]. The compound TiC(δ) melts congruently at 3073± 25 °C and the content is ~44% (at.) C. The homogeneity range of the phase TiC near the solidus is from 32 up to 50% (at.) C. Carbon decreases the melting point of Ti from 1663 to 1653±7 °C - the temperature of eutectic crystallization L ↔ βTi + TiC(δ); a eutectic contains 1.5% (at.) [4]. From a structural point of view, the compound TiC (δ) is a solid solution of nonmetal atoms in octa-interstitial sites of the FCC-lattice of metal [7]. Consequently, in the literature carbide TiC (δ) is often referred to as TiC_x. Structural parameters of the compound TiC_x depending on the carbon content are shown in Table 1.

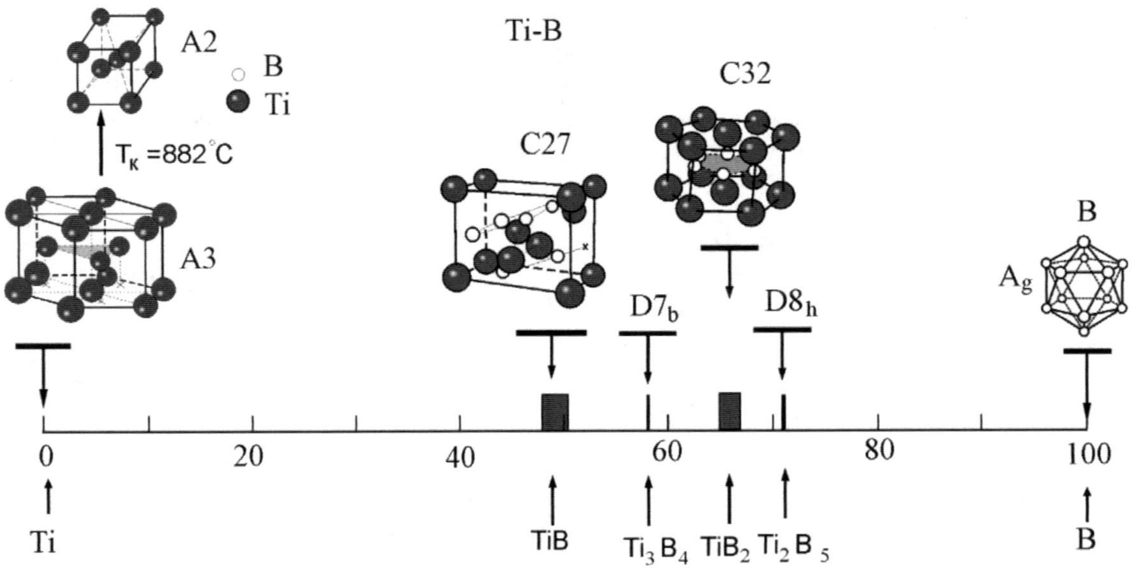

Figure 2. Elementary cells of borides in the binary system Ti-B, depending on the concentration and their homogeneity range.

Table 1. Structural data on ordered phases as a result of PT δ↔δ' [4]

Composition	Syngony	Sp. gr.	Elementary cell parameters	T_C, °C δ↔δ'
Ti_8C_5	trigonal	*R3m*	a = 0.6115 nm, c =1.490 nm in hexagonal axes	
$TiC_{0.67}$	rhombic		$a = 2a_0$	785±5
$TiC_{0.58}$	rhombic		$a = 2a_0$	
$TiC_{0.53}$	rhombic		$a = 2a_0$	
$TiC_{0.46}$ - $TiC_{0.60}$				590

We shall note the following important points that reflect features of the crystalline structure of the phase TiC_x:

1) an important role of the size factor in formation of carbide TiC_x is evident, i.e. arrangement of non-metal atoms in octa-interstitial sites of the FCC-lattice of metal (Fig. 3), which is determined by the well-known Hagg's rule (for TiC_x the size factor is $R_C/R_{Ti} = 0.62$);

2) the presence of a very wide homogeneity range of titanium carbide ($TiC_{0.48}$ - $TiC_{1.0}$);

3) the compound TiC_x belongs to non-stoichiometric compounds of the interstitial phase type (or non-stoichiometric interstitial compounds), the concentration of defects (vacancies) in which provides the interaction among atoms;

4) in the system Ti-C the symmetry of the metallic sublattice of non-stoichiometric compounds is different from the symmetry of the crystalline lattice of corresponding transition metals because titanium carbide crystallizes in the FCC structure in which a succession of layers of Ti atoms is well traced (Fig. 3);

5) at a temperature below 1000 °C the presence of ordering in the carbon sublattice (transition δ↔δ'-transformation) as in conventional solid solutions [4];

6) during the transition δ↔δ'-transformation, as a result of ordering of structural vacancies of the nonmetallic sublattice, a decrease in the symmetry of the space group of the crystal takes place, which results in crystallographically nonequivalent variants of joints of octahedra MS6 (in the structure B1 octahedra are joined by ribs);

7) the solid-solution nature of TiC_x and the structure-forming role of bonds of the nearest neighbors Ti-Ti is confirmed by data value and their physical properties [7];

8) according to modern concepts, a combined covalent-metal-ion type of chemical bond is implemented in non-stoichiometric interstitial compounds MC_x.

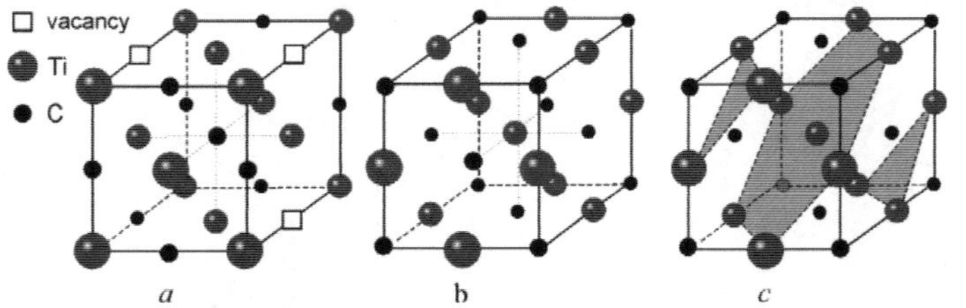

Figure 3. The crystalline structure of carbide TiC_x: *a* – structure B1 with vacancies; *b* –structure B1 without vacancies with a displaced origin of coordinates relative to the represented elementary cell; *a, c* – structure B1 with successively allocated alternating atomic layers of Ti atoms perpendicular to the direction of the type $[111]_{B1}$

Since the compound TiC_x belongs to non-stoichiometric compounds with a wide homogeneity range of the interstitial phase type, the arrangement of structural vacancies leads to formation of a wide variety of crystalline structures. Fig. 4 shows variants of these crystalline structures. It should be noted that the published data on thermodynamic properties of non-stoichiometric titanium carbide is extremely limited. On the other hand, theoretical calculations predict a whole range of crystalline structures based on titanium carbide. Thus, Fig. 5 presents a cubic superstructure of the type M_8X_7. It belongs to a perfectly ordered cubic phase with a space group $R4_332$, which is predicted in [8].

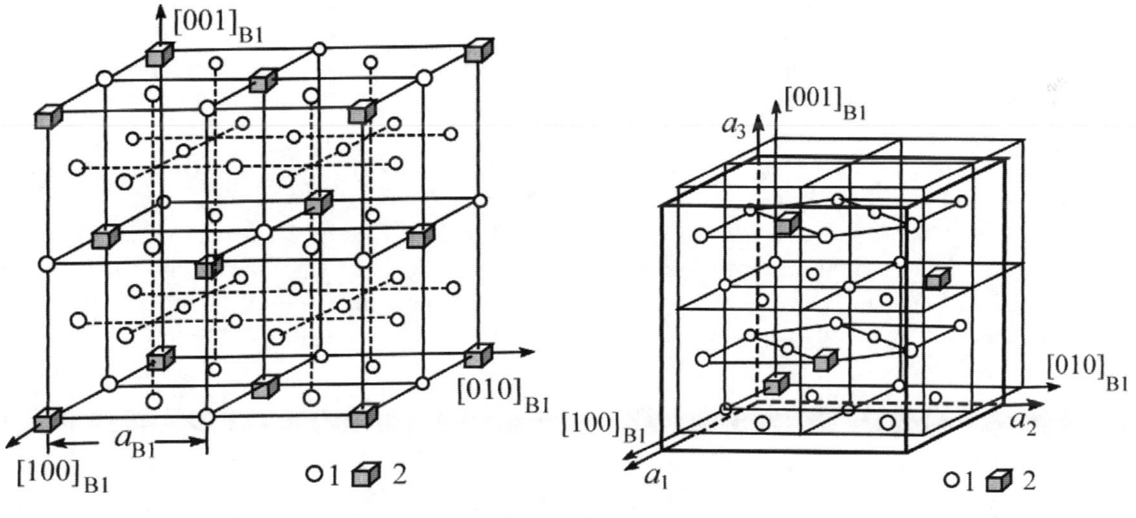

Figure 4. Arrangement of cubic (space group *Fm3m* (*a*) and sp. gr. $P4_332$ (*b*)) elementary cells of the superstructure M_8X_7 in the lattice with a structure Bl (NaCl) and (metallic sublattice is not shown) 1 – interstitial atoms; 2 – vacancies [7].

Fig. 5 shows an isothermal section of the system Ti–B–C. On this diagram it can be seen that the mutual solubility at temperatures below 1400 °C is insignificant. The solubility of boride TiB_2 in carbide TiC_{1-X} is highly dependent on the stoichiometry of the carbide. This is reflected in the form of a non-uniform, in thickness, homogeneity range of carbide TiC_{1-X} on the isothermal diagram. It also finds support on the quasi-binary section between the two coexisting phases TiC_{1-X} and TiB_2 (Fig. 6).

[9] This diagram is characterized by a eutectic transformation at a temperature of 2500 °C and low mutual solubility in TiC_{1-X} of boron atoms and carbon atoms in TiB_2 at temperatures below 2000 °C. In the case of $TiC_{0.95}$ it is nearly 3% (mol.), and in the case of $TiC_{0.68}$ it is nearly 7% (mol.) TiB_2. Furthermore, the given isothermal section of the ternary diagram Ti–B–C indicates the absence of ternary intermediate compounds inside the isothermal triangle (Fig. 5).

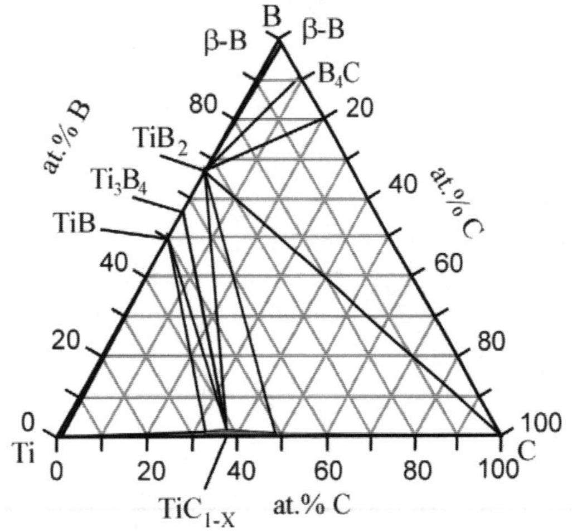

Figure 5. Isothermal section of the system Ti-B-C at 1400 K [9]

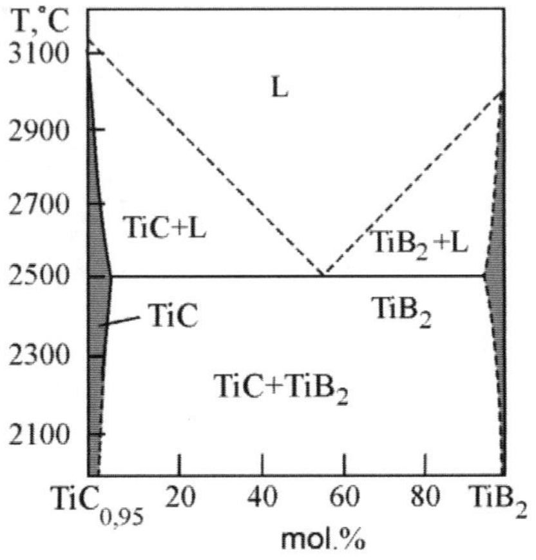

Figure 6. Quasi-binary section $TiC_{0.95}$–TiB_2 of the system Ti-B-C [3]

As previously noted, the system Ti-B-C was formed by doping of commercially pure titanium with plasma formed during an electric explosion of a graphitized carbon fiber, on the surface of which a sample of the nanoscale powder TiB_2 was placed. An additional heat treatment of the modified in such a manner surface layer was carried out by irradiation with an intense pulsed electron beam in the melting mode of the surface layer with a thickness of (20-30) micrometers.

Methods of the X-ray analysis revealed the phase composition of the complex modification of a surface of commercially pure titanium: α–Ti - 17 %, TiC - 60 %, TiB_2 - 14 %, Ti_3B_4 - 2 %, C - 6 %, (B_8C + $B_{25}C$) - 1 %. The total volume fraction of second phases (δ) after EED is 50%; an additional irradiation with an electron beam results in a non-linear change of the value δ: with an increase in the energy density of the electron beam from 45 J/cm^2 to 60 J/cm^2, δ ranges from 25% to 83% reaching a maximum value at the energy density of the electron beam of 60 J/cm^2 (pulse duration of the electron beam exposure is 200 ms).

The study of the phase composition and the defect substructure of the surface layer of titanium modified using the combined method (plasma doping and the subsequent irradiation with an electron beam) was carried out using methods of electron diffraction microscopy of thin foils. It has been established that as a result of the combined treatment in the surface layer of commercially pure titanium a structure of dendritic crystallization based on titanium is formed; the section size of dendritic branches varies from 100 nm to 250 nm. Second-phase inclusions (predominantly titanium carbide of the composition TiC) are located in interdendritic spaces, particle sizes vary from 30 nm to 50 nm.

The structure of the layer located at a depth of 40...50 micron (the intermediate layer separating the melting zone from the heat affected zone) is formed by titanium grains with sizes varying from 80 nm to 200 nm. Second-phase particles with sizes varying from 10 nm to 25 nm are located on grain boundaries and junctions of grain boundaries. Indexing of micro-electron-diffraction patterns obtained from particles allows to conclude that they are, primarily, carbides and titanium borides.

Modification of the structure of the surface layer led to a multiple increase in mechanical and tribological properties of titanium. It has been established that the maximum microhardness achieved in the near-surface layer exceeds the microhardness of the basis by more than 10 times. The thickness of the hardened layer, the micro-hardness of which exceeds the micro-hardness of the basis by ≈5 times, reaches ≈90 microns The wear resistance of the surface layer increases (relative to the wear resistance of the initial material) by ≈8 times. The friction coefficient decreases (relative to the friction coefficient of the initial material) by ≈1.2 times. Having analyzed the results obtained after constructing microhardness profiles it may be noted that one – three maxima of the microhardness, located at the surface and in layers at a depth of 30...50 microns and 80...90 microns, are formed depending on the mode of electron beam treatment.

The carried out studies of the defect substructure and the phase composition of the modified surface layer allow to express opinions about physical mechanisms responsible for a significant increase in mechanical and tribological properties of titanium. It has been established that the main mechanisms responsible for an increase in mechanical and tribological properties of the material are solid-solution (saturation of the crystalline titanium lattice with atoms of carbon and boron); dispersion (allocation of nanoscale particles of carbide, boride, and carbide-boride phases); grain boundary (formation of a grain structure with submicron dimensions).

4. Conclusion

The surface of commercially pure titanium was processed by plasma generated by the electric explosion of the carbon-graphite fibers, the surface of which is placed nanosized TiB_2 powder. Additional heat treatment of the modified in such a manner surface layer was carried out by irradiation with an intense pulsed electron beam in the mode of the surface layer melting. Formation of a multilayer multiphase nanoscale structure with high values of hardness and wear resistance is revealed. It is shown that the maximum micro-hardness achieved in the near-surface layer exceeds the micro-hardness of the basis by more than ≈10 times; the thickness of the hardened layer, the micro-hardness of which exceeds the micro-hardness of the basis by ≈5 times, reaches ≈90 microns. The wear resistance of the surface layer increases (relative to the wear resistance of the initial material) by ≈7.5 times; the friction coefficient decreases by ≈1.15 times. Based on the study results of the phase composition and the defect substructure, physical mechanisms ensuring high strength properties of the modified layer of titanium are proposed.

Acknowledgements

The study is carried out with financial support from the grant of the Russian Foundation for Basic Research (project №13-08-00416_a), this study (research grant No 8.1.42.2015) was supported by The Tomsk State University Academic D.I. Mendeleev Fund Program in 2015 under the project of the Ministry of Education and Science of the Russian Federation and within the framework of the state assignment in the field of scientific activity under the Assignments number 11.351. 2014/K.

References

[1] Ma X, Li C, Du Z, Zhang W 2004 *Thermodynamic assessment of the Ti–B system* J. of Alloys and Compounds **370** 149–158

[2] Karin Frisk A 2003 *Revised thermodynamic description of the Ti–C system* Computer Coupling of Phase Diagrams and Thermochemistry **27** 367–373

[3] Helmut Holleck 1984 *Binare und Ternare Carbid- und Nitridsysteme der Ubergangsmetalle.* (Gebruqer Borntraeger Berlin Stuttgart) 634

[4] Lyakishev N P 1996-2000 *State diagrams of binary metallic systems* (M. Mashinostroenie) **1-3**

[5] Kornilov I I 1964 *Metallides and the interaction between them* (M. Nauka) 180

[6] Serebryakova T I. Neronov V A, Peshev P D 1991 *High-temperature borides* (M. Metallurgiya) 368

[7] Rempel A A 1992 *The effects of ordering in nonstoichiometric interstitial compounds* (Yekaterinburg: Nauka) 226
[8] Lipatnikov V N, Gusev A I 2000 *Ordering in carbides of titanium and vanadium* (Yekaterinburg: UB RAS) 266
[9] Lin Sui R 2013 *Wetting of B4C by molten Ni-Ti alloys*. Journal of Alloys and Compounds. **577**. No15 37-43

Journal of Physics: Conference Series **652** (2015) 012016 doi:10.1088/1742-6596/652/1/012016

Mode Transition of Trichel pulses

Zhang Kexin[1], Piao Yongjun[1], Tang Miao[1], Tang Jingfeng[1], Wei Liqiu[1], Zhang Chaohai[1*]

[1]Harbin Institute of Technology, Heilongjiang, 150001, China

E-mail: zch852@126.com

Abstract. The trichel pulse is a typical kind of negative corona current observed in electronegative gases. In this work, stochastic behavior of the trichel pulse has been investigated. The experiment is performed in a negative corona reactor consisting of a stainless steel pin and a stainless steel powered by a dc high voltage source. The characteristic parameters distributions of corona current pulses, including the amplitude, rise time, half-wave time, and repetition frequency, are analyzed statistically. The results show there is a mode transition during the period of voltage increasing. This transition process happens in a certain voltage region, and change of pulse amplitude is the main difference between the two modes.

1. Introduction

In recent years, the corona discharge presented wide application such as material modification, chemical, electrostatic precipitators and so on. Then, more and more efforts to research on the corona discharge were shown in the related literatures. The corona discharge is usually produced in a strong distortion field by all kinds of discharge electrode structure which has the different characteristics Trichel pulse is a typical kind of negative corona current which was first observed by Trichel in 1934 with a pulsating current mode in electronegative gases and named after him [1]. This kind of corona current consists of a train of pulses with very regular forms, short rise time and durations, and long intervals.

In the past two and half decades both experimental methods and numerical simulation approaches were used to study the characteristics of corona currents. For example, Loeb et al. stated that the Trichel pulses existed only in the electronegative gases [2-3]. They also presented a theory of successive electro avalanches, each giving rise to three successors near the end of its development to explain the characteristics of trichel pulse. Lama et al commented that the charge per pulse of the current was only dependent on the radius of the point electrode but independent of the voltage as well as the gap length [4]. Trinh et al reported that the amplitude of the pulse reduces with the increase of the applied voltage [5-7].

Most of the aforementioned researches were all concentrated on the current-voltage characteristic of Trichel pulses, without investigating the specific characteristics of the pulse including the repetition frequency, the amplitude, the rise time, and the duration of the pulses. In consideration of the randomness of trichel pulse, the relationship between the detailed characteristics of Trichel pulses mentioned above and the discharge condition calls for further was investigated in the paper.

A novel needle-to-plate electrode structure was designed to study the detailed characteristics of Trichel pulses mentioned above. Using the statistical methods, all the detailed characteristics of Trichel pulses mentioned above are analyzed. The results reveal the mutation of the amplitude with the voltage increasing, which can more clearly defined the different modes of Trichel pulse during the process of negative corona discharge developing, and it is significant for the furturefuture research about trichel pulse.

2. Experimental setup

The experimental setup is designed for needle-plate trichel pulse tests, as shown in Figure 1. A needle electrode with a curvature radius of 0.15mm is used as the negative electrode, and connected to a negative dc power supply. A plane with a size of 50mm×100mm and a thickness of 1mm is used as ground electrode and connected with a sampling resistor of 2kΩ. The needle is located with a horizontal distance of 17mm from the plane electrode. A DC generator supplied high voltage for discharges with a continuous regulation ranged from 0 to -20kv.

The voltage between the needle and the round is measured by a DOP2024 type oscilloscope. Meanwhile the current in the circuit can be measured by measuring the voltage on the sampling resistor. All experiments are carried up in ambient air at room temperature without external airflow.

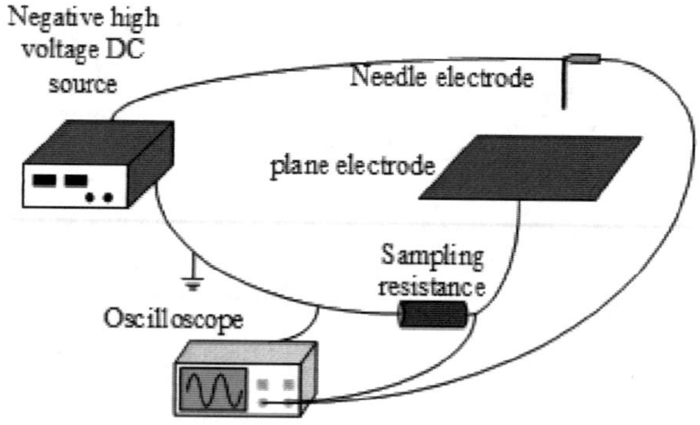

Figure 1. Schematic of the experimental setting

3. Characteristics of trichel pulses

A trichel pulse current measured by a sampling resistor is provided in Figure 2. The typical parameters for the trichel pulse, including pulse amplitude *A*, pulse rise time t_r, and pulse half-wave time t_f are also schematically shown in this figure.

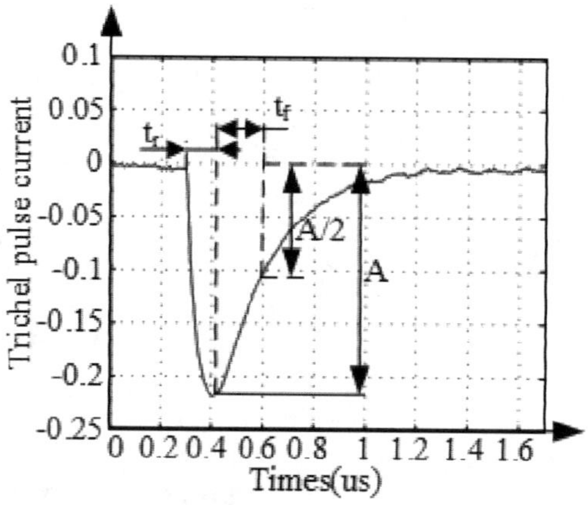

Figure 2. Trichel pulse current measured by a sampling resistor

The applied negative voltage on the needle directly causes a classic trichel corona discharge. On condition that the distance between the needle and the plate is 17 mm, the sampling resistor was 2 KΩ, and the applied voltage of the power supply is ranged from 6000~8000V, the trichel pulse discharges are performed. Figure 3 (a)-(d) show the discharge voltage waveforms of the test resistor at different external power-supply voltages. The total record time is 1ms.

From the figure, it can be found that the discharge voltage waveforms are the typical trichel pulses as well as corona discharges. At the applied power voltage, DC current component does not begin to exist in the Trichel pulses. Compared to Li's works [8], this is mainly due to the applied voltage range just from 6000~8000V. Since DC discharge waveform is the characteristic of the glow discharge, it means that the provided discharge are still trichel pulses now.

At the applied voltage of 6500V, the pulse currents with the amplitude of about 0.2mA always occur. With the applied voltage increasing, the interval time between pulses obviously decreases while more little amplitude pulses obviously appear. At the applied voltage of 8000V, the trichel pulse current is reduced and kept with about 0.1mA level.

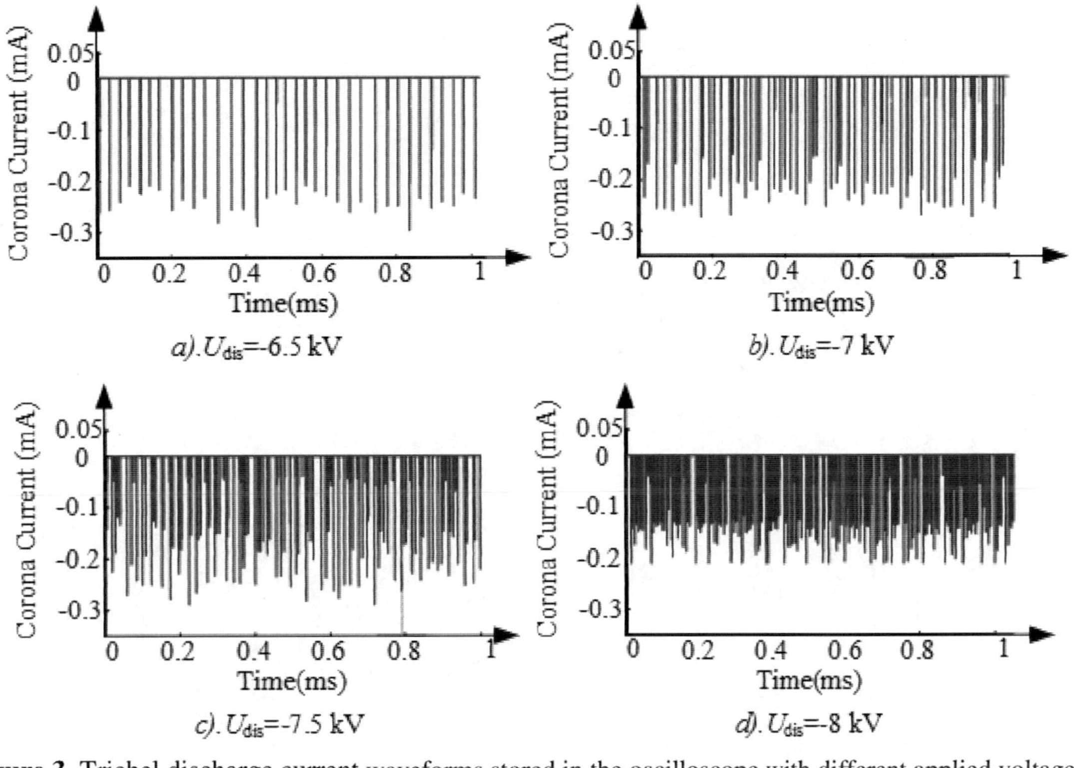

Figure 3. Trichel discharge current waveforms stored in the oscilloscope with different applied voltages.

The characteristics of trichel pulses in a single point-to-plane configuration have been demonstrated. The amplitude and the interval time of these pulses are not generally constant and always fluctuate in time. Since these fluctuations are subject to a certain probability distribution, the statistical processing is used for the analysis of the pulse fluctuations.

All the measured pulses under different applied voltages are used to analyze the probability density distribution of the amplitude. As shown in Figure 4, the amplitude of trichel pulses obeys a normal distribution. Since the pulse amplitude obeys a normal distribution, the probability density function $P(A)$ can be defined as,

$$P(A)=\frac{1}{\sqrt{2\pi\sigma}}e-\frac{(A-\mu)^2}{2\sigma^2} \qquad (1.1)$$

where μ and σ are the mean and variance of corona pulse amplitudes, respectively.

Table 1. The mean and the variance of four pulse parameters

		6.5kv	7kv	7.5kv	8kv
The rising time	Mean(s)	7.9e-8	7.8e-8	7.5e-8	7.2e-8
	Variance(s^2)	8.5e-9	7.9e-9	1.0e-8	8.1e-9
	Maximum of probability density (s)	7.8e-8	7.75e-8	7.47e-8	7.1e-8
The half-wave time	Mean(s)	1.94e-7	2.00e-7	1.88e-7	1.90e-7
	Variance(s^2)	6.00e-8	7.35e-8	3.97e-8	5.78e-8
	Maximum of probability density(s)	1.95e-7	1.94e-7	1.90e-7	1.89e-7
The amplitude	Mean(mA)	-0.2465	-0.2259	-0.1922	-0.1620
	Variance(mA^2)	0.0271	0.0416	0.0410	0.0282
	Maximum of probability density(mA)	-0.23	-0.22	-0.15	-0.16
The interval	Mean(s)	2.86e-5	1.75e-5	1.02e-5	6.26e-6
	Variance(s^2)	3.46e-6	5.88e-5	4.32e-5	2.88e-6
	Maximum of probability density(s)	2.89e-5	1.98e-5	8.92e-6	6.18e-6

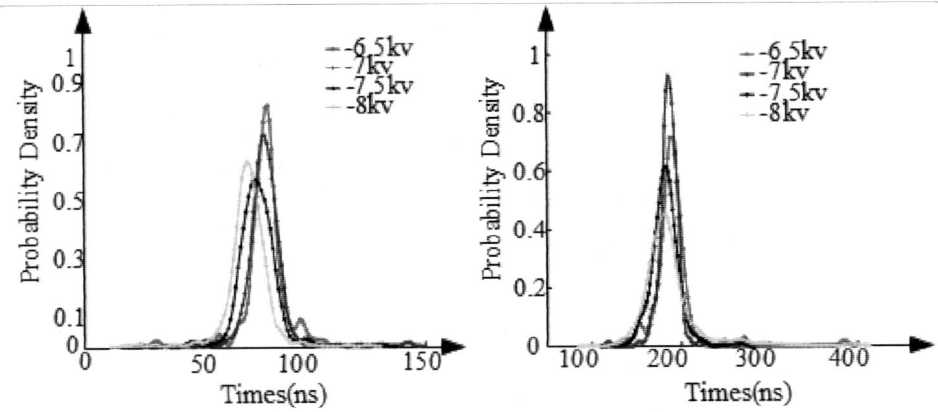

a). the rise time

b). the half-wave time

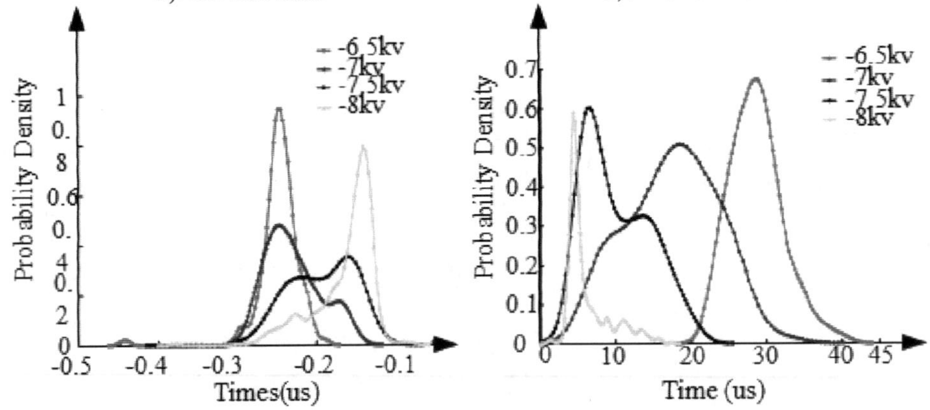

Journal of Physics: Conference Series **652** (2015) 012016 doi:10.1088/1742-6596/652/1/012016

c). the amplitude d). the interval of pulses

Figure 4. Probability distribution of trichel pulse parameters

According to the Table 1 and Figure 4, As the voltage increase, the rising time and falling time change a little as shown. Both rising time and falling time slight reduce as the voltage increase, this characteristic can be seen both in Table 1 and Figure 4a-4b. The rising time is all between 70~80ns, and the falling time is around 180~200ns. When voltage is 6.5kv, there is only one peak in the waveform, which means there is one regime of pulse amplitude, this regime current is about -0.25mA, consistency as Table 1. However, when voltage is 7kv or 7.5kv, there are two regimes of pulse amplitude. As the voltage increasing from 7kv to 7.5kv, the bigger regime transfer from -0.25mA to -0.15mA. As in Table 1, the mean of pulse amplitude is -0.2259 at the voltage 7kv while -0.19 at voltage 7.5kv. according to Figure 4c, it suggest that when voltage is around 7kv or 7.5kv, the amplitude of pulse has both large amplitude and small amplitude, as the voltage increasing, the large amplitude pulse become fewer as the small amplitude pulse become more.

Figure 4d shows the probability distribution of the pulse interval, as the same as Figure 4c, when voltage is 7kv and 7.5kv, there is two regime, contrast to Fig.4c, we suggest that the smaller interval is between smaller pulse.

4. Discussion

According to the results, we believe that during the period of voltage increasing, there are different modes of trichel pulse appeared, in this paper, when voltage is below 6.5kv, we make it mode A, the characteristic of this mode is that the amplitude is bigger, the corona discharge current is more than -0.2mA, and the interval of pulse is all more than 10us. When the voltage is bigger than 8kv, we call it mode B, the characteristic of this mode is small amplitude, high frequency.

According to the previous work [9-11], Trichel pulse is usually attributed to the release of accumulated energy, which depends on the negative ions in the space. When the applied voltage is low, the diffusion motion is responsible for the release of negative ions in the space. As the voltage increases, the negative ions propagate from the negative electrode to the grounded electrode in the electrical field. At this moment, instead of diffusion motion, the release of charges from grounded electrode becomes the main cause. Consequently, the pulse interval increases with the decrease of voltage. Meanwhile, the corresponding energy of one pulse also reduces. That is to say, the change of release method of negative ions in the space results in the mode conversion.

5. Conclusion

In this letter, mode characteristics of needle-to-plate trichel pulse discharge are studied. The experimental results indicate that with the applied voltage increasing, the interval time between pulses obviously decreases while more little amplitude pulses obviously appear. The statistical analysis showed that there is a mutation during the change process of pulses amplitude, this change happened in a increasing zone of the voltage which is 7000V to 7500V under this experimental condition. The mode of trichel pulse discharge has obviously changed as amplitude changing which can be used as mode identification foundation.

References

[1] Trichel G.W. 1938 *Physical Review.* 54 1078-1084
[2] Loeb L B, Kip A F and Hudson G G.1941 *Physical Review.* 60 714-222
[3] Loeb L B. 1952 *Physical Review.* 86 256-527
[4] Lama W L and Gallo C F. 1974 *Journal Applied Physics.* 45 103-113
[5] N G Trinh and J B Jordan. 1968 *IEEE Trans. Power Apparatus Syst.* 87 1207-1215
[6] N G Trinh and J B Jordan. 1970 J.*Appl.Phys.*41 3991-3999
[7] N G Trinh. 1995 *IEEE Electr. Insul. Mag.* 11 23-29

[8] Hua Li, De-cha Zeng, Jun-yu Liu and Zhen-cheng Chen. 2015 *Vacuum*. 113 28-35
[9] Yang Jinji. 1983 *Gas Discharge*. (Beijing: Scientific)
[10] Golden M and Golden A. 1978 Corona discharges. (USA: Academic)
[11] OUYANG Jiting, Zhang Ziliang, Zhang Yu and Peng Zulin. 2014 High Voltage Engineering. 40 1194-1200

Journal of Physics: Conference Series **652** (2015) 012017 doi:10.1088/1742-6596/652/1/012017

Analysis of microwave interference switches with distributed power of switched wave and plasma gas-discharge switching

S N Artemenko[1], V A Avgustinovich[1], S A Gorev[1] and V S Igumnov[1]

[1] National Research Tomsk Polytechnic University, 30, Lenin ave., Tomsk, 634050, Russia

E-mail: snartemenko@mail.ru

Abstract. This paper reports on development of effective cascade type microwave interference switches using waveguide H-tees providing total transition attenuation value of about -50dB at the non-transmitting state, the power losses of -2-3dB at the transmitting state, and the working power compared to the one of a regular waveguide. These switches have different types of connections of two or three H-tees. Different designs provide different distributions of switched wave power between tees and provide different power levels of the switched wave. Characteristics of the switches made of tees matched from their side arms and of unmatched ones were calculated using the scattering matrix method. It was shown the matched tees used in the design allow decreasing by several times the power level of the switched wave and that increases the operating power and stability of the switching process. While some definite combination of unmatched tees allow increasing the power level of the switched wave and may provide effective switching of the transmitted wave at decreased power level. It was shown experimentally the cascade switches could be used as effective energy extraction device in an active resonant microwave compressor also.

1. Introduction

Generation and use of the electromagnetic energy of ultrahigh frequency are always accompanied by processes of controlling the fluxes of the microwave energy. The work of devices controlling the fluxes affects the quality and efficiency of a source of microwave radiation. This is especially true for resonant microwave compressors being sources of pulse microwave radiation [1,2].

Resonant microwave compressors operate on the basis of storing the microwave energy in a cavity and following fast extraction into a load. Most compressors use an interference microwave switch for switching the cavity over to the energy extraction mode. This switch is simple and effective; it is designed as a waveguide H-tee, where one direct arm is connected to the cavity and the other direct arm or the side arm is short-circuited and contains the microwave switch.

At the same time, due to the limited electric strength of waveguides used in tees and their limited cross-section area the switch has a limited operating power. Since microwave switch components used in the device reduce the electrical strength of the microwave switch, its operating power level is usually below this limit. Therefore, improving the quality of the switch requires an increase of the electrical strength. Traditional methods of increasing electrical strength (optimization of the design, use of medium with higher insulating qualities etc.) are almost exhausted.

The new design of the switch for systems of microwave compression was proposed in [3], where the value of the phase shift corresponding to inversion of a switched wave was distributed between

Journal of Physics: Conference Series **652** (2015) 012017 doi:10.1088/1742-6596/652/1/012017

tees in the cascade. The distribution of the phase shift makes the tee switches being subjected to a place of relatively weak electrical field strength, allowing thus increasing the maximum power of the transmitted wave.

The work [4] proposes also microwave compressors comprising the cascade connection of switching tees. The increase of the transmitted wave power in such switches is provided by dividing of the wave power between the tees.

This paper presents the results of the study of the switch with distributed power for the series connection where each consecutive H-tee is connected to the side arm of a preceding one (figure 1) and the series connection of H-tees through their straight arms (figure 2).

2. Analysis

The cascade switches were analyzed by the method of the scattering matrix. Here the operation of such switches is shown using cascades made of two H-tees, generally unmatched from the side arm.

2.1. Cascade of H-tees in the side arm

The scheme of such switch is shown in figure 1.

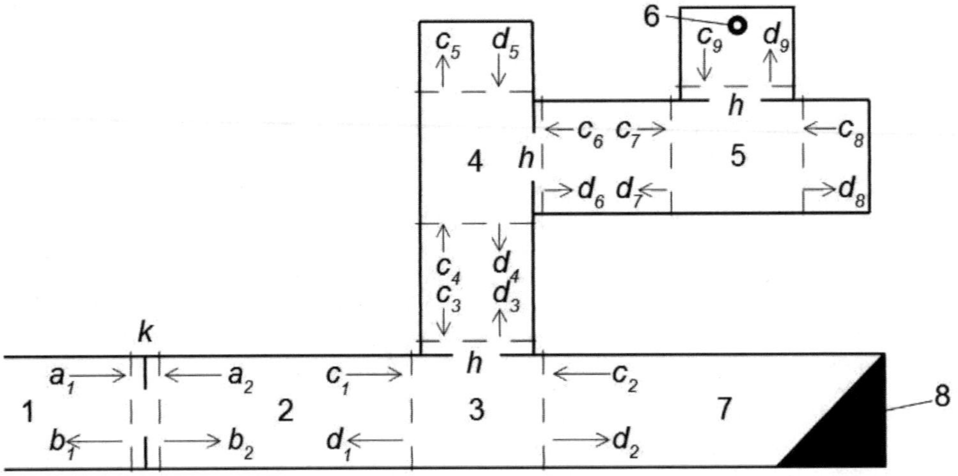

Figure 1. Scheme of compressor with cascade switch in the side arm of H-tee: 1 – input, 2 – cavity, 3-5 – H-tees, 6 – microwave switch, 7 – output, 8 – matched load.

According to the scattering matrix method, the following equations can be written for the wave amplitudes in the studied system:

$$\vec{b} = S_1\vec{a}\ ,\ \vec{d} = S_2\vec{c} \tag{1}$$

where S_1, S_2 are scattering matrices of the input window in the connection of the cavity and waveguide H-tee, and the vectors are constituted by the amplitudes of the incident $\vec{a}$, $\vec{c}$ and reflected $\vec{b}$, $\vec{d}$ waves in the components of the system; k and h are the transmission coefficients of the input window of the cavity in the connection of the cavity and the input window of the side arm of H-tee with straight arms . Using (1) for the cascade of two tees, we can obtain expressions for the wave amplitudes in all components of the system: in the cavity b_2, in the output of the switch d_2, in the side arm of the first tee d_3, in the side arm of the second tee d_6, in the short-circuited straight arm of the first tee d_5.

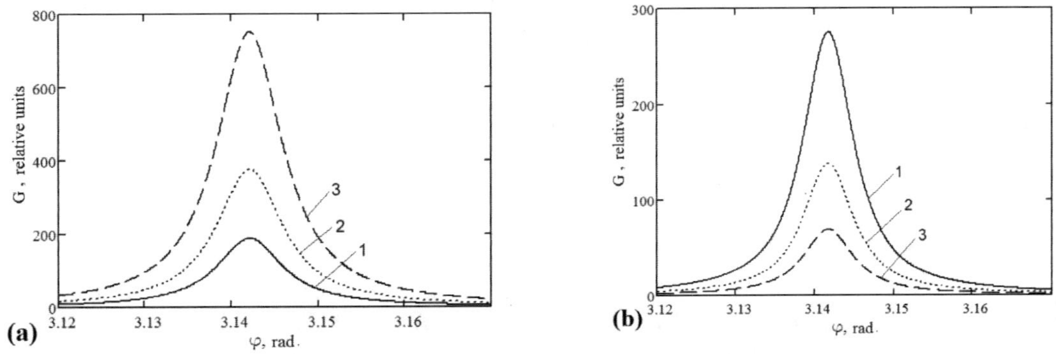

Figure 2. Calculated dependencies of the gain in the components of the compressor with a cascade switch in the side arm of two unmatched H-tees (a, h = 0.8), and tees matched from the side arm (b, h = 1). 1- amplification into the volume of the cavity, 2 – inside of side arm of the first H-tee, 3 – inside of side arm of the second H-tee.

Figure 2 shows the calculated curves of the gain in the components of the compressor with a cascade switch of two conventional unmatched H-tees (fig.2a; h = 0.8), and tees matched from the side arm (fig.2b; h = 1). As follows from the curves, effective matching of tees from the side arm is needed to provide the significant reduction of switched power. For example, it is not difficult to deduce the equality of gain factors in cavity and side arms is achieved at h = 0.95. This value h provides the reflection coefficient $|\Gamma| \approx$ 0.3. Further reduction of gain will need a higher degree of matching. Decreasing h leads to a rapid reduction of the transient attenuation of the switch.

2.2. Cascade of H-tees combined in series

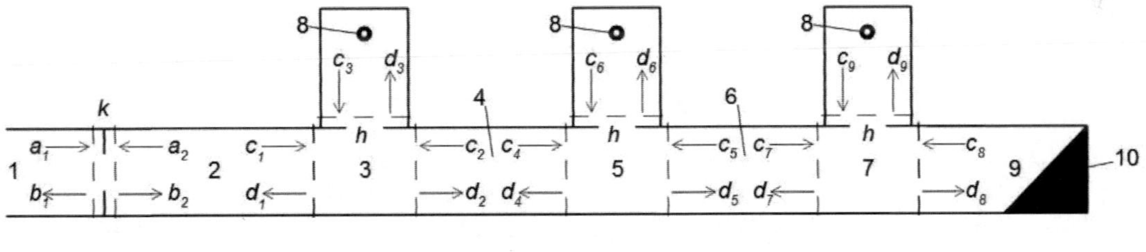

Figure 3. Scheme of compressor with a series cascade switch: 1 – input, 2 – cavity, 3,5,7 – H-tees, 4,6 – common arms of tees, 8 – side arms with microwave switches, 9 – output, 10 – load.

Using (1), the wave amplitudes in all components of the system are expressed in a similar way: in the cavity b_2, in the output of the switch d_8, in the side arm of the first tee d_3, in the side arm of the second tee d_6 and d_9 of the third tee, in the straight arm between the first and second tee d_2 and between the second and third one d_5.

Figure 4 shows the calculated curves of wave amplification in the cavity, side arms and the straight arm between the tees of a cascade of two tees connected in series that were used at experiments. As follows from the curves, the effect of matching is significant.

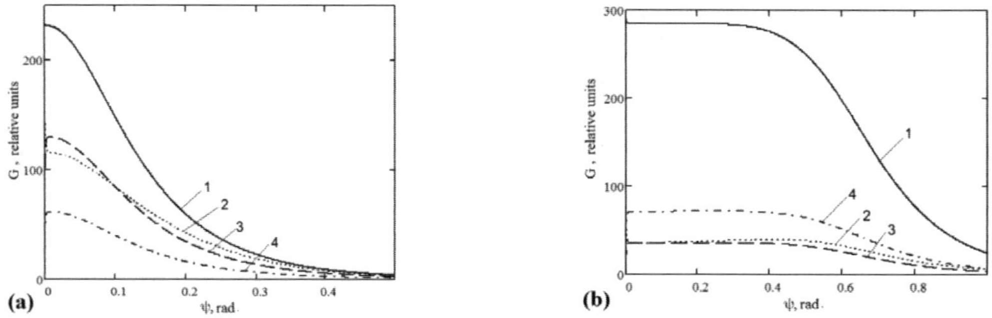

Figure 4. Calculated dependencies of the gain in the components of the compressor with a series cascade switch of two conventional H-tees (a, h = 0.78), and the tees matched from the side arm (b, h = 1). 1- amplification into the volume of the cavity, 2 – in the side arm of the first H-tee, 3 – in the side arm of the second H-tee, 4- in the straight arm between the first and second tees.

The above plots also imply that the unmatched tees form switches with lower quality performance. High quality performance of the cascade switch is obtained when $h > 0.95$, namely, $G < 0.3$. In the case of unmatched tees we will observe stable decrease of the gain in the side arms of tees less than by two times against four times for the matched tees.

It is reasonable to set the switch with a series cascade to the "open" mode by opening the tees in reverse order, starting with the last tee. In this case, as the tees are opened one by one, the gain will increase in steps in the side arm of each subsequent tee. These microwave switches will work at the field level demonstrably higher than the disruptive level, which will provide their more effective functioning.

The gain in side arms is 2 to 3 times less than in the side arm of a conventional switch (the gain in the side arm of a conventional switch is 140; 35 in a cascade of two tees and 16 in a case of three tees). The loss of energy for charging a conventional switch with the travel time along the arms making around 1/3 of the travel time along the cavity is ~ 0.5 dB, for charging the cascade of two tees also ~ 0.5 dB, and the cascade of three tees – less than 1 dB. For the greater the number of tees in a cascade the range of the phase shift ψ values where the gain in side arms is constancy is broader and the value of gain is less.

3. Experimental studies

Experimental studies were carried out with the compressor having the switches of identical S-band H-tees made of circular copper waveguide. Tees were matched from the side of the side arm, their SWR was ~ 1.02. Operating frequency of copper cylindrical cavity was equal to f_0 = 2800 MHz and cavity gain was about 280. The length of input and output pulses was ~ 3 mks and 6-9 ns accordingly.

Plasma microwave dischargers were used as microwave switches. The gas-discharge switches were gas-filled quartz tubes located in the side arms of the tees to the diameter of the waveguide, parallel to field lines of electric field TE_{11} mode [5]. On one end of each tube was the source of illumination.

The first experiments investigated the distribution of gain along the arms of the series switch. The results of these studies confirmed the validity of theoretical analysis data.

In the sequential cascade, after the switch was set on the by-turn spontaneous triggering, one of the switches was set to the controllable triggering mode. In this case, the controllable switch went off first, followed by the spontaneous breakdown of the second switch. This mode of operation was achieved through the distribution of the switched wave power between switches and selecting the composition and pressure of the gas in discharge tubes. The compressor with the switch on the basis of two series connected tees and the oscillogram of output pulses are shown in figure 5a and figure 5b.

.

(a)

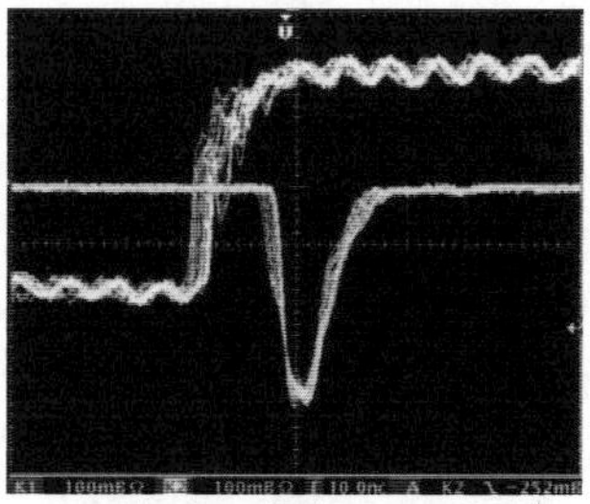

(b)

Figure 5. Resonant microwave compressor (a) with a cascade interference microwave switch based on two H-tees connected in series and the envelope of output pulses (b). (Triggering the first switch of the first tee, followed by the second one, with the interval of time equal to the travel time of waves between the tees (synchronization by laser beam)).

During the experiment with a cascade of tees in the side arm, triggering of switches was controllable. The compressor with such switch and the oscillogram of output pulse are shown in figure 6a and figure 6b.

(a)

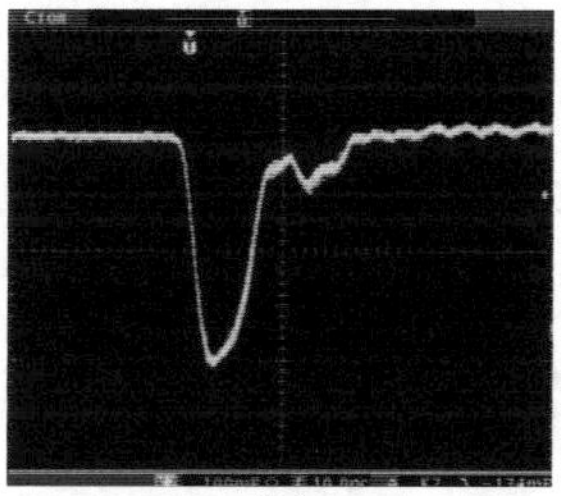

(b)

Figure 6. The compressor with a cascade switch of three tees in the side arm (a) and the oscillogram of output microwave pulses (b).

The gain reached in the compressor with a series cascade was ~ 18.5dB when the first switch of the first tee was triggered. When the switch of the second tee was triggered first, the gain was up to~ 20 dB. This is ~ 1 and 2.5 dB below the gain in the control experiment. Gain fall is explained by the partial loss of the prepulse energy. These losses can be reduced to a negligible level by minimizing the length of the tee arms. Almost similar results were obtained on the compressor with a cascade of tees in the side arm.

Losses in the switches do not affect the gain fall. This follows from the experiments in which the cascade switch provided the output of power without prepulse. This conclusion is realized when the switch in the second tee is triggered synchronously with the incoming wave from the cavity (figure 6b). Here the abrupt increase in the amplitude of the field at the delayed switch facilitates synchronization. In this case, pulses are generated, identical to the pulses of the switch with a conventional switch both in terms of form and gain, which confirms that the losses in a cascade and conventional switches are comparable.

The tubes in switches were filled with a mixture of argon and 1-5 % insulating gas under pressure 0-0.05MPa against 0.3-0.5MPa in a compressor with a conventional switch. The power of input microwave pulses rose to 2 MW, and the power of generated pulses reached 200-250 MW. Moreover, the experiments have shown the possibility of operating the switches filled only with argon at a power level of the input pulses ~ 1.5 MW. This means that the power of input pulses and hence the output pulses can be increased with a corresponding increase in pressure in tubes.

4. Conclusion

The maximum output power of an active compressor can be increased by designing the switch as a cascade of H-tees. It is shown that the level of switched power in such a switch can be reduced by factor of $1/2^n$, where n is the number of tees in the cascade.

The cascaded switches may be used in compressors with solid state switches having relatively low operating power. Increasing the number of tees in a cascade can provide the reduction of switched power to a level acceptable for semiconductor switches.

Acknowledgment

The work has been conducted using the funds of the TPU Competitiveness Program. The work has been also partially supported by the Russian Foundation for Basic Research (Grant No. 15-08-01853). The authors wish to thank Prof. Yuri Yushkov and Prof. Sergei Novikov for useful discussions and some assistance.

References

[1] Didenko A N, Zelentsov V I, Shteyn Yu G, Yushkov Yu G 1972 *Radio Eng. Electron. Phys.* **17** 7 1545-1547 (in Russian)

[2] Alvarez R., Birx D., Lauer E., Scalapino D. 1981 Application of Microwave Energy Compression in Particle Accelerators. *Particle Accelerators*. **11** 3 125-130

[3] Tantawi S G 1999 The design and analysis of multi-megawatt distributed single pole double throw (SPDT) microwave switches. *AIP Conference Proceedings* 472, 959; doi: 10.1063/1.58890

[4] Artemenko S N 2013 Waveguide H-tee T-junction combined in series as the energy extraction device from a cavity. *Izvestiya Vuzov, Fizika* **56** 11/2 238-242. (in Russian)

[5] Avgustinovich V A, Artemenko S N, D'Yachenko V F, Kaminskii V L, Novikov S A, Yushkov Yu G 2009 A study of the switching of the microwave compressor switch in a circular waveguide. *Instruments and Experimental Techniques* **52** 4 547-550

Journal of Physics: Conference Series **652** (2015) 012018 doi:10.1088/1742-6596/652/1/012018

The methods of decrease operating pressure of fast neutrals source

V T Barchenko[1], A E Komlev[2], N A Babinov[1] and M L Vinogradov[1]
[1] Department of Electron Devices and Systems, Saint Petersburg Electrotechnical University "LETI", Professora Popova str., 5, 197376, St. Petersburg, Russian Federation
[2] Department of Physical Electronics and Technologies, Saint Petersburg Electrotechnical University "LETI", Professora Popova str., 5, 197376, St. Petersburg, Russian Federation

E-mail: babinovnikita@gmail.com

Abstract. The fast neutral particles sources are more and more widely used in technologies of surface processing and coatings deposition, especially in the case of dielectric surfaces processing. However for substantial expansion of the sources applications scope it is necessary to decrease the pressure in the vacuum chamber at which they can operate. This article describes the methods to reduce the operating pressure of the fast neutral particles source with combined ions acceleration and its neutralization regions. This combination provide a total absence of the high-energy ions in the particles beam. The main discussed methods are creation of pressure drop between internal and external volumes of the source and working gas preionization which is provided by combustion of auxiliary gas discharge.

1. Introduction

Several decades ago sources of fast neutral particles were poorly studied and its applications were very limited. First of all, this is due to the greater complexity of the fast neutral particles sources compared with ions sources. Moreover there was no significant need for a fast neutral particles processing back at that time.

Active developments of fast neutral particles sources take place in the last decade. These developments occur under influence of two factors: increase of processing quality requirements and increasing demand for high-quality processing of dielectric surfaces. The second factor caused by the fact that compared with ions beam processing the fast neutral particles processing has a significantly high quality especially for dielectric surfaces [1–4]. Thereby at the present time extensive theoretical work is performed and a large number of different designs of fast neutral particles sources are developed.

The difficulty of the fast neutral particles beams generating is the inability of particles acceleration by applied electric field as it is widely performed in ion sources. Nonetheless there are devices in which the direct acceleration of neutral particles is performed [5, 6]. However the fast neutral particles sources that use those principles have very low particles flux and thus low processing performance which lead to low practical value of these sources. Therefore at the present time the sources that use direct acceleration of fast particles are in the stage of laboratory tests and despite certain positive sides of these sources prospects of their future wide uses are not clear.

Due to these most of the developed fast neutral particles sources use the following principle of operation: gas ionization in the plasma volume, acceleration of generated ions by electric field and then its neutralization [7, 8]. A schematic diagram of these sources is shown in figure 1.

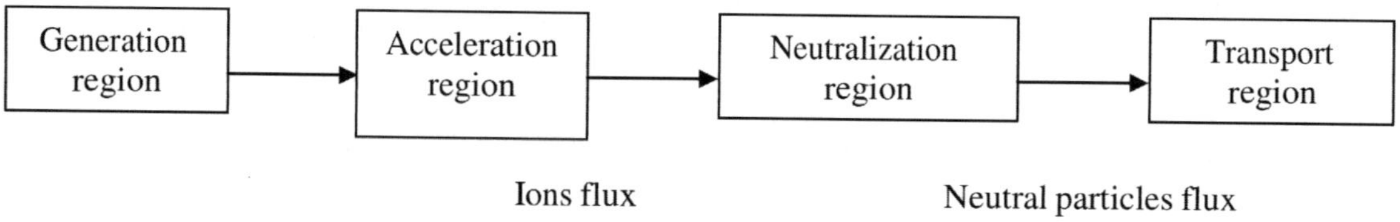

Figure 1. The scheme of fast neutral particles source operation

A significant disadvantage of this operating principle is that due to the non-ideal neutralizer a part of the accelerated ions remains non-neutralized. Since the neutralization occurs in an equipotential space this leads to the fact that some of the ions reach the processed surface while have a high energy. This causes a generation of charge-induced defects on the surface and in some cases is unacceptable.

To remove the high-energy ions out of the beam a magnetic or electrostatic deflection of charged particles can be used. However these methods lead to the considerable complication of the source and increased cost.

Another method to ensure the absence of fast ions in a particles beam is a combination of acceleration and neutralization regions of the source.

2. The fast neutral particles source with combination of acceleration and neutralization regions.

In the work [9] authors describes a fast neutral particles source, its scheme is shown in figure 2. The fast neutral particles source is consists of three electrodes – anode, main and additional cathodes. The main cathode is a set of parallel plates fixed at a certain distance from each other, the anode is a frame wrapped around the cathode. The additional cathode is a hollow electrode, the main and additional cathodes are arranged in a way that they form a quasi-closed volume.

The anode is grounded and the negative voltage of power supply is applied to the main cathode.

The additional cathode is electrically isolated from other electrodes and is under a floating potential.

When the power supply voltage is applied between cathode and anode and the gas discharge is ignited in the quasi-closed volume of the source, the additional cathode potential is determined by equating the ion and electron currents from the plasma surface.

Since the electrons in the glow discharge plasma have much greater mobility than the ions, there is negative potential at the additional cathode that cuts off a part of electron current. Thereby the reflection of the electrons from the additional cathode is possible.

The electrodes design provides an electrostatic confinement of electrons in the quasi-closed volume formed by main and additional cathodes. This leads to the substantial reduction of the discharge voltage and allows the source to operate at the pressure down to about 5 Pa. Moreover, the gas discharge plasma is localized in the source quasi-closed volume and plasma concentration increases.

The fact that additional cathode has floating potential provides a zero current to it, despite the fact that the area of plasma which borders the additional cathode substantially exceed the area of plasma which borders the main cathode. Due to this all the discharge current flows through the main cathode which plays a role of emission grid as well. Thus the current efficiency (the ratio of the equivalent beam current to the discharge current) of the source is close to 100 %. Not 100 % current efficiency is caused only by non-ideal passing of the particles through the cathode.

The ions generated in gas discharge plasma are extracted from the plasma surface and accelerated in the cathode potential drop. Neutralization of the accelerated ions is performed by reflection from the surface of the cathode plates.

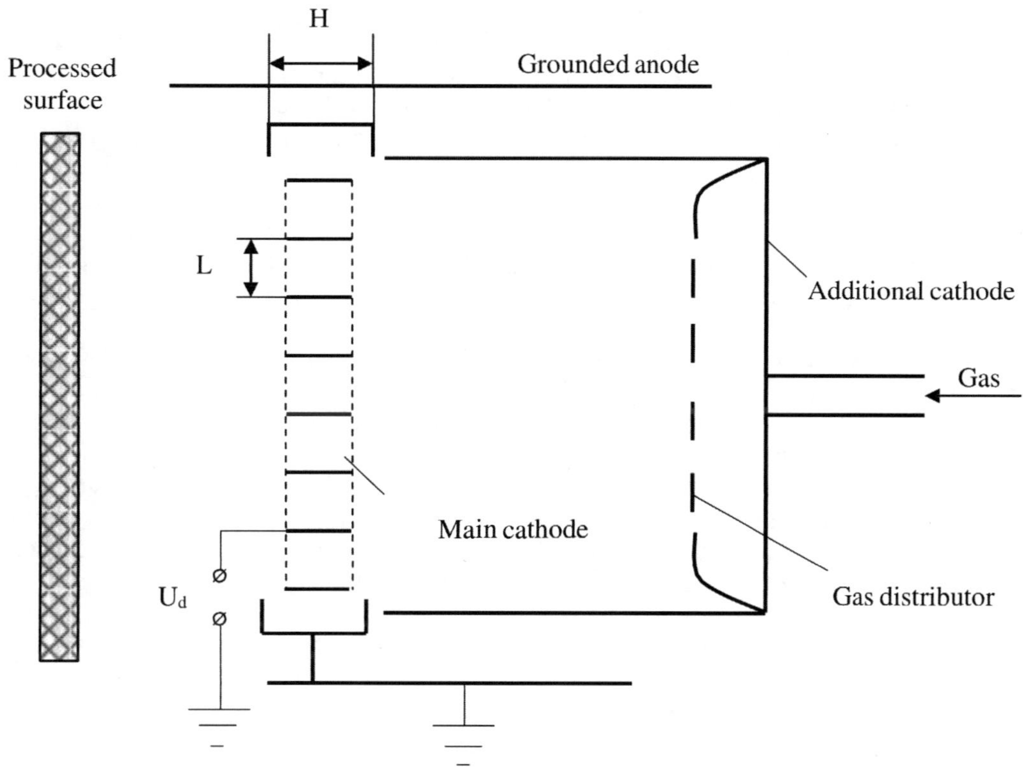

Figure 2. The scheme of fast neutral particles source with combination of acceleration and neutralization regions

Due to the fact that ions neutralization occurs in the region of the cathode potential well, non-neutralized ions cannot leave the near-cathode region and reach processed surface with energy greater than 10…20 eV. The maximum energy of the ions in the beam is determined by plasma potential relative to the anode ("ground") potential. Thus, this source design provides complete absence of fast ions in the particles beam without using additional sources of electric or magnetic fields for deflecting non-neutralized particles from the beam.
However the described source has significant disadvantage – high minimal operating pressure not lower than 5 Pa. Thereby some steps were made to reduce the source operating pressure.

3. The source operating pressure decrease

To reduce the operating pressure of the fast neutral particles source a few methods were used.
One of them was the replacement of the cathode in the form of set of the plates fixed at the rather great distance from each other to the cathode in the form of a thick plate with a plurality of the small diameter holes. The holes in the cathode are angled 10 degrees to the surface thus achieving the conditions for ions entering on the cathode surface at small angles that ensures the highest efficiency of the neutralization and lowest energy losses.
The cathode in the form of a thick plate with holes has substantially greater resistance to gas flow than the cathode plate. This allows creating significant pressure difference between internal and external volumes of the source. Due to this the pressure in the vacuum chamber may be reduced while the pressure in the internal volume of the fast neutral particles source is high enough for gas discharge maintaining.
Moreover, the cathode plate thickness, angle of the holes inclination and its diameter are chosen thus that the plasma is not visible from the processed surface. Thus, the processed surface is isolated from the plasma UV-radiation that is capable to cause a significant damage to the processed surface.

The second method that was used is the separation the source internal volume into two parts with a plate that has approximately ten holes with diameter of 1 mm. The small number of the holes and its small diameter provide a very high pressure drop whereby a gas pressure in one part of the source is high enough and the gas discharge voltage does not exceed a few hundred volts.

Due to this, there is an auxiliary gas discharge that combust at the area with higher pressure (from the gas inlet). The cathode of this discharge is the source additional cathode and the anode is the main discharge plasma surface. The auxiliary gas discharge current flows through the small holes in the plate that indicates that there is a double electrostatic layer near the holes. Maintaining of this layer is possible due to the high gas pressure on the one side of the plate (from the gas inlet) while on the other side of the plate (from the cathode) the pressure is not high enough for double electrostatic layer maintaining.

The auxiliary gas discharge combustion provides preionization of the operating gas, which then enters the main discharge combustion area. Thereby the main gas discharge voltage is reduced and the gas discharge can be maintained at the lower pressure. This leads to the additional decreasing of the source minimal operating pressure.

In addition we assume that creation of a magnetic field to provide a closed electron drift will allow maintaining the gas discharge at a pressure of down to 10^{-2} Pa. Given that the main method of the ions neutralization in the source is the neutralization by reflection from the metal surface that is not influenced by gas pressure this will reduce the operating pressure source to the same value.

Reducing the source operating pressure to that point will significantly expand the scope of the source applications. In particular it makes possible to use the source for film deposition assistance.

Acknowledgments

This work was performed as a part of the Russian Federation government project assignment №8.2456.2014/K. The research was supported by Department of Electron Devices and Systems of Saint Petersburg Electrotechnical University “LETI”.

References

[1] Panda S., Economou D. J., Chen L. 2001. Anisotropic etching of polymer films by high energy (100s of eV) oxygen atom neutral beams *J. Vac. Sci. Tech. A.* **19** 398–404

[2] Wolfe J. C., Craver B. P. 2008. Neutral particle lithography: a simple solution to charge-related artifacts in ion beam proximity printing *J. Phys. D: Appl. Phys.* **41** 024007

[3] Kokoris J., Tserepi A., Gogolides E. 2008. The potential of neutral beams for deep silicon nanostructure etching *J. Phys. D: Appl. Phys.* **41** 024004

[4] B. J. Park, K. S. Min, H. C. Lee et al. 2007. Effect of neutral beam etching of p-GaN on the GaN device characteristics *J. Vac. Sci. Tichnol. B.* **25** № 2 295–298

[5] Giasis K. P., Moore T. A., Minton T. K. 1995. Hyperthermal neutral beam etching *J. Vac. Sci. Technol. A.* **13** 959–965

[6] Suzuki K., Hiraoka S., Nishimatsu S. 1988. Anisotropic etching of polycrystalline silicon with a hot Cl_2 molecular beam *J. Appl. Ph.* **64** 3697–3705

[7] B. J. Park, S. W. Kim, S. K. Kang et al. 2008. Low angle forward reflected neutral beam source and its applications *J. Phys. D: Appl. Phys.* **41** 024005

[8] Ichiki K., Hatakeyama M. 2008. Characterization of neutral beam source using dc cold cathode discharge and its application processes *J. Phys. D: Appl. Phys.* **41** 024003

[9] Barchenko V T, Lisenkov A A, Babinov N A 2014. Module for dielectric surfaces modification by fast neutral particles beams *J. Phys.: Conf. Ser.* **567** 012029

Journal of Physics: Conference Series **652** (2015) 012019 doi:10.1088/1742-6596/652/1/012019

Characteristics of a quasi-stationary non-self-sustained pulsed gas discharge with a hot cathode

D P Borisov, V M Kuznetsov and V A Slabodchikov
Tomsk State University, Tomsk 634050, Russia

E-mail: borengin@mail.ru

Abstract. The paper presents the electrode system and main characteristics of a non-self-sustained pulsed discharge with a hot cathode operating in argon at a pulse repetition frequency of 5 kHz and pulse duration of 50 μs. It is found that during the operation at a pressure of 0.3 Pa, the discharge current can reach amplitudes of up to 300 A and more with a rather high average discharge current of ≈ 100 A and low average discharge operating voltage of less than 20 V. The ion current from the discharge plasma to a Langmuir probe oscillates at frequency of about 100 kHz.

1. Introduction

Increasing the quality and efficiency of vacuum plasma treatment requires new technologies and equipment. In particular, there is a need for efficient plasma sources which provide a plasma density as high as $10^{11}\div10^{12}$ cm^{-3} and allow, for example, high-quality surface cleaning dominated by impurity ion sputtering against adsorption [1]. In the 1990s, a bulk plasma generator based on a thermionic-cathode non-self-sustained arc was designed [2–4] and till now this type of generators, due to its design simplicity and plasma generation efficiency, is one of the most acceptable and reliable technological tools for cleaning, etching, nitriding, doping, and plasma-assisted coating deposition. The generator uses a thermionic or so-called hot cathode, making possible the generation of high-density ($\geq 10^{11}$ cm^{-3}) plasma controllable in a wide pressure range (0.1÷1 Pa). At the same time, the lifetime of a hot cathode is limited (tens of hours) and its use in a plasma generator limits the continuous operation of the source. As has been shown [4], the limiting wear of a hot cathode is defined by the operating voltage of a non-self-sustained arc discharge, and decreasing this voltage from 85 to 28 V in a plasma generator allows a seven-fold increase in the cathode lifetime. The present paper studies the operation of a hot-cathode plasma generator based on a quasi-stationary non-self-sustained pulsed discharge with the aim to decrease the average discharge operating voltage, reduce the loads on the hot cathode, and enhance the efficiency of attaining the desired discharge current and plasma density.

2. Experimental arrangement

An experimental arrangement of the hot-cathode plasma generator based on a quasi-stationary non-self-sustained arc discharge is shown in figure 1.

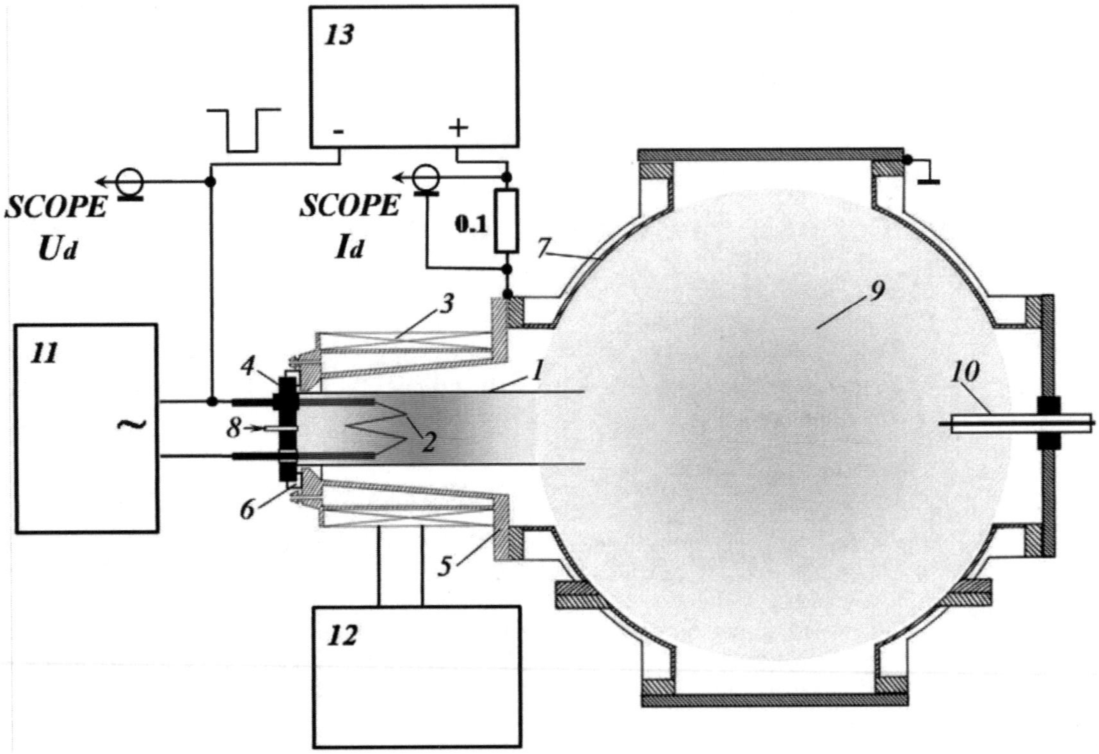

Figure 1. Hot-cathode plasma generator based on a quasi-stationary non-self-sustained pulsed arc.

The main element in the electrode system of the plasma generator is its composite cathode comprising cylindrical hollow stainless steel cathode *1* of diameter 86 mm and length 380 mm with hot tungsten cathode *2* in the form of a spiral (filament) on the inside. The composite cathode is mounted on the water-cooled flange of cathode assembly *4* insulated from water-cooled case *5* of the plasma generator via insulator *6* and is placed in a longitudinal magnetic field of 0.02 T produced by solenoid *3*. Case *5*, which is at anode potential, serves for mounting the plasma generator on vacuum chamber *7* (hollow anode) to produce gas discharge plasma in the chamber through ignition of an arc discharge. The arc is ignited as follows. Alternating voltage and current source *11* provides heating of the hot cathode to the desired thermionic emission current from its tungsten filament. Direct current source *12* powers the solenoid to produce a magnetic field of 0.02 T from the cathode cavity of the plasma generator. Discharge power supply *13* with a no-load voltage amplitude of 90 V produces a train of voltage pulses with a duration of 50 μs and repetition frequency of 5 kHz. As the pulses are applied between the vacuum chamber (anode) and composite cathode (electrically connected hollow and hot cathodes), the electrons emitted by the hot cathode are accelerated along the magnetic field lines toward the anode the role of which is played by the inner walls of the vacuum chamber. Gas supply *8* to the cathode cavity of the plasma generator creates a region of increased pressure near the hot cathode. The region of increased gas molecule concentration and the flow of accelerated electrons whose paths are lengthened by the magnetic field assist the initiation of ionization processes. The electrons reflected from the cathode cavity as a potential barrier execute an oscillatory motion in the cavity, efficiently ionize the working gas molecules, and provide the ignition of a non-self-sustained diffuse arc discharge the operation of which is ensured by thermal electron emission from the hot cathode. The discharge operates between the cathode (electrically connected hollow and hot cathodes) and the anode (inner walls of the vacuum chamber). Gas discharge plasma *9* rather uniformly fills the cavity of the vacuum chamber and provides a medium for surface treatment of objects. The discharge plasma density was estimated by measuring the ion current density from the plasma with Langmuir probe *10* at a distance of 330 mm from the exit aperture of the hollow cathode of the plasma generator.

The characteristics of the plasma generator were studied during its operation with argon. The cylindrical vacuum chamber was 600 mm in diameter and 500 mm in height and was pumped to a residual vacuum of 10^{-3} Pa by a cryogenic pump rated at 1000 l/s. The hot cathode in the experiments was a tungsten (VA) filament of diameter 0.8 mm and working length 150 mm. The filament heating power of the hot cathode P_f was calculated by multiplying the effective values of the alternating filament current I_f and voltage U_f; for example, P_f=56.3 A (I_f) × 16.5 V (U_f)≈ 929 W.

The pulse and average discharge currents I_d were calculated from oscillograms of discharge current signals taken with a shunt (0.1 Ω), as shown in figure 1. The pulse and average discharge voltages U_d were calculated in the same way through direct oscilloscope measurements.

3. Results and discussion

Figure 2 shows typical waveforms of the current I_d (a) and voltage U_d (b) of the quasi-stationary non-self-sustained pulsed arc discharge at an argon pressure of 0.3 Pa and filament power of 580 W. A signal of the saturation ion current density j_p from the argon plasma to the Langmuir probe at a constant negative bias of U_p=200 V with respect to the vacuum chamber (anode) is presented in figure 2 (c). The signal was taken by the oscilloscope in sync with the discharge current and voltage.

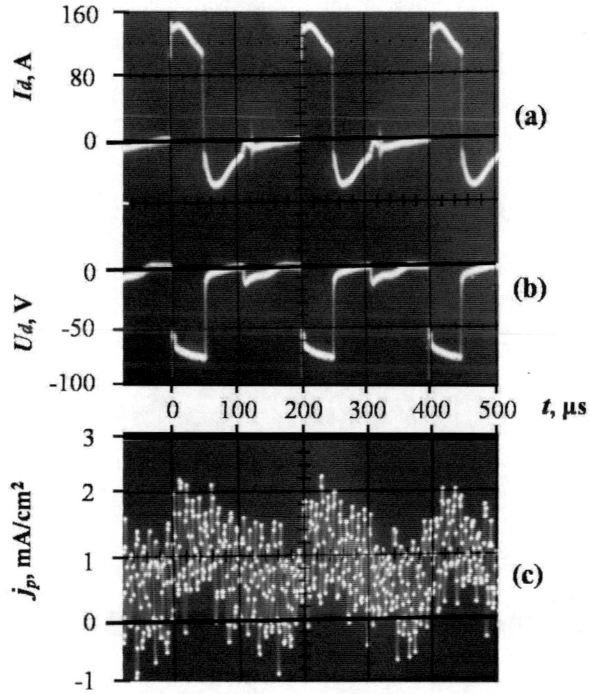

Figure 2. Typical waveforms of the current I_d (a) and voltage U_d (b) of the quasi-stationary pulsed discharge and saturation ion current density j_p (c) from the argon plasma to the Langmuir probe for the indicated discharge current and voltage at an argon pressure of 0.3 Pa and filament power of 580 W.

It is seen from figure 2 (c) that the ion current density from the discharge plasma reveals oscillations with a peak-to-peak amplitude of 1.4 mA/cm^2 which is higher than the average ion current density equal to about 1.2 mA/cm^2 for the indicated values of the discharge current and voltage. The ion current continues to oscillate with the same amplitude even after cessation of the voltage pulses and discharge current. The average ion current density, by estimates from figure 2 (c), can be taken equal to 0.8 mA/cm^2. The pulse discharge current I_d and voltage U_d calculated from the oscillograms in figure 2 (a) and (b) approximate 128 A and 70 V, respectively, and their average values measure about 32 A and 17.5 V. The average discharge current and voltage were determined through dividing the pulse discharge current and voltage by 4, because the discharge pulse duration is 1/4 of the pulse train period.

Figure 2 (c) suggests that during the pulse-to-pulse pause equal to 150 μs, the gas discharge plasma formed by the pulses does not actually decay (recombine) in full, the discharge gap remains filled with the plasma, and the density of the ion current arrived at the probe is about 0.5 mA/cm^2. Thus, it can be concluded that during the pulse-to-pulse interval, the plasma density decreases about 2.4 times due to deionization of the discharge gap in the absence of discharge voltage. During the pulses, the plasma density increases again such that the average ion current density from the plasma measures 1.2 mA/cm^2. This discharge allows maintenance of the quasi-stationary plasma in the vacuum volume with an average ion current density of 0.8 mA/cm^2 (at an average current of 32 A).

The rise of violent oscillations in the ion current to the probe can be due to excitation of plasma oscillations inherent in beam-plasma collective interactions [5]. Really, in our case, the electron beam emitted from the hot cathode of the plasma generator is injected into the gas discharge plasma of the vacuum volume under the action of pulse voltage.

Thus, in our case, it is the excitation of plasma oscillations at ~100 kHz and deep modulation from 0.6 to over 1 in the average ion current during the pulse train period (figure 2) which is responsible for the increase in the ion current density from the plasma if we consider the peak values of this density. The oscillatory character of the ion current density from the discharge plasma can play a positive role in terms of increasing the efficiency and quality of ion plasma treatment and applying the plasma of this type of discharge, similarly to high-frequency discharges, for surface treatment (cleaning, etching) of dielectric materials and products.

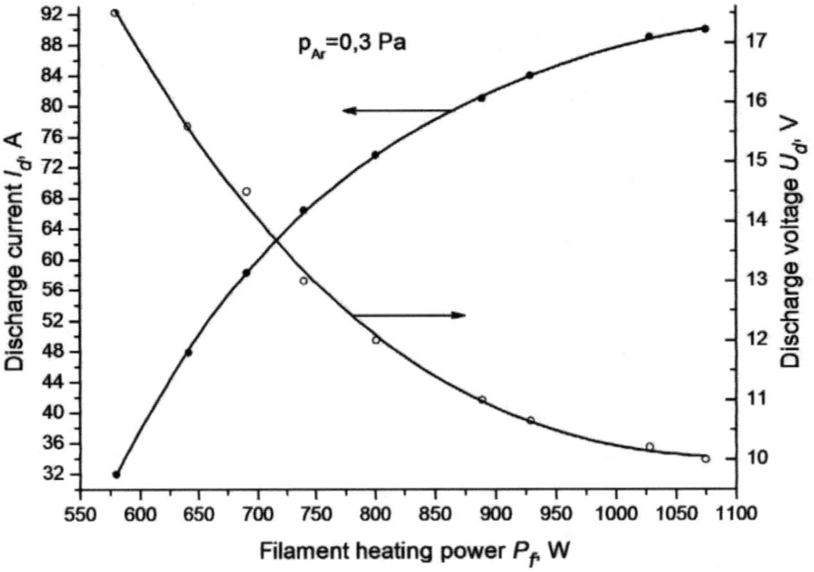

Figure 3. Average current and voltage of the quasi-stationary pulsed discharge vs the filament heating power of the hot cathode.

The discharge system and discharge power supply *13* (figure 1) with a fixed no-load voltage amplitude of 90 V admits control of the discharge current merely by varying the thermionic emission current from the hot cathode of the plasma generator through varying the filament power. Figure 3 shows the dependence of the average current and voltage of the quasi-stationary non-self-sustained pulsed arc discharge on the filament power of the hot cathode at an argon pressure in the vacuum chamber of 0.3 Pa. It is seen from the dependence that the discharge system makes it possible to attain rather high (up to ≈ 90 A) average discharge currents at extremely low (≈ 10 V) average discharge voltages. The average discharge current equal to 32 A for the oscillograms in figure 2 corresponds to the first left point on the dependence in figure 3 (at a filament power of P_f=580 W) and is the minimum value in the range of average discharge currents produced by the plasma generator with the hot cathode comprising a single tungsten filament.

For all discharge currents in the range studied, the ion current from the quasi-stationary argon plasma to the probe at a negative bias of U_p=200 V with respect to the anode (figure 3) also reveals high-frequency modulations like those in figure 2 (c). It should be noted that calculations similar to the calculation from the oscillograms in figure 2 (c) show that the average ion current density from the quasi-stationary pulsed discharge plasma is about three times higher than that from the argon plasma produced at the same argon pressure with the same electrode system, same vacuum chamber, and same positioning of the probe but in the continuous (direct current) mode of the plasma generator [4]. So, for example, for the pulsed operation of the plasma generator at an average current of 32 A (figure 2), the average ion current density to the probe was 1.2 mA/cm^2, whereas that for the continuous operation at 32 A was merely≈ 0.4 mA/cm^2 [4]. For producing the argon plasma with the above ion current density in the continuous mode of the plasma generator at 32 A, the discharge operated at a voltage of 60 V.

The first conclusion on the advantages of the quasi-stationary non-self-sustained pulsed arc discharge system based on a hot cathode is its high efficiency to generate discharge currents compared to the same electrode system but operating in the continuous mode [4]. So, for example, the average discharge current 32 A in the continuous mode is attained at a discharge operating voltage of 60 V, whereas the same current in the pulsed mode is ensured at an average discharge operating voltage of a mere 17.5 V, and this allows an about 3.4-fold decrease in the energy expended for attaining the same discharge current. Also, as shown above, the quasi-stationary pulsed discharge provides a high plasma generation efficiency which, according the estimates from the saturation ion current to the probe, is three times that in the continuous mode of the plasma generator at the same discharge current.

4. Conclusion

Thus, the designed gas discharge system based on a quasi-stationary non-self-sustained arc discharge with a hot cathode demonstrates a high plasma generation efficiency which is several times greater than the efficiency of a dc non-self-sustained arc discharge. The short discharge pulses (50 μs) with short rise times spaced by rather long intervals provide the formation and maintenance of the gas discharge plasma the density of which is about three times greater than that in the stationary discharge. The advantages of the plasma generator based on a quasi-stationary pulsed discharge provide much promise for its use in ion plasma treatment of various materials and products, including dielectrics and products with composite surfaces.

Acknowledgments

This study (state task No.11.1655.2014/K) was supported by the Ministry of Education and Science of the Russian Federation.

References

[1] Vossen J L 1979 *J. Phys. E: Sci. Instr.* **12** 159
[2] Borisov D P, Koval N N and Schanin P M 1994 *Rus.Phys. J.* **3** 295
[3] Borisov D P, Koval N N, Korotaev A D, Kuznetsov V M, Romanov V Ya, Terekhov P A and Chulkov E V 2013 *IEEE Trans. Plasma Sci.* **41** 2183
[4] Borisov D P, Korotaev A D, Kuznetsov V M and Chulkov E V 2014 *Journal of Physics: Conference Series* **552** 012001
[5] Fainberg Ya B 1961 *Atomnaya Energia (Soviet)* **11** 4

Journal of Physics: Conference Series **652** (2015) 012020 doi:10.1088/1742-6596/652/1/012020

Effect of vacuum conditions and plasma concentration on the chemical composition and adhesion of vacuum-plasma coatings

D P Borisov, V M Kuznetsov and V.A. Slabodchikov
Tomsk State University, Tomsk 634050, Russia

E-mail: borengin@mail.ru

Abstract. The paper reports on the chemical composition of titanium nitride (TiN) and silicon (Si) coatings deposited with a new technological vacuum plasma setup which comprises magnetron sputtering systems, arc evaporators, and an efficient plasma generator. It is shown that due to highly clean vacuum conditions and highly clean surface treatment in the gas discharge plasma, both the coating–substrate interface and the coatings as such are almost free from oxygen and carbon. It is found that the coating–substrate interface represents a layer of thickness ≥ 60 nm formed through vacuum plasma mixing of the coating and substrate materials. The TiN coatings obtained on the new equipment display a higher adhesion compared to brass coatings deposited by industrial technologies via intermediate titanium oxide layers. It is concluded that the designed vacuum plasma equipment allows efficient surface modification of materials and articles by vacuum plasma immersion processes.

1. Introduction

Vacuum plasma technologies can open up new possibilities of surface treatment that provides enhanced and even unique functional properties of materials, while providing high ecological safety. However, this type of technologies is rather rare in the market and the industry still gives preference to conventional methods like thermochemical treatment of steels and alloys, electrodeposition of coatings, etc. The cause is that the vacuum plasma technologies available to date fail to meet the desired efficiency and quality of surface modification. The major problem lies in the vacuum conditions which are to be clean and ensure high ultimate residual vacuum with no air leakage into the vacuum chamber, no impurity gas desorption from its walls and other elements, etc. So, for example, due to the presence of oxygen with a partial pressure of 10^{-6} Pa in the vacuum volume, the heated surface of steel is covered with an iron oxide film which impedes the nitrogen penetration into the steel surface and considerably lengthens the nitriding time or even stops the process [1]. The presence of oxygen, carbon, and other impurities in the vacuum chamber not only decreases the rate of nitriding but also results in coatings and their transition layers with poorly controlled chemical compositions and low properties, in particular adhesion.

In this context, it is required to develop new vacuum plasma equipment that would provide highly clean vacuum conditions, high-quality plasma processes in which impurity ion sputtering dominates over residual gas adsorption [2], and increased ion current densities (plasma densities) of inert and reactive gases to a substrate. The paper describes a new plasma vacuum setup designed at Tomsk State University (TSU, Tomsk, Russia) and reports on its efficient use that meets the foregoing requirements and allows deposition of coatings with controlled chemical compositions and enhanced adhesive properties.

2. Experimental equipment, materials, and research technique

The materials under study were TiN and Si coatings obtained on the new vacuum plasma equipment and thin decorative brass (CuZn) coatings with a Zn content of 20 at. % obtained on the industrial equipment of ZCL Nonferrous Casting Plant (ZCL Advanced Technologies, Kurgan, Russia). The ZCL plant produces door furniture (locks, handles, etc.), car furniture, furniture fittings, and other consumer goods from zinc-aluminum alloys deposited with a lacquer undercoating and with decorative brass, aluminum, and nichrome coatings by vacuum plasma processes. For our comparative study we chose brass coatings because of their highest adhesion.

The test CuZn coatings were deposited on polycrystalline Si plates at the ZCL plant. The deposition technology was as follows. The Si substrates were degreased by ultrasonic cleaning in trichloroethylene, rinsed in deionized water, dried, and placed into a vacuum chamber pumped to an ultimate residual vacuum of $6 \cdot 10^{-3}$ Pa for final surface cleaning in a glow discharge. The glow discharge was ignited between the vacuum chamber and magnetron target (titanium) for which a mixture with 60 % of industrial argon and 40 % of carbon dioxide was supplied to the vacuum chamber to a pressure of 1 Pa. The plasma density in the glow discharge was ~ 10^9 cm^{-3}; the treatment time of the Si substrates in the glow discharge plasma was 8 min. The inventors of the technology consider that the ion cleaning in the mixture containing carbon dioxide provides not only removal of impurities (water, organic films, etc.) from the Si substrate surface but also its activation by oxygen and carbon, and this is bound to ensure high adhesion of the intermediate Ti coating formed at the next stage and of the final coating as whole. The Ti undercoating was deposited on the Si substrates cleaned in the glow discharge plasma. The deposition was through magnetron sputtering of VT 1-0 titanium at an argon pressure of 0.3 Pa in the vacuum chamber. The same argon pressure was used to deposit final CuZn coatings through magnetron sputtering. Both the intermediate and final coatings were deposited with no bias potential to the Si substrates.

The other types of coatings, as mentioned above, were TiN and Si coatings deposited using the new vacuum plasma equipment – a SPRUT technological setup designed at TSU [3]. A photo and schematic of the setup are shown in figure 1.

The main plasma device of the setup is its plasma generator based on thermionic or so-called hot cathodes. The cathode units are located on the diametrically opposite flanges of the vacuum chamber measuring 0.7 m^3 in volume. The voltage applied between the hot cathodes and grounded vacuum chamber with gas supply to the cathode cavities at the same flow rate causes the ignition of a non-self-sustained arc discharge the operation of which is ensured by thermionic emission from the hot cathodes. The working volume of the vacuum chamber is thus uniformly filled with "shadow-free" bulk plasma (argon, nitrogen, etc.) which provides efficient treatment such as cleaning, etching, heating, and nitriding of all sides of articles immersed in the plasma. By varying the discharge current from 10 to 250 A, the density of the gas discharge plasma can be varied from 10^8 to 10^{11} cm^{-3} over a wide pressure range of 0.13–0.67 Pa. The setup also comprises four magnetron sputtering systems and two arc evaporators equally spaced over the lateral surface of the vacuum chamber. The vacuum chamber is pumped by a cryogenic pump rated at 5000 l/s.

Once specimens are placed in the vacuum chamber of the SPRUT setup for coating deposition, the chamber is pumped to an ultimate residual vacuum of $6 \cdot 10^{-4}$ Pa and is checked for possible air leak-in and gas desorption from its walls and other elements. For this check, the control system of the setup closes the high-vacuum valve, records the pressure in the vacuum chamber, and starts its timer. Within 30 s after start of the timer, the control system again records the pressure and calculates the leak-in from two pressure values in [Pa·m^3/s] or [sccm]. The leak-in acceptable for further vacuum plasma treatment is defined by specifications of the setup and is no more than $3.4 \cdot 10^{-5}$ Pa·m^3/s (0.02 sccm). If the leak-in exceeds this value, the vacuum conditions are considered to be emergency conditions requiring routine and preventive repair to eliminate the excess leak-in and bring it to the level acceptable for realization of technological processes. Under trouble-free conditions, the technological processes are continued and the specimens are subjected to final surface cleaning for 30 min in the dense gas discharge plasma of high-purity argon (99.998 %) in which the specimens can be heated to the desired temperature by varying both their negative bias and plasma density. Due to the high ion cleaning efficiency in the gas discharge plasma of the setup, the only preliminary treatment of the specimens outside the vacuum is their rinsing in benzine.

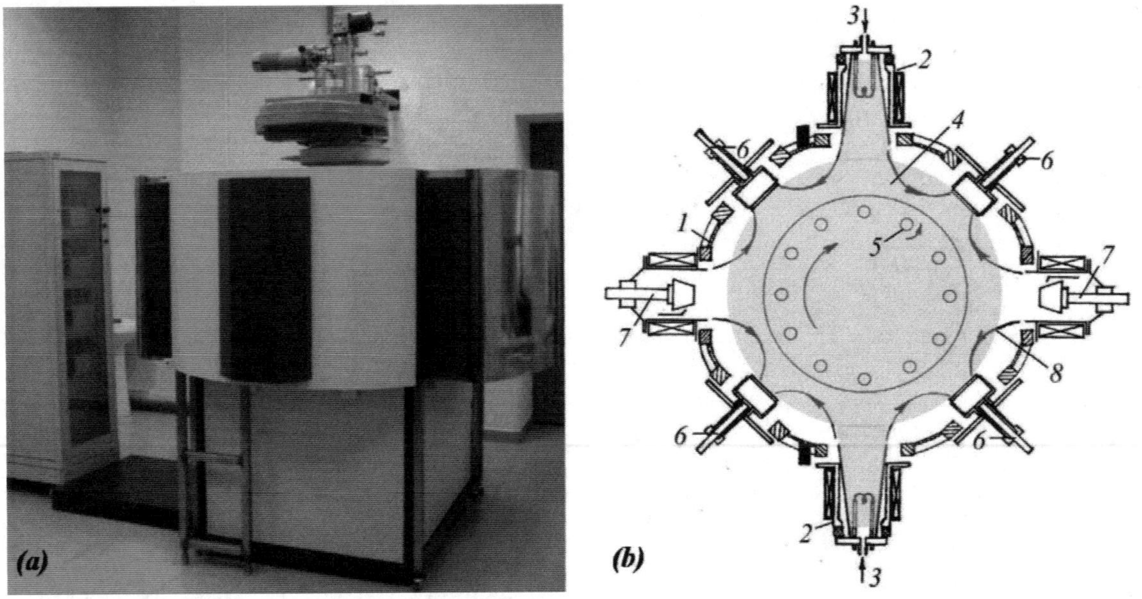

Figure 1. Photo *(a)* and schematic *(b)* of the SPRUT vacuum plasma setup: *1* – vacuum chamber; *2* – cathode units of the plasma generator; *3* – gas supply; *4* – bulk gas discharge plasma; *5* – treated objects; *6* – magnetron sputtering systems; *7* – arc evaporators; *8* – closed magnetic field lines (magnetic wall).

The TiN coatings deposited using the SPRUT setup were of nonstoichiometric composition and were similar in thickness to the polycrystalline Si substrates and in light yellow (brass) color to the decorative CuZn coatings described above. The Si substrates were preliminarily cleaned in the argon plasma of density 10^{11} cm^{-3} with no bias potential applied to them. The coatings were also deposited with no bias potential to the substrates; their floating potential with respect to the vacuum chamber was –20 V. The deposition was realized using arc evaporators of VT 1-0 titanium at a pressure of high-purity nitrogen (99.9997 %) of 0.2 Pa; the density of the nitrogen plasma assisting the deposition was 10^{11} cm^{-3}. The temperature of the Si substrates throughout the deposition was no greater than 150 °C.

As an additional example to demonstrate the efficiency of the SPRUT setup as technological equipment ensuring enhanced control of vacuum conditions and chemical composition of coatings, Si coatings were also deposited on nickel-titanium (NiTi) substrates with a Ni content of 50.9 %. Both ion cleaning and deposition of the coatings were realized in the plasma of high-purity argon with a plasma density of $5 \cdot 10^{10}$ cm^{-3}. The coatings were deposited through sputtering of pure silicon with all four magnetron sputtering systems at a pressure of 0.4 Pa. The bias potential of the NiTi substrates with respect to the vacuum chamber throughout the treatment had a repetitive pulsed form with a pulse frequency of 30 kHz and pulse duration of 17 μs. The pulse amplitude was $\leq$ 160 V and was controlled to hold the substrate temperature at 300 °C.

The chemical composition of all coatings was analyzed by Auger spectrometry. The adhesive properties of the CuZn and TiN coatings on the Si substrates were studied using a CSEM micro scratch tester. In the tester, the surface of a coating is scratched by a special indenter with a tip of radius 20 μm the load on which is increased throughout its motion (track) to record a signal of acoustic emission and its abrupt rise being indicative of separation and fracture of the coating. The critical load for separation (in Newton) or the adhesion strength of each of the test coatings was determined from no less than six scratch tracks by two methods: from acoustic emission signals and from visual observation of fracture in an optical microscope.

3. Results and discussion

Figures 2, 3, and 4 show Auger profiles of the chemical elements in the CuZn coatings (ZCL technology) and in the TiN and Si coatings (SPRUT setup). It is seen in Fig. 2 that the intermediate Ti layer used in brass deposition contains ~ 40 % of oxygen the presence of which, like the presence of carbon, can be due to ion cleaning in the gas mixture containing carbon dioxide. The large amount of oxygen and carbon in the intermediate Ti layer can also be due to the properties of titanium as a good getter which, when sputtered and transported to the substrate, traps oxygen and carbon molecules present in the vacuum chamber because of inadequate vacuum conditions (high residual gas concentration, gas release, and leak-in). The latter explanation is confirmed by the rather high oxygen content≥ 10 % in the brass coating as such (figure 2). According to the ZCL technology, it is likely the formation of titanium oxides and carbides in the intermediate layer between the silicon substrate and brass coating that ensures their high adhesion. The adhesion, in this case, is essentially defined by the adhesion strength of the oxide film to the silicon substrate and outer coating.

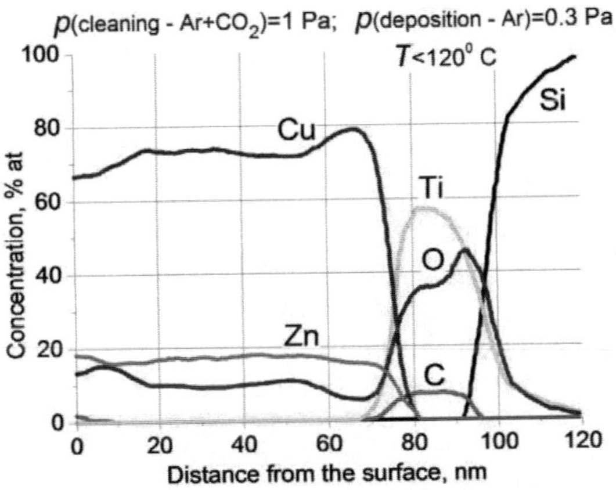

Figure 2. Auger profiles for the CuZn coating on the Si substrate.

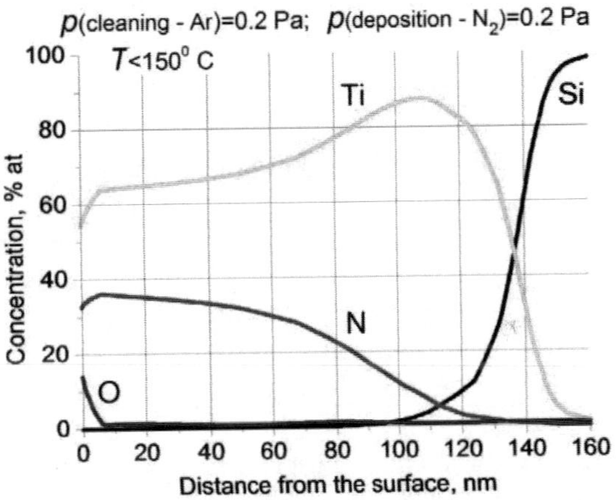

Figure 3. Auger profiles for the TiN coating in imitation of brass on the Si substrate.

Conversely, as can be seen from figure 3, the TiN coating deposited using the SPRUT setup is free from oxygen-containing impurities both at the coating–substrate interface and in the coating itself. The same is true for the Si coating deposited on the NiTi substrate as an additional example. It is seen in figure 4 that the content of oxygen and carbon in the Si coating is clearly negligible compared to their content in the alloy, and this is due to the stringent requirements on vacuum and gas conditions and associated purity of technological vacuum plasma processes realized in the SPRUT setup.

Figures 3 and 4 suggest that the SPRUT setup provides deposition of TiN and Si coatings without any clearly defined coating–substrate interfaces. The interfaces represent transition layers formed through mixing of the substrate and coating materials. The transition layer thickness is 60 nm for the TiN coating deposited on the Si substrate at a temperature of no more 150 ^{0}C and is clearly larger than 100 nm for the Si coating deposited on the NiTi substrate at 300 ^{0}C. Likely the mixing of the coating and substrate materials, along with the absence of oxygen-containing impurities at their interface, ensures an obvious and extremely positive effect on the adhesion of the TiN coating compared to that of the CuZn coating. The results of scratch testing by the procedure described above show that the adhesion strength of the CuZn and TiN coatings is$\approx$ 7 N and$\geq$ 15 N, respectively. Some adhesion ambiguity for the TiN coating is due to surface fracture of the Si substrate along the indenter track at$\geq$ 15 N with no separation of the coating (figure 5). Hence, the load$\approx$ 15 N represents the critical load for fracture of the substrate rather than characterizes the TiN coating and its adhesion the value of which can actually be higher.

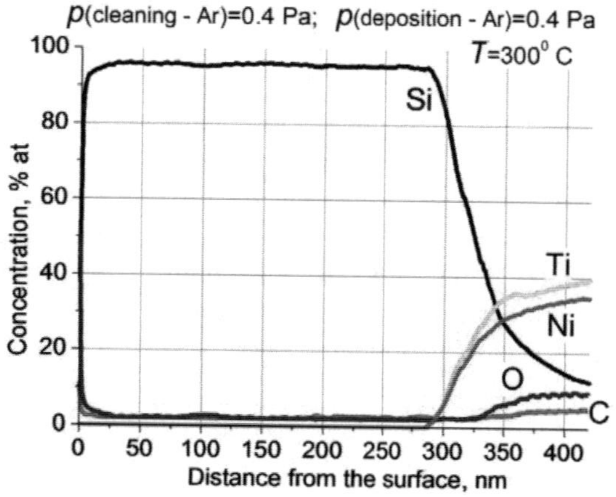

Figure 4. Auger profiles for the Si coating on the NiTi substrate.

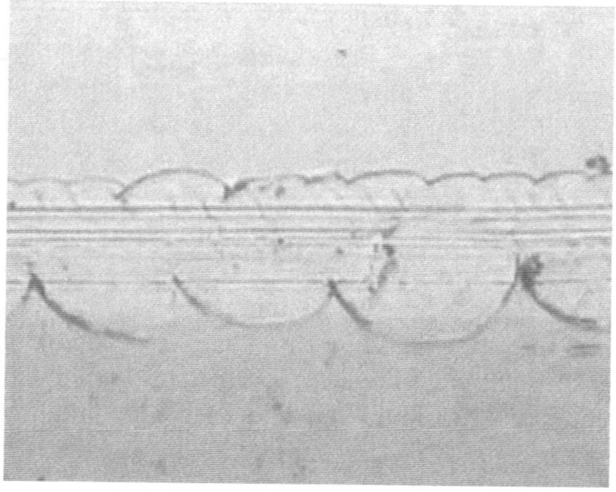

Figure 5. Image of the indenter track for the TiN coating in scratch testing.

4. Conclusion

The SPRUT setup, while providing the generation of dense gas discharge plasma, meets the stringent requirements on vacuum conditions and allows one to enhance the quality of plasma processes, realize the processes with no surface contamination during intense cleaning and activation, and increase the coating adhesion compared to the industrial technology of coating deposition with intermediate oxide layers. The deposition of coatings with the SPRUT setup involves the formation of an extended layer of thickness 60 nm and more through vacuum plasma mixing of the coating and substrate materials. The phenomenon is observed even at low bias potentials of no more than 160 V to the substrate and temperatures of$\leq$ 300 °C and is likely to be due to the absence of impurities on the treated surface. The observed effects give grounds for developing plasma immersion methods based on the new vacuum plasma equipment for surface doping with a wide range of chemical elements in a single vacuum cycle.

Acknowledgments

This study (state task No.11.1655.2014/K) was supported by the Ministry of Education and Science of the Russian Federation.

References

[1] Parascandola S, Knese O and Moller W 1999 *J. Appl. Phys. Lett.* **75** 1851
[2] Vossen J L 1979 *J. Phys. E: Sci. Instr.* **12** 159
[3] Borisov D P, Koval N N, Korotaev A D, Kuznetsov V M, Romanov V Ya, Terekhov P A and Chulkov E V 2013 *IEEE Trans. Plasma Sci.* **41** 2183

Journal of Physics: Conference Series **652** (2015) 012021 doi:10.1088/1742-6596/652/1/012021

Research of transportation efficiency of low-energy high-current electron beam in plasma channel in external magnetic field

E S Vagin, V P Grigoriev

National Research Tomsk Polytechnic University, 634050, Russia, Tomsk, Lenin Avenue, 30

E-mail: VaginEvS@yandex.ru

Abstract. Effective high current (5-20 kA) and low energy (tens of keV) electrons beam transportation is possible only with almost complete charging neutralization. It is also necessary to use quite high current neutralization for elimination beam self-pinching effect. The research is based on the self-consistent mathematical model that takes into account beam and plasma particles dynamic, current and charge neutralization of electron beam and examines the transportation of electron beam into a chamber with low-pressure plasma in magnetic field. A numerical study was conducted using particle in cell (PIC) method. The study was performed with various system parameters: rise time and magnitude of the beam current, gas pressure and plasma density and geometry of the system. Regularities of local virtual cathode field generated by the beam in the plasma channel, as well as ranges of parameters that let transportation beam with minimal losses, depending on the external magnetic field were determined through a series of numerical studies. In addition, the assessment of the impact of the plasma ion mobility during the transition period and during steady beam was performed.

1. Introduction

Growing interest in application-oriented use of low-energy high-current electron beams (LHEB) is predetermined by their unique opportunities for high energy density transportation at rather long distances without essential losses, and also effective transmission of this energy to object of influence. High energy density (up to 20 J/cm^2) and short pulse duration (several μs) make it possible to use electron beams in different technological processes connected with changing the state and properties of material surface [1].

High efficiency of LHEB transportation is possible only with almost full charge neutralization [2]. For this, LHEB should be transported in the pipe with low-pressure plasma or neutral gas (10^{-1} ... 10^{-2} Pa). Besides, self-magnetic field can cause beam pinching, lowering efficiency of transportation. However, sufficient current neutralization and external magnetic field can be used to reduce this effect. Particular issue in such beam systems is potential possibility excitation of instabilities, e.g. Pierce instability, beam-drift instability, current thresholds of which can be below than beam current. Thus LHEB transportation in plasma and external magnetic field with high efficiency represents a sophisticated challenge. The purpose of this work is to determine the optimal parameters for LHEB transportation.

This paper describes mathematical model, equations and results of numerical research of LHEB transportation efficiency in the pipe filled with plasma in the external magnetic field in depending on system parameters.

2. Basic equations of the physical model

The mathematical model of self-consistent dynamics of the beam in the space charge and magnetic fields in the drift area, filled with plasma with the homogeneous density n_p, is based on the description of beam and plasma electrons by macroparticles [2]. The model is generated for the area coinciding with the tube and has dimensionality 2.5 (three-dimensional on dynamics, two-dimensional on fields).

Following assumptions were used in the model:

- Axial symmetry of processes $\frac{\partial}{\partial\theta}=0$;
- Prevalence of the longitudinal current of the beam: $J_z >> J_r, J_\theta$
- The plasma ions have huge mass with respect to electrons that is why they are considered immobile and density homogeneous and constant $n_i=n_p$.

Dynamics of beam and plasma electrons is described by system of relativistic equations in a cylindrical coordinate system:

$$\begin{cases} \frac{d(\gamma_\alpha \dot z)}{dt} = -\frac{e}{m_0}\left(\dot r B_\theta + E_z\right) - \nu_{ef}\dot z \\ \frac{d(\gamma_\alpha \dot r)}{dt} = -\frac{e}{m_0}\left(r\dot\theta B_z^* - \dot z B_\theta + E_r\right) + \gamma_\alpha r\dot\theta^2 - \nu_{ef}\dot r \\ \frac{1}{r}\frac{d\left(\gamma_\alpha r^2\dot\theta\right)}{dt} = \frac{e}{m_0}\left(\dot r B_z^*\right) - \nu_{ef} r\dot\theta \end{cases} \quad (1)$$

where m_0 is an electron rest mass; e is an elementary charge; E_z, E_r, B_θ are components of self electromagnetic field of the beam; B_z*=const is component of an external magnetic field; γ_α is the relativistic factor of particles α; α is plasma and beam electrons; ν_{ef} is effective collision frequency [3]. As shown in calculations, processes of additional gas ionization and plasma recombination can be neglected because they are practically in a dynamic equilibrium at he given parameters.

The fields is described by Poisson equations for the scalar potential Φ and longitudinal component of the vector potential A_z. Densities of charge and a beam current are related by the continuity equation:

$$\mathrm{div}\,\mathbf{J} + \frac{\partial\rho}{\partial t} = 0 \quad (2)$$

The total charge density in equation (2) is presented by the correlation:

$$\rho = \rho_b + \left(\rho_i + \rho_e\right) \quad (3)$$

where ρ_b, ρ_e are charge densities of beam and plasma electrons; $\rho_i = n_p\, e = const$ is charge density of plasma ions. The initial condition for the charge density of beam electrons is set as $\rho|_{t=0} = 0$, which corresponds to the absence of the beam in the drift tube.

Boundary conditions for potentials are set from requirements of ideal conductivity of the surface of tube walls ($r=R$) and potentials continuity on the tube axis ($r=0$) and at tube end faces ($z=0$ and $z=L$):

$$\left.\frac{\partial\Phi}{\partial r}\right|_{r=0} = 0 \quad \Phi|_{r=R} = 0 \quad \Phi|_{z=0} = \Phi|_{z=L} = 0 \quad (4)$$

$$\left.\frac{\partial A_z}{\partial r}\right|_{r=0} = 0 \quad A_z|_{r=R} = 0 \quad \left.\frac{\partial A_z}{\partial z}\right|_{z=0} = \left.\frac{\partial A_z}{\partial z}\right|_{z=L} = 0 \quad (5)$$

Fields components are calculated using formulas of potentials differentiation:

$$E_z = -\frac{\partial \Phi}{\partial z} - \frac{\partial A_z}{\partial t} \quad E_r = -\frac{\partial \Phi}{\partial r} \quad B_\theta = -\frac{\partial A_z}{\partial r} \tag{6}$$

3. Research of transportation efficiency

Numerical simulations of the LHEB transportation were performed with the following parameters that were used in real LHEB guns [4-5]: energy of beam electrons E_0=20..40 keV, beam current I_0=15 kA, plasma temperature 2 eV, gas pressure $p=10^{-1}$ Pa, gas ionization degree 10 %, plasma density $n_p=10^{10}...10^{12}$ cm^{-3}, magnetic field induction B_z = 0..3 kGs. Pipe geometric parameters: L=20 cm, R=10 cm; beam radius: R_b=4.3 cm. The current impulse has leading edge (τ) with the linear increasing, and constant current (I_0) after τ.

Plasma density, external magnetic field induction and leading edge were selected as investigated parameters on which transportation efficiency depends. The relation of the maximum output current to current of pulse (I_b/I_0) was accepted the transportation efficiency criterion.

Dependence of transportation efficiency on plasma density in the case of fixed values of external magnetic field (Bz * = 1 kGs), injection energy E_0=20 keV and pulse leading edge (τ = 300 ns) is demonstrated in Figure 1. The analysis of the results shows that high efficiency is reached at the plasma density equal to or greater than the maximum density of the beam $n_{b0} = I_0 / (I_A S_b r_e \beta_{z0})$, where I_A is Alfevn current, S_b is area of the beam cut, r_e is classical radius of an electron, β_{z0} is velocity of beam electrons relative light speed with initial energy E_0. For the above parameters, the beam density was ~ 2 10^{11} cm^{-3}.

Insufficient efficiency of transportation at a lower plasma density is caused by insufficient charge neutralization for the beam electrons, which leads to the formation of a virtual cathode at the beginning of the drift pipe (Figurc 2).

The presented results for the current differ from approximated estimates on limiting vacuum current [6]. That is due to the fact that in the paper the dynamic task is considered and during beam transportation transverse current density changes from uniform to tubular. As a result of this effect we get higher beam current in comparison with approximate stationary estimates.

Figure 3 shows dependence of transportation efficiency of the beam vs external magnetic flux, in the case of fixed values of plasma density (line 1 - $n_p=2.5\ 10^{11}$ cm^{-3}, line 2 - $n_p=1.8\ 10^{11}$ cm^{-3}) and leading edge (τ = 300 ns). Quite high transport efficiency (90%) is reached at values of the external magnetic field greater than or equal to self-magnetic field of the beam, which can be estimated by the formula:

$$B_b = \frac{1}{4\pi} \frac{2I_0}{cR_b} \sqrt{\frac{\mu_0}{\varepsilon_0}}. \tag{7}$$

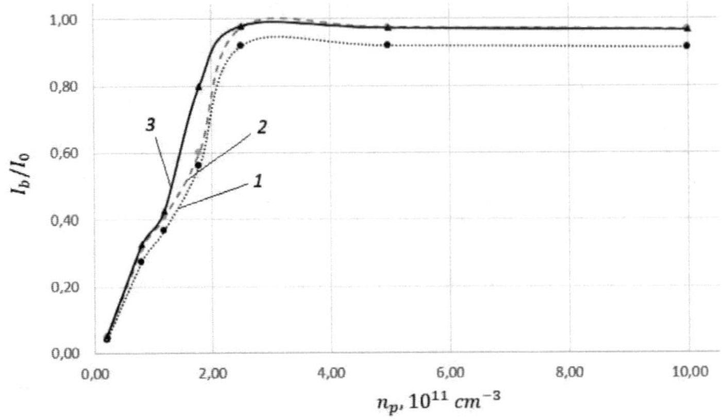

Figure 1. I_b/I_0 vs plasma density: $I_0 = 15\ kA$, $\tau = 300\ ns$, $E_0=20\ keV$, $t = 300\ ns$, *lines:* 1- $B_z=1$ *kGs*, 2- $B_z=1.5$ *kGs*, 3-$B_z=3$ *kGs*

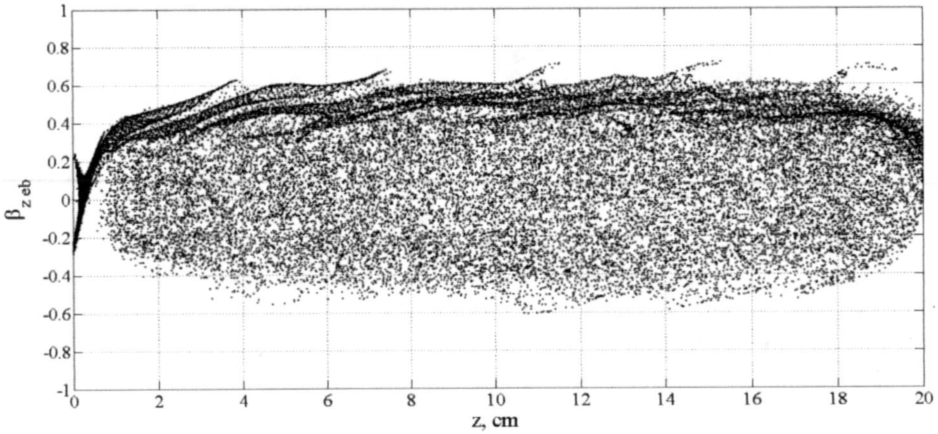

Figure 2. Phase portrait of the beam electrons $I_0 = 15\ kA$, $\tau = 300\ ns$, $E_0=20\ keV$, $B_z = 1\ kGs$, $n_p=1.2\ 10^{11}\ cm^{-3}$, $t = 300\ ns$

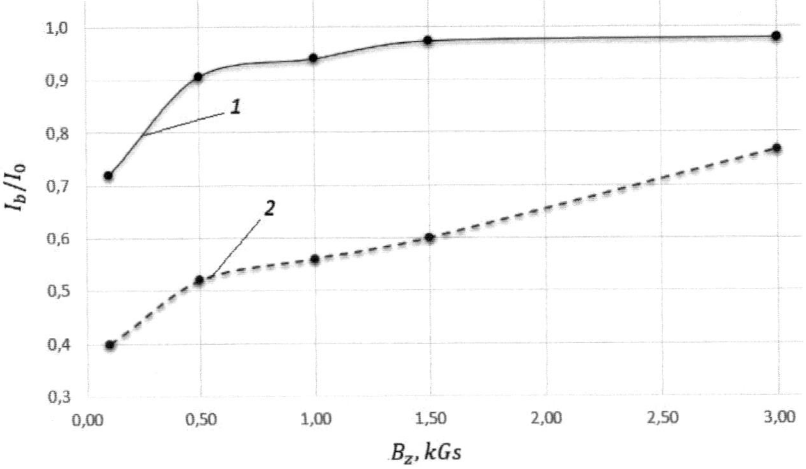

Figure 3. I_b/I_0 vs external magnetic flux: $I_0 = 15\ kA$, $\tau = 300\ ns$, $E_0=20\ keV$, $t = 300\ ns$, *lines:* 1- $n_p=2.5\ 10^{11}\ cm^{-3}$, 2- $n_p=1.8\ 10^{11}\ cm^{-3}$

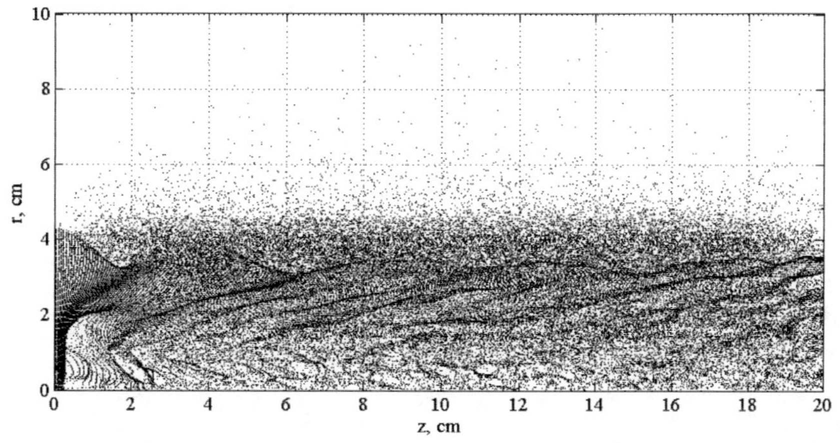

Figure 4. Configuration portrait of the beam electrons: $I_0 = 15\ kA$, $\tau = 300\ ns$, $E_0 = 20\ keV$, $t = 300\ ns$, $B_z = 0.5\ kGs$, $n_p = 2.5\ 10^{11}\ cm^{-3}$

The value of the self-magnetic field for the initial beam parameters was 0.06 T. The lower value of the external magnetic field leads to insufficient weakening of its self-magnetic field and, as a consequence, self-pinching of the beam (Figure 4) and the formation of the virtual cathode.

When assessing the impact of leading edge (Figure 5) on the transportation efficiency, causes sufficiently small values are of the primary interest. The virtual cathode is formed at short leading edge (< 50 ns) even with the sufficient charge and current neutralization, because plasma electrons do not have time to leave the drift area quickly enough. However, virtual cathodc dissipates by forces that extrude plasma electrons from the drift area and as a result the further beam transportation has minimal losses.

The increase of injection beam energy is known to lead to the increase of the limit of vacuum currents. This means that the effective beam transportation requires less plasma density in the drift pipe. It is well confirmed by numerical calculations (see Figure 6).

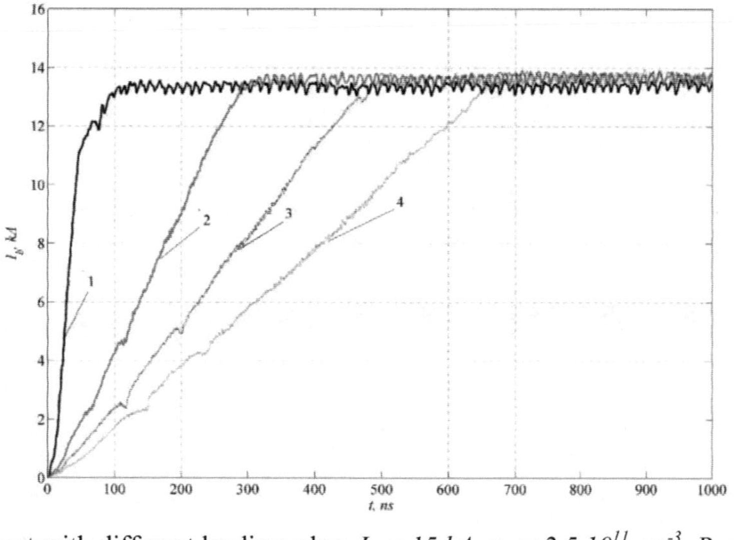

Figure 5. I_b current with different leading edge: $I_0 = 15\ kA$, $n_p = 2.5\ 10^{11}\ cm^{-3}$, $B_z = 1\ kGs$, $E_0 = 20\ keV$ lines: 1 – $\tau = 50\ ns$, 2 – $\tau = 300\ ns$, 3 – $\tau = 500\ ns$, 4 – $\tau = 700\ ns$.

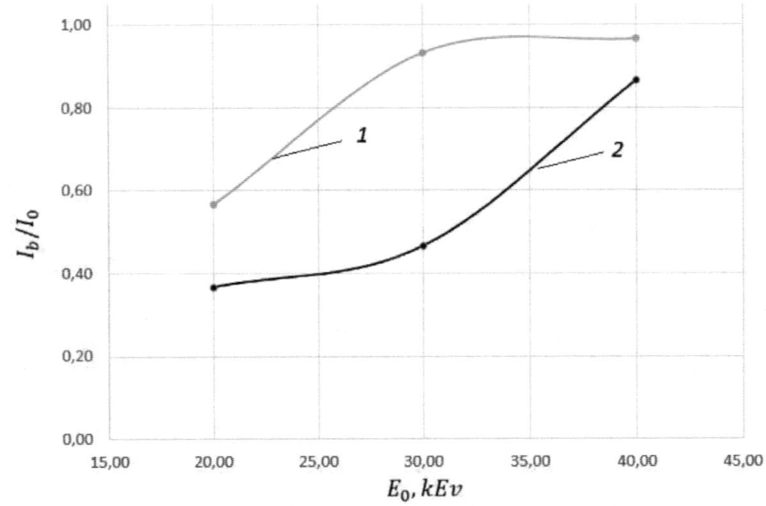

Figure 6. Ib/I0 dependence on injection beam electrons energy E0: I0 = 15 kA, τ = 300 ns, Bz = 1 *kGs*, lines: 1 – $n_p=1.8\ 10^{11}\ cm^{-3}$, 2 – $n_p=1.2\ 10^{11}\ cm^{-3}$.

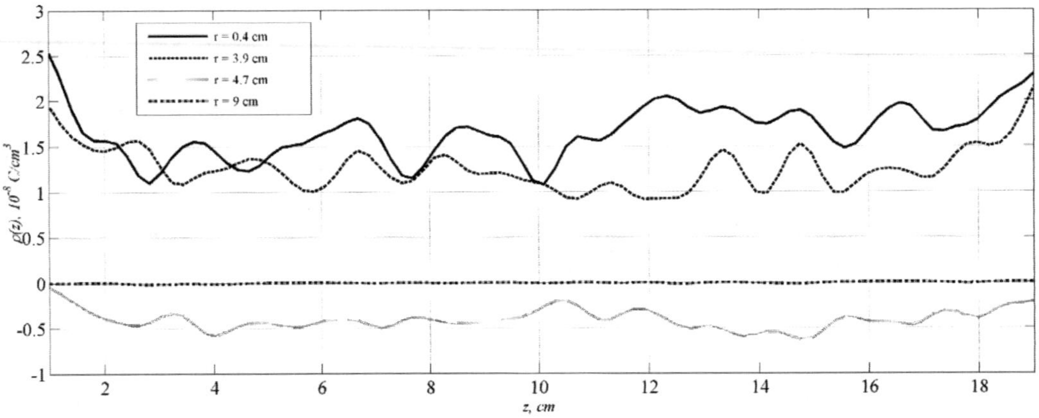

Figure 7. Total charge density $\rho(z)$: $I_0=15\ kA$, $\tau=300\ ns$, $B_z=1.5\ kGs$, $n_p=2.5\ 10^{11}\ cm^{-3}$, $E_0=20\ keV$.

Figures 7 shows typical dependences of total charge density of the system in a longitudinal cut at the time point after beam transient period. There are oscillations of charge density distribution in space and time with frequency of the order of the Langmuir frequency (3-7 GHz) in the system. Due to inertial processes there can appear local areas with prevailing positive or negative charges.

4. Summary

Research of dynamics of plasma electrons leaving the beam drift area shows that the beam is transported in the overcompensation state with a maximum in near-axis area, which is confirmed in experiments [4] (Figure 7). This can be explained by the fact that the exit velocity of the plasma electrons on the initial injection stage exceeds the input electron beam velocity due to its inhibition in the transient mode.

Thus, there is positive scalar potential with a maximum at the drift pipe center (on axis z), which accelerates electrons at the beginning segment of the drift pipe. With increasing time, the flow of the plasma electrons becomes lower than the injected beam current and increases the negative charge influence in the beam drift area. Such dependence of scalar potential is due to boundary conditions at drift pipe end faces. Further research of transportation efficiency in case of different boundary conditions presents interest.

Beam instabilities are not observed apparently because the beam energy dispersion during of the transportation is rather considerable.

Thus as a result of numerical calculations based on the presented model, quantitative estimates of the beam and plasma parameters required to reach the highest efficiency of LHEB transportation can be obtained.

Acknowledgement

This research was supported by RFBR grant 13-08-98066-r_sibir_a.

References

[1] Grigoriev V P, Koval T V, Kuhta V P, Rahadjo P, Uemura K 2008 Transport and focusing of a low-energy electron beam in low-pressure ionized argon. *Journal of Technical Physics* **78, 1** 104-108

[2] Grigoriev V, Vagin E, Ofitserov V 2010 Macroparticles model of charge neutralization of the electron beam at injection in low-pressure plasma. *Bulletin of the Tomsk Polytechnic University*, **316, 2** 85-89

[3] Golant V E, Zhilinskiy A P, Saharov I E 1977 *Fundamentals of Plasma Physics* (Moscow: Atomizdatm, p. 47)

[4] Ozur G E 2012 Low-energy, high-current electron beams for material surface treatment *25th International Symposium on Discharges and Electrical Insulation in Vacuum (ISDEIV)* 580-583

[5] Ozur G E, Proskurovsky D I, Rotshtein V P, Markov A B 2003 Production and application of low-energy, high-current electron beams *Laser & Particle Beams* **21** 157-174

[6] Nezlin M V 1982 *Beam dynamics in Plasma* (Moscow: Energoizdat, p. 102)

Journal of Physics: Conference Series **652** (2015) 012022 doi:10.1088/1742-6596/652/1/012022

Non-equilibrium melting processes of silicate melts with different silica content at low-temperature plasma

V Vlasov[1,2], G Volokitin[1], N Skripnikova[1], O Volokitin[1], V Shekhovtsov[1], A Pfuch[3]
[1]Tomsk State University of Architecture and Building, Applied Mechanics and Material Sciences Department, Tomsk, 634003, Russia
[2]National Research Tomsk Polytechnic University, Tomsk, 634050, Russia
[3]Innovent, Pruessingstraße 27B, D-07745 Jena
E-mail: volokitin_oleg@mail.ru

Abstract. This article is devoted to research the possibility of high-temperature silicate melts producing from different silica content at low-temperature plasma taking into account non-equilibrium melting processes.

1. Introduction

Silicate melting using strong plasma flows is a high-tech process that provides processing of both man-made and natural raw materials. The use of low-temperature plasma (LTP) for the production of high-temperature silicate melts is connected with high temperatures produced by plasma generator much higher than melting temperatures of all silicate materials. This creates consistently high temperatures and energy concentration allowing the transition of processed materials to a molten state. In melting silicate systems using the low-temperature plasma, a range of diverse physicochemical and phase transformations occur, the intensification of which is increased with the increase of temperature.

The original materials for the production of high-temperature silicate melts were as follows:

- Ash from Seversk (Tomsk region, Russia) obtained after coal burning;
- oil shale combustion products from Province Dalian (China).

The chemical composition of these raw materials is presented in Table 1.

Table 1. The chemical composition of original raw materials before plasma treatment.

Raw materials	Oxide content, (wt. %)							
	SiO_2	Al_2O_3	Fe_2O_3	CaO	MgO	K_2O	Δm	M_K
Ash	51.16	35.07	3.62	8.33	0.91	0.23	0.68	9.33
Oil shale combustion products	61.59	23.36	7.91	1.60	1.27	1.34	2.93	29.60

Δm – weight loss, M_K – acidity index, $M_K = \dfrac{SiO_2 + Al_2O_3}{CaO + MgO}$

According to Table 1, the raw materials contain over 50% of SiO_2, which is the main glass former. The acidity index of the used anthropogenic waste is significantly higher than that of the traditional silicate raw materials. The high acidity index has a positive effect on the chemical and thermal resistance of the finished articles. A tendency to crystallization increases with the decrease of the acidity index, and durability decreases.

2. Experimental work

The melting process of raw materials was experimentally carried out in the electro-plasma apparatus for manufacturing a refractory silicate melt [5]. The plasma-supporting gas flows out of the discharge jet of the electro-plasma apparatus and serves as a heating material. The LTP jet enters the water-cooling melting furnace to perform melting of the refractory silicate raw material.

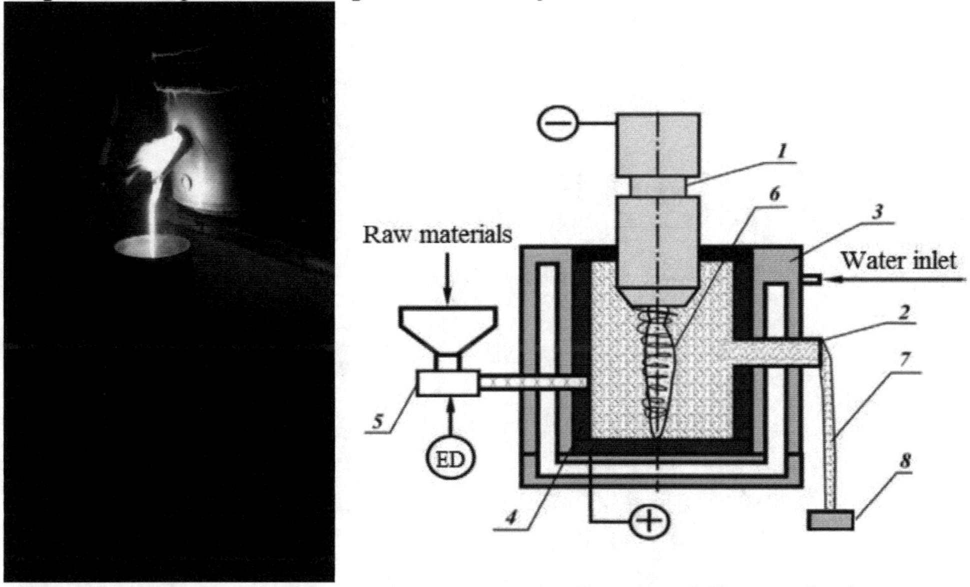

Figure 1. Experimental electro-plasma apparatus for mineral fiber production:
1 – plasma torch; 2 – pouring hole; 3 – water-cooled melting furnace; 4 – graphite electrode; 5 – worm feeder; 6 – plasma jet; 7 – melt; 8 – mold

Operation of the electric-plasma apparatus is based on the interaction between highly concentrated plasma flows and a raw material. Plasma torch 1 is placed in melting furnace equipped with pouring hole. Graphite electrode is installed in the bottom of melting furnace and connected with the positive pole of DC power supply (anode), while the plasma torch is connected with its negative pole (cathode). Once silicate melt reaches the level of the pouring hole, it will be discharged through it into forming hood. By means of batcher mounted into worm feeder 6 driven by electric drive, powder raw material is fed to the melting furnace. Plasma jet is projecting between plasma torch and graphite electrode. A raw material starts to melt and homogenize under a strong thermal effect (3000-5000 °C). The raw material particles are then mixed with a high-temperature silicate melt and melted uniformly. Furthermore, this method of feeding of raw materials makes possible to avoid a loss of fine particles blowing by a LTP jet. This method provides a uniformity and viscosity of low melt within the whole volume of the melting furnace.

It is advisable to identify the behavior of silicate systems under the equilibrium conditions using the state diagram and melting curves reflecting the quantity and composition of the liquid phase at temperature changes [1-4]. The analysis of the equilibrium processes for recalculation to a ternary system showed that the temperature required for the production of 100% melt ranges between 1640-1700°C depending on SiO_2 contained in the raw material.

A study of melting processes of silicate raw materials comprising 50-65% silica includes the analysis and comparison of equilibrium and non-equilibrium processes of the formation of silicate melt mixtures having different SiO_2 content.

3. Results and discussion

The analysis of melt formation based on the actual chemical composition (Table 2) shows that the primary melt is formed at 1170 °C, and comes to 33.33%. These processes are non-equilibrium and reflect the traditional material melting in tanks at running heating rates.

Journal of Physics: Conference Series **652** (2015) 012022 doi:10.1088/1742-6596/652/1/012022

Table 2. The melt amount produced by ash fusibility depending on eutectics melting.

System	Composition, (wt. %)	Temperature, (°C)	Eutectic melt, (%)	Total melt amount, (%)
CaO-Al_2O_3-SiO_2	SiO_2 – 55 Al_2O_3 – 37 CaO – 8	1170	33.33	33.33
FeO-SiO_2	SiO_2 – 38 FeO – 62	1178	4.99	38.32
MgO-Al_2O_3-SiO_2	MgO – 21 Al_2O_3 - 22 SiO_2 – 47	1355	3.90	42.22
CaO-MgO-SiO_2	SiO_2 – 43,6 MgO – 9,2 CaO – 47,2	1400	–	–
CaO-SiO_2	SiO_2 – 63 CaO – 37	1436	–	–
$CaO\cdot Al_2O_3\cdot 2SiO_2$ (anorthite)		1450	–	52
MgO-SiO_2	SiO_2 – 65 MgO – 35	1543	–	–
Melting due to dissolution Al_2O_3 и SiO_2		1550	–	82
Complete melting		1590	–	100

Melting of eutectic comprising 38% of SiO_2 and 62% of FeO, occurs at 1178 °C with the formation of 4.99% of the eutectic melt. Further, the melt is formed due to bonds between MgO and SiO_2 at 1355 °C, the total melt amount is 42.22%. At the further temperature increase, the melt is formed not due to eutectics, but rather anorthite melting and dissolving of silica and aluminum. At temperature of 1590 °C, ash will be completely fused.

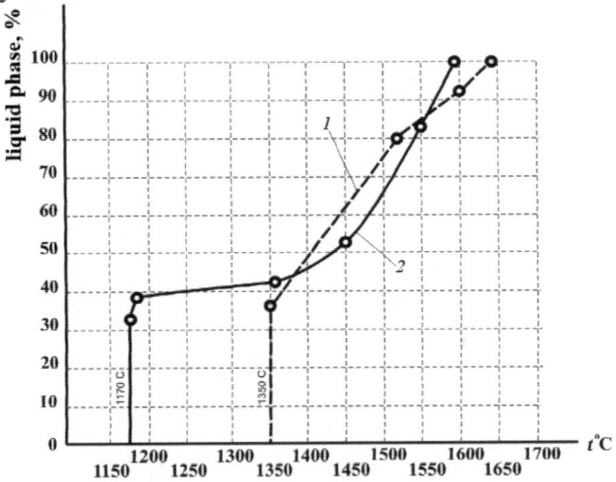

Figure 2. Ash fusibility curves:
1 - equilibrium dependence in CaO-Al_2O_3-SiO_2 system;
2 – non-equilibrium dependence of the actual chemical composition.

The analysis of ash fusibility curves (Fig. 2) shows that the starting temperature of melt formation does not exceed 1350 °C. The same amount of primary melt (approx. 35%) in the real system is formed at 1170 °C, while in the model system it is formed at 1350 °C. A further increase of the liquid phase is similar for equilibrium and non-equilibrium processes. The formation of 100% melt in the model system occurs at temperature of 1640 °C. however, accounting for the real chemical composition, ash will be completely fused at temperature of 1590 °C.

The analysis of the formation of shale combustion products based on the actual chemical composition (Table 3) shows that a relatively small amount of the primary melt is formed at temperature of 1170 °C and comes to 6,4%.

Table 3. The melt amount produced by shale combustion product melting depending on eutectics melting

System	Composition, (wt. %)	Temperature, (°C)	Eutectic melt, (%)	Total melt amount, (%)
$CaO-Al_2O_3-SiO_2$	SiO_2 – 60 Al_2O_3 – 15 CaO – 25	1170	6.4	6.4
$FeO-SiO_2$	SiO_2 – 38 FeO – 62	1178	10.92	17.32
$MgO-Al_2O_3-SiO_2$	MgO – 21 Al_2O_3 - 22 SiO_2 – 47	1355	5.44	22.76
$CaO-MgO-SiO_2$	SiO_2 – 43,6 MgO – 9,2 CaO – 47,2	1400	–	–
$CaO-SiO_2$	SiO_2 – 63 CaO – 37	1436	–	–
Melting due to dissolution Al_2O_3 и SiO_2		1500		43
$MgO-SiO_2$	SiO_2 – 65 MgO – 35	1543	–	–
Melting due to dissolution Al_2O_3 и SiO_2		1550		62
Complete melting		1600	–	100

Melting of the next eutectic comprising 38% SiO_2 and 62% FeO, occurs at 1178°C with the formation of 10.92% of the eutectic melt. Further, the melt formation occurs due to MgO and SiO_2 bond at temperature of 1355 °C. The total melt amount is 22,76%. At further temperature increase, the melt is formed not due to eutectics melting, but rather dissolving silica and alumina oxides, and at 1600 °C the shale combustion product will be completely melted.

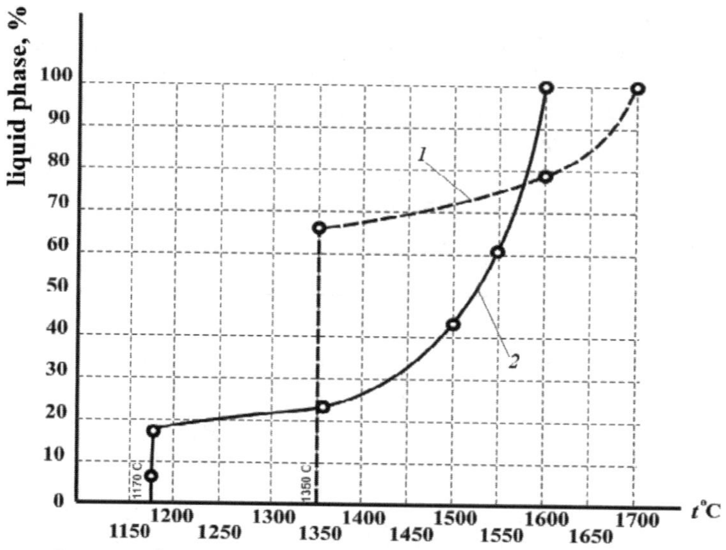

Figure 3. Shale combustion products fusibility curves:
1 - equilibrium dependence in CaO-Al_2O_3-SiO_2 system;
2 – non-equilibrium dependence of actual chemical composition

The analysis of fusibility curves (Fig. 3) shows that the starting temperature for the melt formation of shale combustion product does not exceed 1350 °C. At that, 6,4% of primary melt in the real system is formed at 1170 °C, and in the model system 66% of melt is formed at 1350 °C. The formation of 100% shale combustion product in the model system occurs at temperature of 1700 °C, but accounting for the real chemical composition, the shale combustion product is completely melted at temperature of 1600 °C.

4. Conclusions

Thus, it was found that the melting process of raw materials accounting for their real chemical composition is characterized by 50 °C decrease in the melting temperature in case of ash, and for shale combustion products it is characterized by the decrease of 100 °C. At that, the temperature of the primary melt produced by ash fusibility is formed in the real system at 1170 °C that is 180 °C lower than that in the model system. For shale combustion products, 6.4% of primary melt is formed in the real system at 1170 °C, while in the model system 59.6% of primary melt is formed at 1350 °C.

5. Acknowledgement

The research was financed by the Ministry of Education and Science of the Russian Federation (State Order N 11.351.2014/K).

References

[1] Volokitin O, Timonov E, Volokitin G, Nikiforov A, Chibirkov V RF Patent 2503628 (2014).

[2] Volokitin O, Maslov E, Shekhovtsov V 2014 Plasma technology of silicate melts obtaining in mineral fibers production *Advanced Materials Research:* vol 880 pp 233-236.

[3] Volokitin O, Skripnikova N, Shekhovtsov V 2014 Technology for plasma treatment of molybdenum ore tailings *Applied mechanics and materials* vol 682 pp 515-518.

[4] Volokitin O, Volokitin G, Skripnikova N, Shekhovtsov V 2015 Definition of nanostructural characteristics melting product of quartz sand *AIP Conference Proceedings* vol 1623 pp 655-658.

[5] Volokitin O, Volokitin G, Shekhovtsov V 2015 Physicochemical processes at plasma treatment of quartz-feldspar raw materials *Advanced Materials Research* vol 1085 pp 474-478.

[6] Volokitin G, Skripnikova N, Volokitin O, Iuriev I, Shekhovtsov V 2015 Complex research of molybdenum ore tailings *IOP Conf. Series: Materials Science and Engineering* vol 71 pp 1-5.

[7] Volokitin O, Vereshchagin V, Shekhovtsov V 2015 Melting process of silica sand in low-temperature plasma technology *Chemistry and Chemical Engineering* vol 58 (1) pp 62-64.

Destruction of molecular compounds in gaseous and liquid medium in microwave discharge plasma

A G Zherlitsyn, V P Shiyan, L N Shiyan and S O Magomadova

Tomsk Polytechnic University, Lenin Avenue 30, Tomsk 634050, Russia

E-mail: zherl@tpu.ru

Abstract. The paper presents the results of experimental studies of molecular destruction in gaseous and liquid medium using microwave discharge plasma at atmospheric pressure. As the gas medium hydrocarbon gas is used, the liquid medium were aqueous solutions of methylene blue and more complex organic compound in the form of humic substances. As a result of the destruction of hydrocarbon gas molecules in microwave discharge plasma new products such as hydrogen, ethylene, acetylene and carbon nanostructured material have been formed. In experiments on destruction of molecular compounds in aqueous organic solutions we used air, nitrogen and argon for plasma gases. It is shown that the process of molecular destruction in aqueous organic solutions in the microwave discharge plasma is based on oxidation-reduction reactions. It is found that the maximum efficiency of removal of organic compounds from the solution occurs when using air as the plasma gas.

1. Introduction

It is known that the gas discharge generates reactive plasma, which is capable of providing the necessary conditions for destruction of molecular compounds and synthesis of unique structures. It finds application in development of new technologies for disposal of waste products, purification of products from pollutants, production of new materials, surface modification of materials, etc.

In recent years there have been intensive studies on possibility of using the microwave discharge plasma in plasma-chemical processes. The choice of this type of discharge is based on the fact that plasma-chemical reactions occur with high energy efficiency due to vibrational excitation of molecules in in nonequilibrium plasma of microwave discharge [1, 2]. This is of a great interest primarily for plasma-chemical processes such as reaction with electron-excited particles, dissociation or degradation of molecular compounds of organic substances.

This paper presents the results of experimental studies of destruction of molecular compounds in the microwave discharge plasma at atmospheric pressure in such medium as petroleum gas, aqueous solutions of methylene blue, humic substances.

2. Experimental set-up

The main element of the experimental setup is a microwave plasmatron. The schematic of which is shown in Figure 1.

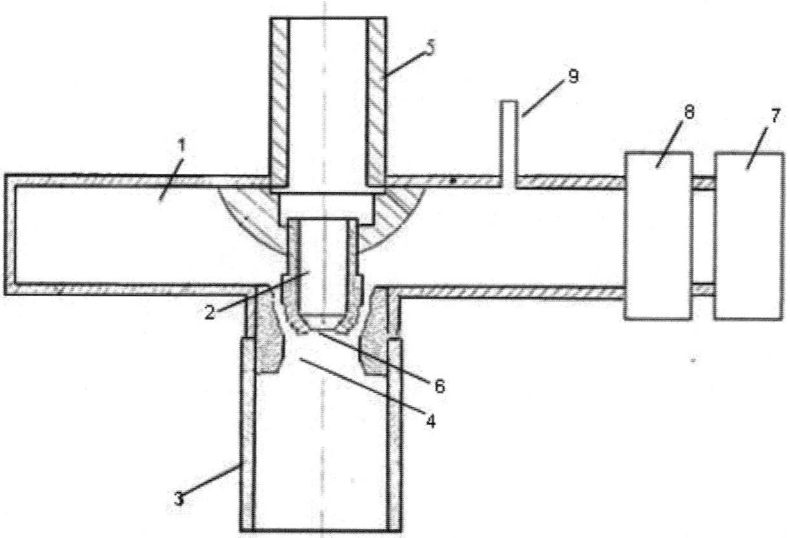

Figure 1. Schematic of a microwave plasmatron
1 - rectangular waveguide; 2 - inner conductor of the coaxial line; 3 - outer conductor of the coaxial line; 4 - discharge chamber; 5 - pipeline; 6 - nozzle; 7 - magnetron; 8 - circulator; 9 – pipe

Structurally, the plasmatron is a waveguide-coaxial junction, consisting of a rectangular waveguide 1 and a coaxial line with a hollow inner conductor 2 and the outer conductor 3, forming a discharge chamber 4. To maintain a stable microwave discharge the plasmatron is provided with an active system of initiation of a discharge, which is described in detail elsewhere [3]. Injection of hydrocarbon gas or aqueous solution into the discharge chamber of the plasmatron is carried out via line 5 through nozzle 6. Power is supplied to the plasmatron from the magnetron 7 with an adjustable output power of up to 2 kW in continuous operation mode with an operating frequency of 2.45 GHz. For breakdown protection of the magnetron a ferrite circulator 8 is used. When working with aqueous solutions the plasma gas is fed into the plasmatron through the inlet 9. The design of the nozzle 6 allows one to spray an aqueous solution, supplied through line 5, in a gas stream similar to an ejector.

The experimental setup works as follows. When working with a hydrocarbon gas, the initiation system of microwave discharge in the chamber 4 is activated first. Then, through line 5 and nozzle 6 the hydrocarbon gas is supplied into the discharge chamber and turns on the magnetron. The microwave energy from the magnetron is supplied into the discharge chamber though waveguide 1 via coaxial-waveguide transition. Near the nozzle 6 due to incident and reflected electromagnetic waves the electric field is increased to the breakdown value. As a result, the microwave discharge is ignited in atmosphere of hydrocarbon gas and nonequilibrium low-temperature plasma is formed. Formation of plasma at atmospheric pressure in hydrocarbon gas proceeds in the absence of oxygen.

When working with aqueous solutions, the solution is fed in the discharge chamber via pipeline 5 through nozzle 6 and the plasma gas is supplied into the plasmatron through pipe 9. In the experiments we studied the following gases: air, nitrogen and argon. The process of microwave discharge and plasma formation proceeds so that the discharge burns in an environment, which contains air and pulverized particles of the aqueous solution.

3. Destruction of molecular compounds in the gaseous medium

The experiments were performed at a flow rate of hydrocarbon gas ranging from 0.4 m^3/h to 1 m^3/h, the level of microwave power delivered to the discharge was varied from 0.8 kW to 2 kW. The degree of destruction of molecular compounds was determined using a chromatographic analysis of gases. Table 1 shows the composition of hydrocarbon gas used in our experimental set-up. Table 2 shows the gas composition after it has passed through the microwave discharge.

Table 1. Composition of the initial hydrocarbon gas

Component	Concentration, %
Methane	93.554
Ethane	3.858
Propane	1.458
Iso-Butane	0.291
Butane	0.262
Isobutyl	0.002
Iso-Pentane	0.027
Pentane	0.013
Carbon dioxide	0.536

Table 2. Composition of the gas after passage through the discharge

Component	Concentration, %
Methane	36.185
Ethane	0.501
Ethylene	1.307
Propane	0.226
Acetylene	2.536
Iso-Butane	0.068
Butane	0.061
Isobutyl	0.000
Cis 2-butene	0.003
Iso-Pentane	0.003
Pentane	0.003
Hydrogen	58.727
Carbon dioxide	0.376

It is seen from Table 1 and Table 2 that in microwave discharge plasma the destruction of the methane and C_{2+} hydrocarbons molecules occurs with the formation of new compounds, such as ethylene, acetylene and hydrogen. Free carbon after discharge settles down in a form of solid particles of carbon material. Analysis of the carbon material using the X-ray diffractometer shows that the material consists of amorphous carbon and carbon nanotubes with a particle size ranging from several nm to several tens of nm.

Figure 2 shows the experimental dependence of the degree of destruction of the gas molecules on the level of microwave power delivered into the discharge. Here, as a parameter, the gas flow rate has been taken. From these results it is clear that the dependence of destruction of molecules on the level of microwave power, delivered into the discharge, has a linear character. The degree of destruction of molecular compounds of hydrocarbon gas increases when adding nitrogen to hydrocarbon gas.

Figure 3 shows the dependence of the degree of destruction K_{distr} of hydrocarbon gas molecules on the relative concentration of nitrogen in hydrocarbon gas β. In [4] authors obtained similar results with methane under the influence of high-frequency discharge. The authors explain this result by a chain mechanism of decomposition of methane molecules, which involves vibrational excitation of nitrogen molecules.

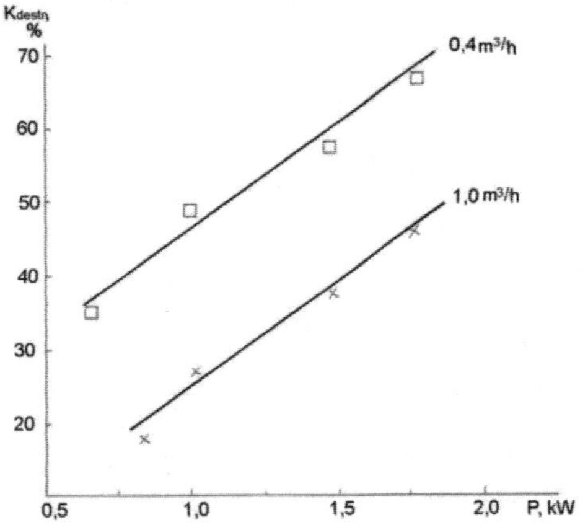

Figure 2. The dependence of the degree of destruction of molecules of hydrocarbon gas on the microwave power level

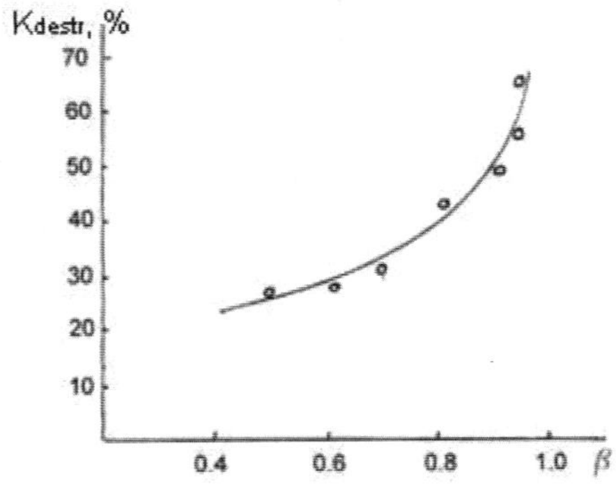

Figure 3. Dependence of the degree of destruction of the hydrocarbon gas molecules on the relative concentration of nitrogen in gaseous medium

4. Destruction of molecular compounds in aqueous solutions

Experiments were conducted at a flow rate of aqueous solution being 3 liters/hour, plasma gas of 3 m^3/h, microwave power level of up to 1.2 kW. In the experiments with methylene blue solution (MB) the concentration of MB in the solution is 4 mg/l, optical density of the solution is 0.35.

Figure 4 shows the optical absorption spectra of the initial MB solution and processed solution with microwave discharge plasma. We observed several absorption maxima in the spectrum: 590nm, 290nm and 200nm. According to the literature [5] absorption at 290nm and 590nm are thought to be due to the chromophore groups of methylene blue and absorption at 200 nm is due to intrinsic absorption of the nuclei of molecules.

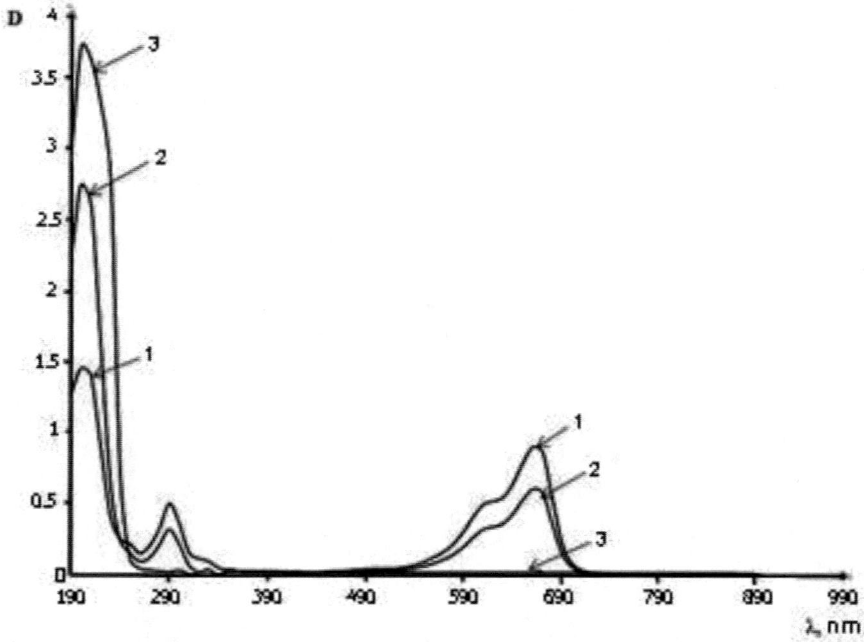

Figure 4. Optical absorption spectra of an aqueous solution of methylene blue.
1 - spectrum of initial solution; 2 - spectrum of the solution after 1 treatment cycle; 3-spectrum of the solution after 2 cycles.

As seen from Figure 4, treatment of MB solution with microwave discharge plasma leads to a reduction in the peak intensity at 290nm and 590nm as compared with the initial solutions spectra (curves 2 and 3). In the area of 200 nm, an increase in the absorption intensity in observed. According to [6], reduction in the intensity of absorption at 290nm and 590nm is due to the destruction of chromophore groups, while an increase in absorption at 200 nm is due to the formation of new compounds according to the schematic shown in Figure 5.

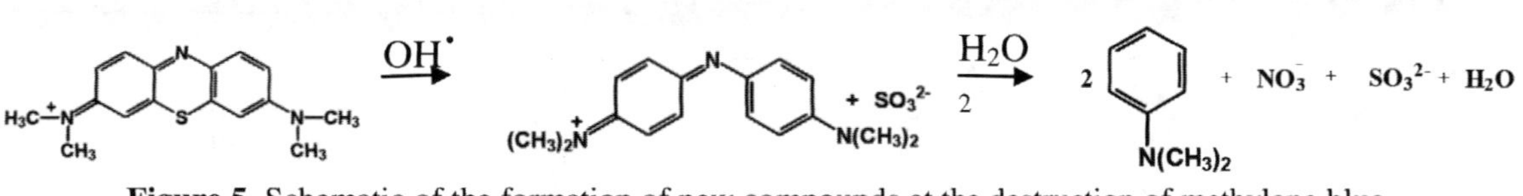

Figure 5. Schematic of the formation of new compounds at the destruction of methylene blue

Along with the change in the absorption intensity there is a decrease in pH. It can be assumed that the reduction in the intensity of the absorption of MB solution is due to the formation of nitrogen-containing compounds, such as HNO_2 and HNO_3 in the solution.

Figure 6 shows the dependence of the decrease in pH (curve 1) and kinetics of the formation of nitrate-ions (curve 2) depending on the number of plasma treatment cycles.

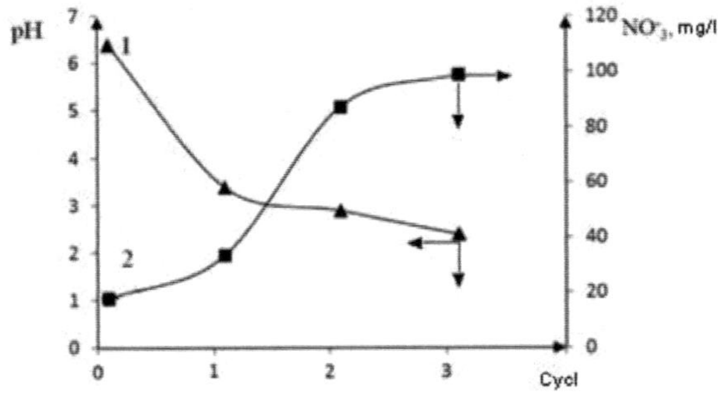

Figure 6. Kinetics of formation of nitrate ions and decrease in pH level

As can be seen in Figure 6, the decrease in pH of the solution is well correlated with the concentration of nitrate ions, which may cause the reduction in the intensity in color of MB.

To exclude the possibility of formation of nitrogen-containing compounds further experiments were performed with argon as plasma gas. Table 3 summarizes the parameters of treated distilled water in air and argon.

Table 3. Parameters of distilled water

Parameters of treated solution	Plasma gas	
	Air	Argon
pH	2.8	6.4
NO_3^-	19.7	0.632
Residual Ozone	2.5	0.71
Permanganate oxidation (PO)	38.6	3.02
Optical density of the solution	0.09	0.11

It can be seen from Table 3 that formation of nitrogen-containing compounds and reduction in pH of the solution is due to the use of air as the plasma gas.

In experiments with aqueous solutions of humic substances we used a model solution with a concentration of humic substances in the distilled water being 8 mg/l, and natural water with a concentration of humic substances being 7.8 mg/l. Table 4 shows the parameters of these solutions before treatment with microwave discharge and after treatment. The plasma gas was air. Each solution was subjected to discharge treatment from 1 to 3 times (1-3 cycles).

Table 4. Parameters of water with humic substances

Tested parameters	Solution of humic substances	Treated and filtered sample		
		Number of cycles		
		1	2	3
pH	5.87[a]	2.83	2.56	2.33
PH, mgO_2/l	2.86[a]	1.07	0.09	0.06
pH	7.10[b]	2.80	2.80	2.80
Fe total., mg/l	8.57[b]	3.30	3.30	3.30
PO, mgO_2/l	2.69[b]	2.10	2.12	2.69
Si, mg/l	11.33[b]	9.40	7.80	6.62
Fe (II), mg/l	$\leq$0.05[b]	3.21	3.17	3.23

[a] Model solution of humic substances
[b] Natural water

Table 4 shows that under the action of microwave discharge there a significant change in the parameters of an aqueous solution of humic substances. Reduction in the value of permanganate oxidation indicates a decrease in the concentration of humic substances in the solution. Concentration of humic substances in the solution, which was determined by measuring the concentration of total organic carbon, was reduced from 7.8mg/l. to 2.35 mg/l. The mechanism of destruction of molecular compounds, shown in Figure 5, is also applicable to solutions of humic substances. Under the action of microwave discharge on complex humic substances degradation of their molecular compounds occurs into smaller organic compounds, such as quinoid structures, carboxylic and phenolic compounds, which are difficultly soluble in water. Chemical analysis carried out on the content of divalent iron ions, showed that divalent iron is formed as a result of the treatment by microwave discharge. This indicates the occurrence of redox processes, which were also observed in the experiments with methylene blue.

5. Conclusion

In this study we demonstrated an efficient destruction of molecular compounds in microwave discharge plasma in carbon dioxide and aqueous solutions of methylene blue and humic substances. In the experiments with hydrocarbon gas we obtained the efficiency of destruction of up to 70%. As a result of gas decomposition several substances are formed, such as: hydrogen, ethylene, acetylene, amorphous carbon and carbon nanotubes. The experiments with aqueous solutions show that the main processes that occur during the treatment of aqueous solutions with discharge plasma are the reduction-oxidation reactions. Under the action of microwave discharge plasma on aqueous solution the destruction of complex organic compounds to simple compounds and new elements take place. The results obtained in this study can be of an interest in the development of new environmentally-friendly technologies for utilization of hydrocarbon gases (natural gas, associated petroleum gas, coal mine methane, biogas) and formation of new useful products for cleaning water from organic substances (humic, petroleum products, etc).

Acknowledgement

This research was supported by the grant for scientific research «Science» from the Ministry of Education and Science of Russia, project # 0.1484.2015 and 7.1504.2015.

References

[1] Eletskii L V, Palkina L A and Smirnov B M 1975 Transport phenomena in a weakly ionized plasma (*Moscow, Atomizdat.*)
[2] Bykov Y V 1984 *High energy chemistry* **18** 347
[3] Korolev Yu D, Frans O B, Landl V, Geyman V G, Zherlitsyn A G, Shiyan V P and Medvedev Yu V 2009 *IEEE Trans. Plasma Sci.* **37** 2298
[4] Savinov S Y, Lee H, Song H K and Na B K 2003 *Plasma Chemistry and Plasma Processing* **23** 159
[5] Parkansky N 2013 *Journal Plasma & Chemistry* **33** 907
[6] Maksimov A, Hlyustova A and Subbotin N 2009 *Chemistry and Chemical Technology* **52** 116
[7] Piskarev I, Solovieva G, Karelin V, Selemir V and Spirov G 2005 *High Energy Chemistry* **39** 228

Magnetron sputtering system for coatings deposition with activation of working gas mixture by low-energy high-current electron beam

N V Gavrilov, A S Kamenetskikh, A I Men'shakov, O A Bureyev

Institute of Electrophysics, 106 Amundsen str., Yekaterinburg, 620016, Russia

E-mail: gavrilov@iep.uran.ru

Abstract. For the purposes of efficient decomposition and ionization of the gaseous mixtures in a system for coatings deposition using reactive magnetron sputtering, a low-energy (100–200 eV) high-current electron beam is generated by a grid-stabilized plasma electron source. The electron source utilizes both continuous (up to 20 A) and pulse-periodic mode of discharge with a self-heated hollow cathode (10-100 A; 0.2 ms; 10-1000 Hz). The conditions for initiation and stable burning of the high-current pulse discharge are studied along with the stable generation of a low-energy electron beam within the gas pressure range of 0.01 – 1 Pa. It is shown that the use of the electron beam with controllable parameters results in reduction of the threshold values both for the pressure of gaseous mixture and for the fluxes of molecular gases. Using such a beam also provides a wide range (0.1-10) of the flux density ratios of ions and sputtered atoms over the coating surface, enables an increase in the maximum pulse density of ion current from plasma up to 0.1 A, ensures an excellent adhesion, optimizes the coating structure, and imparts improved properties to the superhard nanocomposite coatings of (Ti,Al)N/a-Si_3N_4 and TiC/-a-C:H. Mass-spectrometric measurements of the beam-generated plasma composition proved to demonstrate a twofold increase in the average concentration of N^+ ions in the Ar-N_2 plasma generated by the high-current (100 A) pulsed electron beam, as compared to the dc electron beam.

1. Introduction

In the development of coating deposition technology, a significant milestone was reached with the design of magnetron sputtering systems (MSS) with unbalanced magnetic field [1], which have provided an increase in the ion current density on the surface of workpieces and, as a consequence, an increase in the strength of adhesive bonds between coating and substrate and an improved control over the nucleation processes, morphology, chemical composition, and microstructure of coatings as well as over the level of internal stress and the recrystallization processes occurring in the coatings [2]. Also, in order to increase the degree of plasma ionization, initiation of additional discharges in the MSS effective volume has been applied widely [3]. In such ionized physical vapor deposition systems, most frequently used a thermionic cathode discharges, a high-frequency discharges, and a microwave discharges. In the coating deposition technology of magnetron sputtering, an important advance has been made by the application of a high-power impulse discharge mode (HiPIMS mode), which ensures a steep increase in the degree of ionization of sputtered atoms of the cathode materials [4].

Journal of Physics: Conference Series **652** (2015) 012024 doi:10.1088/1742-6596/652/1/012024

The use of a low-energy electron beam for plasma generation, properties of the beam-generated plasma and some applications for material surface modification have been reported in a number of studies, e.g., in [5]. An electron beam (2 keV, 50–100 mA) was generated in a high-voltage glow discharge with a hollow cathode at a pressure of 2–13 Pa in pulse-periodic mode (2 ms, 10–20% duty factors). The ribbon electron beam ((25x1 cm^2) is drifting in a longitudinal magnetic field, thus forming a 'plasma sheet'. Ions arrive at the samples disposed at a distance of 4–5 cm away from the plasma with the energy to be determined by the bias voltage on the samples. A plasma model is reported [6] to have been developed, wherein plasma is generated by an electron beam in low-pressure nitrogen, and a conclusion has been made that the properties and composition of the plasma differ considerably from those of the plasma generated in the gas discharge. An electron source, with a thermionic cathode and an electron optic system, which generates dc electron beam (3.6 A, 140 V) at a pressure of 0.15 Pa, is reported in [7] for steel nitriding application.

The present paper reports about some features of using a broad electron beam in a MSS application, with the beam having a round cross-section (100 cm^2) at an energy 100–200 eV, and a pulse current of 1–100 A. The beam is generated in a plasma source of electrons, which is based on a self-heated hollow cathode discharge. The paper reports on the results of studying the conditions of stable generation of electron beam at elevated pressures (up to 1 Pa) and low accelerating voltages (100 V) in a high-current pulse-periodic mode, on the results of mass-spectrometry of the beam-generated argon-nitrogen plasma, and on the results of using MSS in combination with an electron beam source for the purpose of superhard nanocomposite coatings deposition.

2. Experimental Technique

A schema of a magnetron sputtering system with the gas working medium activated by an electron beam, (MSS EB) is displayed in Figure 1. An electron source is mounted on the working chamber (340 mm in dia, 270 mm in height) and comprises a hollow cathode *1* and a hollow anode *2*, with a grid electrode *3* being a part of it. Once a discharge is initiated and an accelerating voltage is applied between the grid *3* and the walls of a chamber *4*, there appeared a double layer of space charge in the grid area, wherein an electron beam is formed. Four flat magnetrons *5* with a sputtered targets diameter of 75 mm, operating in a pulse-periodic mode (50 kHz, 10 μs) are disposed on the side surface of the chamber. Pulsed discharge current for each of the magnetrons is controlled independently (0.1–2 A). Samples are placed at a manipulator *6* with a planetary gear. A gas flux required for burning the discharge is fed through the hollow cathode made of a tube with an internal

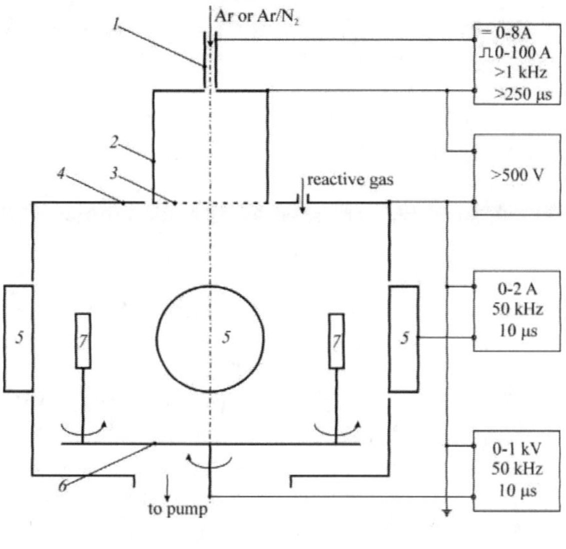

Figure 1. The diagram of magnetron sputtering system. *1* – hollow cathode, *2* – hollow anode, *3* – grid, *4* – working chamber, *5* – magnetrons, *6* – manipulator, *7* – samples.

diameter of 8 mm and a wall thickness of 1 mm. To obtain nitride coatings, a TiN cathode is used, in which case a reaction gas is fed through the cathode. To obtain carbide coatings, acetylene is fed directly into the working chamber.

MSS operating in combination with an electron beam source is featured with a reduction in the threshold value of the working gas pressure and the discharge voltage (Figure 2). When a pulsed beam at a current of 100 A is used, the magnetron discharge current during the beam pulse is approximately twice as great. The described MSS has a specific that, with the use of sputtered cathodes made of different materials, it is necessary to ensure a speed of sample rotation that could eliminate the heterogeneity in the contents of sputtered elements on the coating surface. In our conditions, this was achieved at a sample rotation speed of 10 rpm. The distance between the surfaces of MSS sputtered targets and the samples was 60 mm at an average.

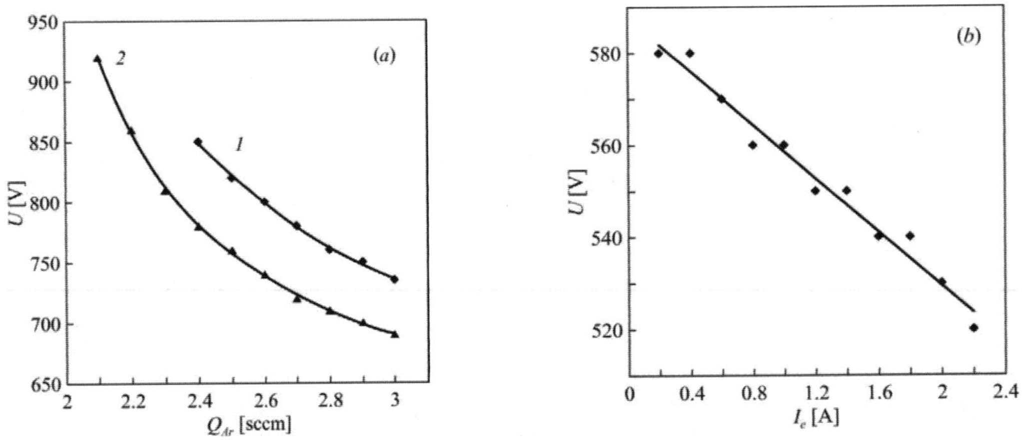

Figure 2. Dependences of magnetron discharge voltage vs Ar flow rate (*a*) and electron beam current (*b*). Electron beam current: *1* – 0, *2* – 2 A. Ar flow rate: *b* – 5 sccm.

The use of an MSS combined with an electron beam source based on a glow discharge with a hollow cathode was reported in [8]. The maximum beam current of this electron source did not exceed 5 A due to glow-to-arc transition, and the maximum ion current density was reported not to exceed a few mA/cm^2. The creation of an electron beam source based on a self-heated hollow cathode discharge (SHHC) has made it possible to increase the beam direct current up to 20 A and to rise the ion current density up to ~10 mA/cm^2 [9]. Further development of MSS EB is associated with the use of a pulse-periodic mode of SHHC discharge. Once a low-current discharge is initiated to provide heating and transition of the hollow cathode into the thermionic emission mode (4-5 A, 50 V), a voltage pulse of 200-600 V in amplitude is applied to the electrodes, which leads to an increase in the discharge current up to a few hundred amperes over a time of about 0,1 ms. By matching the values of pulse repetition rate and current pulse amplitude, such discharge can be sustained in the pulse-periodic mode[10].

2.1. High current electron beam generation

The conditions for stable generation of a high-current electron beam at low accelerating voltages were determined using a double Langmuir probe technique [11]. One of the probes was disposed in the emitting plasma of the electron source, and the other was placed in the beam-generated plasma of the working chamber. Under stable electron beam generation, there exists a double layer of space charge between the plasmas, and the probes' floating potential difference occurs, which value is, to an accuracy of the electron temperature, equal to the potential difference of the plasmas, i.e., is close to the voltage drop on the double layer. Figure 3 displays waveforms recorded under the stable mode of the electron beam generation. Growth of the beam current I_b is delayed relative to the discharge current I_d and reaches saturation in 0.15–0.2 ms, with the floating potential difference U_{dl} being comparable to the accelerating voltage value. In the unstable mode, the voltage drop on the double layer is modulated with oscillations of a frequency of 400 kHz and with the amplitude approaching the sum of the accelerating voltage and the pulse voltage on the discharge gap.

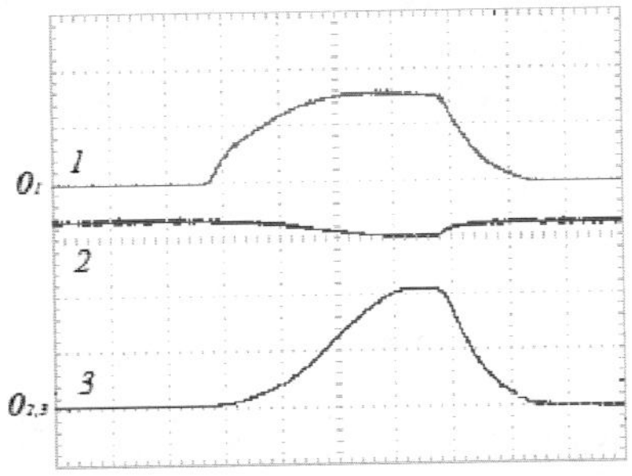

Figure 3. Characteristic waveforms of: *1*-I_d (50 A/div), *2*-U_{dl} (50 V/div), *3*-I_b (40 A/div), 50 μs/div.

Figure 4 shows the curves which divide the regions of stable mode and unstable mode of the electron beam generation; and it becomes evident from the results that the accelerating voltage of at least 160 V is required for stable generation of the electron beam with a current of 100 A, using a fine-grain grid with a mesh size of 1x1 mm. It was established that along with the electron beam instability there occurs the instability of discharge, which is associated with gaseous conditions in the cathode region. To attain the maximum discharge current, it is necessary that a relationship be maintained between the discharge current and the gaseous flux through the cathode, which relationship amounts to ~3 A•min/cm^3, with the discharge burning in nitrogen at a pulse duration of 0.1 ms.

Thermal conditions of the hollow cathode in the pulse mode were estimated with an allowance made for the fact that, during a pulse time, a surface layer of the cathode is heated up, which thickness is small as compared to the cathode wall thickness, and radiation from the surface is absorbed fully by the walls of the closed cathode cavity. Then heat is removed into the cathode bulk and is radiated off the cathode surface. The changes in the surface temperature increment over a pulse at a variable repetition rate and at the same average discharge current were calculated (Figure 5). The lower the frequency, the higher the pulse current is required to sustain stable discharge in the pulse-periodic mode. Calculations carried out for a TiN cathode demonstrate that the thickness of the surface layer, heated during a pulse of 0.1 ms duration, amounts to 30 μm approximately, and the temperature of the cathode emitting surface may rise a few hundred degrees at a pulse power density of ~2 10^4 W/cm^2.

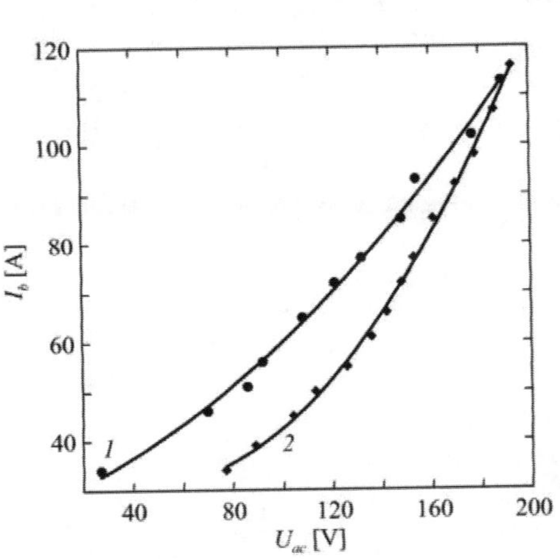

Figure 4. Dependences of maximum electron beam current on accelerating voltage. Gas pressure: *1*-1·10^{-3}, *2*-5,4·10^{-3} Torr.

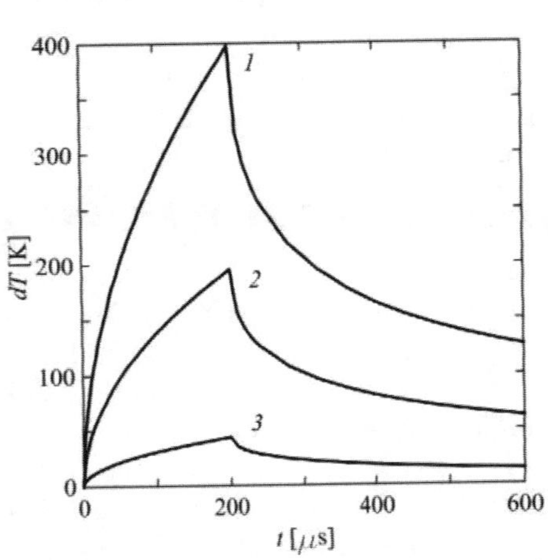

Figure 5. Calculated curves of cathode surface temperature increment in pulse-periodic mode at frequency: *1*-100 Hz, *2*-200 Hz, *3*-800 Hz.

2.2. Mass spectrometry of beam-generated plasma

The experimental technique was described elsewhere [12]. A modified dynamic monopolar mass-spectrometer MX-7304A was used, which had an ion-optic system installed instead of an ionizer. A potential of –300 V was applied on the ion extractor. Accelerated ions arrived into the equipotential space inside the drift tube, and at the input of the analyzer were slowed down to energy of 30 eV ensuring a high resolution of the analyzer. The small size (2 mm) of the aperture at drift tube input and the use of a differential vacuum pumping made it possible to maintain a pressure of 0.03 Pa in the analyzer. Nevertheless, the large lengths of the drift tube and the mass-analysis region entailed the necessity for correction of the measurement results in view of the difference in the charge-exchange cross-sections for atomic and molecular nitrogen ions [12].

Figure 6 shows the resultant curves showing the relative contents of atomic nitrogen ions in the argon-nitrogen beam-generated plasma (at a flux ratio of Q_{N2}/Q_{Ar}=1:1) as a function of the frequency of pulsed electron beam under constant value of average beam current. As compared to the DC electron beam, the high-current pulse electron beam provides approximately twice as great the N^+ ion content in plasma. Measurements of the emitting plasma composition proved the contents of atomic N^+ ions at a level of 8–12% as in the pulsed discharge as in the dc discharge. The content of ions of the hollow cathode material in the beam-generated plasma was below the measurement detection level.

2.3. Coating deposition in MSS EB

Properties of TiN coatings formed in MSS EB have been studied in [13]. Electron beam (100 eV, 0-8 A) was injected into Ar-N_2 gaseous mixture, with the pressure being maintained within 0.15–0.28 Pa by altering the nitrogen flow rate (1-25 sccm). Hardness of TiN coatings deposited without the electron beam reaches maximum value (25 GPa) at nitrogen flow rate of 12 sccm. As the beam current increases up to 5A, the nitrogen content grows in the coatings and the hardness increases to 30–35 GPa at low nitrogen flow rates (~5 sccm). The effect of the electron beam on the gaseous mixture results in increased lattice parameter and the level of microstress, in a reduction of the grain size, and in a change in the coating texture.

Nanocomposite coatings of TiC/a-C:H of 5 μm in thickness were obtained by the reactive magnetron sputtering of Ti in a gaseous mixture of Ar-C_2H_2 which was additionally irradiated with an electron beam (100 eV, up to 1 A). The coating composition was controlled by altering the C_2H_2 flow rate within the range of 1–16 sccm at a constant Ar flow rate of 40 sccm and a constant magnetron current (2 A, 10 μs, 50 kHz). Hardness of the coatings depends nonmonotonically on the C_2H_2 flow rate. The maximum hardness of ~30 GPa was attained at Ti content of ~38at.% in the coating. It was established that irradiation of the gaseous mixture during the coating deposition with electron beam facilitates accelerated acetylene decomposition and allows multiple (from 10 down to 2 sccm) reduction of the C_2H_2 flow rate at which the maximum hardness of the coating can be achieved.

Using the reactive magnetron sputtering with an additional electron beam treatment of the gaseous mixture, nanocomposite coatings of (Ti, Al)/a-Si_3N_4 were obtained. Two cathodes made of Ti (VT1-0, 98.61 – 99.7 %), a Si (99.999 %) cathode, and an Al cathode (A999) were used in the MSS EB. The bias voltage on the samples (50 kHz, 10 μs) amounted to -100 V. In the electron source, a combined mode of discharge was utilized: a low-current mode at a dc (up to 8 A) in combination with a pulsed mode (250 Hz, 0.2 ms, up to 100 A). The pressure of gaseous mixture in the chamber was 0.14 GPa, with the flow rate of argon and nitrogen through the hollow cathode changing in the range of 22 – 34 sccm and 6 - 18 sccm, respectively. The average current density and the pulse current density of ions on the sample were 6 mA/cm^2 and 60 mA/cm^2, respectively.

With the optimal values of element contents in the coating (Si ~5, Al ~15, Ti ~27 at %), the maximum value of hardness H = 41 GPa was obtained, with the elastic modulus E being 590 GPa. Figure 7 shows H and H/E ratio as a function of the discharge current amplitude of the electron source, with the average electron-beam current maintained constant at 7 A. Reducing the average current with decreasing the current amplitude was compensated by an increase in the dc component of the discharge current. As the discharge current amplitude increased from 0 to 100 A, the hardness of coating increased by 10 GPA, with the *H*/*E* ratio increased by 12 % (up to 0.069). A high pulse ion current density (60 mA/cm^2) did not lead to the growth of internal stresses in the coatings.

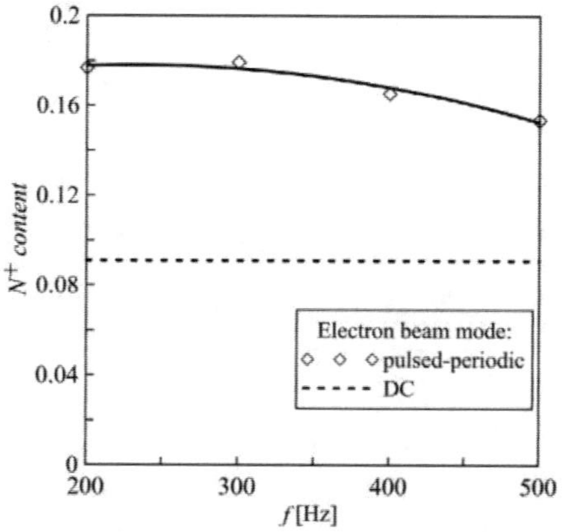

Figure 6. The content of N^+ in the flow of ions from beam plasma vs frequency of beam current.

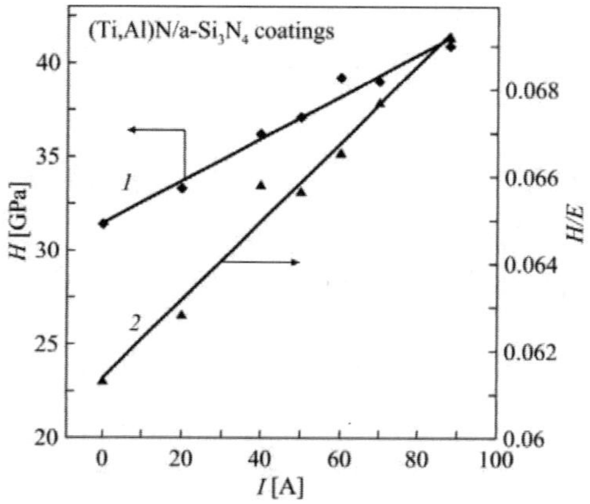

Figure 7. Dependences of (Ti, Al)N/a-Si_3N_4 coatings hardness (*1*) and hardness to Young's modulus ratio (*2*) vs amplitude of discharge current of electron source.

3. Conclusion

Additional ionization and excitation of a gaseous medium with an electron beam during coating deposition using reactive magnetron sputtering makes it possible to attain to high pulsed values of ion current density (up to 0.1 A/cm^2) on the coating surface, which considerably affects adhesion and properties of the coatings.

A plasma source of electrons based on pulse discharge with a self-heated hollow cathode generates an electron beam with a current of 10-100 A and an electron energy of 200 eV at pressures ranging from 0.01 to 1 Pa in a pulse-periodic mode (1000 – 100 Hz, 0.2 ms).

Mass-spectroscopy of Ar-N_2 beam plasma demonstrated that the high-current pulsed electron beam provides twice as great a content of atomic nitrogen ions (up to 18%) as compared to the dc electron beam and gaseous discharge, which can be used in order to improve efficiency of the metals nitriding.

Acknowledgement

The study was carried out within a state program task No. 0389-2014-0006 and was partially supported by the Russian Foundation for Basic Research, project No. 14-08-00249-a.

References

[1] Window B, Savvides N 1986 *J. Vac. Sci. Technol.* **4(3)** 453
[2] Thornton J A 1974 *J. Vac. Sci. Technol.* **11** 666
[3] Musil J, Kadlec S, Münz W D 1991 *J. Vac. Sci. Technol.* A **9(3)** 1171
[4] Kouznetsov V et al. 1999 *Surf. Coat. Technol.* **122(2-3)** 290
[5] Leonhardt D, Walton S G, Fernsler R F 2007 *Physics of Plasmas* **14** 057103
[6] Lock E H, Fernsler R F, Slinker S, Walton S G 2014 *J. of Phys. D: Appl. Phys. 47(42) 425206*
[7] Abraha P, Yoshikawa Y, Katayama Y 2009 *Vacuum* **83** 497
[8] Gavrilov N V, Mamaev A S 2009 *Technical Physics Letters* **35(8)** 713
[9] Gavrilov N V, Men'shakov A I 2011 *Instruments and Experimental Techniques* **54(5)** 732
[10] Gavrilov N V et al. 2014 *Izvestia VUZov. Fizika* **57(11/3)** 209
[11] Gavrilov N V, Kamenetskikh A S 2013 *Technical Physics* **58(10)** 1426
[12] Bureyev O A, Gavrilov N V 2010 *Proc. of 16th SHCE, Tomsk, Russia*, 35
[13] Kamenetskikh A S, Gavrilov N V, Chukin A V 2014 *Izvestia VUZov. Fizika* **57(10)** 161

Journal of Physics: Conference Series **652** (2015) 012025 doi:10.1088/1742-6596/652/1/012025

Changes in the electro-physical properties of MCT epitaxial films affected by a plasma volume discharge induced by an avalanche beam in atmospheric-pressure air

D V Grigoryev[1,2], A V Voitsekhovskii[1], K A Lozovoy[1], V F Tarasenko[2], M A Shulepov[2]

[1] National Research Tomsk State University, 36 Lenin Av., Tomsk, 634050, Russia
[2] Institute of High Current Electronics SB RAS, 2/3 Akademichesky, Tomsk, 634055, Russia

denn.grig@mail.tsu.ru

Abstract. In this paper the influence of the plasma volume discharge of nanosecond duration formed in a non-uniform electric field at atmospheric pressure on samples of epitaxial films HgCdTe (MCT) films are discussed. The experimental data show that the action of pulses of nanosecond volume discharge in air at atmospheric pressure leads to changes in the electrophysical properties of MCT epitaxial films due to formation of a near-surface high-conductivity layer of the n-type conduction. The preliminary results show that it is possible to use such actions in the development of technologies for the controlled change of the properties of MCT.

1. Introduction

The semiconductor narrow-band material $Hg_{1-x}Cd_xTe$ (HgCdTe, MCT) is one of the basic materials for creating sensitive elements of IR photo-receiving devices in the wavelength range from 8 to 14 micrometres [1]. At present the epitaxial MCT films grown by the method of molecular-beam epitaxy (MBE) present a promising material for creating multi-element semiconductor detectors of IR radiation providing signal processing directly in the focal region [1, 2]. Along with the study of the initial properties of epitaxial films of MCT grown by MBE a very urgent task is controlled changing of the parameters of the material in order to obtain the desired semiconductor structures.

Nowadays various discharges and electron beams are widely used for modification of near-surface layers of materials [1]. In [3] the possibility of forming volume discharge at atmospheric pressure in a non-uniform electric field was demonstrated. In air at atmospheric pressure and non-uniform electric field volume discharge forms at both polarities of voltage pulse on the electrode with a small radius of curvature [4]. Under the influence of such discharge it is possible to realize high power density of the energy input (up to 800 MW/cm3 [5]). At the same time runaway electron beam with a current amplitude beyond the foil in the tens to hundreds of amperes is generated from the discharge plasma, and the beam current pulse width at half-maximum is less than 100 ps [6]. Thus, the feature of such discharges is the combined effect of the plasma of dense nanosecond discharge with energy input specific power of hundreds of megawatts per cubic centimeter, ultrashort electron beam with a broad energy spectrum and the optical radiation of different spectral ranges from the plasma discharge.

The aim of this work was to study the effect of volume nanosecond discharge in air at atmospheric pressure on the electro-physical properties of epitaxial HgCdTe grown by MBE.

2. Experimental samples and methods

For experiments the samples of epitaxial HgCdTe films of the p-type conductivity HgCdTe ($p = 1 \div 2 \cdot 10^{16}$ cm^{-3}, $\mu_p = 300 \div 500$ $cm^2V^{-1}s^{-1}$) grown by MBE were prepared at the Institute for Semiconductor Physics of the Siberian Branch of the Russian Academy of Sciences (Novosibirsk). The composition of the working layer of epitaxial films was $x = 0.22$.

Prepared samples were placed in a gas diode on the copper anode. The distance between flat anode and tubular cathode was 8-16 mm. As a source of voltage pulse generator RADAN-220 was used. The generator formed voltage pulses with an amplitude of ~ 230 kV (open circuit voltage), pulse duration at half-height of ~ 2 ns (a matched load) and a rise time of ~ 0.5 ns. The samples were irradiated in the repetitively pulsed mode with a pulse repetition rate of 1 Hz. Exposure was carried out in the range of $100 \div 1200$ pulses. Electro-physical parameters of the MCT samples before and after exposure were determined from measurements of the Hall effect by van der Pauw method. The measurements were performed at a constant current flowing through the sample (I = 1 mA) for the two directions of the current and two directions of a constant magnetic field. Removal of thin layers from irradiated material's surface was carried out in a 0.2 % solution of bromine in dimethylformamide. The surface of the irradiated samples was studied using atomic force microscope (AFM) "Ntegra Prima" (by NT-MDT) and optical profilometer MicroXAM-100 at ambient conditions.

3. Experimental results and discussion

Analysis of the results of measurements of electro-physical parameters of the irradiated samples of epitaxial MCT have shown that after irradiation in the range of $100 \div 1200$ pulses the conductivity of all samples are increases. Moreover, for samples exposed to pulses in the range from 100 to 400, a decrease in the value of the Hall coefficient is observed (figure 1). In this case, on the field dependence of the Hall coefficient shift of the inversion point for the sign of the Hall coefficient in the region of higher magnetic fields (from 0.17 T to 0.28 T) is observed. Increasing the number of pulses of impacts of volume discharge to 600 pulses results in a reversal of the sign of the Hall coefficient (figure 2, curve 2). Further increasing of the number of pulses reduces the Hall coefficient values (figure 2, curves 2, 3, 4), wherein the epitaxial HgCdTe samples are characterized by low electron mobility values $\sim(2 \div 3) \cdot 10^3$ $cm^2V^{-1}s^{-1}$, which is two orders of magnitude lower than the corresponding values for the epitaxial n-type material of high quality.

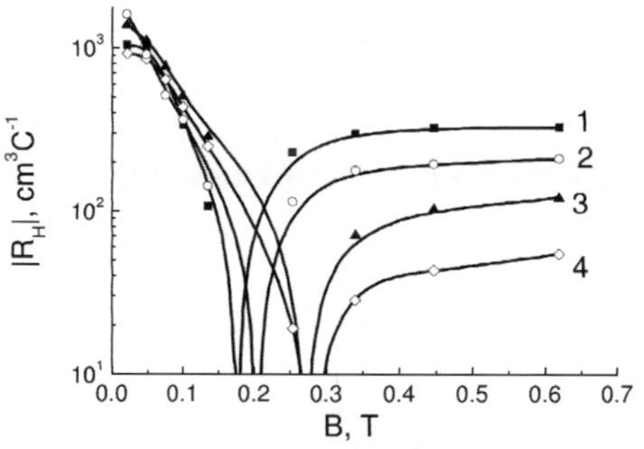

Figure 1. The field dependence of the Hall coefficient for samples of CdHgTe epitaxial films before (1) and after exposure to a volume discharge. The number of exposure pulses: (2) - 100, (3) – 200, (4) – 400.

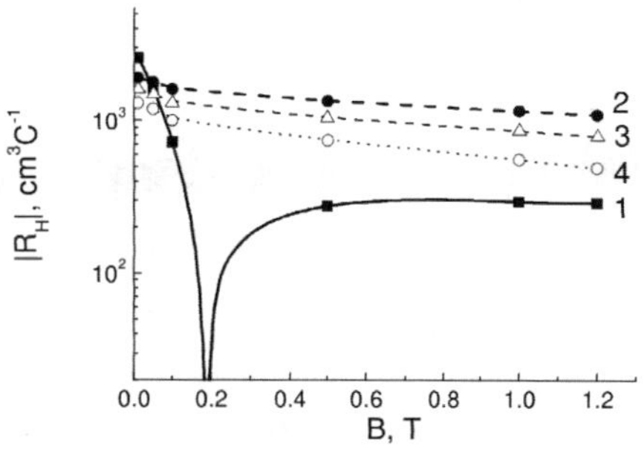

Figure 2. The field dependence of the Hall coefficient for samples of CdHgTe epitaxial films before (1) and after exposure to a volume discharge. The number of exposure pulses: (2) - 600, (3) – 800, (4) – 1000.

The results obtained made it possible to suggest that in the process of impact of volume discharge on the samples of epitaxial films a layer with high electron concentration is formed at the surface or in the surface region of the material. The conductivity of this layer is such that it bypasses the bulk of the epitaxial film in the measurements of the Hall effect.

The study of the surface structure of initial and irradiated epitaxial films with atomic force microscope revealed that the surface quality of the samples after exposure to a volume discharge is not changed. Surface roughness slightly increases from 1.6 to 2.2 nm. Additionally, the distribution profile of the sample surface for which part of the surface was closed with the dielectric plate during irradiation was investigated. Surface profile measurements were carried out with an optical profilometer MicroXAM-100. Analysis of measurement results have shown that the interface irradiated/unirradiated sample has no a characteristic step that leads to the conclusion about the absence of an alien film on the surface of the exposed part of the test sample (figure 3).

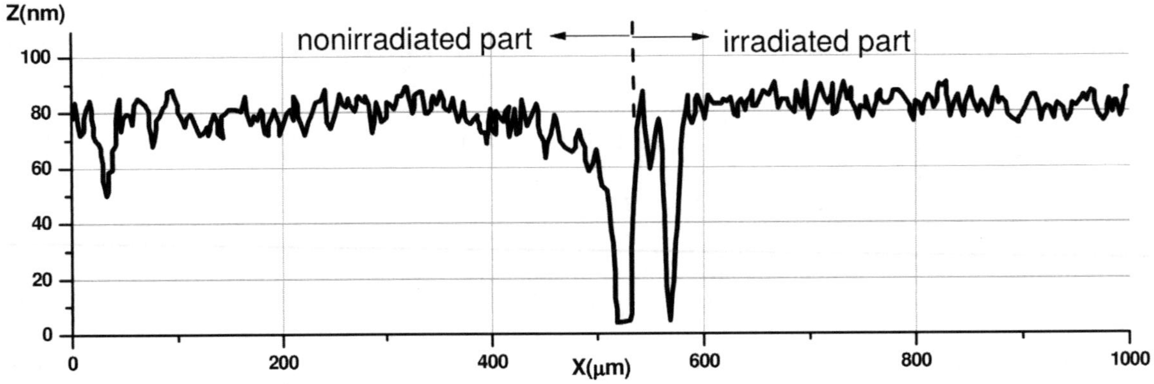

Figure 3. The surface profile of the MCT sample for which part of the surface was closed with dielectric plate during irradiation.

When etching the surface of the irradiated material by 0.1 mm electrophysical parameters of the samples irradiated return to baseline values (figure 4, curve 3). The obtained results allow concluding that during the impact of volume discharge on the samples of epitaxial films high conductivity layer forms in the surface region of the material.

Furthermore, it was found that after treatment of the irradiated samples in a solution of concentrated hydrochloric acid the restoration of the original values of electrophysical parameters of the material is taking place (figure 4, curve 4).

However, hydrochloric acid does not chemically react with initial MCT or with a material subjected to a radiation treatment. In the gas diode during a volume nanosecond discharge in air after a short voltage pulse in the discharge gap plasma consisting of an electron beam of positively and negatively charged ions is formed and the current flows through the diode. In this case, the sample of epitaxial MCT film located at the anode is exposed to complex effect consisting of a beam of electrons and negative ions, which cause the formation of different chemical MCT compounds after the influence on the sample surface.

The obtained data allow concluding that in the surface region of irradiated material the formation of chemical compounds of MCT with oxygen and nitrogen atoms occurs (such as anodic oxide which dissolve in hydrochloric acid). Analysis of published data shows that such chemical compounds contain significant concentration of positively charged centers that lead to the formation of an inversion layer at the interface with the epitaxial MCT film of p–type of conductivity. The formation of such an inversion layer is shown during the study of the properties of the interface between the oxide film and MCT [7].

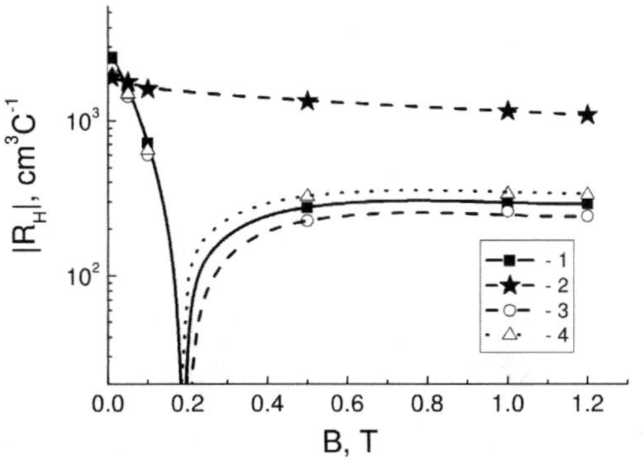

Figure 4. The field dependence of the Hall coefficient for samples of CdHgTe epitaxial films. (1) – before exposure; (2) – 600 volume discharge pulses; after etching of the irradiated sample in a solution of bromine dimethylformamide (3) and hydrochloric acid (4).

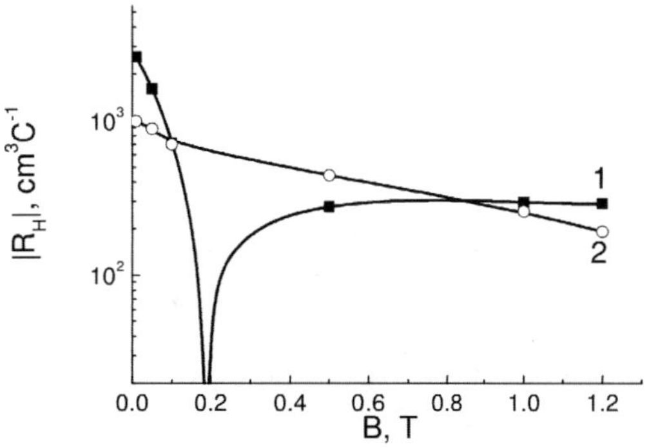

Figure 5. The field dependence of the Hall coefficient for samples of CdHgTe epitaxial films before (1) and after (2) growing of anodic oxide on the surface.

Modeling for this assumption was conducted by growing oxide on the surface of the MCT epitaxial layer by the method of anodic oxidation. After growth of anodic oxide electrophysical parameters of obtained multilayer structure were measured. The measurement results have shown that the observed changes in the field dependence of the Hall coefficient before and after the deposition of anodic oxide are similar to the results observed after irradiating the samples of MCT by volume discharge (figure 5).

4. Conclusion

Thus, the experimental data have shown that the exposure of MCT epitaxial films to the pulse volume nanosecond discharge in air at atmospheric pressure leads to formation of chemical compounds of MCT with oxygen and nitrogen atoms in the surface layer of material. The formed layer has a built positive charge that leads to the formation of an inversion layer at the oxide/MCT interface which "shunts" the rest of the sample so that the measured field dependence of Hall coefficient corresponds to the material of n-type of conductivity.

This research carried out in 2015 was supported by grant (№ 8.2.10.2015) from “The Tomsk State University Academic D.I. Mendeleev Fund Program”.

Acknowledgments

The authors gratefully acknowledge N.N. Mikhailov and V.S. Varavin for providing the MBE HgCdTe epitaxial films.

References

[1] Rogalski A 2012 *Opto-electronics Review* **20** 279
[2] Shulepov M A, Tarasenko V F, Goncharenko I M et al 2008 *Tech. Phys.* **8** 51
[3] Tarasova L V, Khudyakova L N 1969 *Sov. Phys. Tech. Phys.* **14** 1148
[4] Kostyrya I D, Orlovskii V M, Tarasenko V F et al 2005 *Technical Physics Letters* **31** 457
[5] Alekseev S B, Gubanov V P, Kostyrya I D et al 2004 *Quantum Electron* **34** 1007
[6] Tarasenko V F, Shpak V G, Shunailov S A, Kostyrya I D 2005 *Laser Part. Beams* **23** 545
[7] Brogowski P and Piotrowski J 1990 *Semicond. Sci. Technol.* **5** 530

Journal of Physics: Conference Series **652** (2015) 012026 doi:10.1088/1742-6596/652/1/012026

The impact of the plasma volume discharge in the atmospheric-pressure air on the distribution of the surface potential in a V-defect region of epitaxial HgCdTe films

D V Grigoryev[1,2], V A Novikov[1], D A Bezrodnyy[1], V F Tarasenko[2], M A Shulepov[2]

[1] National Research Tomsk State University, 36 Lenin Av., Tomsk, 634050, Russia
[2] Institute of High Current Electronics SB RAS, 2/3 Akademichesky, Tomsk, 634055, Russia

denn.grig@mail.tsu.ru

Abstract. In the present report we demonstrate the experimental data obtained as a result of studying the impact of nanosecond plasma volume discharge in the atmospheric-pressure air on the distribution of the surface potential in the V-defect regions of epitaxial HgCdTe films. The experimental data obtained for the variation of the contact potential difference (ΔCPD) between the V-defect and the main matrix of the epitaxial film show that the mean value of ΔCPD for the original surface differs from the one for the irradiated surface for 55 eV. At the same time the mean value of ΔCPD changes its sign indicating that the original surface of the epitaxial HgCdTe film predominantly contains the grains with increased cadmium content while after the irradiation the grains possess an increased content of mercury. Therefore, during the irradiation process a decrease of the mercury content in the near-surface region of the semiconductor takes place resulting in the alteration of the electrophysical properties in the film's near-surface region.

1. Introduction

The ternary semiconductor compounds $Hg_{1-x}Cd_xTe$ (HgCdTe, MCT) are one of the major materials used for manufacturing the inherent infrared (IR) photodetectors for the wavelength ranges of 3-5 and 8-14 micrometres [1]. The heteroepitaxial HgCdTe films grown by molecular beam epitaxy (MBE) are one of the most promising materials for manufacturing multi-element photodetecting devices capable of the signal processing directly in the focal region (FPA – focal plane array). The investigations of the epitaxial growth process of HgCdTe have shown that without reference to the substrate type there occurs the formation of macroscopic V-defects that represent the inclusions of polycrystalline phase of HgCdTe with material composition different from that of the bulk of the epitaxial film [2, 3]. These defects are mostly impacted on the performance capabilities of photosensitive element arrays in the focal plane. Because of the fact that the spectral range and sensitivity of the final photoelectric cell greatly depends on the material composition one of the important problems is not only to study but also to control the composition of both the individual crystal grains and the overall V-defect.

At the present time the plasma generated by nanosecond pulses have gained wide adoption for using in order to modify the near-surface properties of materials [4]. It was demonstrated that one can form a volume discharge in a non-uniform electric field at atmospheric pressure. In atmospheric pressure air and a non-uniform electric field the volume discharge forms at both voltage pulse directions on the electrode with a small radius of curvature. The characteristics of such discharges represent the complex influence of a dense nanosecond discharge plasma with the specific power of heat input of the order of hundreds of megawatts per cubic centimetre, of an ultrashort electron beam with a wide energy spectrum and of the optical radiation of different spectral ranges from the discharge plasma.

The aim of this work is to study the effect of a nanosecond volume discharge plasma forming in an inhomogeneous electrical field at atmospheric pressure on the distribution of the surface potential in the V-defect regions of epitaxial HgCdTe films.

2. Experimental apparatus and methods

For experiments the samples of epitaxial HgCdTe films of the n-type conductivity grown by MBE on GaAs (013) substrates with ZnTe and CdTe buffer layers were prepared at the Institute for Semiconductor Physics of the Siberian Branch of the Russian Academy of Sciences (Novosibirsk). The content of CdTe in the working layer of epitaxial films was x = 0.22. The width of the upper variband layer was close to 0.4 micrometers while the content of CdTe on the surface was 0.44. The value of the content x and the width of the epitaxial film were controlled by in situ ellipsometric measurements. The content x was controlled based on the optical transmittance spectra at room temperature.

The as-grown samples were placed in a gas diode on a copper anode. The samples were irradiated by 1200 pulses in the pulse-periodic mode at the pulse repetition rate of 1 Hz. The distance between flat anode and tubular cathode was 8-16 mm. The interelectrode voltage was supplied from a pulser of the RADAN-220 type, which generated voltage pulses with an amplitude of ~230 kV (in the open-circuit regime), a FWHM of ~ 2 ns (on a matched load), and a leading front width of ~ 0.5 ns. The discharge current was measured using a shunt composed of chip resistors connected between the foil anode and the discharge chamber housing. The results of measurements showed that the current pulse amplitude for both polarities of the applied voltage pulse was ~3 kA and the total duration of the discharge current pulse was ~30 ns (the first half-period of the discharge current pulse had a duration of ~8 ns). The generator RADAN-220 provided specific power input in gas discharge plasma above 0.8 GW/cm3 under atmospheric pressure of air. The electrodes spacing was 8 mm. The volumetric character of the discharge also remained at a pulse repetition rate of 3 kHz.

For the original surface and the one exposed to discharge the surface potential distribution was studied by the Kelvin Force Probe Microscopy (KFPM). This method is widely used for studying the distribution of the inherent and surface potentials of instrumental micro-and nanostructures [5-7]. The measurements were performed on a commercial atomic-force microscope (AFM) "Solver HV" (produced by NT-MDT). We used polysilicon B-doped probes covered with platinum (probe type NSG11/Pt, manufactured by NT-MDT). To obtain the surface potential profile we measured the contact potential difference (CPD) distribution between the AFM probe and the surface of the epitaxial film. The CPD distribution was measured together with the surface morphology. In order to minimize the influence of the morphology on the CPD distribution we carried out our measurements at distance of sample-probe dz = 50-100 nm.

3. Experimental results and discussion

For the original surface and the one exposed to discharge using the Kelvin force probe microscopy we studied the distribution of the surface potential for at least 10 randomly chosen V-defects. A typical image of the distribution of the CPD is presented on figure 1. One can see that the distribution is significantly inhomogeneous, exhibiting regions where the CPD is either lower or higher than that of the epitaxial film. On the periphery of each individual crystal grain there is a dramatic variation of the CPD, which can indicate the presence of a modified potential.

Using the distribution of the surface potential we derived the difference (ΔCPD) between the CPD value for individual crystal grains of a V-defect and the mean value of the CPD of epitaxial film.

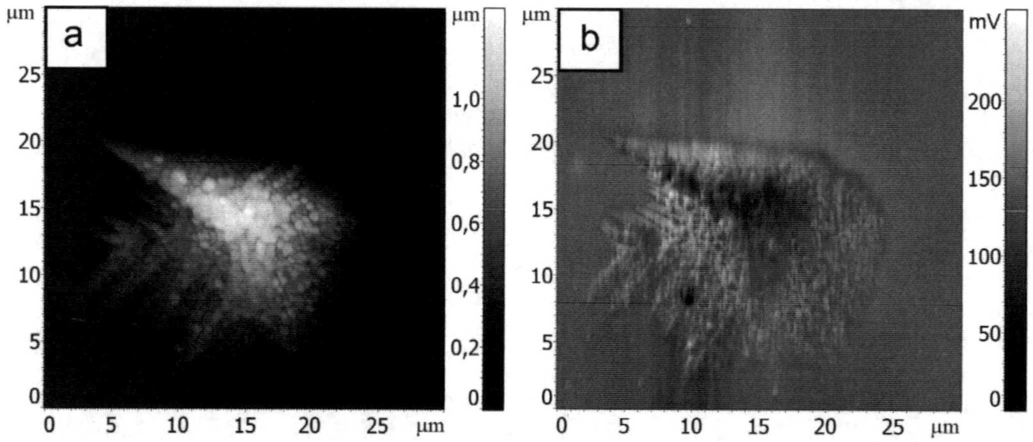

Figure 1. AFM image of a V-defect surface morphology (a) and the distribution of the surface potential or CPD (b).

The distributions of the ΔCPD for a film before and after exposure to radiation are presented on figure 2. The ΔCPD distribution possesses a complex character and can be described by an assemblage of normal distributions. Since the work function of the cantilever needle-point does not change during the measurement process it becomes clear that the observed changes are related to the local changes of the HgCdTe work function, i.e. the changes of the work functions of individual crystal grains of the V-defect are observed. Using the expression for the dependency of the HgCdTe solid solution electron affinity on the component composition [8] one can write the expression for the ΔCPD in the following form:

$$\Delta CPD = \left[1.29 - 7.13\cdot 10^{-4} T\right](x_1 - x_2) - 0.54\left(x_1^2 - x_2^2\right) + 0.56\left(x_1^3 - x_2^3\right), \quad (1)$$

where x_1 is the cadmium content in the epitaxial film and x_2 is the cadmium content in a selected V-defect crystal grain. It is fair to assume that the value of x_1 does not change, in that case one can use the value of the ΔCPD to derive the value of Δx that characterizes the variation of the cadmium content in a selected crystal grain relative to the epitaxial film.

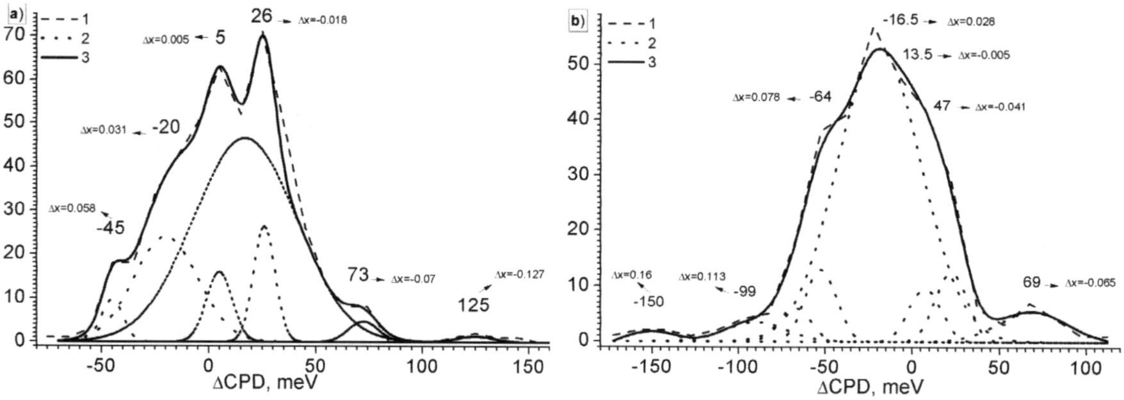

Figure 2. Distribution of the deviation of the CPD value of individual crystal grains of a V-defect versus the CPD of the epitaxial film: a) original film; b) after irradiation, 1 – distribution of ΔCPD, 2 – assemblage of normal distributions, 3 – combined distribution; the arrows indicate the variation Δx of the CdTe content for the given value of ΔCPD.

Taking into account the expression (1) and the data presented on the figure 2 one can show that before and after the exposure to radiation individual crystal grains of a V-defect may possess a higher mercury content ($\Delta x < 0$) as well as a higher cadmium content ($\Delta x > 0$). In the original unirradiated sample the crystal grains with a higher mercury content prevail while in the irradiated sample dominate the grains with a higher cadmium content.

Given the complex character of the ΔCPD distribution consider the mean values of this parameter for the original (27 mV) and the irradiated (-28 mV) surface of HgCdTe. These mean values differ for 55 eV. Note also that the mean value of the ΔCPD changes its sign. Using the expression (1) it can be shown that the original surface of the HgCdTe epitaxial film mainly contained the crystal grains with an increased mercury content while the irradiated surface mainly contains the crystal grains with an increased cadmium content. It follows that during the irradiation process the decrease of the mercury content in the semiconductor near-surface region takes place and, hence, occur the changes of the electrophysical properties in the near-surface region of the film.

4. Conclusion

Using the Kelvin force probe microscopy we showed that the exposure to the radiation of nanosecond volume discharge leads to changes of the material composition of the solid solution of individual V-defect crystal grains. As the decrease of the mercury content is expected to be observed not only in the crystal grains but also in the overall epitaxial film, one can infer that the mercury desorption rate is higher for the crystal grains than for the film itself.

It should also be noted that such a situation is expected to hold true for other triple and quadruple semiconductor compounds with second-phase inclusions.

This research carried out in 2015 was supported by grant (№ 8.2.10.2015) from "The Tomsk State University Academic D.I. Mendeleev Fund Program".

Acknowledgments

The authors gratefully acknowledge N.N. Mikhailov and V.S. Varavin for providing the MBE HgCdTe epitaxial films.

References

[1] Rogalski A 2011 *Infrared Detectors* (CRC Press, Boca Raton)

[2] Sidorov Yu G, Dvoretskii S A, Varavin V S, Mikhailov N N, Yakushev M V and Sabinina I V 2001 Semiconductors **35** 1045

[3] Sabinina I V, Gutakovsky A K, Sidorov Yu G and Latyshev A V 2005 *Journal of Crystal Growth* **274** 339

[4] Shulepov M A, Tarasenko V F, Goncharenko I M et al 2008 *Tech. Phys.* **8** 51

[5] Chun-Sheng Jiang, Moutinho H R, Friedman D J, Geisz J F and Al-Jassim M M 2003 *Journal of Applied Physics* **93** 10035

[6] Cai L X and Wang H M 2004 *Applied Surface Science* **235** 501

[7] Katzer Kl D, Mertin W, Bacher G, Jaeger A and Streubel K 2006 *Applied Physics Letters* **89** 103522

[8] Voitsekhovskii A V, Gorn D I, Izhnin I I et al 2013 *Russian Physic Journal* **55** 910-16

Journal of Physics: Conference Series **652** (2015) 012027 doi:10.1088/1742-6596/652/1/012027

Plasma-chemical method for producing metal oxide powders and their application

I Zhukov[1], **S Vorozhtsov**[1], **V Promakhov**[1], **I Bondarchuk**[1], **A Zhukov**[1], **A Vorozhtsov**[1]

[1] National Research Tomsk State University, Lenin av., 36, Tomsk, Russia

E-mail: gofra930@gmail.com

Abstract. Structure and properties of ZrO_2 and Al_2O_3 powders produced using plasma chemical technique were studied in the framework of this research. Obtained Al_2O_3 powder was used for reinforcement of Al alloy. Improvement of mechanical properties of Al alloy associated with introduction of alumina particles into the melt was demonstrated.

1. Introduction

Micro- and nanoparticles of Al_2O_3, ZrO_2, AlB_2, SiC etc. are used as reinforcing ones providing reduction of development of dislocations in accordance with Orowan's mechanism [1] in the process of alloy deformation which improves strength characteristics. Nowadays, the most promising methods of production of oxide nanopowders with high capacity are plasma chemical method. Plasma-chemical systems are characterized with high energy intensity of the heat flux: the temperature of plasma heat transfer media comprises ~ 10^4 K while their enthalpy comprises $(1\text{-}2)\times10^2$ kCal/mol [2], which provides high specific production capacity of equipment.

It is known that morphology, particle size distribution, phase composition and structure parameters of powders are determined by the method of production. Thus, properties of the powders used for alloy reinforcement are to be studied carefully in order to provide target properties of materials produced on their basis.

The objective of this research is to study structure and properties of ZrO_2 and Al_2O_3 powders produced using plasma chemical technique and their use for modification of properties and structure of Al alloys.

2. Experimental

Alumina and zirconia powders were produced from sprayed water solutions of salts in air flow heated to low-temperature plasma state using a 70 kW high-frequency unit. The diagram of the unit is given in Figure1. Sprayed solution is supplied to plasma jet of the heat transfer medium generated by plasmatron 3 in the plasma chemical reactor 2 via nozzles 1. The powders produced are separated by filter 5 and supplied to collector 9. Vapor-gas flow after cooling down in apparatus 6 is separated from the liquid (condensate is collected in tank 10) and prior to emission to atmosphere is cleaned in scrubber 7. In order to obtain metal oxides initial nitrate solutions of corresponding metals of these compounds were prepared with stoichiometric ratios of components and optimal temperature mode of the reactor of plasma-chemical unit was selected. The plasmatron was remotely activated 3-5 seconds after application of voltage to ignition electrode from the inductor at the pressure level in discharge chamber close to atmospheric pressure or higher. The principle of operation of plasma chemical unit consists in the following: the prepared initial solution with specified composition is sprayed by a spraying device – a pneumatic atomizer – forming a gas-droplet mix spray cone which is supplied to cylindrical reactor with gas (heat transfer medium) flow, in our case it is air heated up to 5000-6000K in a low-temperature plasma generator (plasmatron). Droplet size comprises 50-100 μm. Gas droplet mixture is heated by a plasma jet generated in induction plasmatron powered by a high-frequency generator. Heating process takes place in the plasma chemical reactor having a direct-flow cylindrical scheme, which provides reliable, steady and safe operation in various modes. Solvent is evaporated here from solution droplets with subsequent decomposition of nitrates of Zr, Al etc. to oxides. Powders produced are separated in containers.

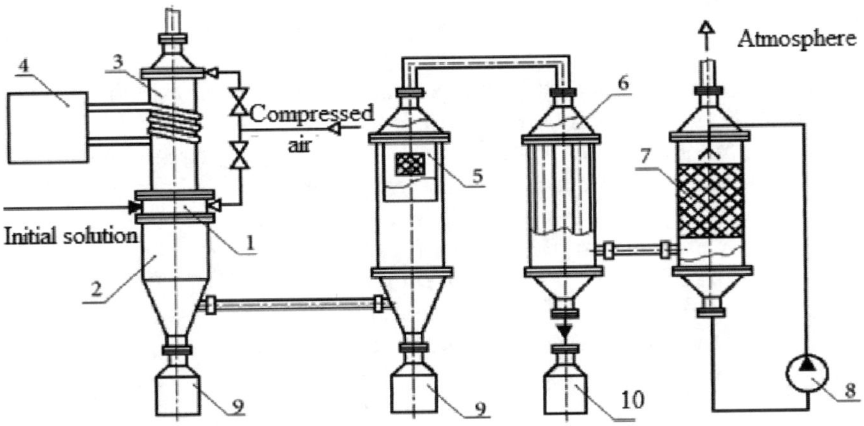

Figure 1. The diagram of plasma chemical powder synthesis unit

The structure of powders was studied using a Philips SEM 515 scanning electron microscope (SEM). The average size of powder particles was measured using the random linear intercept method [3]. The study of the phase composition and structural parameters of powders was performed using an X-ray diffractometer with Cu*K*α radiation. Calculation of coherent scattering region (CSR) and microdistortion of the crystal lattice $<\varepsilon^2>^{1/2}$ was performed using Williamson–Hall method [4] where all broadenings in the X-ray diffraction (XRD) profile were used for calculation. Phases were identified by comparing peaks of X-ray spectra with ASTM (American Society for Testing and Materials) card file. Specific surface area of powders was measured by BET method using a Sorbi analyzer with the measurement error not exceeding 3%.

3. Result and discussion

Preliminary study of kinetics of thermal decomposition using the method of derivatography was performed for Zr nitrates [5]. It was found that the process of decomposition has 8 stages with different kinetic characteristics. Results of this study are given in Table 1.

As it can be seen from the Table 1 water solutions of Zr nitrates are decomposed to Zr oxides at the temperatures higher than 1053 K. The study of Zr nitrate decomposition in a plasma chemical reactor show that the process of oxide formation from nitrate solutions at temperature levels in plasma chemical reactor is complete.

Table 1. Characteristics of the process of thermal decomposition of Zr nitrates.

Temperature range (K)	Mass reduction, (g.)	Chemical formula of the sample at the end of the process
353-443	0.030	$Zr(NO_3)_4 \cdot 8\text{-}9H_2O$
443-583	0.052	$Zr(NO_3)_4 \cdot 7H_2O$
583-738	0.075	$Zr(OH)(NO_3)_3 \cdot 5H_2O$
738-813	0.140	$Zr(OH)_2(NO_3)_2 \cdot 5H_2O$
813-893	0.180	$Zr(OH)_3 \cdot NO_3 \cdot H_2O$
893-913	0.200	$Zr(OH)_3 \cdot NO_3$
913-973	0.222	$Zr(OH) \cdot NO_3$
973-1053	0.268	ZrO_2

Powders represented a structure formed by individual spherical particles and bulk aggregated loose substance consisting of particles having irregular form (Figures 2, 3). According to the results of XRD and electron microscopic studies the powders represent an loose micro-aggregate system. Multiple morphology of powder particles produced using plasma chemical methods can be explained by the following circumstances. As the drops of solution come into contact with a heat transfer medium which has a high temperature intensive evaporation of solvent from droplet surfaces takes place, this leads to a sharp saturation of surface layer of the liquid relative to the salt dissolved in it. As a result the shells are formed on the surface of droplets with liquid solution or vapor-gas mixture inside. Under the influence of internal pressure the vapor-gas mixture breaks through these shells which harden after quick cooling down in reactor chill zone producing hollow porous spheres as well as fragments of broken shells [5]. The BET surface of plasma-chemical powder has been measured to amount 60 m^2/g. Phase analysis has indicated that alumina powder is in a highly nonequilibrium state and contains 5 structural modifications: rhombic (α), cubic (γ), tetragonal, hexagonal (ε) and monoclinic (θ). According to XRD analysis data the size of coherent scattering regions of alumina powder is approximately 30 nm. The crystal lattice micro-distortion is 5×10^{-3}.

Figure 2. SEM image powder Al_2O_3

Electron microscopic studies have indicated that the average sphere size(diameter) comprises 1.2 µm, grain size (grain size is considered equal to CSR size) comprises 20nm. As zirconia powder was obtained using a modifying additive(yttrium oxide), the powder contained tetragonal and monoclinic phases of zirconia, Figure 3b.

Journal of Physics: Conference Series **652** (2015) 012027 doi:10.1088/1742-6596/652/1/012027

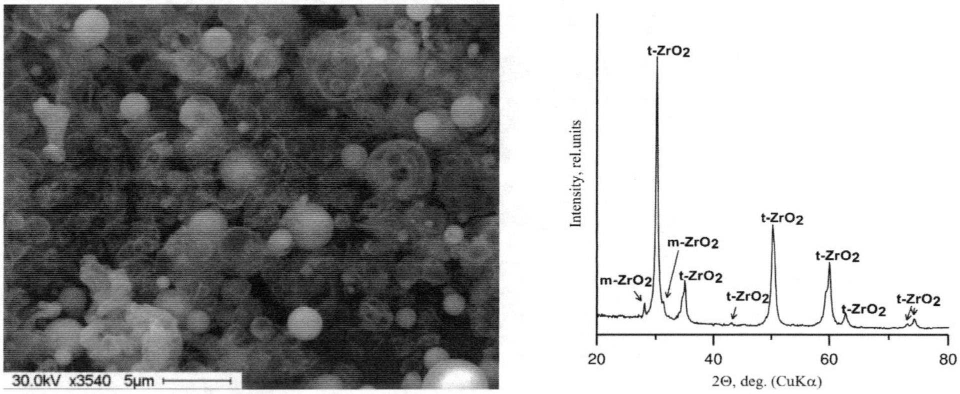

a) b)

Figure 3. Powder $ZrO_2(Y_2O_3)$. SEM image (a); X-Ray pattern (b)

The tetragonal phase of zirconia prevailed, its share in the powder being 90%. The crystal lattice micro-distortion value was 2.4×10^{-3}; and the lattice parameters were: c = 5.1749 Å, a = 5.1028 Å. The comparison of ZrO_2 tetragonal phase lattice parameters with the values given in work [6] showed that the values of lattice parameters for zirconia powder correspond to the values reported for tetragonal phase in the powders composed of ZrO_2 + 3 mol. % Y_2O_3.

Al_2O_3 - particles were used as reinforcing particles for introduction into an Al alloy. Densely compacted master alloy containing 10 wt% of nanoparticles was introduced into the Al melt with simultaneous ultrasonic processing. The microstructure of an Al-4 wt% Cu–0.1 wt% Al_2O_3 alloy is given in Fig. 4. The stress-strain tensile curves for produced samples are given in Figure 4b. As it can be seen, the introduction of 0.1% Al_2O_3 nanoparticles into the alloy leads to simultaneous improvement of yield strength, ultimate tensile strength and ductility. Moreover, alloy hardness also increases from 570 to 710 MPa.

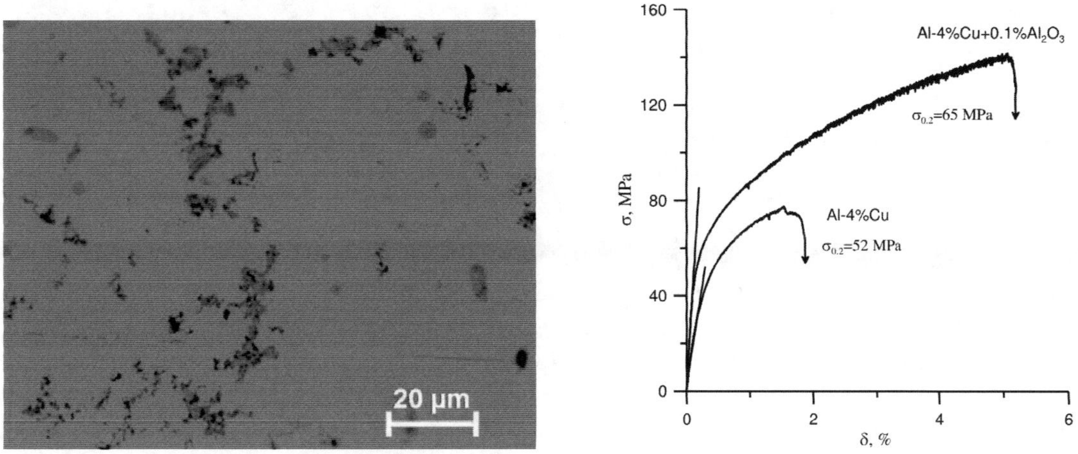

a) c)

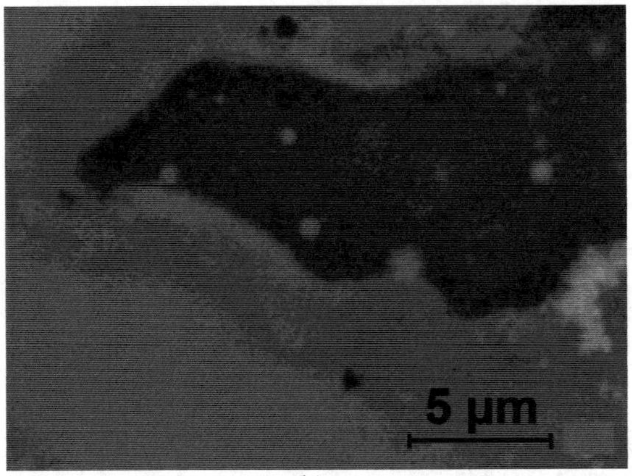

b)

Figure 4. Microstructure of the as-cast composite material Al–4 wt% Cu+0.1 wt% Al_2O_3 (a, b) and stress-strain diagrams of this composite material in comparison with the matrix (c).

4. Conclusion

It was found that morphological structure of zirconia and alumina powders produced using plasma-chemical technique is a microaggregate system consisting of hollow spherical particles. Possible application of plasma chemical technique for production of metal oxides to be used for improvement of properties of Al alloys was illustrated using the example of alumina powders. Authors plan to synthesize AlN powders using plasma-chemical technique in order to use them as modificators in nonferrous metallurgy and production of ceramics.

Acknowledgments

This research was supported by the Ministry of Education and Science of the Russian Federation under the Federal Target Program "Research and development on priority directions of scientific-technological complex of Russia for 2014-2020", the agreement # 14.578.21.0034, unique identifier PNI RFMEFI57814X0034. “The Tomsk State University Academic D.I. Mendeleev Fund Program” grant # 8.2.46.2015 and Grant President RF # MK – 5681.2014.8, 3 MK –5883.2014.8 is gratefully acknowledged.

5. References

[1] Polak L S, Ovsyannikov A A, Slovetsky D I, Wurzel F B 1975 *Theoretical and applied plasma chemistry (Moscow: Nauka)* p. 304 (in Russian).
[2] Orowan E 1963 Classic and dislocational theory of brittle fracture, in: Atomic Mechanism of Fracture [Russian translation] *M.: Metallurgy* 170-184.
[3] Gokhale A M, ASM Handbook, 2004, Metallography and Microstructures, Metals Park (OH): *ASM International* **9** 428–447.
[4] Williamson G K, Hall W H 1953 *Acta Metallurgica.* **1** 22–31.
[5] Andriets S et al. 2008 *Russian Journal of Non-Ferrous Metals.* **49(3)** 205–211.
[6] Ingel R P, Lewis D 1986 *Journal of the American Ceramic Society.* **69** 325–332.

Structure-phase states of silumin surface layer after electron beam and high cycle fatigue

S V Konovalov[1], K V Alsaraeva[1], V E Gromov[1], Yu F Ivanov[2,3,4]

[1] Siberian State Industrial University, Novokuznetsk, Russia
[2] Institute of High Current Electronics Siberian Branch of Russian Academy of Science, Tomsk, Russia
[3] National Research Tomsk State University, Tomsk, Russia
[4] National Research Tomsk Polytechnic University, Tomsk, Russia

E-mail: konovalov@physics.sibsiu.ru

Abstract. Modification of eutectic silumin surface has been implemented by high-intensity pulsed electron beam. The irradiation mode has been revealed; it allows increasing silumin fatigue life in more than 3.5 times. It has been established that the main reason of this fact is the formation of a multiphase submicro- and nanosized structure. It has been elicited that the most danger stress concentrators are large silicon plates situated on the surface and near-surface layers.

1. Introduction

Currently, in various branches of industry aluminum alloys are becoming increasingly popular. The most common of them is an aluminum alloy with silicon – silumin. This is due to its relatively low cost and low specific gravity. However, the relatively low strength properties of silumin narrow significantly its scope of application. Silumins are not strengthened by heat treatment, because of the small differences in the solubility of silicon at high and low temperatures. Therefore, the most important method of improving their mechanical properties is the modification [1]. The most effective method of the specified modification is the treatment of material surface by high-intensity pulsed electron beam. It allows modifying the structure of the surface layer with a thickness of tens of micrometers, turning it into a multi-modal structural-phase state and almost without changing the structural-phase state of the bulk of the alloy [2]. As shown in works [3-7], the formation of such surface structural-phase states contributes to the improvement of steel fatigue life of different structural classes in 2-3.5 times.

The aim of this work is to study the structural-phase states, which are formed in the surface layer of silumin, subjected to electron beam irradiation and high-cycle fatigue testings up to the fracture.

2. Materials and research methods

Silumin of a grade AK12 was used as a material for study. Fatigue tests, as in [3-7], were carried out on a special installation according to the scheme of an asymmetrical cantilever bending. The samples had the shape of a parallelepiped with dimensions 8x14x145mm. Simulation of crack was carried out by a cut in the form of a semicircle of radius 10 mm. Test temperature is of 300 K, the frequency of sample loading by bending – 15 Hz with a load of 10 MPa. Irradiation of the sample surface, prepared for fatigue testing was carried out at the device "SOLO" [2] with the following parameters: electron energy of 16 keV; the pulse frequency 0.3 s^{-1}; duration of the electron beam pulse of 50 μs and 150 μs; the energy density of the electron beam (10...25) J/cm^2; the number of the exposure pulses – 1, 3, 5. The front surface of the samples was irradiated, that is, the sample surface located over the incision simulating the crack. Each irradiation mode tested at least 5 samples. Examination of electron-beam exposure and fracture surface was carried out by methods of optical and scanning electron diffraction microscopy.

3. Results and their discussions

The structure of silumin before electron beam irradiation (the structure of the original state) is characterized by the presence of relatively large (from tenths to tens of micrometer) inclusions of silicon predominantly of a lamellar morphology (Fig. 1).

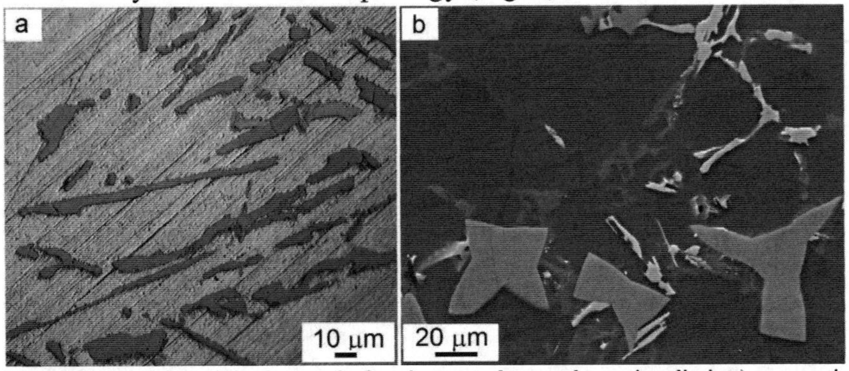

Fig. 1. Structure of silumin of the original state (before intense electron beam irradiation); a – optical microscopy; b – scanning electron microscopy.

In the fatigue tests of silumin samples a non-monotonic dependence of the number of cycles up to fracture on the irradiation mode by high-intensity pulsed electron beam has been received (Table 1). In the initial state silumin samples withstood N=1.3 10^5 the number of cycles up to the fracture. The greatest increase in fatigue life (~3.5 times) has been provided by the irradiation mode No. 4.

Table 1. Dependence of a circle number up to the fracture on the irradiation mode by electron beam.

Number of the mode	Beam energy density, E_S, J/cm^2	Exposure time, τ, μs	Number of pulses, n, pulse	Total power density, W*n, 10^6, Wt*pulse/cm^2	<N>, 10^5, cycles
1	20	150	1	0.13	1.32
2	15	150	3	0.30	1.80
3	25	150	3	0.50	2.70
4	20	150	5	0.67	5.17

5	10	50	5	1.00	2.09

It is evident that the fatigue life of silumin is determined primarily by the structure of the surface layer, modified by electron-beam processing. For structural studies of the irradiation surface of silumin the samples have been selected; they have showed minimal (for mode No. 2) and maximum (for mode No. 4) fatigue life.

The irradiation of silumin surface with pulsed electron beam depending on the energy density of the electron beam is accompanied either by melting the surface of the sample (Fig. 2, a, b), or the melting of the surface layer of the material of a certain thickness (from one to tens of micrometers) (Fig. 2, c, d). In the first case, the fatigue life in some cases has been below the fatigue life of the initial samples, and in the second one it has exceeded the fatigue life of the initial material in more than 3.5 times.

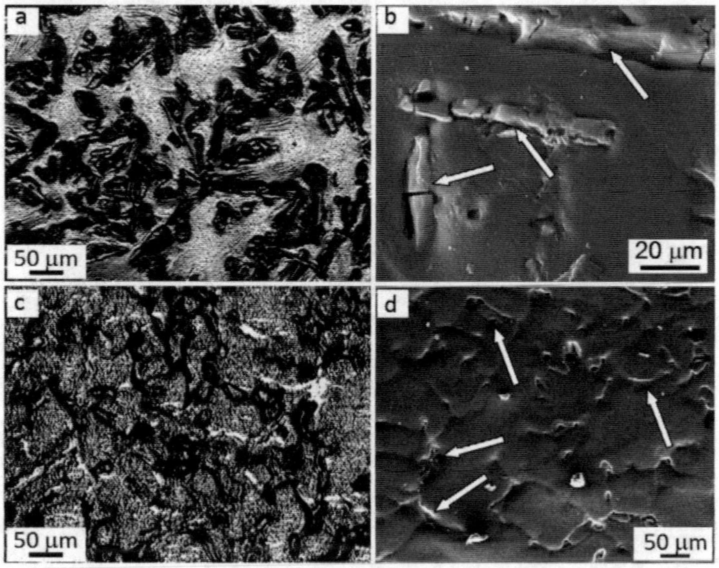

Fig. 2. Surface structure of silumin irradiated by electron beam in modes No. 2 (a, b) and No. 4 (c, d); a, c – optical microscopy; b, d – scanning electron microscopy; silicon particles are arrowed on (b, d).

The irradiation of silumin surface for mode No. 2 leads to the partial melting of the excess silicon inclusions (Fig. 2, a). In the surface layer numerous micropores along the boundary between the plate/matrix and microcracks are formed; they are located in the silicon plates, which weaken the material. On the image of fatigue fracture surface of the silumin sample (Fig. 3) it can be clearly seen that the fatigue crack is formed on the sample surface (Fig. 3, a). The reasons for the formation of fatigue cracks are coarse inclusions of silicon (Fig. 3, b), which are stress concentrators. As a result, the fatigue tests lead to the fracture of plates and the formation of long microcracks.

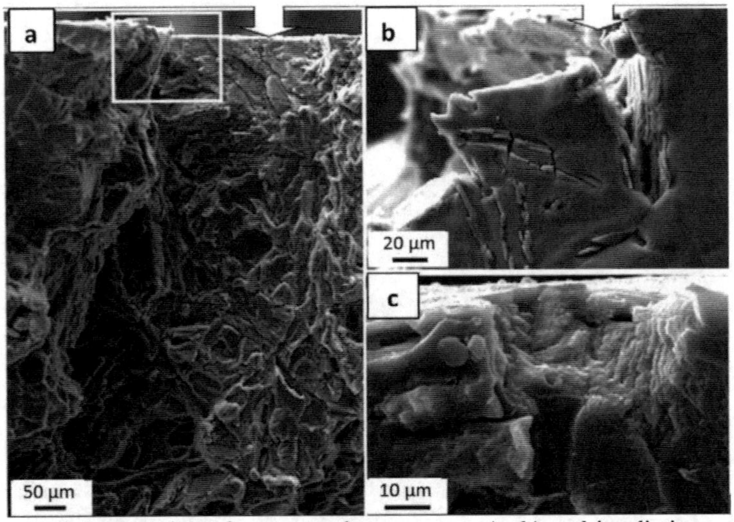

Fig. 3. Electron microscope image of the fracture surface structure (a, b) and irradiation surface (b) of silumin, processed by electron beam according to mode № 2. On (a) the frame singles out a fatigue crack formation range; the pointers indicate: on (a) – irradiation surface, on (b) – place of origin of fatigue crack.

During silumin irradiation with high-intensity electron beam in mode No. 4 the structure of the surface layer by morphological characteristic differs significantly from the structure of the original sample (Fig. 1) and the sample irradiated in the mode of the surface melting (Fig. 2, a, b). On the irradiation surface the homogeneous structure of a grain type is formed (Fig.2 (c, d), and the thickness of the molten layer varies in the range up to 20 µm (Fig. 4, d).

Fatigue fracture surface analysis of the sample irradiated by the electron beam in mode No. 4, showed that the thickness of the melted layer varies in the range up to 20 µm (Fig. 4, a). Subsequent to the melting high-speed crystallization leads to the formation of multimodal structures represented at the macro level by the aluminium based grains, sizes of which varies within 30...50 µm with located on the boundary silicon particles, whose dimensions do not exceed 10 µm (Fig. 2, c, d). Two-phase (silicon and solid solution on the basis of aluminum) crystallization cells detected on the fatigue fracture surface form meso-level of modified layer. Crystallization cell sizes vary from 100 nm to 250 nm (Fig. 4, b). This reflects the submicrocrystalline structure of the near-surface layer. It is important that the stress concentrators, which can be a source of fracture of silumin samples at this mode of irradiation, on the edge of a fracture are not revealed. Apparently, the concentrators, which caused the fracture of the sample, are situated below the surface, most likely at the interface of the liquid and solid phases. It can be assumed that one of the possible reasons for the decline in the number of cycles up to the fracture to the initial state after irradiation of the silumin samples according to mode No. 5 can be a powerful residual thermal stresses, emerging in the modified surface layer, as in [8].

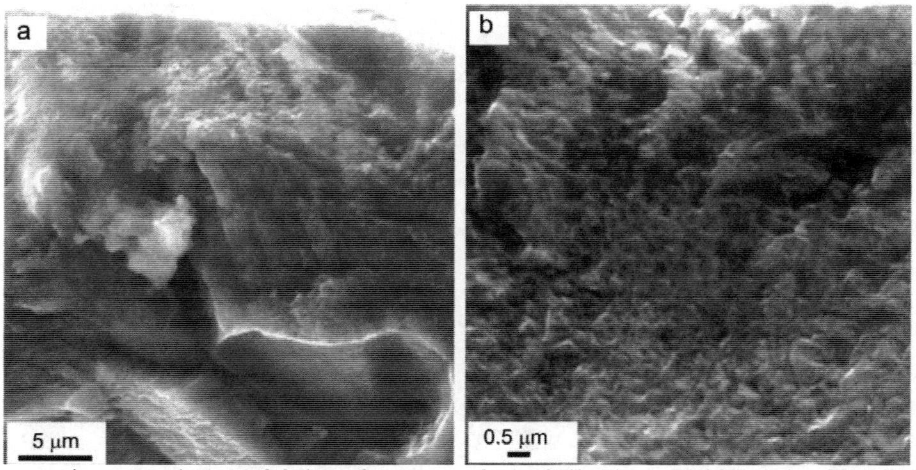

Fig. 4. Electron microscope image of fatigue fracture surface of silumin treated in mode of irradiation by high intensity pulsed electron beam No. 4.

The formation of multi-level structural-phase state determines the development of the damping properties in the surface layer of the modified silumin with respect to the main material by mechanical and thermal external influences. This prevents the premature birth and spreading of fragile microcracks from the surface into the bulk of the material. These cracks can lead to the formation of main cracks and to the fracture of the basic material [9]. Thus, the formation of a submicro-and nanosized multiphase structure at the irradiation of silumin according to mode No. 4 is also the defining reason assisting a multiple increase of its fatigue life.

4. Conclusion

Multicycle fatigue tests up to the fracture of silumin samples subjected to the electron-beam processing have been carried out. The irradiation mode, that promotes the multiple increase (in more than 3.5 times) of fatigue life of silumin, has been identified.

The researches of an irradiation surface structure and the surface of fatigue failure of silumin in an initial state and after modification states with intense pulsed electron beam according to different regimes have been carried out. It is shown, that in mode of partial melting of the irradiation surface modification process of silicon plates is accompanied by the formation of numerous large micropores along the boundary plate/matrix and microcracks located in the silicon plates. A multi-modal structure (grain size within 30...50 μm with located on the boundaries silicon particles up to 10 μm) is formed in stable melting mode, as well as subgrain structure in the form of crystallization cells in size from 100 μm to 250 μm).

It has been established that the main cause of multiple increase of silumin fatigue life, processed with a pulsed electron beam, is the formation of the nanoscale multiphase structure in a modified surface layer.

Acknowledgement

This work has been supported by a grant of the President of the Russian Federation for state support of young Russian scientists – Doctors of Sciences (project MD-2920.2015.8) and state task № 3.1496.2014/K.

References

[1] Zolotorevskiy V S and Belov N A 2005 *Metal science of casting aluminum alloys*, (Moscow, Publishing House MISiS).

[2] Laskonev A P, Ivanov Yu F, Petrikova E A *et al* 2013 *Structure and properties modification of eutectic silumin by electron-ionic-plasma processing* (Minsk, Belaruskaya nayka).

[3] Ivanov Yu F, Koval N N, Gorbunov S V, Vorobyov S V, Konovalov S V and Gromov V E 2011 *Rus. Phys. J.* **54** 575-583.

[4] Sizov V V, Gromov V E, Ivanov Yu F, Vorob'ev S V and Konovalov S V 2012 *Steel in Translation* **42** 486-488.

[5] Gromov V E, Ivanov Yu F, Sizov V V, Vorob'ev S V and Konovalov S V 2013 *J. of Surf. Investigation. X-ray, Synchrotron and Neutron Techniques* **7** 94.

[6] Grishunin V A, Gromov V E, Ivanov Yu F, Teresov A D, Konovalov S V 2013 *J. of Surf. Investigation. X-ray, Synchrotron and Neutron Techniques* **7** 990-995.

[7] Grishunin V A, Gromov V E, Ivanov Yu F, Volkov K V and Konovalov S V 2014 *Steel in Translation* **43** 724-727.

[8] Gromov V E, IvanovYu F and Grishunin V A 2013 *Progress in metal physics* **14** 67-80.

[9] Psakhie S, Ovcharenko V, Baohai Yu *et al* 2013 *J. Mater. Sci. Technol.* **29** 1025-1034.

Synthesis of Polycrystalline Diamond Films in Abnormal Glow Discharge and their Properties

A V Gaydaychuk, S A Linnik, A V Kabyshev, F V Konusov and G E Remnev

National Research Tomsk Polytechnic University, 30, Lenin Avenue, 634050, Tomsk, Russia

E-mail: konusov@hvd.tpu.ru, stepan_lin@mail.ru

Abstract. The optical and electrophysical properties of polycrystalline diamond films (PDF) deposited from the abnormal glow discharge have been studied. The dominating mechanisms of absorption and charge carrier transfer and the energy spectrum of the localized states (LS) of defects which determine the properties of the films have been specified. The parameters of the interband absorption and electrical conductivity are determined by the continuous energy distribution in the band gap (BG) of the states of defects of different nature. The absorption edge of the crystalline phase of the films is separated from the absorption zone determined by the electron transitions between LS defects. The width of BG is narrowed to 0.2–0.5 eV from the quantity typical to the diamond. An additional film absorption edge is formed in the energy interval 1.2–3.3 eV, where Urbach rule is fulfilled and the interband absorption is realized at direct transitions through the optical gap 1.1–1.5 eV. The average width of BG is 2.6–3.24 eV estimated within semiclassical interband model. The interaction of the parameters of the interband and exponential absorption is determined by the crystal lattice static disorder. The dominating n-type of the activation component of the electrical conductivity is complemented by the hopping mechanism with the participation of the localized states of the defects distributed near the Fermi level with a density $5.6\cdot10^{17}$–$2.1\cdot10^{21}$ $eV^{-1}\cdot cm^{-3}$.

Keywords: polycrystalline diamond films, absorption, electrical conductivity, photoconductivity, localized states of defects

1. Introduction

The unique chemical, mechanical, electrical and thermophysical, optical and photoelectrical properties of a diamond contribute to its wide application in high-frequency and high-temperature electronics [1–5]. Due to the limited potentials for application of diamond single crystals and epitaxial diamond films caused by their high cost, the most appropriate instruments are the polycrystalline diamond films (PDF), which are obtained by well-studied methods of the gas-phase deposition [1–9]. PDF are successfully used for the production of the stable radiation-resistant UV detectors and ionizing radiation, as well as for laser and photodiode structures [1,4–7]. Depending on the characteristics of PDF polycrystalline structure, the content of alloying impurity atoms and defects in the PDF crystal structure, their optical and electrical characteristics vary widely. It stimulates the study of the electrophysical and optical properties of PDF, the investigation of the mechanisms determining these properties and the influence of deposition conditions on the change in the characteristics. The method

Journal of Physics: Conference Series **652** (2015) 012029 doi:10.1088/1742-6596/652/1/012029

of the film deposition from abnormal glow discharge (AGD) plasma stands out among various methods of PDF deposition – using microwave discharge plasma, hot filament, and plasma-arc CVD. Each of the methods has a lot of disadvantages: high cost and/or limitations in maximum deposition area; low rate and the necessity to repeated replacement of filaments; the drawback of a plasma-arc CVD method is a small deposition area and high gas consumption. The efficiency of the method of PDF deposition from AGD plasma compared to the alternative ones is in its simplicity and a high growth rate.

The aim of the work is to determine the optical, electrical, and photoelectrical characteristics, the energy spectrum of the defect levels and their nature in PDF deposited from AGD plasma.

2. Research methods

The methods of film deposition from AGD and the study of the film structural features are described in detail in [2, 3]. A new system of diamond film deposition from AC of AGD has been presented. In the process of deposition the discharge in the form of the plasma filament is realized between two small electrodes, and the distance between electrodes reaches 20 cm. The electrodes are removed from the substrate, and the material of the electrodes does not contaminate the film [2]. The absorption spectra of $\alpha(h\nu)$ were calculated from the diffuse reflection spectra and were approximated by the Urbach rule to determine the characteristics of the localized states (LS) of defects distributed in the band gap (BG) of the films and by the laws for the interband absorption of crystalline and amorphous semiconductors [10]. The dark surface conductivity σ and photoconductivity $\Delta\sigma_{ph}=\sigma_{ph}-\sigma$ (σ_{ph} is conductivity under lighting), photosensitive $K(U, T, h\nu)=\Delta\sigma_{ph}/\sigma$ were measured at a constant voltage across the electrodes U=0.01–300 V, at temperature T=300–700 K, and photon energy $h\nu$=1.5–4.0 eV. The electrodes were deposited on the surface by rubbing the soft graphite. The temperature dependences of σ, $\Delta\sigma_{ph}(T)$ were approximated by the equation for the activation mechanism

$$\sigma_a(T) = \sigma_0 \times e^{(-\varepsilon_\sigma / k \cdot T)} \quad (1)$$

(σ_0 is a pre–exponential factor, ε_σ is an activation energy, k is the Boltzmann constant) and by the equation for the hopping mechanism of transfer between LS near the Fermi level E_F within the Mott model

$$\sigma_p(T) = \sigma'_0 \times e^{(-(T_0/T)^{0.25})}, \quad (2)$$

(σ'_0 is a pre–exponential factor, T_0 is activation energy) [11, 12]. The LS density $N(E_F)$ near E_F was calculated from T_0 according to [11, 12]. The sign of the dominant charge carriers was determined by the photostimulated current amplitude $I_{PhTSC}(T, h\nu, U=0)$ [11].

3. Results and discussion

The $\alpha(h\nu)$ spectra indicate the heterogeneity of defect pattern with LS when $h\nu$=1.1–5.5 eV (figure 1). The concentration of LS defects vary from $N=2\cdot10^{19}$ to $3\cdot10^{20}$ cm^{-3} across the surface of the film deposition and over their thickness (figure 1). The surface roughness has a greater effect on the properties than the size of the crystallites [1]. The accumulation of LS with an energy of $h\nu$=1.1–1.9 and 2.5–4.5 eV is higher from the side of the substrate than from the growth side (figure 1). The characteristics of $\alpha(h\nu)$ spectra indicate the dominant defect agglomeration near the intercrystalline boundaries of PDF and the impact of the amorphous hydrogenated carbon a–C:H by analogy with [1, 6–9]. A sharp rise of α when $h\nu$>5.0 eV responds the fundamental absorption in crystallites (figure 1). The width of BG films E_g=5.0–5.27 eV for direct transitions when $h\nu$=5.2–5.6 eV is close to E_g=5.45 eV of diamond and E_g=5.0–5.5 eV of CVD films [1, 4–5]. The absorption in PDF crystallites ($h\nu$=5.0–5.6 eV) is separated from the absorption zone ($h\nu$=1.1–5.0 eV) caused by defect LS (figure 1). BG narrowing by 0.1–0.2 eV is more explicitly from the side of the substrate to the edge of deposition. The spectra indicate the realization within the interval $h\nu$=1.1–5.0 eV of several types of absorption. Within the intervals $h\nu$=1.18–1.3, 1.3–1.8, and 1.8–2.3 eV the spectra $\alpha(h\nu)$ obey the Urbach rule

$$\alpha(h\nu) = \alpha_0 \times e^{(-h\nu / E_U)} \quad (3)$$

with the PDF typical parameters α_0 and E_U (E_U is Urbach energy) (figure 1). In the intervals hν=1.18–1.3 eV the value E_U=0.05–0.2 eV is close to the value E_U in a–C:H [8]. The big values E_U>1.6 eV within the intervals hν=1.3–1.8, 1.8-2.3 eV in the films are typical to nanocrystalline diamond films (NCDF) [1, 5, 6, 9, 13, 14]. Increasing E_U with the depth of the level in BG is determined by the high overlapping of exponentially distributed LS and enables the realization of interband absorption at the distance from the edge of BG. The characteristics of exponentially distributed levels in BG are determined by a high impact of graphite-like phase disordered by the defects [1, 6–9].

The spectra α(hν) are also approximated by the law for the interband absorption

$$\alpha \times h\nu = B'(B'') \times (h\nu - E'_g(E''_g))^m, \quad (4)$$

where m=1/2 and 2 for direct and indirect allowed transitions through the optical bands E'_g and E''_g, respectively. For the indirect transitions realized within the interval hν=1.2–2.4 eV, $E''_g \lesssim$ 0 eV, and for the direct within the interval hν=1.5–3.4 eV the BG width is E'_g=1.1–1.6 eV (figure 2). It indicates the probable overlapping of LS with the allowed bands caused by the increase in the content of sp^2 clusters and their sizes by analogy with [1]. In NCDF the close interrelations (E''_g=1.0 eV, E'_g=2.2 eV) are determined by the defect LS near the grain boundaries [15, 16]. In the films a–C and a–C:H E''_g=1–3 and 1.5–3.5 eV [8]. In PDF E''_g is narrower than in the diamond-like carbon films (DLCF) and ultradispersed nanodiamond films (UDNDF), where E''_g=3.3–3.5 eV. The interval overlap of the interband and exponential absorption suggests that a new edge of absorption is formed in PDF biased to the zone hν=1.2–3.3 eV, where the Urbach rule is followed and the interband absorption is realized. The decrease in E_g with E_U growing is typical to LS with hν=1.18–1.3 and 1.8–2.3 eV (figure 3). The similar interrelation E''_g and E_U is typical to the films a–C, a–C:H [8] and to the materials with a high concentration of defects [10].

The effect of the structural disorder in PDF is estimated and within the semiclassical model for the interband absorption of amorphous or highly defect materials [1, 10]. Here the spectra obey the law

$$\alpha(h\nu) = \alpha''_0 \times e^{(-(h\nu - E_{g0})^2 / 2 \cdot \sigma^2)}, \quad (5)$$

where α'_0 is a multiplier, E_{g0} is a mean BG width (E_g obeys the Gaussian distribution relative to E_{g0}, σ^2 is the dispersion of the value E_g determined by the disorder induced by defects) (figure 1, curve 2). The dispersion σ is close to the value E_U for the defects localized when hν=1.3–1.8 eV. The values E'_g and E_{g0} correlate due to the equality of both interband absorption models (figure 3). Since the overlapping degree of the defect levels is growing (~E_U), the mean BG width E_{g0} is decreased (figure 3). The values α_0'' and B' (proportional to the density of states bound by the interband transitions through the optical gap) vary in the correlation when E_{g0} is changed (figure 4). The values E_{g0} in PDF are close to the optical gap of DLCF E''_g=3.77 eV and UDNDF E_g=3.3–3.5 eV. The PDF spectra α(hν) correspond to the transition model between the states π–π* determined by the border inclusions a–C, through the pseudogap E_{g0}=3.0–3.5 eV [1, 6, 7, 9, 17]. The correlations between the parameters E_g, E_{g0} and σ, E_U are typical to the materials with the static and/or dynamic disorder of the crystal lattice (figures 3 and 4) [10, 14–16]. It indicates the impact of the absorption in crystallites on the properties, and it suggests that PDF is the material with the BG width for direct transitions 1.1–1.6 eV and the absorption edge determined by the defect LS with ε=1.2–3.3 eV.

Spectrum factorization on the elementary Gaussian curves $\alpha \sim \exp(-(h\nu - E)^2/(0.833 \cdot \gamma^2))$ (E is the center, γ is half-width) identify the local defect levels: **1**–E=1.17 eV, γ=0.1 eV, N=$2 \cdot 10^{19}$ cm^{-3} (V^+ is the center [14]); **2**–1.4 eV, 0.3 eV, $1.2 \cdot 10^{20}$ cm^{-3}; **3**–(1.8–2) eV, 0.4 eV, $7 \cdot 10^{20}$ cm^{-3} (GR1, I^0 are the centers [18]); **4**–(2.1–2.2) eV, 0.2 eV, $2 \cdot 10^{19}$ cm^{-3} (sp^2(π→π*) [1, 7]); **5**–(2.6–2.7) eV, 0.5 eV, $2 \cdot 10^{20}$ cm^{-3} (sp^2(π→π*) [1, 7]); **6**–(3.2–3.4) eV, 0.4 eV, $3 \cdot 10^{20}$ cm^{-3} (V^-, sp^2(π→π*) [1, 6]); **7**–(3.9–4.1) eV, 0.4 eV, $2 \cdot 10^{20}$ cm^{-3} (N4–6, sp^2(π→σ*) [1, 7]); **8**–(4.4–4.6) eV, 0.3 eV, $7 \cdot 10^{19}$ cm^{-3} (Σ→Σ*) [1, 6, 7]).
The defects are distributed both in crystallites and at the intercrystalline PDF boundaries.

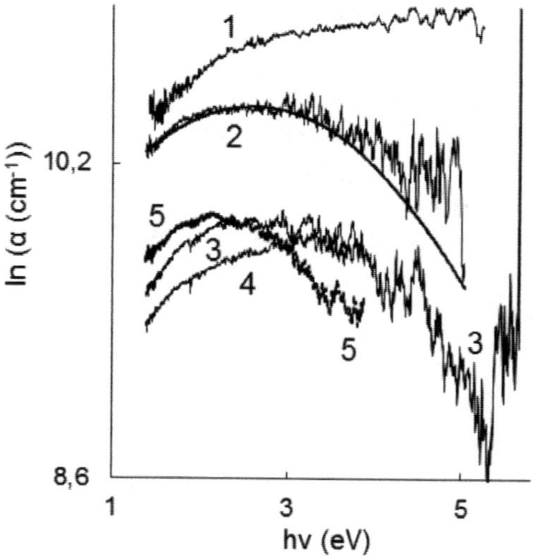

Figure 1. The absorption spectra α(hν) of PDF with the side of the substrate (1–3) and growth (4, 5) at the edge (1, 2, 5) and in the deposition center (3, 4) and their approximation (2) with the equation (5) at E_{g0}=2.6 eV and σ=1.8 eV.

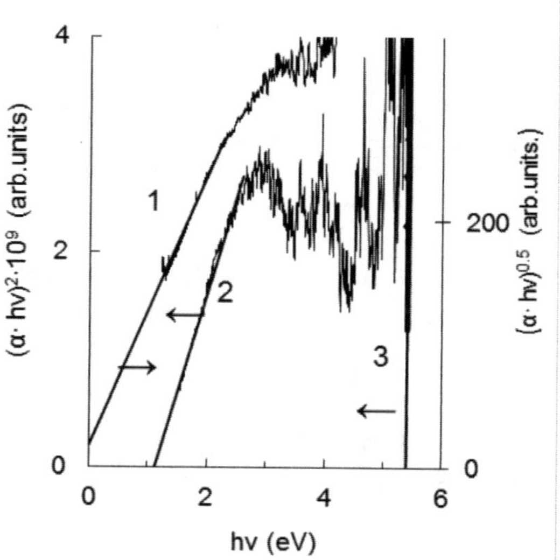

Figure 2. The spectra of interband PDF absorption with indirect (1) and direct allowed transitions (2, 3) through the optical gaps E''_g and E'_g, respectively.

The photosensitivity spectra K(hν) are similar to the film absorption spectra and typical to the diamond films [1, 4, 9, 11]. The spectra K(hν) are determined by the electron transitions between the valence band and defect LS. Some definite contribution to K(hν) is made by the transitions between the "tails" of the LS of the allowed bands and the levels associated with the graphite-like phases [1, 11]. The oxygen molecules adsorbed on the film surface facilitate the increase in UV photosensitivity due to their passivating action on defect LS [11].

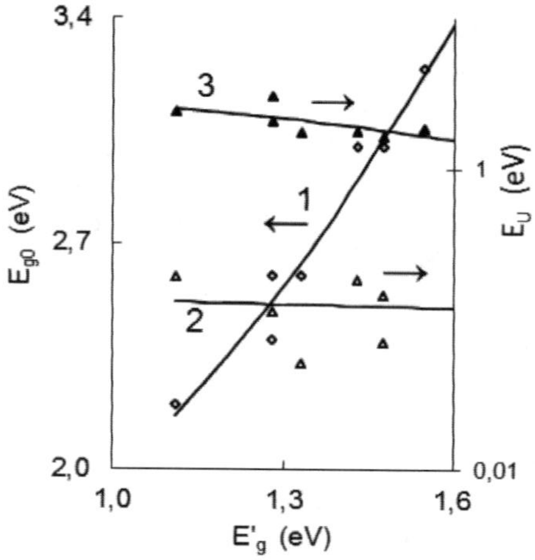

Figure 3. $E_{g0}(E'_g)$ (1) vs. $E_U(E'_g)$ (2, 3) in PDF. The value E_{g0} is calculated within the interval hν=1.2–5 eV, E'_g within the interval hν=1.5–3.4 eV, and E_U within the interval hν=1.18–1.3 (2) and 1.8-2.3 eV (3)

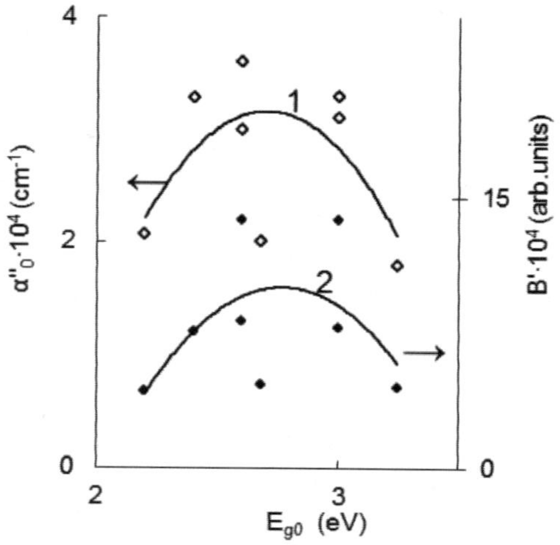

Figure 4. $\alpha''_0(E_{g0})$ (1) vs. $B'(E_{g0})$ (2) in PDF. The value E_{g0} is calculated within the interval hν=1.2–5 eV, and the value B′ within the interval hν=1.5–3.4 eV.

Volt-ampere characteristics, field dependence of photosensitivity K(U), energy and kinetic characteristics of electrotransfer ε_σ, σ_0, T_0, σ'_0, $N(E_F)$, I_{PhTSC} vary in the thickness and in the PDF deposition plane caused by the heterogeneity of their properties [11]. The temperature dependencies of σ, $\Delta\sigma_{ph}(T)$ within the interval T=300–600 K are determined by the thermally stimulated electron exchange between the shallow donor levels with an energy ε_σ=0.007–0.21 eV and the conduction band (CB) [11]. The dominating n-type of σ and $\Delta\sigma_{ph}$, as the dependencies $I_{PhTSC}(T, h\nu)$ show [11]. We cannot exclude the effect of the holes exchange between defect LS and the valence band on the characteristics of the films. The electrical parameters indicate the effect on the transport of the hopping conductivity that is verified by the approximation $\sigma(K)$ by the equation (2) within the interval ΔT=300–600 K [11]. The LS density calculated from the dependencies σ, $\Delta\sigma_{ph}(T)$ is $N(E_F)=5.6\cdot10^{17}$–$1.5\cdot10^{21}$ $eV^{-1}\cdot cm^{-3}$ in PDF with ε_σ=0.007–0.15 eV. In PDF deposited by the alternative methods, the density of LS exponentially distributed near E_F is $N(E_F)=(4–8)\cdot10^{19}$ [7, 12] and 10^{15}–10^{18} $eV^{-1}\cdot cm^{-3}$ [19]. The value of the most probable length of the hop calculated using the values $N(E_F)$ is R=1.1–8.1 nm and is decreased at lightning to 1.0–6.0 nm under the increase in the density of the states by 3–5 times for the films with electrical conductivity $\sigma<10^{-10}$ S [11].

The correlations between σ_p, σ_a and K enable to make the dominance of one of the transport mechanisms [11]. The dominance of σ_a is verified by the dependence $N(E_F)(\varepsilon_\sigma)$, which is typical to the noncrystalline semiconductors, which have a high concentration of defects (figure 5) [7, 11, 12, 19]. In PDF, the hopping LS conductivity near E_F predominates over the activation one at T=300–400 K. When T=400–600 K, the PDF characteristics are determined by $\sigma_a(T)$ involving LS defects distributed at the grain boundaries and localized in the BG in the region of the "tails" of the allowed bands [11]. When T>600 K, the effect of the energy barrier between crystallites on the transport is increased [11]. The established interrelation between the characteristics σ, ε_σ, and E'_g agree with the models of the electrotransfer typical for amorphous and highly defect materials [10, 11]. The accumulation of the concentration of absorption centers is related with the decrease in the LS density which participate in the hopping of σ_p from $N(E_F)=10^{19}$ to $5\cdot10^{17}$ $eV^{-1}\cdot cm^{-3}$ in the correlation with the increase in α from $2\cdot10^4$ to $5\cdot10^4$ cm^{-1} (figures 1 and 5). The decrease in σ and $N(E_F)$ correlate with the increase in ε_σ and E_U for LS at hν=1.15–1.3 eV and 1.6–3.0 eV (figures 5 and 6). The growth of E_U from 0.07 to 0.15 eV, as σ is decreased from $\sigma=10^{-4}$ to 10^{-11} S and ε_σ is increased from 0.007 to 0.16 eV, is determined

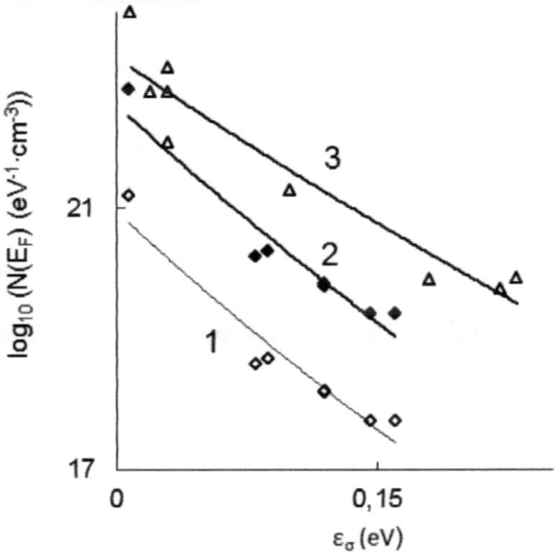

Figure 5. Interrelation between LS density and activation energy ε_σ in the deposited films (1, 2) and according to [7] (3). Attenuation constant $\alpha=5.7\cdot10^6$ (1) and $2\cdot10^7$ cm^{-1} (2).

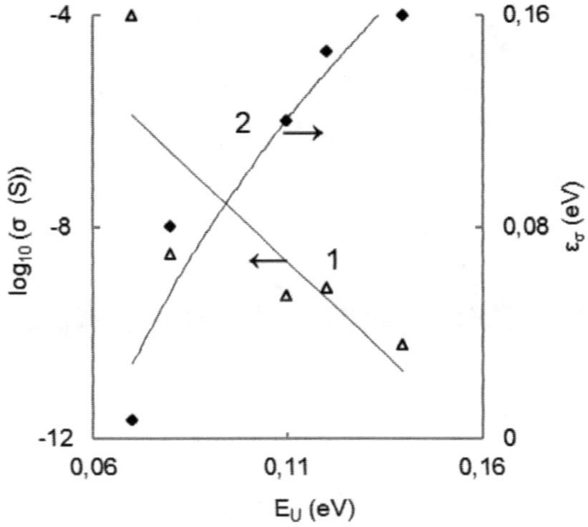

Figure 6. Interrelation between conductivity σ, Urbach energy E_U (1) and activation energy ε_σ and E_U (2) in the deposited films. The values E_U are calculated within the interval hν=1.15–1.3 eV.

by the impact of the structural disorder induced by defects in the film material and the impact of the disordered graphite-like phases [10, 11]. In the film material BG, the LS energy exponential distribution is formed, LS are induced by the growth defects [1, 10, 11].

4. Conclusion

The properties of the films deposited from AGD do not yield to those of the diamond films obtained by the alternative methods. PDF absorption deposited from the AGD plasma is determined by the continuous LS spectrum induced by different imperfections. The edge of PDF absorption when 5.0–5.6 eV is separated from the absorption zone within the interval 1.1–5.0 eV induced by defect LS. The BG width is narrowed to E_g=5.0–5.3 eV. The absorption in the energy interval of 1.0–5.0 eV is realized with the transitions between defect LS and interband transitions which are typical to the crystalline and amorphous materials. A new edge of absorption is formed in the films in the energy interval of 1.2–3.3 eV, where the Urbach rule is followed the interband absorption is realized by the direct and indirect transitions. The electrical characteristics of the films are determined by the structural disorder induced by the growth defects in crystallites and by the presence of graphite-like phases in the films. The n-type conductivity activation component is realized when the electrons are exchanged between CB and donor levels with an energy $\lesssim$ 0.21 eV and is amplified by the LS hopping mechanism near the Fermi level with a density of $5.6 \cdot 10^{17}$–$2.1 \cdot 10^{21}$ $eV^{-1} \cdot cm^{-3}$.

Acknowledgments

The paper has been completed within megaproject of National Research Tomsk Polytechnic University (VIU_IHTPh_85_2014) «Materials for Extreme Conditions».

References

[1] Williams O A 2011 *Diamond and Relat. Mater.* **20** 621
[2] Linnik S A and Gaydaychuk A V 2013 *Diamond and Relat. Mater.* **32** 43
[3] Linnik S A and Gaydaychuk A V 2012 *Tech. Phys. Lett.* **38** 258
[4] Ralchenko V, Pimenov S and Konov V 2007 *Diamond and Relat. Mater.* **16** 2067
[5] Willems B, Tallaire A and Achard J 2014 *Diamond and Relat. Mater.* **41** 25
[6] Varga M, Remes Z, Babchenko O and Kromka A. 2012 *Phys. Stat. Sol.* **B249** 2635
[7] Achatz P, Garrido J A, Stutzmann M, Williams O A, Gruen D M, Kromka A and Steinmüller D 2006 *Appl. Phys. Lett.* **88** 101908
[8] Fanchini G and Tagliaferro A 2004 *Diamond and Relat. Mater.* **13** 1402
[9] Vanesek M, Rosa J, Nesladek M and Stals L M 1996 *Diamond and Relat. Mater.* **5** 952
[10] Konusov F V, Kabyshev A V and Remnev G E 2011 *J. of Surface Invest.* **5** 228
[11] Konusov F V, Kabyshev A V, Linnik S A, Gaydaychuk A V and Remnev G E 2014 *J. of Phys.: Conf. ser.* **552** 012046
[12] Gan L, Bolker A, Saguy C, Kalish R, Tan D L, Tay B K, Gruen D and Bruno P 2009 *Diamond and Relat. Mater.* **18** 1118
[13] Nebel C E, Reisel R and Stutzmann M 2001 *Diamond and Relat. Mater.* **10** 639
[14] Sharda T, Rahaman M M, Nukaya Y, Soga T, Jimbo T and Umeno M 2001 *Diamond and Relat.Mater.* **10** 361
[15] Nagano Akira, Yoshitake Tsuyoshi, Hara Takeshi and Nagayama Kunihito 2008 *Diamond and Relat. Mater.* **17** 1199
[16] Franta D, Zajíčková L, Karásková M, Jašek O, Nečas D, Klapetek P and Valtr M 2008 *Diamond and Relat. Mater.* **17** 1278
[17] Mathioudakis C, Kopidakis G, Patsalas P and Kelires P C 2007 *Diamond and Relat. Mater.* **16** 1788
[18] Jones R 2009 *Diamond and Relat. Mater.* **18** 820
[19] Nath S and Wilson J I B 1996 *Diamond and Relat. Mater.* **5** 65

The Effect of Doping on the Electrophysical Properties of Polycrystalline Diamond Films Deposited from an Abnormal Glow Discharge

A V Kabyshev, F V Konusov, S A Linnik and G E Remnev
National Research Tomsk Polytechnic University, 30, Lenin Avenue, 634050, Tomsk, Russia

E-mail: konusov@hvd.tpu.ru, kabyshev@tpu.ru

Abstract. The paper is focused on the study of the boron doping effect on the electrical characteristics, on the mechanism of charge carrier transfer, and on the energy spectrum of the localized defect states in the polycrystalline diamond films (PDF) deposited from an abnormal glow discharge. PDF doping enables to form the semiconductor layers of p-type conductivity, which have as good properties as those of PDF produced by the alternative methods. The doping reduces the degree of disorder in the film material brought by the growth defects, which determine the film electrical characteristics and electrotransfer mechanism. The PDF electrical characteristics and electrotransfer mechanism are determined by the defects of different nature, whose band gap energy levels have a continuous energy distribution. A p-type activation component is realized in the exchange of charge carriers between the valence band and shallow acceptor levels with the activation energy of 0.013–0.022 eV. Doping increases the effect of the hopping mechanism of the conductivity involving the localized states with a density of (1–6)$\cdot 10^{20}$ $eV^{-1}\cdot cm^{-3}$ distributed near the Fermi level, which is in the low half of the band gap.

Keywords: polycrystalline diamond films, electrical conduction, photoconductivity, localized states of defects

1. Introduction

The unique chemical, mechanical, electrical and thermophysical, optical and photoelectrical properties of a diamond contribute to its wide application in high-frequency and high-temperature electronics [1–5]. Due to the limited potentials for application of diamond single crystals and epitaxial diamond films caused by their high cost, the most appropriate instruments are the polycrystalline diamond films (PDF), which are obtained by well-studied methods of the gas-phase deposition [1–9]. Not only dielectric PDF are used for these purposes (electrical conduction $\sigma=10^{-15}$–10^{-11} $S\cdot cm^{-1}$), but also high-conductivity films of n– and p–type obtained by doping during deposition with dopant atoms of nitrogen [1, 7, 10–12] and boron [1, 2, 13–16]. Doping with B atoms with a concentration of 10^{17}–$3\cdot 10^{21}$ cm^{-3} causes the high-conductivity layers of p–type with $\sigma=10^{-10}$–$3\cdot 10^{2}$ $S\cdot cm^{-1}$ [1, 7, and 13–16] and with the altered optical properties [1, 2, 17]. The use of such doped layers makes it possible to significantly increase the photoelectrical characteristics of the devices [1, 17–20]. The features of the formed polycrystalline structure of the films and the content of the growth defects influence the electrical characteristics of the deposited semiconductor layers. It stimulates the study of the impact of the deposition conditions on the electrical and photoelectrical properties of PDF. The method of the film deposition from abnormal glow discharge (AGD) plasma stands out among various methods of PDF deposition – using microwave discharge plasma, hot filament, and plasma-arc CVD. The efficiency of the method of PDF deposition from AGD plasma compared to the alternative ones is in its simplicity and a high growth rate [3, 4].

The aim of the work is to study the effect of boron doping on the electrical characteristics, on the dominant transfer mechanism, and on the energy spectrum of the defect localized states in PDF deposited from AGD plasma.

2. Research methods

The methods of film deposition from AGD and the study of the film structural features are described in detail in [3 and 4]. The dark surface conductivity σ and photoconductivity $\Delta\sigma_{ph}=\sigma_{ph}-\sigma$ (σ_{ph} is conductivity under lighting), photosensitive $K(U, T, h\nu)=\Delta\sigma_{ph}/\sigma$ were measured at a constant voltage across the electrodes U=0.01–300 V, at temperature T=300–700 K, and photon energy $h\nu$=1.5–4.0 eV [18]. The electrodes were deposited on the surface by rubbing the soft graphite. The temperature dependences of σ $\Delta\sigma_{ph}(T)$ were approximated by the equation for the activation mechanism

$$\sigma_a(T) = \sigma_0 \times e^{(-\varepsilon_\sigma / k \cdot T)}, \qquad (1)$$

where σ_0 is a pre–exponential factor, ε_σ is an activation energy, k is the Boltzmann constant and by the equation for the hopping mechanism of transfer between the localized states (LS) near the Fermi level E_F in the band gap (BG) in the Mott model

$$\sigma_p(T) = \sigma_0' \times e^{(-(T_0/T)^{0.25})}, \qquad (2)$$

where σ'_0 is a pre–exponential factor, T_0 is an activation energy [10, 11, 18]. The LS density $N(E_F)$ near the Fermi level was calculated from T_0 according to [18]. The sign of the dominant charge carriers was determined by the photo- and thermostimulated current amplitude $I_{PhTSC}(T, h\nu, U=0)$ [18].

3. Results and discussion

Current voltage characteristic (CVC) I(U), field dependence of K(U) and carrier transfer energy characteristics ε_σ, σ_0, T_0, σ'_0, $N(E_F)$, and I_{PhTSC} in the undoped low-conductivity ($\sigma=10^{-14}$–10^{-4} S) films deposited on the silicon substrate or separated from it vary in thickness and PDF deposition plane, due to the inhomogeneity of the PDF properties (figure 1). The variation in the electrical parameters of the deposited films may be caused by the heterogeneity in the crystallite size distribution and by the growth defects in their volume and at the boundaries or by the influence of hydrogen impurity in the surface layers that is proved by the analysis of [2, 12, 14–18, 20]. The differences of I(U) and K(U) are recorded from the growth smooth and the opposite rough sides (figure 1) [18]. The conductivity in the PDF separated from substrate from the growth side is higher than from the side of the substrate and the ratio between K is opposite (figure 1) [18]. In the films with $\sigma \lesssim 10^{-4}$ S, the deviation from the linear law is fixed: CVC obeys the exponential equation $I \propto \exp(\beta U)$ (β=0.02–0.06) or the power law $I \propto U^\alpha$ (α=1.2–3.3) (figure 1, curves 5, 7–9). This indicates the current effect limited by the space charge as in [12, 17, and 18]. With σ decreasing from 10^{-7} to 10^{-14} S, the parameters of nonlinearity increase from α=1.2 to α=3.3 and from β=0.02 to β=0.06. This indicates the increase in the variation in LS population caused by the increase in the content of defects in PDF. The value of the photoconductivity (K=0.1–1.0) and the nature of its field dependence K(U) strengthen the ratios between CVC [18]. The CVC and K(U) characteristics are determined by the inhomogeneous spatial distribution of the energy barrier that exists on the border of PDF with the electrodes and on the intercrystallite boundaries [8, 9, 12, 17, and 18]. Both in the center of the deposition and from the growth side, the CVC of the films with a conductivity $\sigma \gtrsim 10^{-4}$ S are almost linear (α=1.05–1.15) (figure 1, curve 6), photosensitivity of these films is low (K=0.003–0.05), and the field dependence of K(U) is low as shown in [18].

Regarding the substrate, the undoped film on the Si substrate with $\sigma=(1-2)\cdot 10^{-5}$ S and the

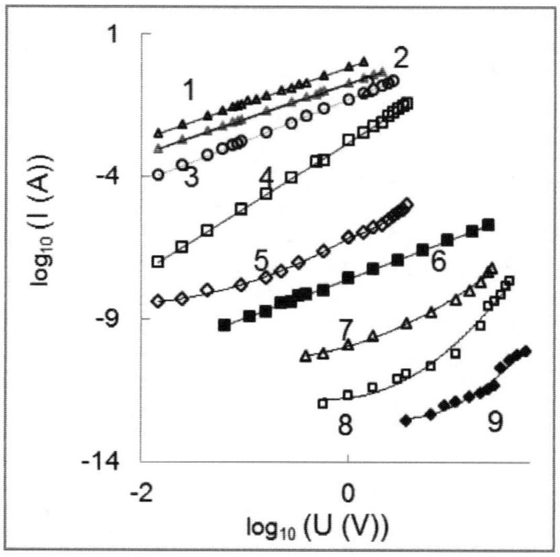

Figure 1. CVC of the films doped with boron and deposited on Si (1 and 2), Si substrate (3); CVC of the undoped PDF on Si (5) and Si substrates (4); CVC of the films separated from the substrate (6–9) for the surface (1, 3–9) and bulk conductivity (2) from the growth side (1, 5, 6, 8) and from the substrate side (7, 9) in the center (5–7) and on the edge of deposition (8, 9).

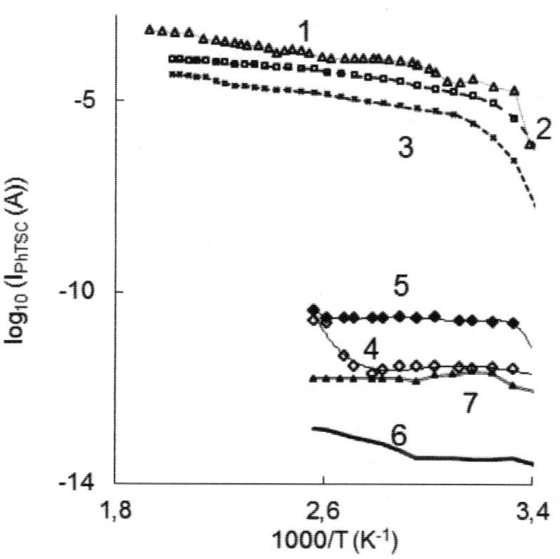

Figure 2. Temperature dependencies $I_{PhTSC}(T)$ in the dark (1–3, 4, 6) and under lighting (5, 7) in the doped PDF on Si in consecutive measurements (1, 2), in the substrate (3), in the undoped PDF on Si (4, 5) and in PDF separated from the substrate (6, 7).

coefficient α=1.85 of CVC possesses a lower conductivity σ=2·10^{-7} S and CVC higher nonlinearity (α=2–2.3, β=0.3–0.4) (figure 1, curves 4 and 5), as well as unstable photoconductivity [18]. The ratio between CVC and K(U) of the film and the substrate indicates the independence of the PDF and substrate properties that is determined by the decrease in σ by 2–4 orders from the layers from the growth side to the layers near the substrate, as for the most of the films separated from the substrates (figure 1, curves 6–9). In the film doped with boron on the ohmic Si substrate (α=1), CVC obeys almost linear law with α=1.05–1.1 for the bulk and surface conductivity, that reveals a weak influence of various heterogeneities, which are typical to the undoped films, on the conductivity (figure 1, curves 1–3). This conclusion is verified by a very low photosensitivity of the film K$\lesssim 10^{-4}$.

In the most low-conductivity undoped films with $\sigma \lesssim 10^{-7}$ S, n-type of σ and σ_{ph} dominates, as thermo and photostimulated currents $I_{PhTSC}(T)$ show [18]. Temperature dependences of σ, $\sigma_{ph}(T)$ within the interval T=300–720 K are determined by the thermostimulated electron exchange between shallow donor levels with the activation energy $\varepsilon_\sigma \lesssim$ 0.21 eV and the conduction band (CB) (figure 2 and 3) [18]. Donor domination in the electrotransfer is most likely caused by the effect of the nitrogen-containing centers formed in PDF involving residual impurity of nitrogen in the reactor. The values of ε_σ in the deposited PDF indicate a weak population of electrically active LS, which is in 10^3–10^6 times lower compared to single crystals and epitaxial PDF. The ε_σ activation levels which are typical to diamond do not occur, as in [2, 7–11, 14, 16, 20]. We cannot exclude the effect of the holes exchange between LS of the defects and the valence band (VB) on the CVC, σ(T) and K(U, hν) of the undoped films [18, 20]. It is verified by thermo- and photostimulated currents (figure 2, curves 4, 6 and 5, 7). The comparison of $I_{PhTSC}(T)$ for the films on the substrate and the films separated from it enables to eliminate the impact of the substrate on the thermoactivation of the holes (figure 2). In the doped film, σ(T) is determined by the thermoactivated hole exchange between VB and shallow acceptor levels with an energy of $\varepsilon=\varepsilon_v$+(0.013–0.022) eV ($\varepsilon_v$ is a top of VB) (figures 2 and 3). It is

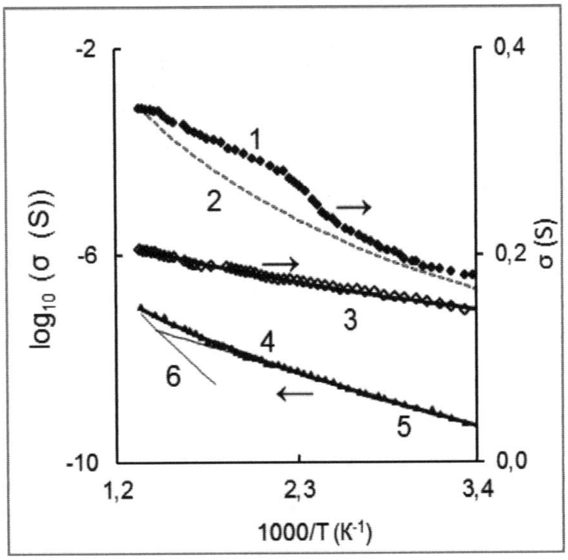

Figure 3. Temperature dependences of σ(T) of doped PDF on the Si substrate before (1), after annealing to 730 K (3) and undoped PDF on the Si substrate (4) and their approximation with the equations of hopping (2) (2, 3) and activation conductivity (1) (5, 6) at $T_0^{0.25}$=23.7 $K^{0.25}$, σ'_0=943 S (2); 16.01 $K^{0.25}$, 121 S (3); ε_σ=0.19 eV, σ_0=8.7·10^{-7} S (5), 0.6 eV, 9·10^{-4} S (6).

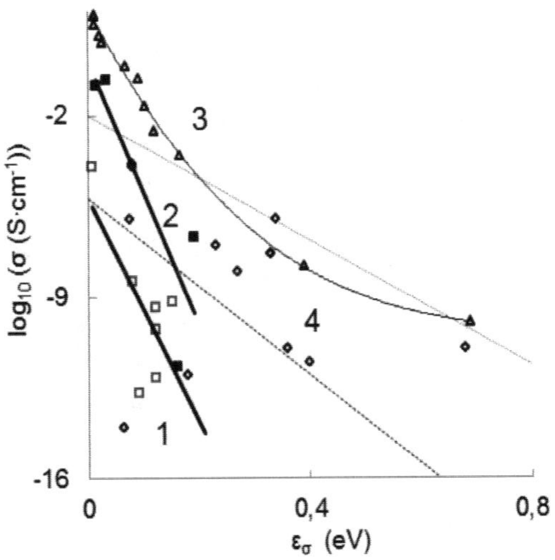

Figure 4. Interrelation between conductivity σ of the films and its activation energy ε_σ in the deposited films separated from the substrate (1) (□), in the films on the silicon substrate (2) (■) and in the other PDF reported by [16] (3) (Δ) and [2, 7, 10, 11, and 14] (band 4) (◇).

typical for PDF doping by boron impurity, as shown in [1, 13, 14, 16, 17]. The conclusion is verified by the near values of ε_σ for σ(T) and I_{PhTSC}(T) (figures 2 and 3).

The electrical parameters and the ratio between them indicate the effect on the hopping conductivity transfer that is verified by the approximation σ(T) by the Eq. (2) within the interval T=300–700 K (figure 3) is typical for PDF [7–11, 17, 18]. LS density calculated from the dependence σ(T) for the undoped PDF is $N(E_F)=10^{17}–10^{21}$ $eV^{-1}\cdot cm^{-3}$ [18]. The typical LS density for the other PDF is $N(E_F)=(4–8)\cdot 10^{19}$ [10, 11] and $10^{15}–10^{18}$ $eV^{-1}\cdot cm^{-3}$ [17]. The Fermi level is located in a lower part of BG close to the VB by analogy with the data in [1, 13, 17]. The value of the most probable jump distance calculated according to [10, 11, and 18] with the values $N(E_F)$ is R=1–10 nm. Boron doping enhances the approximation of σ(T) by Eq. (2), that indicates the increase in the effect of σ_p to the electrotransfer (figure 3). The density of LS distributed near E_F after doping is $(1–6)\cdot 10^{20}$ $eV^{-1}\cdot cm^{-3}$, and a jump distance R=1.4–2.1 nm.

The dependences σ(T) and parameters K, ε_σ, σ_0, T_0, σ'_0, $N(E_F)$ indicate the presence of the activation and hopping mechanisms, whose effect on the electrotransfer varies during the variation of deposition conditions (figures 3–6). In the undoped PDF on the Si substrate, σ(T) is approximated separately by the Eq. (1) or (2), but more accurately by the sum of two activation components $\sigma(T)=\sigma_{a1}(T)+\sigma_{a2}(T)$ or by the sum of the activation and hopping components $\sigma(T)=\sigma_p(T)+\sigma_{a2}(T)$ (figure 3). The decomposition of curves σ(T) for the sum of components confirms the impact on the properties of the deep levels with $\varepsilon_\sigma \gtrsim 0.6$ eV induced by the more stable defects. In the PDF doped with boron after heating up to 720 K, the nonequilibrium component of conductivity disappears due to the annihilation of unstable acceptor defects, LS density being increased from $1.2\cdot 10^{20}$ to $5.9\cdot 10^{20}$ $eV^{-1}\cdot cm^{-3}$, meanwhile a jump distance decreases from 2.1 to 1.4 nm (figure 3).

The correlation between the parameters of the activation and hopping components enable to additionally specify a predominant transfer mechanism. The interrelations between the photosensitivity K, conductivity σ, multiplier factor σ_0, and activation energy ε_σ indicate a significant

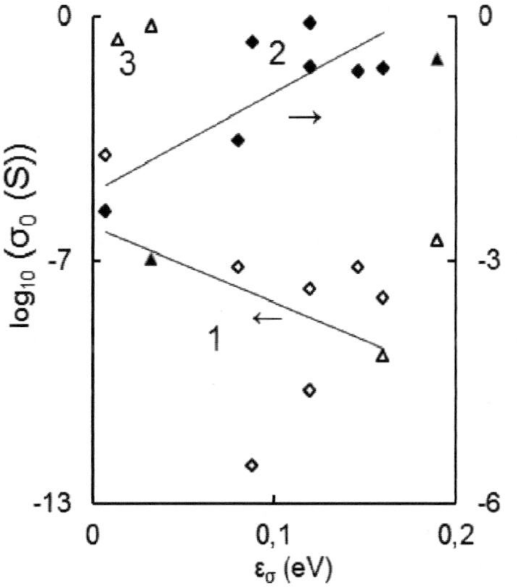

Figure 5. σ_0 vs. ε_σ (◇, Δ) (1, 3) and K vs. ε_σ (♦, ▲) (2) in the films separated from the substrate (◇,♦) and on the silicon substrate (Δ ▲).

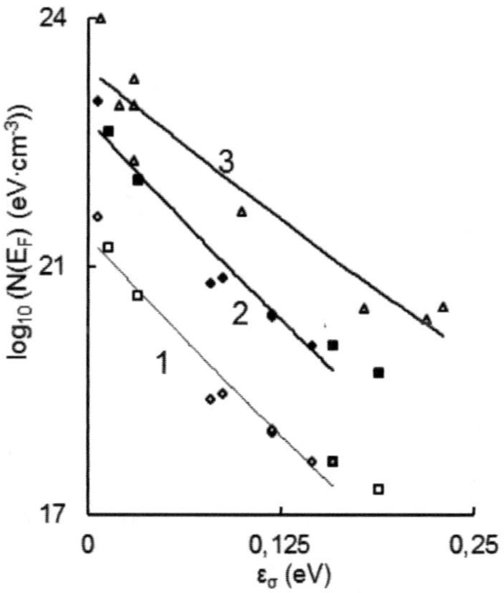

Figure 6. $N(E_F)$ vs. ε_σ in PDF separated from the substrate (◇, ♦) and on the Si substrate (□, ■) (1, 2). The data are given for UNCD (Δ) (3) [10]. Attenuation constant α=5.7·10^6 (1) and 2·10^7 cm^{-1} (2).

role of activation transport in PDF (figures 4 and 5). The role of the hoping component of the electrotransfer in PDF deposited from AGD is higher than in the other CVD films, as the dependencies $\sigma(\varepsilon_\sigma)$ show (figure 4). The correlation between σ and ε_σ for the AGD films on the Si substrate and the films separated from it is slightly different. The values σ and ε_σ boron-doped films are separated from those in the undoped and nitrogen-doped PDF (figure 4). In the boron-doped PDF deposited from AGD on the Si substrate, the values σ and ε_σ are located near these values (curve 2) that corresponds to the concentration of boron atoms in the films (0.3–3)·10^{20} cm^{-3} according to [16]. The multiplier factor σ_0 significantly exceeds its values for the undoped films due to the increase of the σ_p effect (figure 5, curve 1 and points 3). The correlation K(ε_σ) (figure 5, curve 2) indicates the predominance of the activation mechanism of photoconductivity [18]. Besides, this correlation shows that the content of photosensitive defects in the undoped PDF on the substrate is lower than in the films separated from it. In the doped films this transfer mechanism of the light-activated holes is lacking, as shown by the low photosensitivity $K \lesssim 10^{-4}$ and curves $I_{PhTSC}(T)$ (figure 2).

The correlation $N(E_F)(\varepsilon_\sigma)$ in PDF deposited by AGD and by the other methods is typical for noncrystalline semiconductors possessing a high concentration of defects (figure 6) [10, 11, 17, 18]. The effect of the substrate is insignificant. The data for the doped PDF form a total array with the undoped films. By analogy with the approach developed in [1, 8–11, and 13–17], we concluded in [18], that in the undoped PDF hopping transfer σ_p on the LS near E_F dominates over the activated σ_a within the interval T=300–500 K. The invert correlation in favour of the activation component involving LS defects in the region of the "tails" of the allowed bands occurs when T=500–700 K. During the film doping, the range of the σ_p component dominance is expended to T=300–700 K, as shown by the approximation σ(T) (figure 3) and the correlation $\sigma(\varepsilon_\sigma)$ (figure 4). When T>600 K, the effect of the potential barrier between crystallites is increased [18]. The correlation between the values

σ, ε_{σ}, and BG width, calculated form the absorption spectra agree with developed transfer models, the transfer being realized through the charge carrier exchange between BG and LS through the mobility gap [1 and 18].

4. Conclusion

The doping of the polycrystal diamond films with the impurity atoms of boron during the deposition from abnormal glow discharge enables to form the semiconductor layers with p-type of conduction on the silicon substrate that have as good properties as PDF obtained by the alternative methods. The bottom low-conducting PDF layers isolate the upper conducting layers from the influence of the substrate. The vacuum annealing of the films up to 720 K stabilize the electrical characteristics. The doping diminishes the degree of film material disordering by the growth defects compared to the undoped films that affects the electrical and photoelectrical characteristics and the conductivity mechanism. The activation component of the p–type conductivity is realized in the exchange of charge carriers between the VB and shallow acceptor levels with the activation energy of 0.013–0.022 eV. Doping increases the effect of the hopping mechanism of the conductivity involving the LS defects with a density of up to $(1–6)\cdot10^{20}$ $eV^{-1}\cdot cm^{-3}$ near the Fermi level located in the low half of the BG.

Acknowledgments
The work is performed within the megaproject of Tomsk Polytechnic University (VIU_IFVT_85_2014) “Materials for Extreme Conditions”

References

[1] Williams O A 2011 *Diamond and Relat. Mater.* **20** 621
[2] Polyakov V I, Rukovishnikov A I, Avdeeva L A, Kun'kova Z E, Varnin V P, Teremetskaya I G and Ralchenko V G 2006 *Diamond and Relat. Mater.* **15** 1972
[3] Linnik S A and Gaydaychuk A V 2013 *Diamond and Relat. Mater.* **32** 43
[4] Linnik S A and Gaydaychuk A V 2012 *Tech. Phys. Lett.* **38** 258
[5] Bevilacqua M and Jackman R B 2009 *Appl. Phys. Lett.* **95** 243501
[6] Ueda K, Kawamoto K, Soumiya T and Asano H 2013 *Diamond and Relat. Mater.* **38** 41
[7] Chiquito A J, Berengue O M, Diagonel E and Galzerani J C 2007 *J. Appl. Phys.* **101** 033714
[8] Kopylov P G, Lotonov A M, Apolonsky I A and Obraztzov A N 2009 *Vestn. of MSU.Ser.3(in russian)* 2 54
[9] Trucchi D M, Cappelli E, Conte G, Mattei G, Gramaccioni C and Ascarelli P 2005 *Diamond and Relat. Mater.* **14** 575
[10] Achatz P, Williams O A, Bruno P, Gruen D M, Garrido A and Stutzmann M 2006 *Phys. Rev.* **B 74** 155429
[11] Gan L, Bolker A, Saguy C, Kalish R, Tan D L, Tay B K, Gruen D and Bruno P 2009 *Diamond and Relat. Mater.* **18** 1118
[12] Conte G, Rossi MC, Spaziani F and Arcangeli R 2005 *Diamond and Relat. Mater.* **14** 570
[13] Hikavyy A, Clauws P, Maes J, Moshchalkov V V, Butler J E, Feygelson T, Williams O A, Daenen M and Haenen K 2006 *Phys. status solidi* **A 203** 3021
[14] Muret P and Saby Ch 2004 *Semicond. Sci. and Technol.* **19**1
[15] Hubik P, Mares J J, Kozak H and Kromka A 2012 *Diamond and Relat. Mater.* **34** 63
[16] Gajewski W, Achatz P, Williams O A, Haenen K, Bustarret E, Stutzmann M, and Garrido J A 2009 *Phys. Rev.* **B 79** 045206
[17] Nath S and Wilson J I B 1996 *Diamond and Relat. Mater.* **5** 65
[18] Konusov F V, Kabyshev A V, Linnik S A, Gaydaychuk A V and Remnev G E 2014 *J. of Phys.: Conf. ser.* **552** 012046
[19] Chaudhary A, Welch J O and Jackman R B 2010 *Appl. Phys. Lett.* **96** 242903
[20] Ri Sung-Gi, Takeuchi Daisuke, Kato Hiromitsu, Ogura Masahiko, Makino Toshiharu, Yamasaki Satoshi, Okushi H, Rezek B and Nebel Ch E 2005 *Appl. Phys. Lett.* **87** 262107

Journal of Physics: Conference Series **652** (2015) 012031 doi:10.1088/1742-6596/652/1/012031

Plasma treatment of heat-resistant materials

V A Vlasov[1,2], P V Kosmachev[1], N K Skripnikova[1], K A Bezukhov[1]
[1]Tomsk State University of Architecture and Building, Tomsk, 634003, Russia
[2]National Research Tomsk Polytechnic University, Tomsk, 634050, Russia

E-mail: pvkosm@gmail.com

Abstract. Refractory lining of thermal generating units is exposed to chemical, thermal, and mechanical attacks. The degree of fracture of heat-resistant materials depends on the chemical medium composition, the process temperature and the material porosity. As is known, a short-term exposure of the surface to low-temperature plasma (LTP) makes possible to create specific coatings that can improve the properties of workpieces. The aim of this work is to produce the protective coating on heat-resistant chamotte products using the LTP technique. Experiments have shown that plasma treatment of chamotte products modifies the surface, and a glass-ceramic coating enriched in mullite is formed providing the improvement of heat resistance. For increasing heat resistance of chamotte refractories, pastes comprising mixtures of Bacor, alumina oxide, and chamot were applied to their surfaces in different ratios. It is proved that the appropriate coating cannot be created if only one of heat-resistant components is used. The required coatings that can be used and recommended for practical applications are obtained only with the introduction of powder chamot. The paste composition of 50% chamot, 25% Bacor, and 25% alumina oxide exposed to plasma treatment, has demonstrated the most uniform surface fusion.

1. Introduction

The use of plasma technologies in different industries is being constantly widened due to a range of advantages over the traditional treatment of materials. Thanks to plasma generation units, it has become possible to apply the powerful plasma energy to surfaces that, in turn, intensifies the chemical reaction rate. It was proved that plasma treatment enhances the existing properties of materials and adds new ones. The description of the low-temperature plasma (LTP) technique used for the surface modification of diverse materials found in the literature, presents the results of obtaining protective and decorative properties of plasma-treated surface of materials [1-12].

The importance of heat-resistant materials for the industrial use is constantly grows. The quality of these materials mostly depends on the development of productions, especially, ferrous and nonferrous metallurgy, electrical power engineering, and chemical, gas, and glass industries. Heat-resistant materials can be used at various temperature conditions and with various media. Some of them contact with different types of melts, such as glass, sludge, metal and are subjected to both physical and chemical corrosion. Therefore, the density, strength, a certain chemical composition are the main requirements for heat-resistant materials. Others are used only in gaseous atmosphere, i.e. in less aggressive medium. Some materials are intended for a multiple alternating heating and cooling, and thus, must possess a high thermal resistance.

The material morphology (presence and nature of porosity) and the surface condition are also of great importance. The heat resistance of a porous material, especially when pores are big, open and interconnected, may turn to be several dozens of times lower than that of a dense and non-porous material. The proper grain composition and the increase of burning temperature of heat-resistant materials result in pore size reduction and increase of the material corrosion resistance. Non-uniformities on the surface and cracks inside the material facilitate its corrosion. Still, there are no heat-resistant materials equally combining all operating characteristics necessary for a failure-free operation under any conditions. Each type of material is characterized by its own intrinsic properties that define its efficient use and scope of applications. The material is considered to be in operation at the furnace firing-up after its building or repair. Herewith, a temperature differential occurs in the body of material causing the respective mechanical stresses. In case the heating rate exceeds the allowable one for the given material and conditions, the material starts to crack and even crash. Upon completion of the furnace firing, heat-resistant materials are still experience the effect of high temperatures.

The reconstruction and construction of glass-, copper-, steel-melting and blast furnaces as well as cement kilns require a great number of diverse heat-resistant materials (refractories). Most of refractories are small piece materials which are used for hand-laid masonry of different configuration. In order to industrialize and optimize the period of brick-laying and lining of thermal generating units, the production of large refractory blocks, panels and unmolded refractories is being expanded. As is known, the production of small piece refractory materials and lining therefrom is rather labour-intensive and difficult to mechanize. The increase of refractory and heat-insulating material resistance enables the increase of the service life, decrease of thermal generating unit downtime due to repair of refractory lining, and increase of their productivity. Hence, the increase of the refractory service life by at least two times will allow their significant saving.

Refractory lining of thermal generating units and furnaces is subjected to chemical attack of fuel combustion products, melts, and materials that are melt or burned in furnaces. A degree of fracture of heat-resistant materials depends on the chemical composition of the medium, process temperature, and porosity of the material itself.

This paper mainly focuses on the creation of protective coating for heat-resistant chamotte products using the LTP technique.

2. Experimental

Chamotte refractory products were used in this experiment. Surfaces of specimens produced from the given heat-resistant materials were treated by LTP at different process conditions. The LTP treatment was provided by the laboratory plasma generator installed at the Department of Applied Mechanics and Material Sciences of Tomsk State University of Architecture and Building, Tomsk, Russia. Figure 1 shows the schematic layout of this plasma generator.

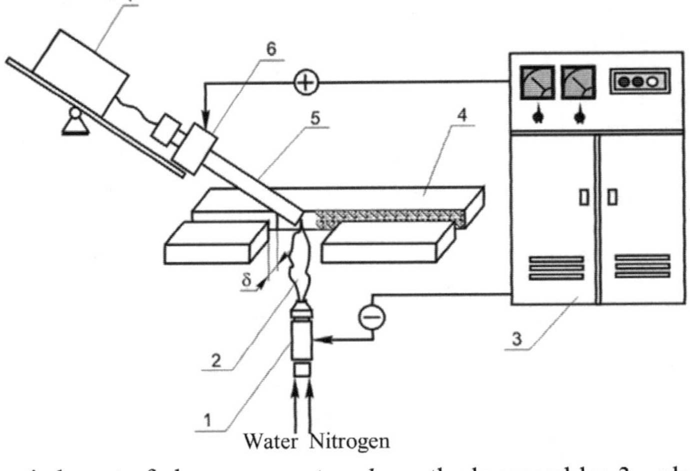

Figure 1. Schematic layout of plasma generator: *1* – cathode assembly; *2* – plasma arc; *3* – power source; *4* – workpieces; *5* – graphite anode; *6* – cooling unit; *7* – electric drive of electrode feeding.

At a short-term LTP exposure, the melt is formed on the surface of heat-resistant materials and then instantly frozen forming either a vitreous or a glass-ceramic coating. This process can be referred to plasma-chemical processes of non-equilibrium and local nature. Plasma-chemical processes differ from common thermal processes by the reaction behavior, namely, missing or unifying some of phases [13].

The surface of workpieces was melted at different rate and power of the plasma generator [14-15]. Table 1 presents the quality of the modified surface depending on fusion modes.

Table 1. The quality of coatings depending on fusion modes

Type of material	Fusion mode		Porosity (%)	Outcome
	Power (kW)	Fusion rate (m/s)		
Chamot	60	0.13	1.6	Underburning
	80	0.13	1.2	Overburning, roughness
	80	0.17	0.8	Good-quality glass-ceramic coating
	80	0.19	1.1	Underburning

As Table 1 shows, that in treating chamotte products, the good-quality, i.e. low porosity coating is formed at 0.17 m/s fusion rate and 80 kW power of plasma generator.

With a view to increase the heat resistance of chamotte products, heat-resistant pastes having 1-1,5 mm thickness comprising a mixture of Bacor, alumina oxide, and chamot are applied to surfaces in different ratios. The preliminary investigations of different refractory compositions applied to chamotte products show that no coating is created in case if only one heat-resistant component (Bacor, alumina oxide, zirconium dioxide, mullite or zircon) is used. Also, one more regularity is observed. Coatings that can be recommended for practical applications are obtained only after the incorporation of broken chamotte in the paste composition. This provides a better adhesion between the coating and the base. Seven compositions are suggested for further research, comprising two or three components one of which is powder chamot (Table 2).

Table 2. Properties of modified surface of chamotte products depending of paste composition

No	Paste composition (wt.%)	Coating thickness (mm)	Adhesive strength (MPa)	Modified surface properties
1	Chamot -50 Bacor-25 Alumina oxide-25	1.5-2	3.32	Grass-ceramic uniform coating with vitreous drops
2	Chamot -25 Bacor -50 Alumina oxide -25	1-2	2.28	Non-uniform, porous coating with base visible
3	Chamot -50 Bacor -50	1-2	3.02	Rough coating; glass phase coagulation
4	Chamot -75 Bacor -25	1-2	2.08	Non-uniform, porous coating with base visible
5	Chamot -50 Bacor -75	1-2	2.01	Non-uniform, porous coating with base visible; glass phase coagulation
6	Chamot -50 Alumina oxide -50	1-2	-	No coating formation
7	Chamot -25 Bacor -25 Alumina oxide -50	1-2	-	No coating formation

It is shown that the most uniform coating is obtained at fusion of 50% chamot, 25% Bacor, 25% alumina oxide paste composition. At a higher content of Bacor in the paste composition, the glass phase coagulation is observed and the base of the workpiece becomes visible, while at a higher content of alumina oxide, the coating formation is not achieved.

Research results show that phosphate-based pates are well-applied to the surface and adhere to it very strongly. However after the plasma treatment, coagulation of this coating is observed in the form of separate drops.

The strongest adhesion between the coating and the base is observed in the liquid glass-based coating. The experimental results show that the thickness of the paste to be applied must not be over 1 mm. Otherwise, it does not adhere to the base in the proper way, and the coating is cracked thereby decreasing the adhesive strength.

3. X-ray phase analysis

The structural and physicochemical transformations of the surface and contact layers were studied by the X-ray phase analysis. The analysis shows that chamotte product base is presented by high quartz, aluminum silicate $Al_2O_3{\cdot}SiO_2$ and mullite $3Al_2O_3{\cdot}2SiO_2$ (d = 0,539; 0,377; 0,339; 0,270; 0,254; 0,221 nm) (Fig. 2).

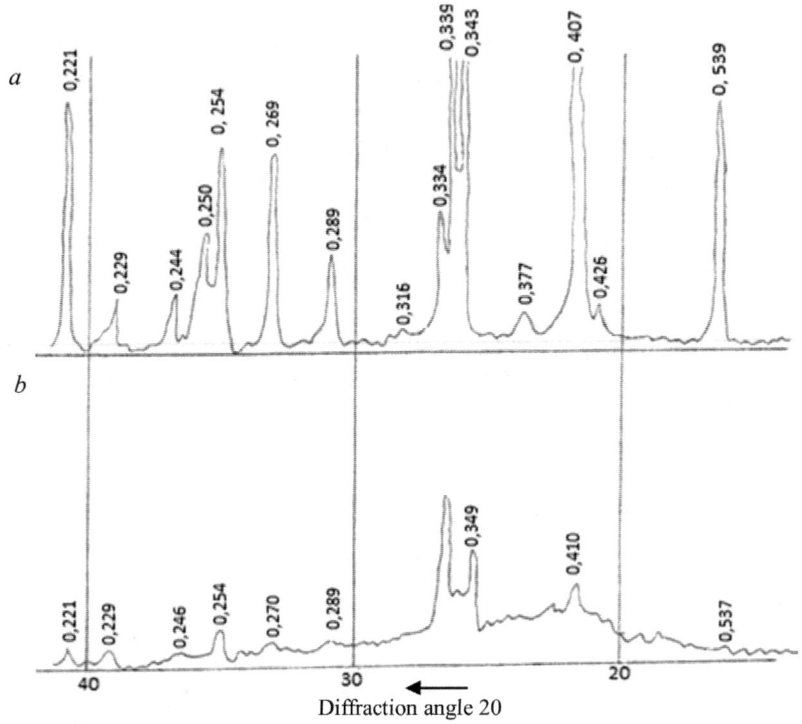

Figure 2. XRD pattern of chamot: *a* – chamot base material; *b* – fused coating

As a result of chamotte surface fusion as shown in Figure 2, a portion of the base material is transformed to the X-ray amorphous state. The decrease of mullite diffraction maximums indicates that these phases are transformed to the amorphous state, however, mullite has partially retained its crystallinity (d = 0.339; 0.349).

As for aluminum silicate $Al_2O_3{\cdot}SiO_2$ (d = 0.244; 0.289; 0.377 nm), the decrease of diffraction maximums indicates a transformation of this composition to a vitreous state. α-quartz (d = 0.229; 0.407; 0.426 nm) incorporated in the base material after fusion partially has transformed to the glass phase.

The X-ray phase analysis shows that after the fusion of chamot base material, a glass-ceramic coating enriched in mullite, is formed on its surface indicating that its heat-resistance will be higher than that of the base material. As is known, the heat-resistance of chamot is in the order of 1600 °C, while the melting temperature of mullite is over 1800 °C.

4. Conclusion

Experimental findings showed that the plasma treatment of heat-resistant specimens resulted in the formation of the glass-ceramic coating enriched in mullite that improves heat-resistance of the given products.

In order to increase heat-resistance of chamotte products, special paste compositions are required, namely comprising different ratios of Bacor, alumina oxide, and chamot. It was proved that the appropriate coating could not be created if only one of heat-resistant components is used. It was stated that paste composition of 50% chamot, 25% Bacor, and 25% alumina oxide produced the most uniform surface fusion.

Acknowledgement

The authors acknowledge the Ministry of Education and Science of the Russian Federation for the financial support (Project # 11.351.2014/K).

References

[1] An'shakov A S, Volokitin O G and Urbakh E K 2012 *Rus. Phys. J.* **2.** 5–7.
[2] An'shakov A S, Volokitin G G, Volokitin O G and Skripnikova N K 2014 *Elektrotekhnologicheskie ustanovki dlya plazmenno-termicheskoi obrabotki materialov* [Electrical engineering units for plasma treatment of materials]. Tomsk: TSUAB Publ.
[3] Shershnev A A, Gaenko I S, Savel'chikova I L, Men'shelova G G and Borovskii Yu F 1987 *Refractories* **2**.105.
[4] Fridment M G, Pershin V A, Gaenko I S and Karasev V P 1986 *Refractories* **6**. 121.
[5] Sun Y, Liang Q, Chi H, Zhang Y, Shi Y, Fang, D, Li F 2014 *Fib. Pol.***15** (1). 1-7.
[6] Ye Y, Xu Y, Liu Y, Tu W, Liao J 2013 *Pol.Mat. Sc. Eng.* **29** (11). 187-190.
[7] Kusano Y 2014 *J. Adh.* **90** (9). 755-777.
[8] Yoshida S, Hagiwara K, Hasebe T, Hotta A 2013 *Surf. Coat. Tech.* **233**. 99-107.
[9] Zille A, Oliveira F R, Souto P A P 2015 *Pl.Proc.Pol.* **12** (2). 98-131.
[10] Peltekis N, Kumar S, McEvoy N, Lee K, Weidlich A, Duesberg G S 2012 *Carbon* **50** (2). 395-403.
[11] Vesel A, Mozetic M 2012 *Vacuum* **86** (6). 634-637.
[12] Yun J W 2013 *Plasm. Sc.Tech.***15** (6). 521-527.
[13] Volokitin G G, Skripnikova N K, Shilyaev A M and Petrochenko V V 2005 *Plazmennye tekhnologii v stroitel'stve* [Plasma technologies in construction]. Tomsk: TSUAB Publ.
[14] Kosmachev P V 2014 *Vest. TSUAB* **6**.141–147.
[15] Skripnikova N K, Kosmachev P V and Bezukhov K A 2015 Modifitsirovanie poverkhnosti shamotnykh ogneuporov nizkotemperaturnoi plazmoi [Surface modification of fireclay refractory by low-temperature plasma] Novosibirsk: Constr Mat. – 4S: Coll Papers. 211–214.

Antiemissive coatings

N Z Vetrov[1], D K Kostrin[2], A A Lisenkov[3] and M S Popova[2]
[1] JSC «S.E.D.-SPb», 199178, Morisa Teresa pr. 64, Saint-Petersburg, Russia
[2] Saint-Petersburg State Electrotechnical University, 197376, prof. Popova str. 5, Saint-Petersburg, Russia
[3] Institute of Problems of Mechanical Engineering, Russian Academy of Sciences, 199178, Bolshoy pr. V.O. 61, Saint-Petersburg, Russia

E-mail: lisran@yandex.ru

Abstract. High demands, the implementation of which determines the quality of the products in whole, are presented to the grid electrodes of the generating lamps as the elements that determine parameters and characteristics of electronic devices. The grid is a complex structure with a certain degree of transparency. The choice of material of the grid electrodes is determined by the conditions of their operation and design features of the device. Grid nodes are located in the area of the electron beam and during operation of a lamp they should have high heat dissipation to maintain their geometric characteristics and possess a low coefficient of secondary emission.

1. Introduction

The application of circuitry decisions in radio-transmitting devices allows to increase overall performance of the powerful generating lamps. However, the use of the principles of pulse-width modulation in modulator and radio-frequency cascades caused the essential revision of the requirements to their active elements – powerful generating and modulator lamps. Obvious need in increasing the intensifying parameters of powerful generating lamps (steepness, stability, frequency range and bandwidth) led to transition from the triode to tetrode devices, as more adapted for the specified requirements. The shielding grid electrodes preserve considerable electronic loadings in the conditions of pulse amplification. To achieve their maximum efficiency a problem of thermal stability is needed to be solved.

The grid electrodes of powerful generating lamps are made from molybdenum, tungsten, or a tungsten alloy with rhenium addition. Oxidic or thoriated and carbonized cathodes (figure 1) are used as a source of electrons. The electronic structures of the corresponding lamps providing effective switching of high currents (hundreds of amperes) at the voltage of 20…45 kV must have thin and precise structure working in the conditions of high temperatures (up to 2000 K) and carry out functional tasks of the opposite plan – to provide the highest emission ability of the cathode in hundreds of amperes and the minimum emission ability of both grids at the level of no more than units of milliampere. For powerful generating lamps the norm of undesirable thermionic emission is about 10^{-5} A/cm^2.

Constructive solutions that reduce the temperature of the grid electrodes by increasing their working surface or distance between the grid and the cathode are generally not acceptable because they decrease such technical characteristic of the lamps as steepness; also they increase interelectrode capacities, dimensions and weight. The problem of improving the properties of antiemissive coatings is researched in many countries, but the main decisions for powerful generating lamps with filamentary thoriated and carbonized cathode only result in using multilayer coatings with end use of platinum (Pt – order number 78; atomic weight 195.08; configuration of the outer electron shells $5d^9 6s^1$; face-centered structure; melting point 2041.4 K; boiling point 4098 K; work function 5.3 eV; effective atomic radius 1.387 Å; ionization potentials 9.0, 18.56 and 23.6 eV). Platinum as an element of the eighth group in the chemical compounds is able to exhibit multiple valences: 0, +2...+6, +8.

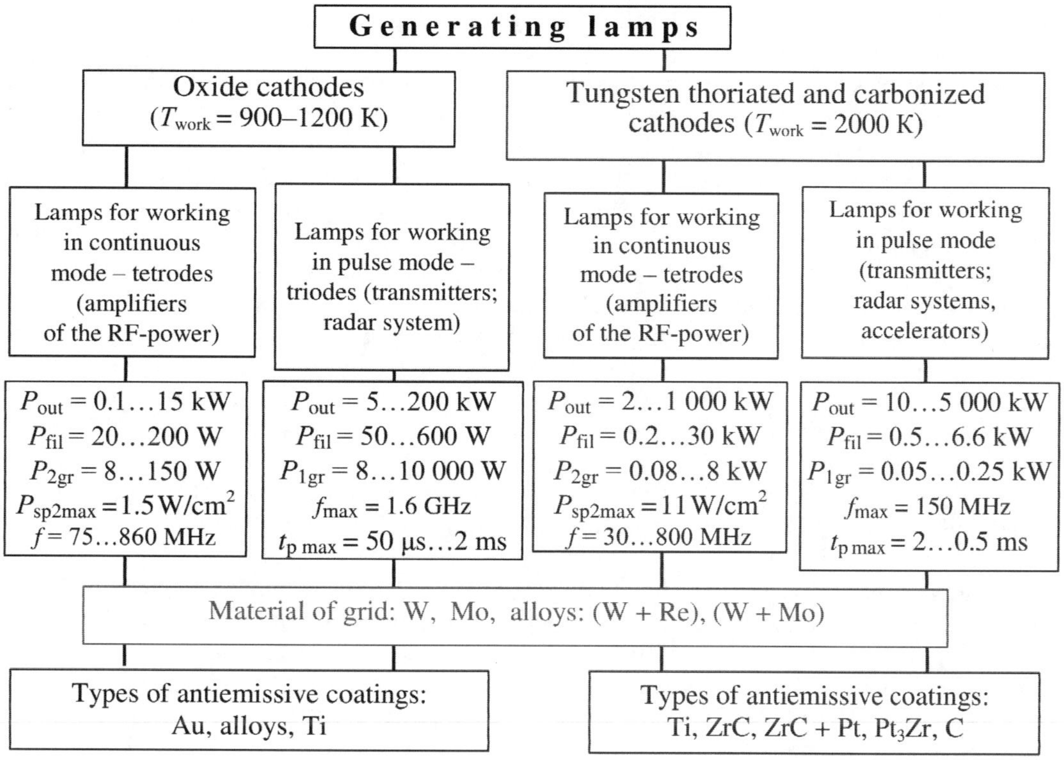

Figure 1. Operating parameters and materials of the grid electrodes.

Platinum and intermetallic compounds of platinum group metals make interest while creating high-temperature alloys and a compound based on platinum and zirconium (Zr) has proven its effectiveness when used as antiemissive coating of the grid electrodes of powerful generating lamps [1]. In such devices antiemissive coating consists of a layer of platinum and a layer of zirconium carbide (ZrC), acting as a diffusion barrier between the platinum and the core grid. To reduce the thermionic emission, the grid electrode is coated with a metal of the VIII group of the periodic system, in particular platinum. To reduce the diffusion of platinum into the core grid between the base metal and platinum an intermediate layer consisting of compounds Zr–Pt or Ti–Pt is applied.

2. Technology of receiving antiemissive coatings

The technology of obtaining antiemissive coatings consists of the following technological operations: by the method of cataphoresis a powder intermetallic compound Zr–Pt or Ti–Pt with a thickness of 5…10 μm is deposited onto the grid, after which it is baked for 20 minutes at a temperature of 1773…1873 K in an inert atmosphere or in vacuum; the next three microns layer of Pt is formed by electrolysis, after which the grid is again annealed in vacuum. By adjusting the particle size of the powder of the intermetallic compounds the surface roughness can be accurately determined and thereby secondary emission of grid electrode can be purposefully influenced.

Journal of Physics: Conference Series **652** (2015) 012032 doi:10.1088/1742-6596/652/1/012032

The grid electrodes of generating lamps with a big capacity of thermal load made from Mo, W or Ta have the antiemissive coating, including the intermediate layer formed by metalloid and metal of which the grid is made, and a platinum blanket. Preferable thickness of an intermediate layer with a small roughness makes about 15 % of width of a metal wire. Thickness of a superficial metal layer of platinum reaches 30 μm.

In [2] possibility of receiving an antiemissive intermetallic coating of Pt_3Zr is shown. In this technological process for deposition of carbide of zirconium coating the plasma arc method is used, which is carried out by a plasmatron with self-established arc length in argon with a pressure of 10 kPa.

In recent time improving the parameters of the devices is only possible by the use of more efficient technological processes and by achieving new performance properties of the used materials by creating complex systems of multilayered structures of alloys or chemical compounds. The most perspective method for surface treatment of materials is the method of using the fluxes of metallic plasma generated by vacuum arc discharge with integrally cold cathode. This type of technological equipment provides high efficiency when receiving the ionized and high-speed flow of substance, opens the possibility to control technological process of deposition from plasma of various pure metals, implement plasmachemical synthesis of simple and complex compounds, reproduce alloys and obtain complex combined coatings. The technological process in vacuum ensures the purity of the obtained coatings and high adhesion on materials with various physical and chemical parameters.

The formation of intermetallic compound Pt–Zr occurs at high temperature in vacuum on a refractory substrate (Mo, W) in the presence of a stable layer structure of ZrC in the existence of platinum. Unlike the solid solutions intermetallic compounds usually have a complex crystalline structure different from the structures of initial materials and properties unlike to the original components. The composition of the intermetallic compounds does not always obey the rules of valency and therefore any formed compound is characterized by its crystal structure and its individual properties.

For modifying the phase structure of the processed surface and formation of a ZrC coating vacuum arc plasma sources of coaxial and extended designs were used. The zirconium carbide coating (with thickness 5…8 μm) was synthesized in a flux of metal plasma in the environment of carbon-containing gas (C_6H_6). Platinum with thickness of 10 μm was deposited in an electrolytic solution containing hexachloroplatinic acid ($H_2(PtCl_6)\cdot 6H_2O$), disodium hydrogen phosphate (Na_2HPO_4) and diammonium hydrogen phosphate ($(NH_4)_2HPO_4$) at a current density of 0.015…0.045 A/cm^2.

Chemical heat treatment was carried out in vacuum at a temperature of 1300…1700 K. In the created system reactions of interaction proceed in a certain functional order. As a result of the high-temperature diffusive processes, defining kinetics of change of materials properties and nature of phase transformations, and also exchange chemical reactions, occurs an interaction of elements forming intermetallic connection

$$\mathrm{ZrC} + \mathrm{Pt} \xrightarrow{T} (\mathrm{Zr} + \mathrm{C}) + \mathrm{Pt} \xrightarrow{T} \mathrm{Pt}_x\mathrm{Zr}_y + \mathrm{C}\,.$$

The composition of the formed intermetallic compound depends on the quality and thickness of the formed layer of zirconium carbide, the thickness of the layer of platinum, the surface properties of the substrate (the presence of a transition layer and barrier layer), as well as on time and temperature of vacuum annealing.

3. Characteristics of intermetallid Pt–Zr coatings

The X-ray phase analysis showed that at the chosen temperature of formation of an intermetallid Pt–Zr varying the time of vacuum annealing allows to change structure of the formed coating. It was found that each couple of elements (in certain limits of structure and temperature) is capable to form some types of nonstoichiometric compounds, in each of which in the field of homogeneity a number of transitional phases is observed. Structure of an intermetallid with 15…20 at. % Zr is characteristic for Pt_4Zr compound; with 20…30 at. % – Zr–Pt_3Zr; with 60…65 at. % – Zr–Pt_3Zr_5; compound PtZr in high-temperature (β-PtZr, structure of CsCl) and low-temperature (α-PtZr, structure of CrB) phases contains 47…54 at. % Zr. At temperatures of annealing of 1400…1600 K the main phase as a part of a coating is intermetallic connection Pt_3Zr which in the range of working temperatures is capable to chemically connect thorium (Th).

In a crystal lattice of a metallid Pt_3Zr atoms of metals hold strictly certain positions, as if creating some sublattices inserted one into another. In these sublattices there is a significant amount of unoccupied sites that provides creation of connections with high valency. Intermetallic compounds based on platinum retain stable properties in a narrow temperature interval, which is associated with high polarizing effect of platinum and effect developing when heating its compounds that results in destruction of chemical bonds and the restoration of the metallic state. This leads to the gradual decomposition of Pt_3Zr followed by the formation of compounds $Pt_{11}Zr_9$ and Pt_9Zr_5. Thus the processes of counter diffusion which are characterized by different coefficients of diffusion of atoms (atoms of free platinum in a substrate and substrate material in a coating) migrating towards each other start proceeding actively (Kirkendall's effect). This effect is caused by the chaotic thermal movement of atoms and existence of the gradient of concentration directed towards decrease of this concentration.

Due to the high diffusion mobility of the components the diffusion zone reaches significant proportions: molybdenum is fixed throughout the depth of the formed coating Pt_3Zr; and the diffusion of platinum, passing along the edges of the crystals, is fixed at a depth of 20 μm or more (figure 2).

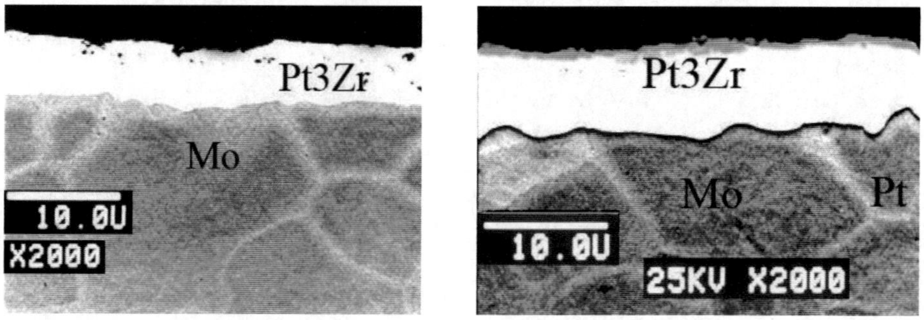

Figure 2. Metallographic sections of an intermetallic coating Pt_3Zr on a molybdenic substrate with traces of diffusion of platinum in molybdenum (scale: 10.0U = 10 μm).

Diffusion along the grain boundaries occurs many times (sometimes by several orders of magnitude) more intensively than in the bulk of grains due to the lower activation energy of the diffusion. Observation of such large grains of molybdenum of variable size indicates that sample is in the temperature region above the recrystallization temperature. Grain growth occurs by the addition of atoms from the surrounding deformed volume to the crystal lattice of the growing grain. Because of the relatively small differences in the sizes of atoms of molybdenum and platinum (Mo – 1.45, Pt – 1.39 Å) in the diffusion region, in a considerable range of concentrations the unlimited formation of solid solutions and such metal compounds as MoPt and Mo_2Pt is typical.

With increasing temperature and time of operation the factor of dissolution diffusion of the intermetallic coating into the substrate material becomes dominant. Increase of the initial thickness of the coating is not able to stop a counter-diffusion of the substrate material into the coating. For inhibition of these processes it is necessary to use an additional barrier layer. When choosing a solution it was assumed that the rate of diffusion in multilayer coatings is significantly lower than in single layer, so any phase boundary represents the energy barrier for the diffusing atoms. Thus the rate of diffusion will be lower the more densely structure is packed and the higher is the value of the bond energy between atoms. Previously it was shown that carbides of transition metals are able to perform the barrier layer function.

4. Deposition of a carbide layer

With regard to the mesh material of which grid is made, the following technological processes of carbidization were used: for items made of pure molybdenum and tungsten or a combination of molybdenum and tungsten direct carbidization in the atmosphere of carbon-containing gas was used; for items, in manufacture of which niobium was used, with the aim of reducing the time of interaction of niobium with hydrogen, was used carbidization in vacuum with a partial supply of carbon-containing gas into the working volume. Direct carbidization of molybdenum and tungsten was carried out in a mixture of the hydrogen stream and carbon-containing gas on the technological flow-type installation. To exclude the possibility of oxidation of the substrates and their decarbidization during the heat treatment, the hydrogen was carefully purified from oxygen and water vapor. During annealing of the substrate at a temperature of 1500 K and above in the region adjacent to its surface, hydrocarbon is thermally decomposed with the formation of atomic carbon, that reacts and forms carbide compounds [3]; gaseous components are removed from the chamber and neutralized ($Me + C_nH_n \rightarrow MeC + H_2\uparrow$). The optimum condition for the formation of carbide phases and avoiding the formation on the surface and in the volume of liquate free carbon is to achieve a balance between the speed of the process of formation of atomic carbon and its rate of diffusion into the material.

Compounds of carbon with hydrogen differ in the nature of bonds, number of atoms in molecules and molecular composition. With increasing number of carbon atoms in the homologous series of hydrocarbons their melting and boiling points usually increase. Thus, the composition of the feed gas determines the yield of carbon from the compound and speed of reaction. As a carbon-containing gas, depending on the chosen method of carbidization, natural gas or benzene was used. To obtain stable results in the process of carbidization it is necessary to provide a constant feed rate of hydrocarbon into the working volume.

When determining the thickness of the carbide layer the fact that one of the characteristics of the resulting carbides is emerging fragility of the material increasing with the growth of the carbide layer was taken into account. The thickness of the carbide layer was estimated by measuring the electrical resistance before and after the process of carbidization, and also by studying metallographic sections [4]. As a result of carbidization the resistance of a material changes, which allows to assess the degree of carbidization γ with a method of measuring the electrical resistance of the metal before (R_{Me}) and after (R_{MeC}) carbidization: $\gamma = [\rho_{MeC} / (\rho_{MeC} - \rho_{Me})][\gamma' / (1+\gamma')]$, where ρ_{MeC} and ρ_{Me} are the electrical resistivity of the carbide and pure metal respectively; $\gamma' = (R_{MeC} - R_{Me}) / R_{Me}$.

The advantage of using vacuum arc plasma sources is the ability to consistently obtain in a single technological cycle the carbide substrate material (MoC, WC) and zirconium carbide [5]. In this case the formation of the coating is carried out by deposition of ions with high kinetic energy. Changing the value of the accelerating potential U allows to adjust the energy of ions and hence to control the flow of the process of coating formation. Preliminary condensation of sputtered cathode material (Mo, W) to the edge of related metal helps to ensure a structurally oriented coating that in a subsequent reaction of diffusion reduces interfacial tension caused by differences in the crystal lattices and improves the strength of the adhesive connection.

Because of the peculiarities of the electronic structure and complex covalent-ionic-metallic nature of the interatomic interactions carbide phase diagram of systems of Mo and W have several structural modifications which are stable in different temperature and concentration intervals [1]. For the selected conditions of carbidization layers were obtained, consisting of the phases of molybdenum carbide – MoC and the tungsten carbide formed from the inner layer of semicarbide W_2C and surface monocarbide WC layer with hexagonal structure. With increasing time of carbidization thickness of W_2C layer increased, while the thickness of the surface WC layer has changed slightly.

For coating deposition on a grid with height more than 150 mm the plasma source with extended design with length of 450 mm was used, operating in a pulsed mode and generating a directional tape flux. Developed plasma sources provide a lower operating temperature of the cathode, thereby achieving a reduction in the number of droplets generated into the plasma flux and increase in the quality of the coating. For molybdenum grids not exceeding a height of 150 mm vacuum arc plasma sources of coaxial structure with the diameter of the cathode 100 mm were used [6]. The presence in the technological installation of two evaporators with molybdenum and zirconium cathodes allows conducting a consistent deposition onto the grid of a sublayer of molybdenum carbide and then a coating of zirconium carbide.

For the synthesis of carbide in vacuum arc plasma source a wide range of gases was used – from methane (CH_4) to cyclohexane (C_6H_{12}). Experiments have shown that with increasing relative molecular weight of hydrocarbon molecules the yield of carbon increases. Aromatic compounds give a higher percentage yield of carbon compared to the naphthenic molecules (cyclohexane C_6H_{12}) or paraffin (*n*-heptane). The most effective yield of carbon is achieved by using methane, which has the lowest relative molecular mass of these compounds, while the hydrocarbons of high molecular weight have an increased output of carbon black [7]. During experiments Pt_3Zr coating obtained on a refractory substrate (Mo) with a barrier layer of molybdenum carbide MoC (system: $Me_{subst}(Mo,W) - Me_{subst}C - ZrC - Pt$), was subjected to heat treatment and the changes in the composition of the coating were monitored [8].

5. Results and discussion

The phase composition of coatings was studied by X-ray diffraction analysis. Studies were conducted on X-ray device DRON-2 in the $Cu_{K\alpha}$ radiation at a voltage of 50 kV and using the four slotted nickel monochromator. The elemental composition of the coatings was determined by electron probe microanalysis, based on the comparison of the characteristic X-ray spectrums of the test sample with standards of known composition. The integral sensitivity of the method is approximately 0.5 weight %.

On the obtained diffraction patterns all the lines specific to the compound, but of lesser intensity compared to the original were recorded, while no other connections or loose components were observed. The plot of one of the diffraction patterns of the Pt_3Zr coating has the following registered lines: (201) – 2.3622 (ASTM-2.3601); (004) – 2.3076 (ASTM-2.3061); (202) – 2.1616 (ASTM-2.1616); (203) – 1.9125 (ASTM-1.9120); (205) – 1.4741 (ASTM-1.4719); (206) – 1.3026 (ASTM-1.3009).

Analysis of metallographic sections of the coatings showed the presence in all of the studied samples of a thin edge carbide layer of the substrate material with a thickness of about 1 μm and in some cases it was observed the presence of characteristic contrasting areas saturated with carbon, also reproduced on the distribution profiles of the elements. For comparison figure 3 presents the distribution of molybdenum and platinum on samples with a barrier layer (figure 3, a) and on similar samples, but without carbide layer of the substrate material (figure 3, b).

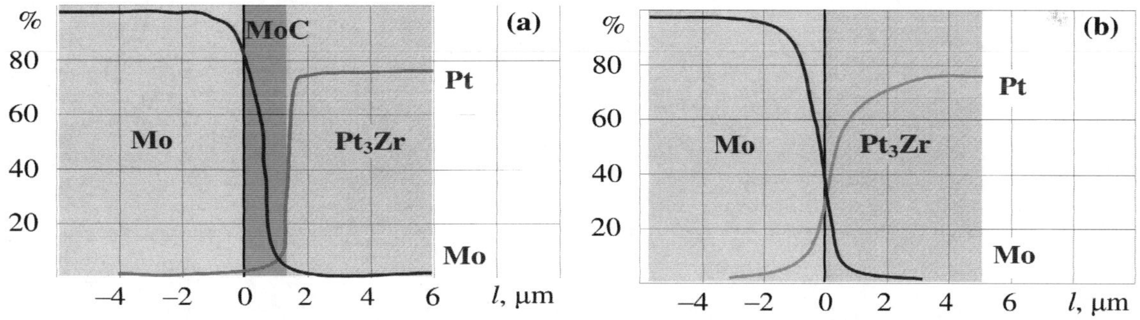

Figure 3. Distribution of elements across the depth of the metallographic section of the Pt_3Zr coating on a molybdenum substrate with (a) and without (b) the underlayer of the molybdenum carbide after 75 h of operation.

In the presence of the barrier layer of the carbide of substrate material on the phase boundary it is observed an abrupt drop in the concentration of diffusing metals in the transition from one layer to another (figure 3, a). In molybdenum, base metals included in the composition of antiemissive coating were not observed and molybdenum was not found in the coating.

Thus, the proposed technology allowed obtaining a multilayer antiemissive coating containing as the underlayer carbide of refractory material, which creates favorable conditions for obtaining a high quality intermetallic compound Pt_3Zr and prevents the counter flow diffusion processes; and improves the lifetime performance properties of antiemissive grid coatings in powerful generating lamps.

References

[1] Barchenko V T, Bystrov Yu A, Lisenkov A A, Prilutsky V S and Shanurenko A K 2009 *Modern technological processes in production of powerful generating lamps* (Saint-Petersburg: Saint-Petersburg State Electrotechnical University Press)

[2] Vildgrube V G, Tserpitsky B D, Sharonov V N and Shatalov S M 1989 *Electronic equipment. Ser.: Electrovacuum and gas-discharge devices.* **125** 43–51

[3] Prilutsky V S 2001 *Tungsten thoriated and carbonized cathode* (Moscow: Ore and metals)

[4] Bystrov Yu A, Vetrov N Z and Lisenkov A A 2010 *Technical Physics Letters* **6** 570–573

[5] Bystrov Yu A, Vetrov N Z and Lisenkov A A 2008 *Technical Physics Letters* **9** 734–736

[6] Bystrov Yu A, Vetrov N Z and Lisenkov A A 2013 *Technical Physics Letters* **10** 914–916

[7] Bystrov Yu A, Vetrov N Z and Lisenkov A A 2011 *Technical Physics Letters* **8** 707–709

[8] Bystrov Yu A, Vetrov N Z and Lisenkov A A 2014 *Technical Physics Letters* **12** 1126–1129

Journal of Physics: Conference Series **652** (2015) 012033 doi:10.1088/1742-6596/652/1/012033

The optical emission spectroscopy of pulsed and pulse-periodic discharges initiated with runaway electrons

M Lomaev[1,2], **D Sorokin**[1] **and V Tarasenko**[1,2]

[1]Institute of High Current Electronics, Akademichesky Ave. 2/3, 634055 Tomsk, Russia
[2]National Research Tomsk State University, Lenina Ave. 36, 634050 Tomsk, Russia

Lomaev@loi.hcei.tsc.ru

Abstract. We report on the results of measurements of an electron *Te* and a gas *Tg* temperatures as well as a reduced electric field strength *E/N* in the plasma of a high-voltage nanosecond discharge initiated with runaway electrons in a gap with a strongly nonuniform electric field distribution. The foregoing plasma parameters were determined with optical emission spectroscopy techniques. The possibility of using the method for determining *Te* and *E/N* in thermodynamically nonequilibrium plasma, which is based on a determination of a ratio of a peak intensities of the ionic (λ =391.4 nm) and molecular N_2 (λ= 394 nm) nitrogen bands, is proved. To measure a gas temperature the optical emission spectroscopy technique based on the measurement of a relative radiation intensity of rotation structure of electronic-vibrational molecular transitions was used, as well.

1. Introduction

Currently a lot of attention is given to study diffuse pulse and pulse-periodic discharges in dense gases preionized with runaway electrons (REP DD) under conditions of an inhomogeneous electric field distribution [1-5]. The main feature of such discharges is a generation in a gap of runaway electrons and X-ray, affected on a breakdown. These processes was shown to provide the formation of diffuse discharges at the excitation of gas medium by high-voltage nanosecond pulses of both polarities.

Determination of the main parameters of REP DD plasma is of both scientific and practical interest. The knowledge of the density *Ne* and electron temperature *Te*, gas temperature *Tg*, and reduced electric field strength *E/N*, where *E* is the electric field strength and *N* is the particle density, is important for general plasma characterization and verification of theoretical models of the REP DD plasma. Reliable data on these parameters can be used for calculating the rates of different plasma chemical processes and estimating the amplitude-time, spectral, and energy characteristics of the REP DD plasma light emission.

In practice, diffuse discharges at atmospheric gas pressure can be used for generation of chemically active plasma. Such plasma is in demand, for example, in surface modification of solids, surface cleaning and sterilization, decomposition of organic compounds, and waste processing. The data on the plasma parameters are also important for developing other modern technologies based on homo- and heterophase plasma chemical processes.

At present, it is known about several publications concerning the determination of REP DD plasma parameters [6-10]. The results of measurements of *Ne* in the cathode spot by the method of Stark

Journal of Physics: Conference Series **652** (2015) 012033 doi:10.1088/1742-6596/652/1/012033

broadening of spectral lines were reported in [6]; this method was used in [7, 8] for measuring the values of *Ne* in the plasma of the diffusive discharge in helium, as well. It should be noted that the values of *Ne* were measured in [7] in various regions of the discharge volume, the dependence of *Ne* on a gas pressure and its dynamics were determined. In addition, the value of *Te* in the plasma of the atmospheric pressure discharge was estimated. The method of Stark broadening of spectral lines was used to *Ne* measurement in hydrogen [9] and nitrogen [10].

It should be noted that the REP DD plasma is thermodynamically nonequilibrium. This excludes the possibility of using the methods for *Te* measurement, which are applicable where local thermal equilibrium or partial local thermal equilibrium conditions are fulfilled [11]. In nitrogen and nitrogen mixtures, in particular in air, *Te* and reduced electric field strength *E/N* in the REP DD plasma can be determined by means of the procedure based on the radiative-collisional plasma model [12, 13]. Its physical substantiation is that the ratio $R_{391/394}$ of peak radiation intensities of the N_2^+ ion (transition $B^2\Sigma_g^+$ (v = 0) $\rightarrow X^2\Sigma_g^+$ (v = 0), λ = 391.4 nm) and N_2 molecule (transition $C^3\Pi_u$ (v = 2) $\rightarrow B^3\Pi_g$ (v = 5),
λ = 394.3 nm) bands is a function of electron temperature $R_{391/394}$ = f(*Te*) or reduced electric field strength $R_{391/394}$ = f(*E/N*). The values of *Te*, *E/N* in the discharge plasma can be determined from the ratio of peak radiation intensities in bands of the first negative (1^-) ($B^2\Sigma_g^+ \rightarrow X^2\Sigma_g^+$) and second positive ($2^+$) ($C^3\Pi_u \rightarrow B^3\Pi_g$) nitrogen systems. The procedure is applicable providing fulfillment of the following conditions:

(I) Maxwellian electron velocity distribution;

(II) Excitation of the upper states N_2^+ $B^2\Sigma_g^+$ and N_2 $C^3\Pi_u$ is mainly by direct electron impact from the ground state of nitrogen molecule.

The optical emission spectroscopy technique can be used to measure the gas temperature *Tg* in plasma of REP DD in nitrogen, as well. It is based on a measurement of a relative radiation intensity of rotation structure of electronic-vibrational molecular transitions [11, 14].

This study aims at determining the values of *Te*, *Tg* and *E/N* in the plasma of REP DD operated in atmospheric-pressure nitrogen in a pulsed and pulse-periodic (2 kHz) modes.

2. Experimental setup and measurement

In the experiments, two setups were used; their block diagrams are shown in figure 1. Both setups comprised a system for pump-out and supply of gases. On setup No. 1 (figure 1*a*), the integral values of plasma parameters within a pulse were measured. The peaks radiation intensities of the N_2^+ ion (transition $B^2\Sigma_g^+$ (v = 0) $\rightarrow X^2\Sigma_g^+$ (v = 0), λ = 391.4 nm) and N_2 molecule (transition $C^3\Pi_u$ (v = 2) $\rightarrow$ $B^3\Pi_g$ (v = 5), λ = 394.3 nm) bands were recorded with a PI-MAX 2 CCD camera (Princeton Instruments) installed at the exit of a MDR-23 monochromator. In the plane of its rectangular entrance slit, an image of the discharge plasma was formed with a lens, while the monochromator exit slit was demounted. On the matrix of the CCD camera in the exit slit plane, an image of the entrance slit was formed in monochromatic components. One of the directions (along the matrix row) coincided with the dispersion direction. The second direction (vertical along the column) coincided with the direction of the entrance slit, which made possible determination of peaks radiation intensities in sites corresponding to different zones along the vertical line. All the forgoing along with possibility to horizontally shift the image of the discharge plasma with respect to the entrance slit allowed spatial-resolved measurements of plasma parameters in two directions.

The photodetector used on setup No. 2 (figure 1*b*) was an EMI-9781B photomultiplier mounted on a VM-502 monochromator (Acton Research Corp.). The time behaviour of the peaks wavelength intensity of the N_2^+ ion (transition $B^2\Sigma_g^+$ (v = 0) $\rightarrow X^2\Sigma_g^+$ (v = 0), λ = 391.4 nm) and N_2 molecule (transition $C^3\Pi_u$ (v = 2) $\rightarrow B^3\Pi_g$ (v = 5), λ = 394.3 nm) bands were recorded by the EMI-9781B photomultiplier. An overview radiation spectrum of the plasma was recorded using an EPP2000C-25 spectrometer (StellarNet Inc.) with known spectral sensitivity in relative units in the range 200–850 nm. This ensured that the spectral bands required for calculations was present in the radiation spectrum and was not overlapped by other spectral components.

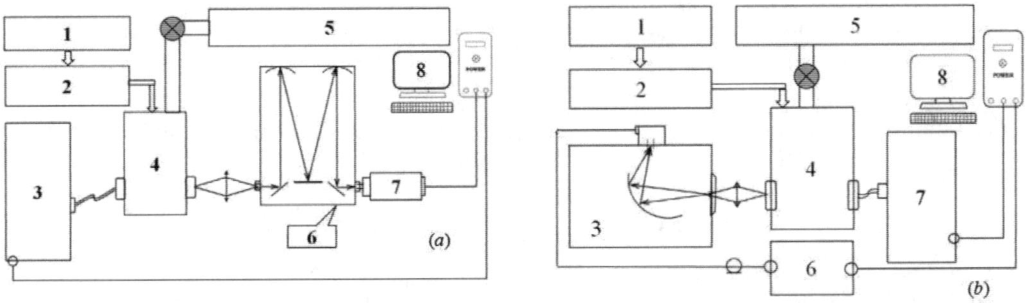

Figure 1. Blok-scheme of experimental setup: (a) - for time-integrated measurements: 1 - master oscillator BNC, 2 – pulser, 3 – spectrometer EPP2000C-25 (StellarNet Inc.), 4 – discharge chamber, 5 – pumping system, 6 – monochromator MDR-23, 7 – CCD-camera PI MAX 2, 8 – PC.
(b) - for time-resolved measurements: 1 - master oscillator BNC, 2 – pulser, 3 – monochromator VM-502, 4 – discharge chamber, 5 – pumping system, 6 – oscilloscope TDS-3054B, 7 – spectrometer EPP2000C-25 (StellarNet Inc.), 8 – PC.

For each of the setups, the instrument function and normal slit width were determined using a low-pressure mercury lamp. For setup No. 1 comprising the MDR-23 monochromator with a diffraction grating (2400 grooves/mm), the instrument function was ~0.24 Å at a normal slit width of ~10 μm. For setup No. 2, the instrument function and normal slit width were ~3.0 Å and ~30 μm, respectively.

The high-voltage pulsers RADAN-220 and FPG-60 were used for REP DD excitation in pulsed and pulse-periodic modes, respectively. In the idle running regime, RADAN-220 formed voltage pulses in the discharge gap with an amplitude up to ~250 kV and the half-amplitude duration of ~2 ns for a matched load with a front duration of ~0.5 ns in the transmission line. High-voltage pulser FPG-60 formed voltage pulses with an amplitude on a high resistance load up to 60 kV and voltage pulse rise time of 2÷3 ns at a pulse repetition rate up to 2 kHz.

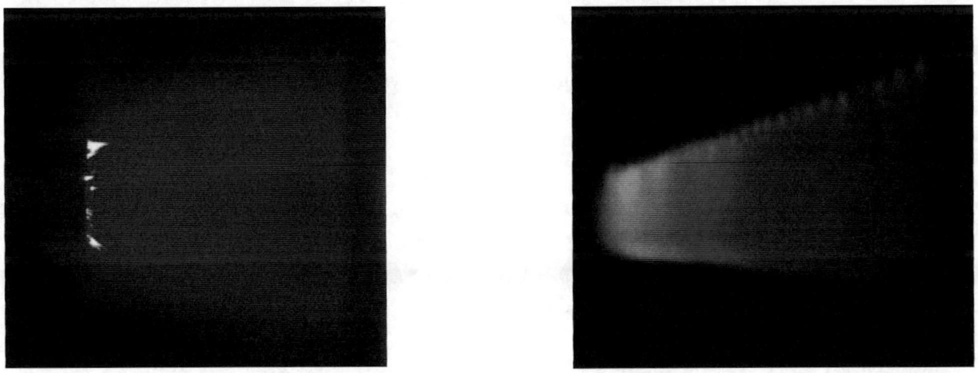

(a) (b)

Figure 2. REP DD glowing in nitrogen at pressure of 1 atm at pulsed (*a*) and pulse-periodic (*b*) modes.

At the pulsed mode, the electrodes of discharge chamber were a cathode of small curvature radius and a plane anode. The cathode was made of steel foil of thickness ~100 μm in the form of a tube of diameter ~6 mm. The interelectrode gap was ~12 mm. The radiation was extracted through side windows of the discharge chamber. At the pulse-periodic mode, the discharge gap consisted of electrodes with length of ~ 40 cm. The typical photos of glowing of REP DD plasma in nitrogen at pressure of 1 atm at pulsed and pulse-periodic modes are shown on the figure 2.

3. Experimental results and discussion

To substantiate the possibility of using above-mentioned technique for *Te*, *E/N* measurements, it is necessary to prove that the Maxwellian velocity distribution sets in by the instant at which the intensity of nitrogen bands used in this technique is sufficient for recording. In view of the effective energy transfer during the electron–electron collisions, the characteristic time τ of stabilization of the Maxwellian distribution can be estimated as the quantity reciprocal to the electron–electron collisions frequency (1):

$$\tau = \frac{1}{N_e v_e \sigma_{ee}} \qquad (1)$$

In (1), *Ne* is the electron density; *Ve* is the relative electron velocity; σ_{ee} is the electron-electron collisions cross-section. Estimation showed that at conditions under study, the values of τ are from fractions of nanosecond to several nanoseconds and this suggests that condition (I) is fulfilled. To substantiate that condition (II) is valid for the PER DD plasma in nitrogen, it is necessary to show that the main channel of population of the N_2^+ ($B^2\Sigma_g^+$) and N_2 ($C^3\Pi_u$) states is collision of a nitrogen molecule in the ground state with an electron. It is known that excitation of the resonant $C^3\Pi_u$ state of a nitrogen molecule by direct electron impact prevails over other excitation channels at high electron energies when the reduced electric field strength *E/p* is close to values optimal for excitation of this state: *E/p* ~ 150–200 (V·cm^{-1}·Torr^{-1}), which corresponds to a short phase of voltage drop across the gas discharge gap after its breakdown [15]. The fact that the excitation of the N_2^+ ($B^2\Sigma_g^+$) and N_2 ($C^3\Pi_u$) states by direct electron impact from the ground state of a nitrogen molecule dominates under the experimental conditions is confirmed by calculation and comparison of the excitation rates of the N_2^+ ($B^2\Sigma_g^+$) and N_2 ($C^3\Pi_u$) states due to the direct and stepwise excitation process. The calculation showed that at conditions under study the direct excitation process was markedly higher as compare to the stepwise one.

The measurements of *Te* and *E/N* in plasma of REP DD were carried out at nitrogen excitation at pressure of 1 atm both in pulse-periodic (~2 kHz) and pulsed modes. In the first case the time average values of *Te* and *E/N* was found to be of ~2 eV and ~270 Td, respectively. As well, dynamics of these parameters was determined. Values of the ratio $R_{391/394}$ monotonically decreased of about two times during ~ 10 ns. It means that values of *Te* and *E/N* are reduced by ~1.5 times (from 3 to 2 eV for *Te*, from 400 to 270 Td for *E/N*).

For discharge excitation in the pulsed mode, the measurements of *Te* and *E/N* values were carried out both in time-integrated and time-resolved modes, as well. In the first case, the spectrum of plasma emission radiation in the spectral range of 390-395 nm was registered with a monochromator MDR-23 and CCD-camera PI-MAX 2. The average values of *Te* and *E/N* corresponding to this spectrum was found to be of ~ 2 eV and ~ 240 Td, respectively. The dynamic of these parameters was determined by the registration of the time behaviour of N_2^+ (λ = 391.4 nm) and N_2 (λ = 394.3 nm) bands' peaks intensity with the monochromator and photomultiplier tube. The values of *Te* and *E/N* was obtained to decrease monotonically in the ranges of 3 to 2 eV and of 400 to 260 Td, respectively.

A value of *Tg* is known to link to a value of rotational temperature *Tr* due to the relaxation process of inelastic collisions between molecules. The relaxation number, which is the average number of inelastic collision required for the exchange of translational and rotational energy, was shown to be of 4-6 [16]. The mean time of N_2 molecules elastic collisions $\bar{\tau}_{Coll}$ can be calculated according to (2):

$$\bar{\tau}_{Coll} = \frac{1}{[N_2]\cdot\sqrt{2}\cdot\bar{v}\cdot\sigma}, \qquad (2)$$

where $[N_2]$, $\bar{v}$, σ - density, thermal velocity and gas-kinetic cross section of N_2 molecules, respectively. Since the value of $\bar{\tau}_{Coll}$ in atmospheric pressure nitrogen at room temperature is ~ 0.13 ns, the rotational relaxation time is about 0.65 ns. This is more than one order of magnitude shorter than the duration of REP DD under study. The relation of *Tg* and *Tr* of the $C^3\Pi_u$ state is as follows [17]:

$$T_g \approx T_r \cdot \frac{B_e^0}{B_e^*} \approx 1.09 \cdot T_r \quad (3)$$

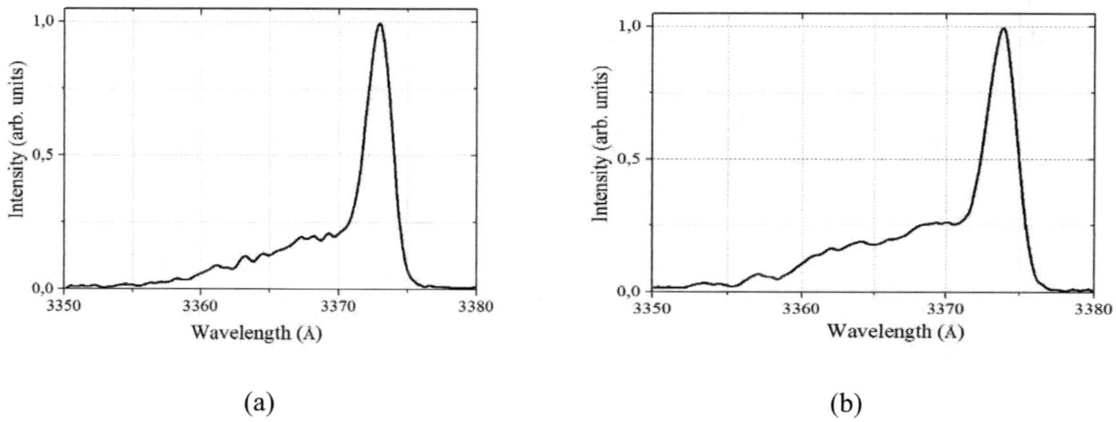

Figure 3. Spectrum of N_2 ($C^3\Pi_u$ (v=0)→ $B^3\Pi_g$(v=0)) transition in atmospheric-pressure nitrogen at pulsed (*a*) and pulse-periodic (*b*) excitation modes.

where B_e^0, B_e^* - rotational constants of $X^1\Sigma_g^+$ and $C^3\Pi_u$ states of N_2 molecule, respectively. The measurements of *Tg* were carried out both at pulsed and pulse-periodic modes. The typical unresolved rotational emission spectra of N_2 ($C^3\Pi_u$ (v = 0) →$B^3\Pi_g$ (v = 0)) transition are presented on the figure 3. The value of *Tg* at pulsed mode was found to be of ~ 380 °K. In the case of pulse-periodic mode, the value of *Tg* calculated was found to be of 820 °K.

4. Conclusion

Electron *Te*, gas *Tg* temperatures and reduced electric field strength *E/N* of gas discharge plasma under REP DD excitation of atmospheric-pressure nitrogen at pulsed and pulse-periodic (2 kHz) modes were determined by optical emission spectroscopy techniques. We have performed calculations confirming the applicability of the technique based on the radiative-collisional plasma model [12, 13] to determine *Te* and *E/N* at condition under study.

Acknowledgments

The work was supported by the grant from the Russian Science Foundation, project №14-29-00052.

References

[1] Babich L P, Loiko T V and Tsukerman V A 1990 *Sov. Phys. Usp.* **33** 521
[2] Babich L P 2003 *High-Energy Phenomena in Electric Discharges in Dense Gases: Theory, Experiment and Natural Phenomena* (ISTS Science and Technology Series, Vol. 2. Arlington, Virginia: Futurepast) p 353
[3] Tarasenko V F and Yakovlenko S I 2004 *Physics-Uspekhi* **47** 887
[4] Levko D, Krasik Y E and Tarasenko V F 2012 *International Review of Physics* **6** 165
[5] *Runaway Electrons Preionized Diffuse Discharges* ed by V F Tarasenko (New York: Nova Science Publishers, Inc.) 2014 p 598
[6] Babich L P, Bereznin I A, Loiko T V and Tarasov M D 1982 *Izv. Vyssh. Uchebn. Zaved., Radiofiz.* **25** 1131
[7] Sorokin D A, Lomaev M I and Krivonogova K Yu 2010 *Izv. Tomsk. Politekh. Univ.* **316** 80

[8] Yatom S, Stambulchik E, Vekselman V and Krasik Ya E 2013 *Physical Review E* **88** 013107
[9] Yatom S and Krasik Ya E 2014 *J. Phys. D: Appl. Phys.* **47** 215202
[10] Sorokin D A, Lomaev M I , Banokina T I, Tarasenko V F 2014 *Technical Physics* **59** 1119
[11] *Plasma diagnostics* ed by W Lochte-Holtgreven (Amsterdam: North-Holland publishing company) 1968
[12] Paris P, Aints M, Valk F, Plank T, Haljaste A, Kozlov K V and Wagner H-E 2005 *J. Phys. D: Appl. Phys.* **38** 3894
[13] Britun N, Gaillard M, Ricard A, Kim Y M, Kim K S, Han J G 2007 *J. Phys. D: Appl. Phys.* **40** 1022
[14] Ochkin V N 2009 *Spectroscopy of Low Temperature Plasma* (Wiley-VCH) p 609
[15] Godarg B 1974 *IEEE J. Quantum Electron.* **10** 147
[16] Phillips D M 1975 *J. Phys. D: Appl. Phys.* **8** 507
[17] Otorbaev D K, Ochkin V N, Rubin P L, Savinov S Yu, Sobolev N N and Tskhai S N 1985 *Electron-excited molecules in non-equilibrium plasma* (Moscow: Nauka) p 187

Growing aluminum nitride films by Plasma-Enhanced Atomic Layer Deposition at low temperatures

V A Tarala, A S Altakhov, V Ya Martens and S V Lisitsyn

North-Caucasus Federal University, 2 Kulakov Ave., 355029, Stavropol, Russia
E-mail: vmartens@yandex.ru

Abstract. Aluminum nitride films have been grown by Plasma-Enhanced Atomic Layer Deposition method. It was found that at temperatures of 250 °C and 280 °C increase of the plasma exposure step duration over 6 s, as well as increase of reactor purge step duration over 1 s does not affect the growth rate, however, it affects the microstructure of the films. It was found that crystalline aluminum nitride films deposit with plasma exposure duration over 10 s and the reactor purging over 10 s. When the temperature drops the increase of reactor purge step duration and plasma exposure step duration over 20 s is required for crystalline AlN film growth.

1. Introduction

Aluminum nitride (AlN) is a material with a large band gap (6.2 eV) [1], high chemical inertness and thermal conductivity (320 W/m·K) [2]. AlN thin films are promising for the production of cold cathodes [3-5], buffer layers [6], dielectric and passivation coatings [7-9] as well as in the creation of gas sensors [10, 11], UV LEDs and photodetectors [12, 13].

Various methods of synthesis are used for AlN growth: chemical vapor deposition (CVD) [14], plasma-enhanced chemical vapor deposition (PECVD) [15], magnetron sputtering (MS) [16], molecular beam epitaxy (MBE) and atomic layer deposition (ALD) [17].

Plasma-Enhanced Atomic Layer Deposition method (or PEALD) is particularly interesting because it allows high precision control of film thickness and drastic reduction of the synthesis temperature by increasing the reactivity of the reagents. Furthermore, with regard to the deposition of aluminum nitride films, PEALD allows usage of nitrogen-hydrogen plasma as a nitrogen source, as alternative to ammonia, which is toxic and explosive gas.

Conventionally, atomic layer deposition method is utilized to grow amorphous films of aluminum nitride. The aim of our study was to obtain crystalline films at substrate temperatures less than 300 °C, which is important for creating passivation coatings for elements of micro- and nanoelectronics.

2. The experiment

PEALD TFS-200 device, manufactured by Beneq (Finland) was used for the deposition of AlN films. The operating frequency of the RF generator was 13.56 MHz. All experiments were performed at HF radiation power (W) of 200 W. Films were grown on a monocrystalline silicon substrate (111).

After loading the substrates into the reactor, a roughing pump was used to decresase pressure below 10 Pa. Then, nitrogen feed was turned on for ventilation of reactor chamber and the reactor itself. The nitrogen flow rate through the reactor chamber (R_k) was 200 cm^3/min, and 300 cm^3/min through a reactor (R_r).While maintaining constant values of R_k and R_r, the reactor was heated up to temperatures (T) ranging from 150 to 280 °C. 60 minutes after reaching the temperature, the surface of the substrate was treated by the plasma of mixture of hydrogen (R_{H2} = 80 cm^3/min) and nitrogen (R_{N2} = 20 cm^3/min) gases for 1 minute. The plasma of this gas mixture was used as the nitrogen source in a cyclic film deposition process. This is followed by a cyclic process of feeding the reactants in the reactor (Figure 1). Trimethylaluminum (TMA) was used as aluminum source. Evaporator with TMA was kept at a temperature of 18 °C. TMA feed pulse duration (τ_{TMA}) of experiments in "F-series" was 0.3 s, and in "M-series" – 0.05 s. Duration of reactor purging phase after the step of treatment the substrate with the vapor of TMA (τ_{PUR}) varied from 1 to 30 s, and the duration of exposure of the substrate stage to plasma gas mixture of hydrogen and nitrogen (τ_{PE}) varied from 3 to 30 s. Duration of reactor ventilation phase after plasma exposure (τ_{PR}) in all experiments was 10 s.

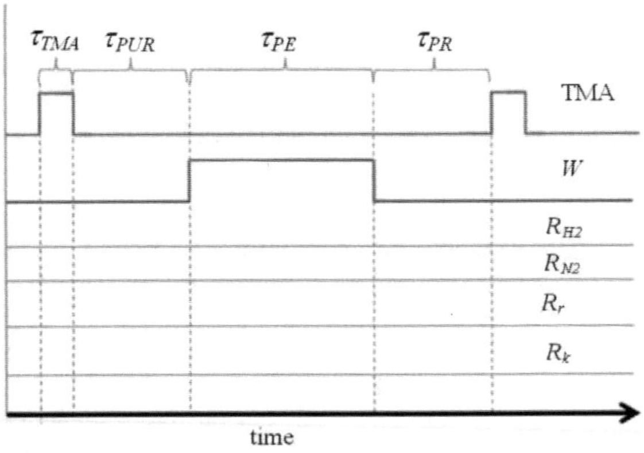

Figure 1. Illustration of the deposition process.

AlN film refractive index (n) and thickness (h) was measured by spectral ellipsometer SE-800 (Sentech, Germany).The growth rate of the film was evaluated as the ratio of its thickness to the quantity of deposition cycles.

Chemical composition of the grown film was evaluated infrared absorption spectra obtained using FTIR spectrometer FSM-1201 (Infraspek Ltd., Russia).

Film phase composition was studied using X-ray diffractometer Smart Lab (Rigaku, Germany).

3. Result discussion

Studies of aluminum nitride film deposition processes at low temperatures (150 °C) shown (Figure 2) that, ceteris paribus(τ_{PE} = 10 s, τ_{PR} = 10 s) increasing the duration of reactor purging step (τ_{PUR}) leads to a decrease in film growth rate (V).The most significant changes of V are observed at τ_{PUR}<10 s, after which the dynamics of the growth rate declination becomes slower. Presumably, growth rate is reduced due to efficiency of physically adsorbed TMA molecules removal from the substrate surface.

This assumption is supported by the fact that V drops not only because of τ_{TMA} decrease, but also due to an increase in temperature of the substrate.

At low temperatures (150 °C), the dependence of the growth rate on the duration of the plasma exposure is of extreme nature (Figure 3). Presumably, increasing value of τ_{PE} from 3 to 10 s raises the proportion of TMA radicals and molecules adsorbed on the substrate surface and engaged into chemical interaction with plasma particles; as a result, the growth rate increases. In our opinion, increase in growth rate at temperatures of 250 and 280 °C is also caused by these reasons.

However, the deposition process at temperatures well below the decomposition temperature of $Al(CH_3)_3$ (230 °C), in particular at a temperature of 150 °C, complete chemical conversion of TMA does not occur. Therefore, some of the CH_x (where $0 \leq x \leq 3$) radicals can be overgrown into the bulk of the film, making it amorphous and less dense.

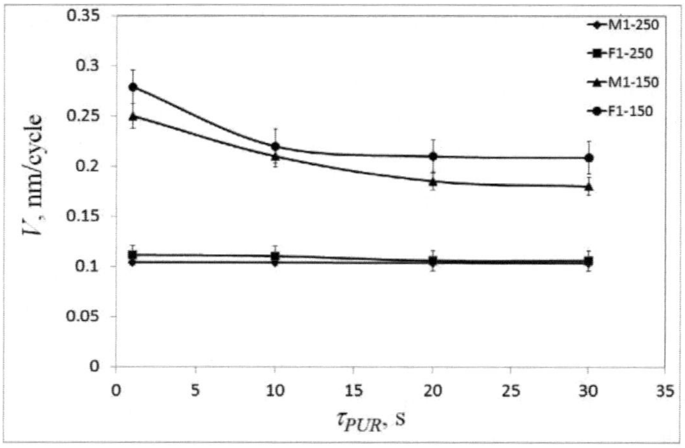

Figure 2. Dependence of film growth rate on the duration of post TMA treatment purge: τ_{PE} = 10 s; τ_{PR} = 10 s; F1-150 – T = 150 °C, τ_{TMA} = 0.3 s; F1-250 – T = 250 °C, τ_{TMA} = 0.3 s; M1-150 – T = 150 °C, τ_{TMA} = 0.05 s; M1-250 – T = 250 °C, τ_{TMA} = 0.05 s.

With increase of plasma exposure duration from 10 to 30 s efficiency of plasma exposure continues to increase, resulting in a reduction of CH_x radical fraction and in increase of density of the film. As a result of density increase, the rate of film growth decreases.

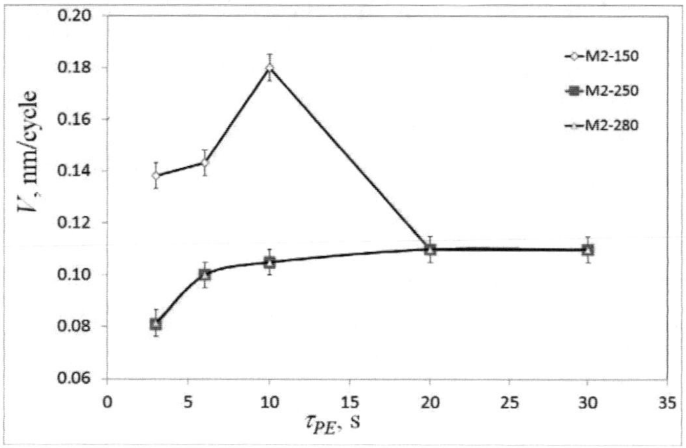

Figure 3. The dependence of film growth rate on the duration of plasma exposure step: τ_{TMA} = 0.05 s, τ_{PUR} = 30 s, τ_{PR} = 10 s; M2-150 – T = 150 °C, M2-250 – T = 250 °C, M2-280 – T = 280 °C.

As a confirmation of the density change assumption, we can employ the results of the study of dependence of refractive index (n) on the film deposition conditions. It is known that for AlN film refractive index value varies within the range from 1.9 to 2.1; it decreases as crystallinity and, therefore, the density, lowers.

As shown in Figure 4, in case of high-temperature processes (250 and 280 °C), increase of plasma exposure to 6 s leads to the increase of n from 1.9 to 1.97, after which significant changes of the refractive index substantially does not occur. This fact indicates that the most significant changes in the microstructure of AlN films occur at $\tau_{PE} < 6$ s. Under the low-temperature process conditions (150 °C), with the duration of plasma exposure of 10 s deposited films have low values of refractive index, which indicates their amorphousness and low density. Increasing τ_{PE} to 20 s and more allows growing films with a refractive index of about 1.9, which can be regarded as films of aluminum nitride, in contrast to those obtained at lower values of the plasma exposure duration.

It is important to note that the samples grown at 150 °C and τ_{PE} < 20 s and τ_{PUR} < 30 s almost completely lack IR absorption band with maximums at 670-675 cm^{-1} and 600-610 cm^{-1} [15], responsible for the vibrations of the Al-N.

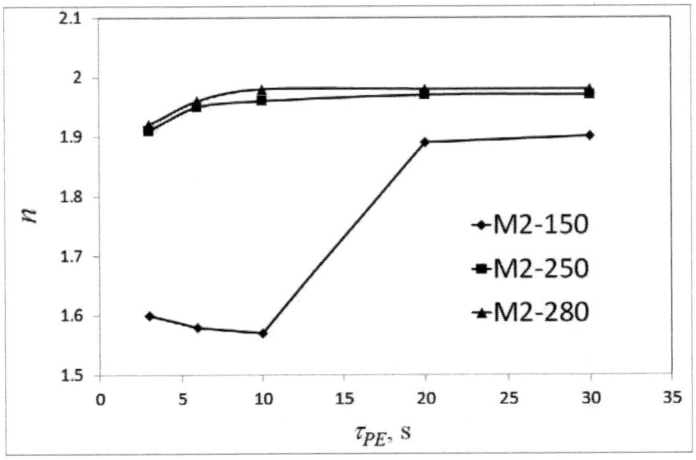

Figure 4. Dependence of the refractive index on the plasma exposure stage duration: τ_{TMA} = 0.05 s, τ_{PUR} = 30 s, τ_{PR} = 10 s; M2-150 – T = 150 °C, M2-250 – T = 250 °C, M2-280 – T = 280 °C.

As shown in Figure 5, the maximum of the IR absorption spectrum in the wavenumber range of 400-1000 cm^{-1} increases with increase in substrate temperature and duration of purging and plasma exposure steps. These results correlate well with changes in the values of the refractive index of the deposition conditions, and show the way to an increase in the crystallinity of the AlN films.

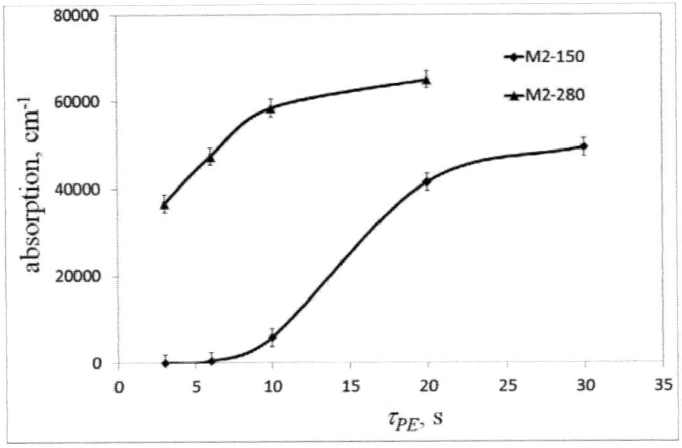

Figure 5. The dependence of the infrared absorption on the plasma exposure duration: τ_{TMA} = 0.05 s, τ_{PUR} = 30 s, τ_{PR} = 10 s; M2-150 – T = 150 °C, M2-280 – T = 280 °C.

In X-ray diagrams of samples with IR absorption coefficient of more than 55 000 cm^{-1} distinct reflexes, typical of aluminum nitride, were present (Figure 6). It should be noted that the crystallites in aluminum nitride films synthesized at a temperature of 280 °C, predominantly had a hexagonal polytype. At the same time, films obtained at a temperature of 250 °C, were formed by crystallites of the cubic modification. Samples with IR absorption coefficient of less than 55 000 cm^{-1} no X-ray reflections was observed.

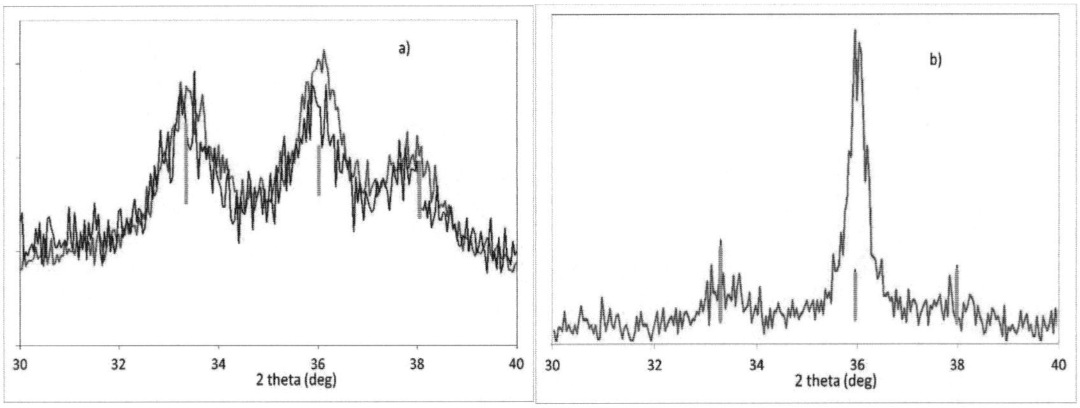

Figure 6. The XRD patterns of aluminum nitride film samples grown at temperatures of 280 °C (a) and 250 °C (b); τ_{TMA} = 0.05 s, τ_{PUR} = 20 s, τ_{PE} = 20 s, τ_{PR} = 10 s.

Thus, it was found that the aluminum nitride films that have a refractive index greater than 1.9 and having a spectrum of infrared absorption bands responsible for Al-N bonds, deposit at a rate not more 0.10±0.02 nm/cycle. Herewith, it was found that at temperatures of 250 °C and 280 °C increase of the plasma exposure step duration over 6 s, as well as increase of reactor purge step duration over 1s does not affect the growth rate; however, it affects the microstructure of the films. It was found that crystalline aluminum nitride films deposit with plasma exposure duration over 10 s and the reactor purging over 10 s. When the temperature drops the increase of τ_{PE} and τ_{PUR} over 20 s is required for crystalline AlN film growth.

4. Conclusion

The results of the investigation of the dependence of the refractive index, coefficient of IR absorption, XRD, and the growth rate on the deposition conditions showed that crystalline AlN film can be obtained by plasma-enhanced atomic layer deposition at temperatures less than 300 °C, and the film growth rate does not exceed 0.12 nm / cycle.

Acknowledgments

This work was supported by the RF Ministry of Education and Science (state research target no.2014/216, project no. 2516) and was carried out using equipment at the Shared Facilities Center, North-Caucasus Federal University.

References

[1] Silveira E, Freitas J A, Schujman S B, and Schowalter L J AlN bandgap temperature dependence fromits optical properties 2008 *J. Cryst. Growth* **310** 4007–4010.
[2] Junior A F and Shanafield D J Thermal conductivity of polycrystalline aluminum nitride (AlN) ceramics 2004 *Ceramica* **50** 315 247–253.
[3] Sowers A T, Christman J A, Bremser M D, Ward B L, and Davis R F Thin films of aluminiumnitride and aluminum gallium nitride for cold cathodeapplication 1997 *Appl. Phys. Lett.* **71** 16 2289–2291.
[4] Nikiforov D K, Korzhavyi A P and Nikiforov K G Modeling of charge carrier injection and emission processes in aluminum nitride-based nanostructures 2012 *Proc. INTERMATIC Int. Sci.Technol. Conf.* **2** 58–60.

[5] Shi S C, Chen C F, Li H Y, Lo J T, Lan Z H, Chen K H and Chen L C Field emission from quasialigned aluminium nitride nanotips 2005 *Appl. Phys.Lett.* **87** 7 3109–3112.
[6] Chen Z, Newman S, Brown D, Chung R, Keller S, Mishra U K, Denbaars S P and Nakamura S High quality AlN grown on SiC by metalorganic chemical vapor deposition 2008 *Appl. Phys. Lett.* 93 191 906.
[7] Bosund M, Mattila P, Aierken A, Hakkarainen T, Koskenvaara H, Sopanen M, Airaksinen V M and Lipsanen H GaAs surface passivation by plasma enhanced atomic layer deposited aluminum nitride 2010 *Appl. Surf. Sci.* **256** 24 7434–7437.
[8] Chen C, Chen D J, Xie Z L, Han P, Zhang R, Zheng Y D, Li Z H, Jiao G and Chen T S Effects of an AlN passivation layer on the microstructure and electronic properties of AlGaN/GaN heterostructures 2008 *Appl. Phys. A* **90** 3 447–449.
[9] Sen Huang, Qimeng Jiang, Shu Yang, Zhikai Tang, and Chen K J Mechanism of PEALD grown AlN passivation for AlGaN/GaN HEMTs: compensation of interface traps by polarization charges 2013 *Electron Device Lett.* **34** 2 193–195.
[10] Ivaldi P, Abergel J, Arndt G, Robert P, Andreucci P, Blanc H, Hentz S and Defay E 50 nm thick AlN resonant microcantilever for gas sensing application 2010 *Frequency Control Symposium (FCS)* 81–84.
[11] Samman A, Gebremariam S, Rimai L, Zhang X, Hangas J and Auner G W Platinum–aluminum nitride–silicon carbide diodes as combustible gas sensors 2000 *J. Appl. Phys.* 87 3101–3107.
[12] Taniyasu Y, Kasu M and Makimoto T An aluminum nitride light emitting diode with a wavelength of 210 nanometers 2006 *Nature* 441 325–328.
[13] Dung-Sheng Tsai, Wei-Cheng Lien, Der_Hsien Lien, Kuan-Ming Chen, Meng-Lin Tsai, Senesky D G,Yueh-Chung Yu, Pisano A P and Jr Hau He Solarblind photodetectors for harsh electronics 2013 *Sci. Rep.* 4 2628.
[14] Kakanakova-Georgieva A, Nilsson D and Janzйn E High quality AlN layers grown by hot wall MOCVD 1 52–56.
[15] Bouchkour Z, Tristant P, Thune E, Dublanche-Tixier C and Jaoul C Aluminum nitride nanodots prepared by plasma enhanced chemical vapor deposition on Si(111) 2011 *Surf. Coat. Technol.* 205 S586–S591.
[16] Pat S and Kokkokoglu M Characterization of deposited AlN thin films at various nitrogen concentrations by rf reactive sputtering 2010 *Optoelectron. Adv. Mater. Rapid Commun.* **4** 6 855–858.
[17] Yong Ju Lee and Sang-Won Kang Growth of aluminum nitride thin films prepared plasma enhanced atomic layer deposition 2004 *Thin Solid Films* **446** 2 227–231.

Translated by M Yu Shevchenko

High-voltage electrode optimization towards uniform surface treatment by a pulsed volume discharge

A V Ponomarev[1,4], M S Pedos[1], S V Scherbinin[2], Y I Mamontov[2] and S V Ponomarev[3]

[1] Institute of Electrophysics, 106 Amundsena Street, Ekaterinburg, 620016, Russia
[2] Ural Federal University, 19 Mira Street, Ekaterinburg, 620002, Russia
[3] Tomsk Polytechnic University, 30 Lenina Street, Tomsk, 634034, Russia
[4] Author to whom any correspondence should be addressed

E-Mail: apon@iep.uran.ru

Abstract. In this study, the shape and material of the high-voltage electrode of an atmospheric pressure plasma generation system were optimised. The research was performed with the goal of achieving maximum uniformity of plasma treatment of the surface of the low-voltage electrode with a diameter of 100 mm. In order to generate low-temperature plasma with the volume of roughly 1 cubic decimetre, a pulsed volume discharge was used initiated with a corona discharge. The uniformity of the plasma in the region of the low-voltage electrode was assessed using a system for measuring the distribution of discharge current density. The system's low-voltage electrode – collector – was a disc of 100 mm in diameter, the conducting surface of which was divided into 64 radially located segments of equal surface area. The current at each segment was registered by a high-speed measuring system controlled by an ARM™-based 32-bit microcontroller. To facilitate the interpretation of results obtained, a computer program was developed to visualise the results. The program provides a 3D image of the current density distribution on the surface of the low-voltage electrode. Based on the results obtained an optimum shape for a high-voltage electrode was determined. Uniformity of the distribution of discharge current density in relation to distance between electrodes was studied. It was proven that the level of non-uniformity of current density distribution depends on the size of the gap between electrodes. Experiments indicated that it is advantageous to use graphite felt VGN-6 (Russian abbreviation) as the material of the high-voltage electrode's emitting surface.

1. Introduction

A pulsed volume discharge is of great interest for the creation of significant volumes of low-temperature plasma within gaps up to tens of centimeters [1]. Processes in the pulsed volume discharges are similar to processes in the classical stationary glow discharge, where cathode and anode fall off potentials, Faraday dark space and positive column exist. At the same time, the volume discharge possesses some significant differences from the classical glow discharge. One of them is that to materialize such a discharge a pre-ionization of gas is needed, which can be performed with auxiliary discharges such as DBD, corona, spark. Another difference is that this type of discharge is not stationary. If its duration won't be limited artificially, it will certainly constrict into a single conducting channel. Also, one more difference is its specific peak power which can exceed 1 MW/cm^3, that significantly exceeds the similar characteristics of a classical glow discharge. Unlike the classical low-pressure discharge, the pulsed high-pressure discharge was given the name transient glow discharge [2] but it most often referred to as pulsed glow discharge. It may be that this discharge would be more properly referred to as a high-pressure glow discharge, but the term 'volume discharge' had come into wide use before their kinship was established. The adjective 'volume' emphasizes the fact that the characteristics of this discharge are determined primarily by the process in the discharge itself and not related to its interaction with the chamber walls, as is the case with classical glow discharge. Volume discharge has wide practical use. These discharges are used for laser pumping for various technological applications. Furthermore, it is used to construct plasma-chemical reactors and also to initiate and support various processes on the surface of solid bodies [3].

Regardless of the majority of work in the field of pulsed volume discharge being done in late last century, the interest towards this type of discharge remains high even today. Research continues to study its characteristics [4], and systems are constructed on its basis. Reference [5] proposes a system based on pulsed volume discharge for decontamination of surfaces.

Pulsed volume discharge has a diffuse nature and enables to treat larger surfaces per pulse when compared to other discharge types. But effective treatment requires a uniform distribution of discharge current density on the entire surface of the object being treated. Research of current density distribution in pulsed volume discharge is presented in this paper. Factors affecting current distribution on the surface of the low-voltage electrode are discussed. The shape and material for the emitting surface of the high-voltage electrode in a gas discharge system are presented.

2. Experimental setup and methods

The set of equipment used for studying the pulsed volume discharge parameters consisted of a high-voltage generator DOPG-0.7 [6] and a system to measure the uniformity of current density distribution in sections [7].

The generator is a modular structure consisting of 2 separate parts: power and high-voltage modules. The high-voltage module contains a high-frequency transformer, whose windings ratio defines the output voltage and inductivity of windings - the rate of high-frequency oscillations. The output transformer allows easy alignment the output impedance of the generator with the impedance of various loads in a wide range for effective energy transfer into them. To change the parameters of a discharge or its configurations it is enough to change either the high-voltage module itself or the transformer located in it. The power module remains unchanged in all configurations.

The generator is capable of forming pulses of damped harmonic oscillations with a frequency up to 1 kHz. Herewith, the rate of high-frequency oscillations themselves can reach units of megahertz. The maximum energy of the output pulse at optimal resistive load at frequency of 1 MHz reached 0.4 J. The advantage of this generator type is that for one power pulse it is capable of repeatedly generating a plasma in the discharge gap. This results in the increase of the plasma interaction time with the treated object, with the preservation of all advantages of pulsed operation mode. In addition, it increases the actual repetition rate of discharge pulses that will be equal to the multiplication of generator operation rate by the number of plasma generation moments per one pulse.

The discharge in the configuration "multi-needle high-voltage electrode – grounded plate" was ignited within gaps from 55 to 125 mm. Changing the gap between electrodes changed the strength of the electric field in the gap. This was the only possible way to change the strength of the electric field in our case. The generator's output parameters are constant and depend on the high-voltage transformer's parameters only.

A high-speed multi-channel measurement system of the current density distribution of a completed pulsed corona discharge has been used based on a 32-bit ARM™ microcontroller.

A photograph of a single pulsed volume discharge in air at atmospheric pressure and outside view of the measuring system are provided as figures 1(a) and 1(b), respectively.

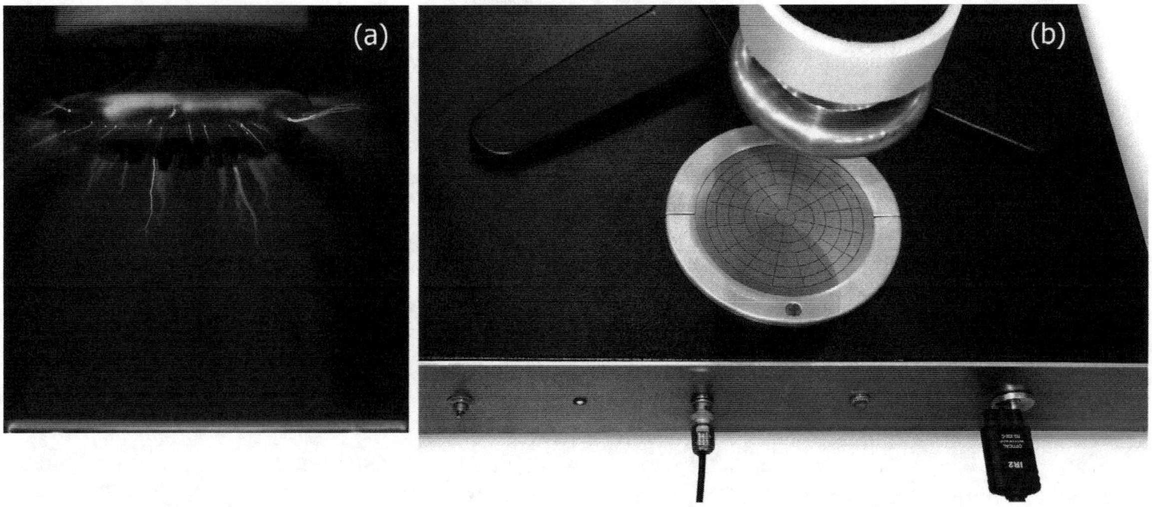

Figure 1. A photograph of a single pulsed volume discharge in air at atmospheric pressure (a); outside view of the measuring system (b).

The system allows the current density distribution on a grounded electrode – collector to be determined for both a negative and positive current. The collector consisted of a nickel-plated double-sided PCB-plate of 100 mm in diameter, the outer track of which was divided by gaps of 0.5 mm width into 64 radially located segments of equal surface area. Discharge current passing through the circuits of each segment induced a charge in the measuring capacitor. The voltages of 64 measuring capacitors were directly proportional to the induced charges and were measured by an automated system. Discharge current of only one polarity could be measured in one experiment.

The device provides the ability to measure the current density distribution as a single pulse and to operate in the accumulation mode of measurement results. The sampling time per measurement channel does not exceed 1 μs, which allows for a pulse repetition frequency of a current in units of kilohertz with 64 channels used in the work.

The results of the measurements are transmitted to a personal computer, where they are processed and rendered. To facilitate the interpretation of results obtained, a computer program was developed to visualise the results. The program provides a 3D image of the current density distribution on the surface of the low-voltage electrode. The images provided as column graphs were used as a basis when assessing the uniformity of the distribution of discharge current.

A TDS 684B oscilloscope with a 1 GHz bandwidth and Barth Electronics wide band signal attenuators were used for recording the current pulses.

3. Experimental results

3.1. Shape and material of the high-voltage electrode

As indicated above, gas ionisation is a necessary preparation for the ignition of a pulsed volume discharge. In our case, a corona discharge was used as an auxiliary discharge. Our choice was based on the fact that a high-strength electric field in the gap is needed for the ignition of both the auxiliary and the main discharge. The source of high-voltage used, i.e. the generator DOPG-0.7 enables to achieve an electric field with the strength of roughly 1 MV/m in the first half-period of the oscillation, if the discharge gap is 0.1 m.

One of the traditional methods of facilitating the ignition of a corona discharge is to use sharp-pointed electrodes that increase the electric field's strength due to their heterogeneous geometry. To form a discharge with a larger surface or a longer duration, multi-needle electrodes are used. In previous studies [8], the authors used a multi-needle electrode made of copper wire. The emitting surface of such a high-voltage electrode had a diameter of 60 mm and consisted of ~8,000 wires of 0.13 mm in diameter and roughly 20 mm in length. The main disadvantage of this design is its inconvenience of technological manufacture. Other disadvantages are the impossibility of precise formation of the emission surface shape as well as the difficulty of keeping its shape unchanged during its subsequent use.

A more advantageous material for the emitting surface of the high-voltage electrode was found to be graphite felt VGN-6 (Russian abbreviation). It is a flexible sheet inorganic conductive material. Felt consists of individual thin hairs, which provide significant number of micropoints on the surface. Micropoints in its turn made the corona discharge ignition easier.

In the course of research it was found that the material possesses a very convenient property of self-formation of its surface. When working with a freshly made electrode at the discharge currencies of about one ampere, individual hairs excessively projecting out of the electrode's surface were gradually burned out. Being made of carbon, the hairs burned off in air and the electrode's surface become uniform with only a short training.

The system's operation was affected not only by the shape and material of the high-voltage electrode's emitting surface. High-strength electric field for efficient ignition of a pulsed volume discharge cannot be achieved without careful design of the high-voltage electrode as a whole. It is determined that the most efficient form of the electrode is an axially symmetrical rotational body. Other forms, including sharp edges, brought along non-uniform distortions in the electric field and the formation of intensive streamers in those regions of the surface. This led to reduced impedance of the system and, in turn, to decreased output voltage of the generator and a decreased strength of the electric field in the gap. A drawing of an optimum shape of the high-voltage electrode to treat the surface of roughly 1 dm^2 is provided as figure 2(a). The entire electrode except the emitting surface is made of aluminium alloy. A photograph of the shape as viewed from the side of the emitting surface is provided as figure 2(b). This is the electrode that was used in all experiments described in this paper.

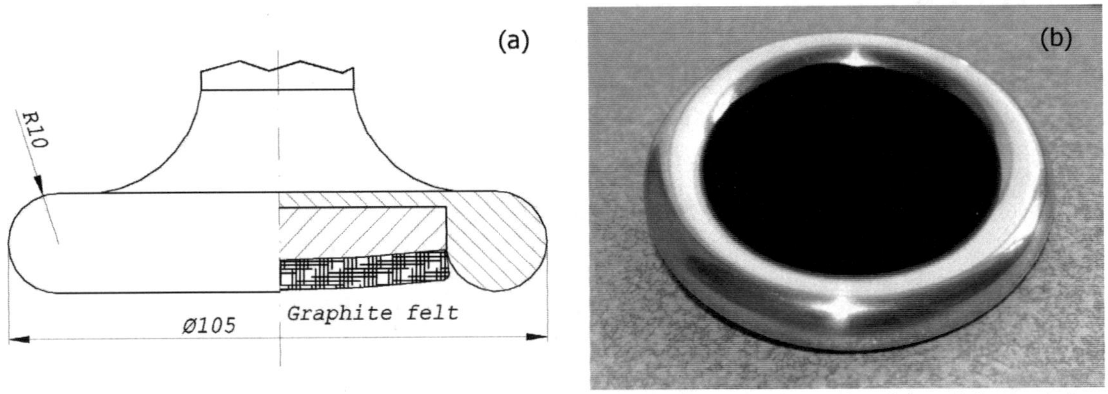

Figure 2. Design of the high-voltage electrode (a) and its outside view from the side of the emitting surface (b).

3.2. Research of discharge current density distribution on the surface of the low-voltage electrode

We assume that the objects to be treated by pulsed volume discharge will be acting as the discharge system's low-voltage electrode. Therefore, the uniformity of the discharge current distribution on the surface of the low-voltage electrode is the most important parameter determining the effectiveness of our system functioning as a whole.

With the help of the system described above we obtained data about the uniformity of the discharge current distribution at various gap widths between the electrodes. In those experiments, we also measured the form of the current on the low-voltage electrode – the collector. Measurements were taken both in single pulse mode and in repetitive mode with accumulation of 100 discharge pulses results. In single pulse mode, the system performed 10 preparatory pulses and measured the parameters of the 11th discharge pulse. This was done to ensure identical measurement conditions in both single pulse and repetition modes. The data presented were measured at the pulse frequency of 100 Hz.

Figure 3 shows four groups of data obtained with the discharge gaps of 55, 75, 95 and 115 mm. For each group, two 3D-distributions of the induced current are presented: from a single discharge pulse and normalised across 100 discharge pulses. Axes X and Y represent the collector's dimensions in mm. The readings on axis Z are presented in arbitrary units proportional to the charge induced by the flow of discharge current in the circuit of each segment. The segments in the figure correspond to the size and shape of the actual measuring segments. For each group an oscillogram of the collector's single pulse current is provided, measured for the respective group of distances between the electrodes.

4. Discussion and conclusion

The main purpose of the research was to optimise the shape and material of the high-voltage electrode of a gas discharge system. The assessment criterion used was the uniformity of discharge current density distribution on the surface of the low-voltage electrode – the collector. It is clear that the measuring system enabled only coarse measurements of that uniformity due to the finite number of measuring segments and their significant size. For a more detailed determination of the level of uniformity of discharge current, the size of the measuring segments must be reduced and their number increased. Still, we assume that the system used by us can be successfully utilised for a qualitative assessment of the level of uniformity of discharge current. The following conclusions were made in the course of using that measuring system:

- In single pulse mode, treatment with pulsed volume discharge can be performed only at small gaps between the electrodes. But even at the minimum gap, there are observable regions on the surface of the low-voltage electrode where no current runs. With the gap widening, the number of such segments increases.
- If assessing the effects of 100 discharge pulses, current was registered in all measuring segments of the collector at small gaps between the electrodes. This may be due to shifting of emission centres along the electrode's surface. This allows for a conclusion that in repetitive mode, the entire surface of the low-voltage electrode will be treated by discharge.
- Combining the data of the distribution of currents on the surface of the low-voltage electrode and the data of the current oscillograms leads to a conclusion that the surface area of the discharge's impact on the electrode and the amplitude of the discharge current are in direct correlation to each other. From this point of view, the situation can be compared to classic glow discharge where an increase of the current leads to an increase of the area of emitting surface.
- High stability of the summary charge induced by the discharge current across pulses could be observed. The data of 10 experiments at the electrode gap of 55 mm were reviewed. For example, the mean value of all summary charges from a single pulse was 65.306 arbitrary units. The minimum and the maximum values were 62.924 and 68.017, respectively. Standard deviation of the charge value was 1.545.
- The shape of the high-voltage electrode has a significant impact on the distribution of discharge current density.

A discussion of the shape and behaviour of plotted curves of the current on the collector is beyond the scope of this paper. It must be noted that the non-symmetrical nature of the current upon symmetrical alternating voltage applied to the discharge gap is the result of non-symmetrical configuration of electrodes used. With a symmetrical configuration of electrodes, the current on the collector is also symmetrical.

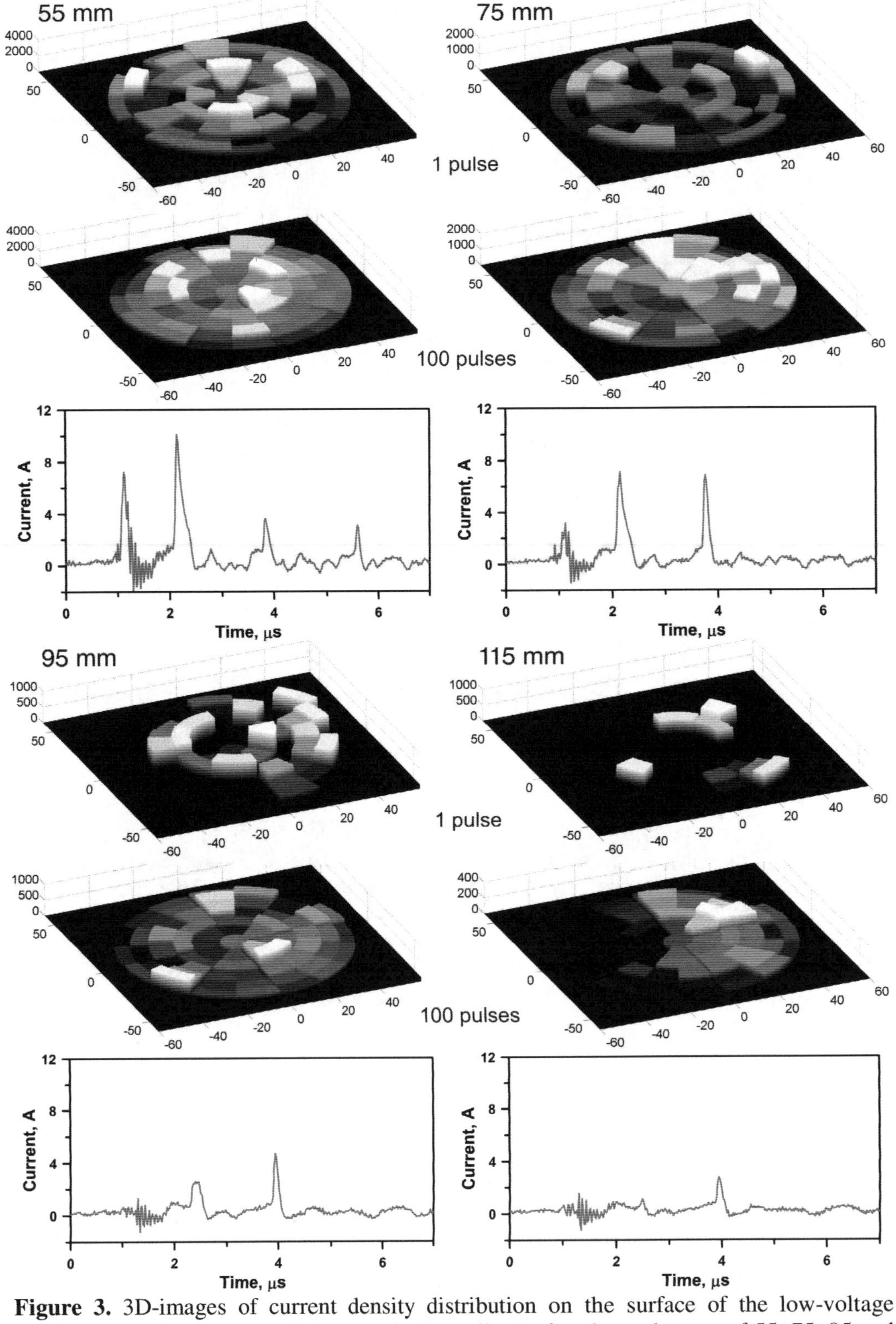

Figure 3. 3D-images of current density distribution on the surface of the low-voltage electrode, and oscillograms of currents in the collector for electrode gaps of 55, 75, 95 and 115 mm.

5. Acknowledgments
This work was supported by the Russian Foundation for Basic Research, project no. 15-08-01707a.

References
[1] Korolev Yu D and Mesjac G A 1998 *Physics of Pulsed Breakdown in Gases* (Yekatriburg: URO-PRESS)
[2] Chalmers I D 1971 The transient glow discharge in nitrogen and dry air *J. Appl. Phys.* **4** 1147
[3] Osipov V V 2000 Self-sustained volume discharge *J. Phys.-Uspekhi* **43** 221-41
[4] Elistratov E A, Kuznetzov A P, Maslennikov S P, Protasov A A and Shkol'nikov E Ya 2012 Measuring parameters of volume nanosecond pulsed discharge in air at atmospheric pressure *J. Technical Phys. Lett.* **38** 793
[5] Zhuravlev M V, Slobodyan M S and Shubin B G 2012 System for atmosphere HF volume discharge *J. Izvestiya Vischih Uchebnih Zavedeniy: Phizika* **55** 456
[6] Ponomarev A V, Pudikov A S, Gusev A I and Pedos M S 2014 A solid-state generator with pulsed excitation of the oscillating circuit *J. Instrum. Exp. Tech.+* **57** 135
[7] Ponomarev A V, Pedos M S, Mamontov Y I, Gusev A I and Scherbinin S V 2015 System for measuring the current density distribution of a completed pulsed corona discharge *J. Instrum. Exp. Tech.+* in print
[8] Ponomarev A V, Pudikov A S and Ignatenko Y G 2014 Damping oscillation pulse generator for atmospheric-pressure gas discharge applications *J. Izvestiya Vischih Uchebnih Zavedeniy: Phizika* **57** 28-31

Journal of Physics: Conference Series **652** (2015) 012036 doi:10.1088/1742-6596/652/1/012036

Simulation of Intense Electrohydrodynamic Flow Based on Dielectric Barrier Discharge

I E Rebrov, V Yu Khomich and V A Yamshchikov
Institute for Electrophysics and Electric Power RAS, Dvorzovaya Naberezhnaya 18, 191186, Saint- Petersburg, Russia

E-mail: re_i@rambler.ru

Abstract. This paper presents a numerical analysis of the intense EHD gas flow velocity distribution at atmospheric pressure. The system consists of a coaxial plasma emitter and a collector grid. The ion source is a dielectric barrier discharge distributed over the surface of the emitter. The computational results coincide with the experimental data which allow us to predict the maximum flow rate produced by the experimental setup.

1. Introduction

An electrohydrodynamic (EHD) flow is a physical phenomenon of a gas collective motion caused by the drift of charged particles in a high electric field. The promising atmospheric plasma technology with many potential applications in aeronautics, which allows controlling the airflow over the surface of the wing, reducing drag and impeding the transition to turbulence [1], is related to electrohydrodynamic interaction of ions and neutral particles.

The EHD flow systems are used to intensify the heat dissipation in modern high-performance computers and as an electrical pump of a gas mixture in lasers, electric filters for gas media cleaning, and surface modification applications of materials [2-4]. It is possible due to the generation of high speed gas flows without using traditional electromechanical systems, such as turbines and fans. An advantage of electric pumping is the absence of moving parts and related failures due to the rotor wear or the thermal and mechanical fatigue of fans [5-7].

2. Numerical simulation

As it was offered in [8] the high frequency barrier discharge distributed over the dielectric surface was used as an ion source for the EHD flow system for a gas mixture circulation in electric-discharge lasers. We have constructed a device with the three tubular plasma emitters (1) with optimal spacing, a grid collector (2) with transparency T, a DC high voltage supply and a high frequency high-voltage solid-state pulse generator (HFHVPG) (Figure 1). The positive polarity output pulses generated by the HFHVPG were quasirectangular with a voltage amplitude of 0 – 12 kV, a rise time of 100-200 ns, a variable duration of 500 ns –10 μs, and a repetition rate of 1 – 25 kHz [9].

Such a system has a lot of variables and to achieve maximum efficiency it requires multiparametric optimization. Since analytical solutions were obtained for the very simple cases [8], the numerical model has been constructed to design systems with a complex geometry. The intense EHD flow generated by the plasma discharge on the surface of the dielectric tube was simulated in the air at atmospheric pressure depending on the transparency of the collector grid. To achieve an acceptable computational time the model was simplified to a two-dimensional case in the X-Y plane and was solved for a single emitter.

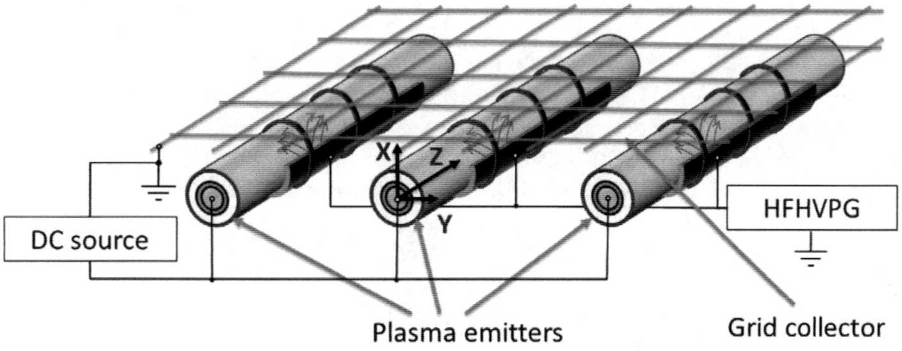

Figure 1. Schematic of the intense EHD flow system.

The model based on the system of equations connects several areas of physics [10, 11].
The electric potential U is determined by the Poisson equation:

$$\nabla^2 U = -\frac{q}{\varepsilon_0}, \tag{1}$$

where q is the space charge density and ε_0 is the dielectric permittivity of free space.
The electric potential associated with the electric field strength is given by:

$$\boldsymbol{E} = -\boldsymbol{\nabla} U \tag{2}$$

The electric current in the drift zone of charged particles is the result of the three processes: an electric drift (ion motion in the environment under the influence of an electric field), advection (movement of charged particles in the gas stream), and diffusion. Thus, considering the above, the current density can be written as:

$$\boldsymbol{j} = \mu_i \boldsymbol{E} q + \boldsymbol{V} q - D \boldsymbol{\nabla} q \tag{3}$$

where μ_i is the ion mobility in an electric field, $\boldsymbol{V}$ is the fluid velocity, and D is the diffusion coefficient of ions in gas. The convection term of equation is negligible as $\boldsymbol{V} \ll \mu_i \boldsymbol{E}$ (typical gas flow velocity is about 1-5 m/s whereas the drift velocity of ions reaches hundreds of meters per second) [12].
The continuity equation for the current is given by:

$$\nabla \boldsymbol{j} = 0 \tag{4}$$

The volumetric electrodynamic force in a gas with dielectric permeability ε, density ρ, and temperature t is expressed by:

$$\boldsymbol{F} = q\boldsymbol{E} - \frac{1}{2} E^2 \boldsymbol{\nabla} \varepsilon + \frac{1}{2} \boldsymbol{\nabla} \left[E^2 \left(\frac{\partial \varepsilon}{\partial \rho} \right)_t \rho \right] \tag{5}$$

The first term in the equation is the Coulomb force acting on a free charge in the gaseous medium, the second and the third terms are dielectrophoretic and electrostrictive forces, respectively. The second and the third terms are important in the phase transition, which is great fluctuation of permeability at the liquid-gas boundary. The permeability in air is almost constant and hence the Coulomb force [13] is the dominant term.
The model hydrodynamic part is described by the Navier-Stokes equations for a steady state incompressible gas flow:

$$\rho(\boldsymbol{V} \cdot \nabla)\boldsymbol{V} = -\boldsymbol{\nabla} p + \mu \nabla^2 \boldsymbol{V} + \boldsymbol{F}, \tag{6}$$

where ρ is the mass density, p is the gas pressure and μ – is the viscosity.
The mass continuity equation is given by:

$$\nabla \boldsymbol{V} = 0 \quad (7)$$

The system of equations (1)–(7) describes the numerical model of the EHD device in this study with the boundary conditions as follows. The DC voltage of 20 kV was applied to the emitter and the ground potential was set at the collector grid. A space charge density of q was imposed on the tubular ion emitter corresponding to ion concentration of 10^{10} cm^{-3} [14]. Open boundary with no viscous stress was applied to all boundaries and wall with no-slip condition was set to the collecting grid wires and the emitter tube.
The numerical simulation was based on the finite element method. The simulation space was discretized by the inhomogeneous grid consisting of 130,000 triangular elements. The minimum size of the finite element defined on the surface of the collector electrode is 50 μm and the maximum size of 400 μm is localized in the region center.

3. Numerical results

The numerical simulations for the EHD flow circulation system are presented for a device electric field profile, a charge distribution and an output flow velocity profile. The typical intensity distribution in the X-Y plane of the amplitude of the EHD flow speed at 0.7 grid transparency is shown in Figure 2.

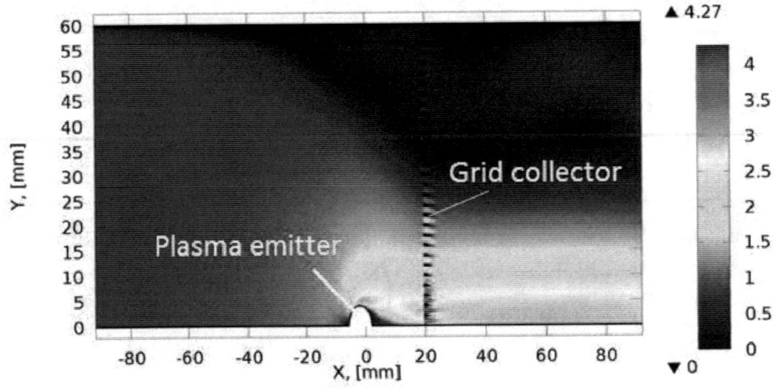

Figure 2. Numerical simulation results: air velocity as coloured surface map with units in m/s.

Flow velocity V was simulated depending on transparency T of the mesh collector at optimal power supply parameters [4] of HVHFPS and the DC supply, which are taken into account at the boundary conditions. The velocity distribution profiles along the Y coordinate, depending on the spacing of the grid is shown in Figure 3. In the figure an increase of V with increase of T is demonstrated. Transparency was varied by changing the spacing and the wire radius using a commercially available selection of grids.

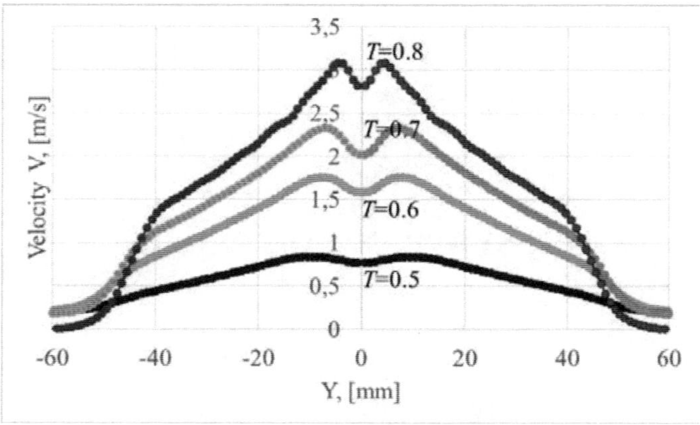

Figure 3. Velocity profile at X=70 mm depending on collector transparency *T*.

The computational results show that the grid electrode has a strong deceleration of the EHD flow, so optimizing its design provides significant potential to raise the speed and the flow quantity. There is a limit to increase in *T* which results to high field strength near the surface of the wire mesh. This field causes an opposite ion drift decelerating the main flow. Therefore, along with the influence of the geometry of the electrode gap and the space charge, the effect of grid parameters should also be taken into account.

On the basis of this simulation optimal transparency of the grid collector of 0.77 has been selected and an airflow speed up to 3 m/s was achieved. The results of computations coincide with the experimental data [4] which allow us to predict the maximum flow rate produced by the experimental setup and improve its efficiency.

4. References

[1] Guo S 2013 *An Investigation of Dielectric Barrier Discharge Based Plasma Actuator Designs With Enhanced Performance In Active Flow Control* (Minnesota: University of Minnesota)

[2] Go B D, Garimella S V_and Fisher T S 2007 *J. of Appl. Phys.* ***102*** 053302

[3] Kuryachii A P, Rusyanov D A, Skvortsov V V 2011 *TsAGI Sci. Journal* **42** 227

[4] Khomich V Yu, Moshkunov S I, Nebogatkin S V et al. V A 2011 *Quant. Elect.* **41** 1093

[5] Cooperman P, 1960 *Trans. of the Commun. and Electr.* **79** 47

[6] Wolf R A 2012 *Atmospheric Pressure Plasma for Surface Modification* (New York: Wiley–Scrivener)

[7] Ong A O 2013 *Electrohydrodynamic Microfabricated Ionic Wind Pumps for Electronics Cooling Applications* (Cleveland: Western Reserve University)

[8] Khomich V Yu, Malashin M V, Moshkunov, Rebrov I E 2013 *EPE Journal* **23** 51

[9] Moshkunov S I, Rebrov I E, Khomich V Yu 2013 *Advances in Appl. Phys.* **5** 630

[10] Karpov S V, Krichtafovitch I A. 2005 *COMSOL Multiphysics User's Conference*

[11] Honer K A, Jewell-Larsen N E, Ran H, Zhang Y 2008 *ESA Annual Meeting on Electrostatics*

[12] Mahmoudi S R, Adamiak K, Cstle G S P 2011 *ESA Annual Meeting on Electrostatics*

[13] Kogelschatz U 2003 *Plasma chemistry and plasma processing* **23** 46

[14] Patel V 2012 *Chemical Kinetics* (Rijeka: InTech)

Generation of Shock-Wave Disturbances at Plasma-Vapor Bubble Oscillation

N S Kuznetsova, A S Yudin and N V Voitenko
National Research Tomsk Polytechnic University, Institute of High Technology Physics, 30 Lenin avenue, Tomsk, 634050, Russian Federation

E-mail: tevn@hvd.tpu.ru

Abstract. The complex physical and mathematical model describing all steps of plasma-vapor bubble evolution in the system of the water–ground condensed media is presented. Discharge circuit operation, discharge plasma channel expansion, its transformation into the vapor-plasma bubble and its pulsation, pressure wave generation and propagation of the mechanical stress waves in the ground are self-consistently considered in the model. The model allows investigation of the basic laws of stored energy transformation into the discharge plasma channel, next to the plasma-vapor bubble and transformation of this energy to the energy of pressure wave compressing the surrounding ground. Power characteristics of wave disturbances generated by gas-vapor bubble oscillation in liquid depending on the circuit parameters are analyzed for the prediction of the ground boundary displacement. The dynamics of the shock-wave propagation in water–ground condensed media depending on the rate of the plasma channel energy release is investigated. Simulation of the shock-wave phenomena at a plasma-vapor bubble oscillation in condensed media consecutively describes the physical processes underlying technology for producing piles by electro-discharge stuffing. The quantitative model verified by physical experimental tests will allow optimization of pulse generator parameters and electrode system construction of high-voltage equipment.

1. Introduction

Shock-wave disturbances generated by the plasma-vapor bubble expansion and oscillation in different condensed media at electro-blast are widely used in engineering and technology due to ability to manage their properties over a wide range by varying both the parameters of RLC-circuit and the geometrics of electro-discharge gap [1–3]. Short time period (in the range of 10–100 μs) and the high rate of electrical energy release into the discharge channel in liquid cause the high plasma pressure in it (up to $3\cdot10^9$ Pa) and the formation of intensive compression waves in the surrounding liquid. The resulting power in the discharge channel leads to the mechanical work both above the liquid, and in the movement and deformation of the surrounding medium (ground) [7–8]. The correlation of contributions of shock waves at the discharge channel expansion and at the plasma-vapor bubble the expansion and oscillation are not defined at present. The lack of experimental information about the general laws determining the amplitude and profile of the pressure wave generated by plasma channel expansion does not allow estimation of the generated pressure in the condensed media on the basis of measured electrical characteristics of the explosion (current and voltage). Some attempts have been made by the authors of monographs [2, 3, 9], but formulations of their practical tasks did not assume the recording of the shock wave profile.

The experimental studies of the fast processes in the water–ground media are difficult and complicated technically. Investigation of the mechanisms of formation, evolution and dissipation of mechanical perturbation pulses in this case is only possible with the use of mathematical modeling. A comprehensive physical and mathematical model describing the plasma-vapor bubble expansion, generation of the shock-wave disturbances, the dynamics of elastoplastic waves, the formation of fields of mechanical stresses and deformations in solid medium is necessary for studying the basic laws of electrical energy conversion into the mechanical work of plasma-vapor bubble expansion and oscillation.

2. The Physical and Mathematical Model of Electro-Blast Action on the System of Water-Ground Condensed Media

The development of shock-wave disturbances in liquid at electro-blast can be divided into three stages. The first – the plasma channel expansion due to high pressure in it leading to the shock waves generation with a sound and supersonic speeds, pulse movements of liquid volumes with the speeds reaching the hundreds of meters per second [6]. The large resistance to the expansion from the liquid surrounding the channel leads to the sharp pressure rise in the channel, whereby the phenomenon takes on the character of the explosion. The second – the discharge channel degeneration into the plasma-vapor bubble filled with the liquid degradation products, vapors under the different thermodynamic conditions and plasma. The diameter of the spherical (or near-shaped) cavity depends on the released energy and the external hydrostatic pressure in the liquid. The third - the subsequent pulsation of a plasma-vapor bubble with generation of the secondary pressure waves. The plasma-vapor bubble expansion extends to the maximum radius at which all the kinetic energy is transferred to the water. The increased potential energy of the water again transferred to the bubble, causing its compression to a minimum sizes. This cycle is repeated until the bubble energy decreases due to the emission of acoustic shock waves, as well as due to condensation and viscous losses in the medium. Thus, due to the bubble formation and oscillation in liquid there is the hydrodynamic pulse pressure on the wellbore walls, as estimated in [7], it is greater than 50 MPa at a distance of 10 cm from the source of pulsations and decreases to zero at a distance of 0.6–0.8 m.

The current pulse generator operation at its discharge on the gap located in a hole filled with water is analyzed. Formation of plasma channel occurs, its expansion generates a pressure wave in the liquid at electro-blast. The propagating wave is refracted into the ground, is reflected in the liquid, returns to the channel and interacts with it. Due to the energy release in the discharge gap the discharge channel degenerates into a plasma-vapor bubble which under the influence of the changing pressure begins to pulsate. The plasma-vapor bubble oscillation leads to the generation of secondary pressure waves propagating in the liquid and soil. As a result, in the ground-water media a complex, fast-changing wave pattern is formed, which results in the displacement and deformation of the borehole wall (figure 1).

The model is a consistent solution of the transient process equations for the pulse generator's circuit (figure 1), the energy balance equation for the discharge channel, the hydrodynamic equations for description of wave disturbances and state equations for plasma channel and surrounding medium [10].

Electrodynamic oscillating processes in the discharge circuit were characterized by the voltage balance equation:

$$L\frac{di(t)}{dt} + (R_{ch} + r)i(t) + \frac{1}{C}\int_0^t i dt = U_0 \, , \tag{1}$$

where L is the circuit inductance, μH; R_{ch} is channel resistance,Ω; i(t) is the circuit current, A; U_0 is the charge voltage of capacitor bank, V; r – the active resistance of the generator, Ω.

Electrical plasma channel energy conversion into the plasma energy and mechanical work, performed by expanding channel and pulsating plasma-vapor bubble:

$$\frac{1}{\gamma-1}\frac{d}{dt}(P_{ch}V_{ch}) + P_{ch}\frac{dV_{ch}}{dt} = i(t)^2 R_{ch} \, , \tag{2}$$

where γ is the effective ratio of specific heats; P_{ch} is the channel pressure, Pa; $V_{ch} = \pi r_{ch}(t)^2 l_{ch}$ – channel volume, m^3; r_{ch} and l_{ch} are the channel radius and length accordingly, m.

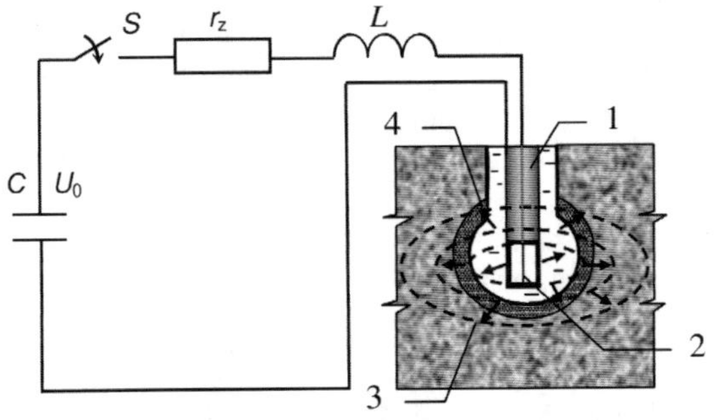

Figure1. The scheme of shock-wave action on the system of water–ground materials at electro-blast: 1 – coaxial cable, 2 – electro–discharge system, 3 – area of medium deformation (packed ground), 4 – shock-wave disturbances, generated by plasma-vapor bubble expansion and oscillation.

The time dependences of the current and voltage have been calculated according to Kirchhoff's laws along with the empirical equations for plasma channel resistance. The plasma channel resistance $R_{ch}(t)$ is determined through the integral of current action in the form of Vaicel-Rompe [11]:

$$R_{ch}(t) = A_{mid} \cdot l_{ch} \left(\int_0^t i^2(t)dt \right)^{-\frac{1}{2}}, \tag{6}$$

where A_{mid} is a spark constant (in case of water A_{mid}=120 V·s$^{0.5}$·m^{-1} [12]).

Wave propagation is described in the hydrodynamic approximation. The description of the channel expansion and plasma-vapor bubble oscillation in a liquid is based on a system of equations including the differential equations of laws of mass conservation, momentum, and energy in the adiabatic approximation, the equations of media and plasma state, the correlation of the stress tensor, spherical tensor and deformation deviator, equations of wave dynamics [13]. Kirchhoff's equations and power balance equation of the discharge channel were jointly solved with the state equations of water and soil, the equations of continuity, internal energy, rations for the components of the stress deviator. The system of equations describing the evolution and dissipation of shock-wave disturbances was solved self-consistently and approximated by the difference scheme according to Wilkins method [14]. On the basis of the submitted equations the numerical algorithm and software were developed that allows to carry out the electro-blast computer experiments in the water–ground condensed media.

3. Simulation Results of Shock-Wave Dynamics at Plasma-Vapor Bubble Oscillation

The results of computer investigations of electro-blast in the water–ground media by means of the presented model under different conditions of energy release into the plasma channel are presented below. Figure 2 shows the main electrical parameters and the plasma–vapor bubble dynamics at the different modes of electrical energy release into the channel: I – U_0=10 kV, C=1200 μF, r_z=114 mΩ, L=11.12 μH; II – U_0=20 kV, C=168 μF, r_z=20 mΩ, L=1.06 μH; the discharge gap length l_{ch}=50 mm. To simulate the well depth of about 7 meters the initial channel pressure was assumed to be 50 MPa.

Figure 2 shows that during the electro-blast in the water at the expansion stage the velocity of plasma–vapor bubble expansion reaches the values of about 800 m/s and 1100 m/s respectively for I and II modes of energy release into the plasma channel, and the pressure are in the range of 1.5 GPa to 3.4 GPa, which causes a rapid growth of the bubble parameters. During the active phase of the discharge the pressure is sharply reduced up to 50 MPa, the velocity of plasma–vapor bubble expansion reaches extremum during the first third of the current oscillation period, and then reduced to zero.

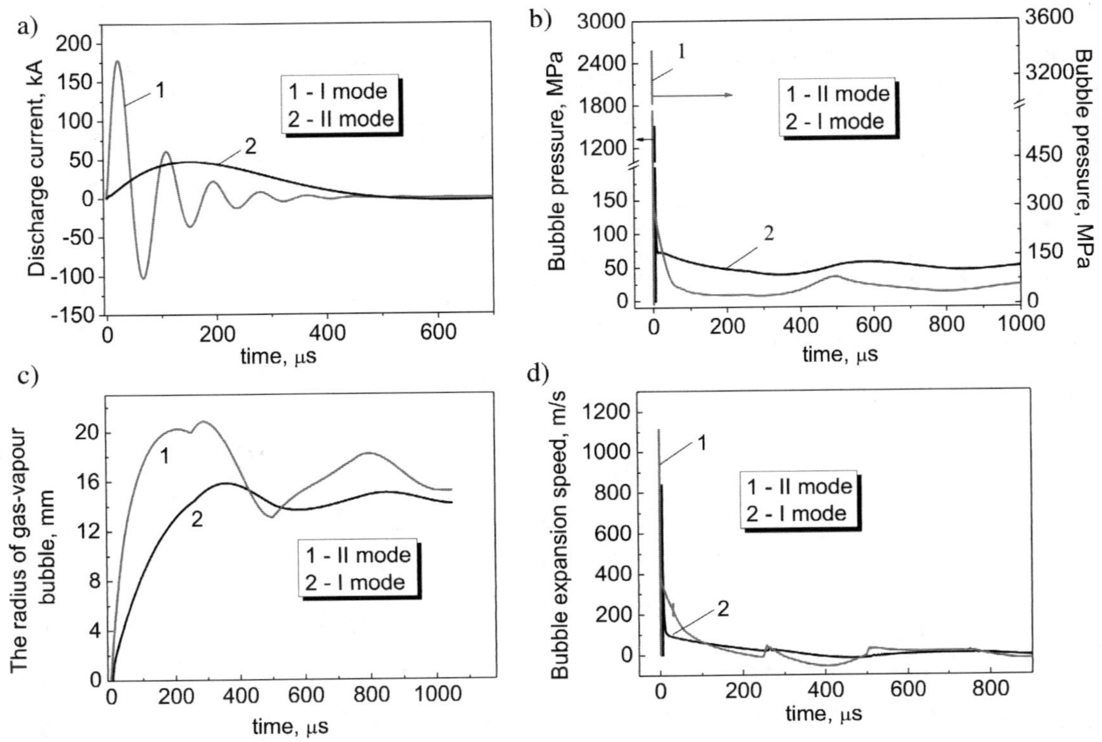

Figure 2. Time dependence of the discharge current (a), the plasma–vapor bubble pressure (b), the plasma–vapor bubble radius (c) and its expansion speed (d) in water.

The plasma–vapor bubble radius increases up to 200–300 microseconds, after that the bubble is compressed with the same speed and pressure as in the expansion stage. When the plasma–vapor bubble pressure becomes higher than the pressure in the surrounding liquid, the repeated growth of its radius occurs (figure 2, c). Pulsation period and the maximum bubble radius increases with the rise of power of energy release into the plasma channel. The maximum plasma-vapor bubble radius reaches 20 mm at the power of the channel energy release of about 400 kW (II mode). These values are in a good agreement with the experimental results [15], which confirms the adequacy of the model.

Due to the plasma–vapor formation and oscillation in liquid the pressure of about 57 MPa and 80 MPa effects on the borehole wall respectively for I and II modes of electrical energy release, figure 3, a.

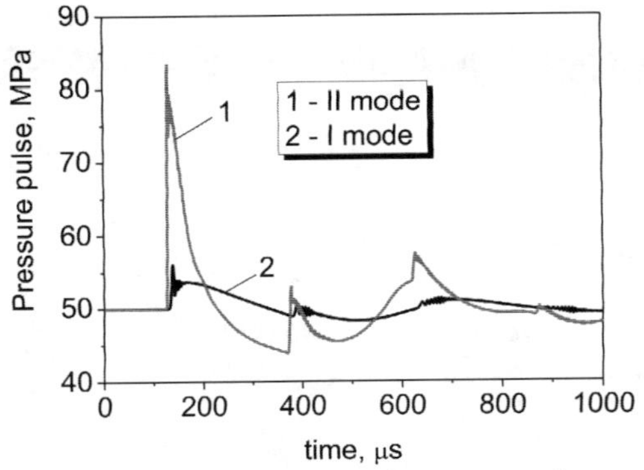

Figure 3. Pressure versus time waveforms, produced on the borehole walls of diameter of 400 mm for different modes of electrical energy release.

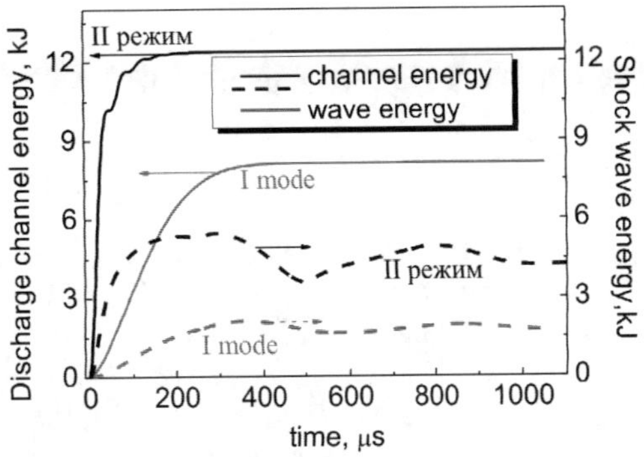

Figure 4. Dynamics of channel and wave energy for different modes of electrical energy release.

The secondary pressure waves were caused by the processes of wave reflection from the walls of gas-vapor bubble and borehole. The displacement of the borehole wall was in the first case – in order of 1 mm, in the second – 1.8 mm. Comparison of the results of numerical modeling with the experimental data [16] (the dependence on the bore-hole radius from the discharge conditions) is showed a high degree of agreement. The energy of shock-wave disturbances is thus less than 4 % from the storage electrical energy at I mode of energy release, and about 16 % in the II mode (figure 4).

The plasma–vapor bubble expansion and its subsequent pulsation result to the formation of mechanical compression stresses in the ground. The dynamics of mechanical stresses under the different modes of energy release is shown in figure 5.

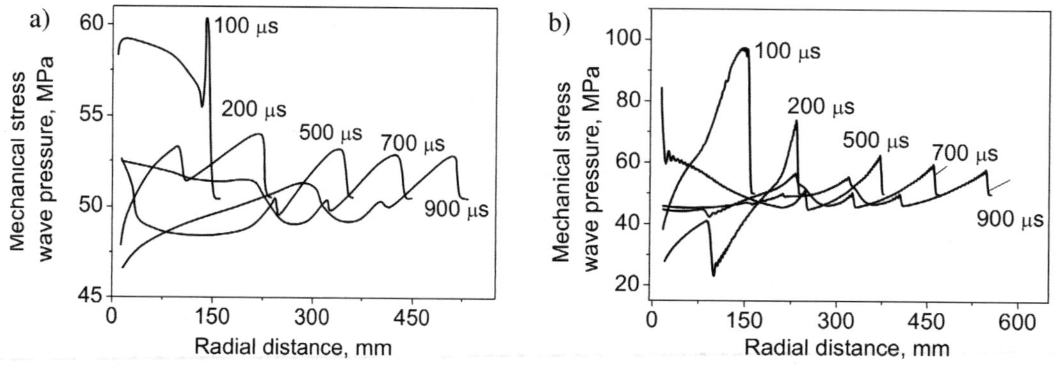

Figure 5. Distribution diagrams of mechanical stress wave profiles in the ground at plasma-vapor bubble pulsation in liquid: a) – I mode of energy release, b) – II mode of energy release.

The profiles of mechanical stress waves for two modes of the channel energy release are different as for the amplitude parameters and for the propagation velocity. In the first case (figure 5, a) the wave velocity is about 385 m/s, in the second (figure 5, b) – 462 m/s.

4. Conclusion

The obtained results of the gas-vapor bubble dynamics have been testified to the determining role of the influence of the electrical energy release mode into the plasma channel on the generated pressure wave in the water–ground system. It is shown that the presented complex physical and mathematical model of the plasma-vapor bubble dynamics in the condensed media allows to estimate the radius and the expansion rate of gas-vapor bubble, its pressure, the pressure wave profile on the borehole wall, the wave pattern evolution in the ground. The model is applicable in a wide range of parameters and gives a good qualitative and quantitative agreement with the experimental results.

Acknowledgements

This work was supported by the Federal Target Program, project No. 14.575.21.0059 (RFMEFI57514X0059).

References

[1] Bluhm H, Frey W, Giese H, Hoppe P, Schultheis C and Sträbner R 2000 *IEEE Trans. On Dielecrics and Elecr. Insul.* **7** 625–636
[2] Naugolnyh K and Roy N 1971 *Electric discharges in water* (Moscow: Nauka) p 238
[3] Krivitskij E *Dynamics of electro explosion in a liquid* (Kiev: Naukova Dumka) p 208
[4] Okun I 1971 *J. Tech. Phys.* **41**, 2 292–300
[5] Vinogradov B and Fedin D 2009 *Mathematical modeling* **1** 23-26
[6] Yutkin L 1986 *The electrohydraulic effect and its application in industry* (St. Petersburg: Engineering) p 253

[7] Bakholdin B and Dzhantimirov H 1998 *Grounds, foundations and geotechnics* **4** 47–52
[8] Bunttsen R 1965 *Electric wire explosion* (Moscow: Mir) p 360
[9] Stolovich N 1983 *Electro-explosion energy converters* (Minsk: Science and Technology), p 151
[10] Burkin V, Kuznetsova N and Lopatin V 2009 *J. Tech. Phys.* **5** 42–48
[11] Rompe R and Weizel W 1944 *Zs. Physik B* **122** 9–12
[12] Sjomkin B, Usov A and Ziniviev N 2000 *Transient processes in electro-discharge installations* (St. Petersburg: Nauka) p 276
[13] Burkin V, Kuznetsova N and Lopatin V 2009 *J. Phys. D: Appl. Phys.* **43** 185204
[14] Wilkins M 1964 *Calculation of elastic-plastic flow, methods in computational physics* (New York: Academic) 211–264
[15] Ivanov V, Shvets I and Ivanov A 1982 *Underwater sparks* (Kiev: Naukova Dumka) p 192
[16] Kurets V, Tarakanovskiy E, Filatov G and Yushkov A 2004 *Electronic material processing* **6** 70-74.

Journal of Physics: Conference Series **652** (2015) 012038 doi:10.1088/1742-6596/652/1/012038

Low-current electric arc in the open air between the end of the cathode and long vertical anode

Kh K Tazmeev[1] and B Kh Tazmeev[2]

[1]Branch of Kazan Federal University in Naberezhnye Chelny, 10A Sujumbike av., Naberezhnye Chelny, 423812, Russia
[2]Naberezhnye Chelny Branch of Kazan National Research Technical University named after A.N. Tupolev, 1 Academician Korolyov str., Naberezhnye Chelny, 423814, Russia

E-mail: tazmeevh@mail.ru

Abstract. The electrical discharge was investigated experimentally in the open atmosphere at currents less than 1 A. The cylindrical rods and flat plates were used as anode. They were installed in front of the cathode rod vertically or obliquely. The modes in which the discharge makes regular self-oscillations were detected. Electric field intensity in the discharge column and the electron temperature were calculated.

1. Introduction

The first experiments were carried out with an electric arc in a high combustion mode (Petrov, Davy and Ritter) [1, 2]. However, despite a long history, the study of open high-current arc is not complete. A more detailed study opens its new properties. In particular, high-speed imaging methods for recording shows that the geometrical form continuously changes, including in the case of horizontally oriented electrodes [3]. The low-current modes of combustion are studied in less detail. Currently, they are virtually ignored. Some characteristics of the low-current modes of combustion are given in the research [4]. This work can be attributed to the attempt to fill this gap in the study of low-current electric arc.

2. Experiment

Copper rods of different diameters (4-12 mm) and a copper plate with dimensions of 150x150 mm were used as anode. They were installed either vertically or inclined at a certain angle. The cathode was a hafnium rod with a diameter of 6 mm. It was mounted in a copper frame and located horizontally at a predetermined distance l from the anode. In the case of the inclined anode l was calculated along the horizontal line from the upper extremity of the cathode as shown in Figure 1.

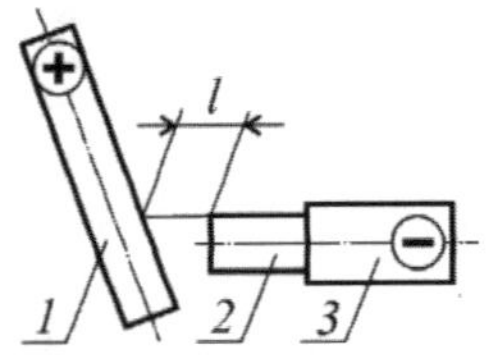

Figure 1. Electrodes assembly: *1* – anode; *2* – cathode; *3* – barrel.

A full-wave rectifier which was connected to the secondary windings of a three-phase step-up transformer was the power source. Pulsations of the rectified voltage were smoothed with *C-L-C* filter. The output voltage amounted to 1760 V. The discharge current is changed by varying the ballast resistor in the range of 1 to 11 kΩ. Oscillograms of current and voltage were recorded with a digital oscilloscope AKIP-15/1. The voltage drop across the shunt with a resistance of 0.5 ohms was current signal, and the signal of voltage was by ohmic divider 6kΩ/12MΩ.

Snapshots of discharges were taken by using high-speed video camera Videoscan-401. The emission spectra were recorded by high-speed fiber-optic spectrometer AvaSpec-3648 in the wavelength range of 484-708 nm with a resolution of 0.15 nm (diffraction grating 1200 lines/mm, the input optical gap of 10 microns).

3. The results of experiments, their processing and discussion

The feature of the electric arc in the present configuration of the discharge was anode spot, which makes a forced move to a considerable distance in the vertical direction when l exceeds a certain threshold value $l_{threshold}$. In the current ranges of 0.1-0.9 A numeric values $l_{threshold}$ were in the range of 3-4 mm.

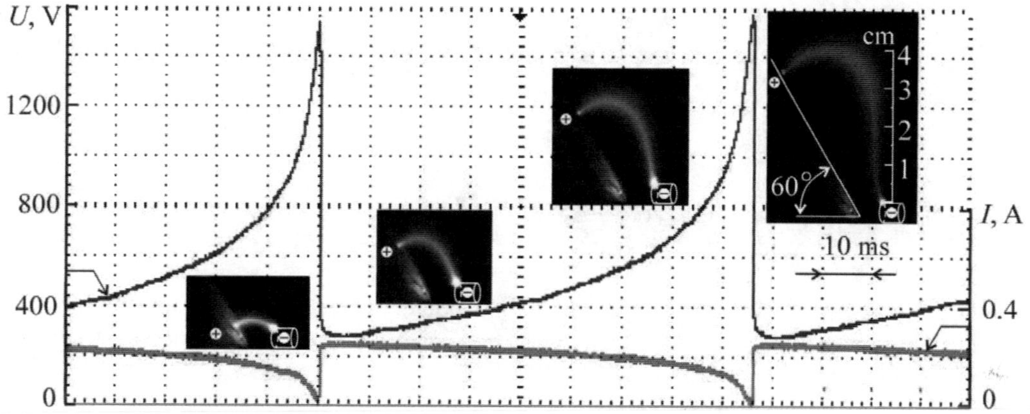

Figure 2. Oscillograms of voltage and current. Anode - copper plate. Exposure of snapshots 2 ms.

When anode spot moves up, the length of the discharge channel l_{arc} was increasing, voltage arc U was increasing and the discharge current I was decreasing (figure 2). The higher was the initial current, the higher was rise the anode spot. This behavior of the arc is explained by increasing buoyancy force, defined as

$$F_{\text{Arch}} = \rho_{\text{air}} g V_{\text{arc}} , \tag{1}$$

where ρ_{air} – air density, V_{arc} – volume of the discharge channel of the arc. With an increase current discharge channel expands, the volume of the arc increases and, consequently, according to (1) increases the buoyancy force.

The movement of the anode spot broke off by shunting. After shunting anode spot again moved up, and thus appeared the pulsating combustion. The pulsations frequency was varying, and the amplitude of the voltage (and current, respectively) was not constant too (figure 3a). With increasing l pulsations became ordered and discharge switched in a mode of regular self-oscillations at practically constant frequency and almost unchanged values of voltage and current amplitudes (figure 3b). A further increase of l resulted to a violation of the regularity of oscillations (figure 3c). Pulsations occurred less frequently. There were long periods of stable combustion, which was amounted to tens of seconds. Stable regimes were established with different voltages and, respectively, the currents (figure 3d). To do this, the rise of the discharge column upward was limited by foreign refractory object which was placed on top of the discharge area.

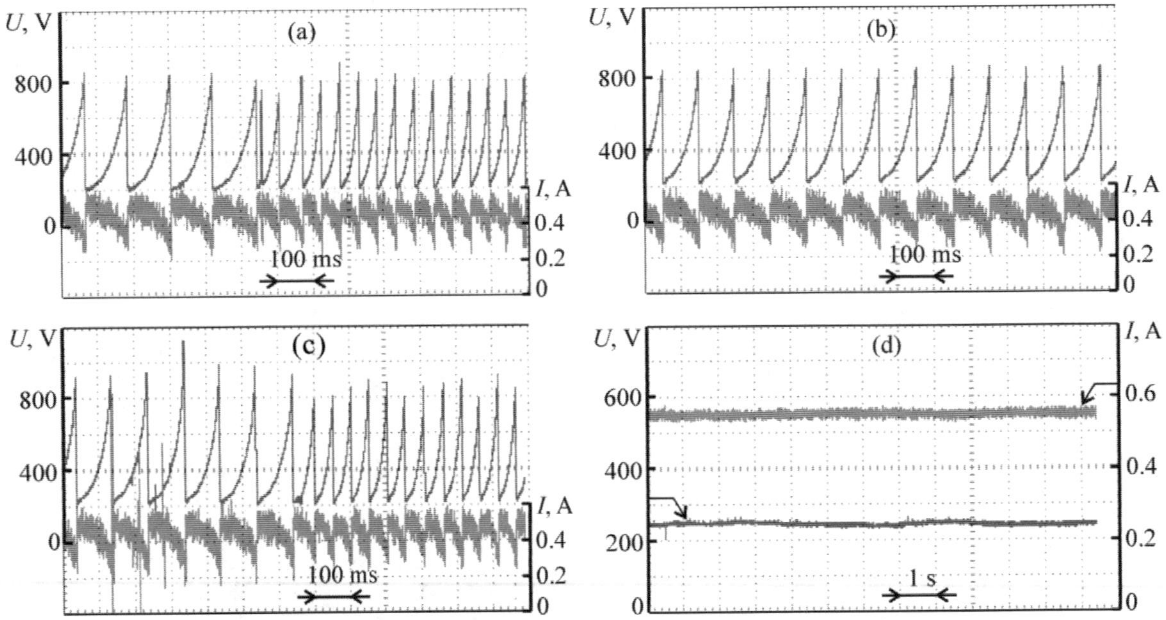

Figure 3. Oscillograms of voltages and currents in oscillatory (a, b, c) and stable (d) combustion modes. Anode - copper rod Ø4.5 mm. (a) – l = 5.0 mm; (b) – 6.0; (c) – 7.5; (d) – 8.0.

Continuous stretching of the discharge channel without shunting was terminated after reaching some maximum l value of l_{max}. It amounted 12-13 mm in the interval of currents of 0.8-0.9 A. l_{max} reduced when current decreased.

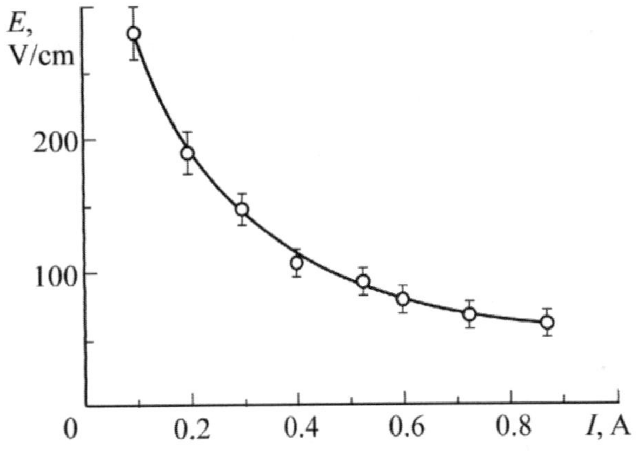

Figure 4. Electric field intensity in the arc column.

Registration of voltage waveforms in the mode of stable combustion which was performed simultaneously with the photographing, allowed to calculate the intensity E of the electric field in the arc column by the formula

$$E = \Delta U / \Delta l_{arc} , \qquad (2)$$

where ΔU – the voltage difference determined from the oscillograms, Δl_{arc} - change the length of the discharge channel, as measured by the corresponding photographs. Results of calculation of E in graph form are shown in figure 4.

This dependence of E on I allows us to give an explanation of a sharp increasing U, arising out of the discharge pulsations (figure 2). Discharge channel elongates and the current decreases in the final phase of pulsations. Both of these factors contributed to an increase of voltage and if they were acting together they accelerated its growth.

The high intensity E of the electric field in the discharge column can be noted as one of the features of the low-current mode of combustion. For example, in a open carbon arc at I = 200 A, they are in the range of 10-20 V/cm [2].

Anode region was painted in greenish color on the instant photos of discharge. The copper vapor had this characteristic. Spectral studies confirmed the receipt of the vapors of copper in the discharge region. In the visible range, the spectral lines of copper was the most intense.

Radiation spectra of copper atoms are used to determine the electronic temperature T_e of the relative intensities of the lines in the approximation of LTE [5]. In this study, T_e was calculated using three pairs of lines: CuI 515.32 and CuI 578.21; CuI 521.82 and CuI 578.21; CuI 510.55 and CuI 521.82. These modes in which the spatial position of the anode spot can be considered stationary (anode spot was moving insignificantly and the current was changing slightly) were chosen. The values of T_e which was obtained by the three pairs of lines of copper atoms at currents of more than 0.3 A, were comparable within the error of calculations. Thus, the acceptability of this method of calculation of T_e were determined by the condition $I \geq 0.3$ A. The average value T_e was accepted as the final result. In the current range of 0.3-0.8 A was a slight increase of T_e with increasing current I. The average value of T_e was within in the range of 5100-5500 K.

4. Conclusions

In the low-current combustion mode, open horizontally oriented electrical discharge had a number of features. The most notable of these were: 1) the length of the discharge channel can many times exceed electrode spacing; 2) the current range of 0.2-0.5 A and the interval of 5-7 mm inter-electrode distance was a favorable factor for the emergence of regular self-oscillations; 3) the strength of the electric field in the discharge column was much higher than in high-current open arc; 4) anode material is heated to high temperatures and erodes in the root spot at currents 0.3 A or more. As a side note it should be mentioned that in this research the electrical discharge was investigated in the transition from arc to the glow discharge and the name "low-current arc" in relation to him was conditional. This was also stipulated in paper [4].

In practice, the combustion mode with regular self-oscillations can be used in processes of surface modification of metals and other solid conductors, in particular graphite.

References

[1] Raizer Y P 2001 *Gas Disharge Physics* (Berlin: Springer)
[2] Drummond J E 2013 *Plasma Physics* (New York: Dover Publications)
[3] German O V, Glinov A P, Golovin A P, Kozlov P V and Lyubimov G A 2013 *Plasma Physics Reports* **39** 1142
[4] Korolev Y D, Frants O B, Landl N V, Bolotov A V and Nekhoroshev V O 2014 *Plasma Sourses Science and Technology* **23** (5) 054016
[5] Fujimoto T 2004 *Plasma spectroskopy* (Oxford: Clarendon Press)

Journal of Physics: Conference Series **652** (2015) 012039 doi:10.1088/1742-6596/652/1/012039

Surface hardening of stainless steel by runaway electrons preionized diffuse discharge in air atmosphere

M V Erofeev[1,2], M A Shulepov[1], K V Oskomov[1] and V F Tarasenko[1,2]

Institute of High Current Electrons, 2/3 Akademichesky ave., Tomsk, 634055, Russia
National Research Tomsk Polytechnic University, 30 Lenin ave., Tomsk, 634050, Russia

E-mail: mve@loi.hcei.tsc.ru

Abstract. In this paper we present microhardness measurements of stainless steel surface treated by diffuse discharge in air atmosphere. The cleaning from carbon in comparison to the initial sample was observed at a depth exceeding 20 nm. The oxygen concentration was also increased in comparison to that in the initial sample at a depth of up to about 50 nm. Comparative analysis shows that after treatment the microhardness of stainless steel surface increased in 2 times due to interaction of near-surface layers with product of plasma chemical reactions produced in diffuse discharge.

1. Introduction

In recent years, diffuse discharges with runaway electron preionization have attracted much interest because of the possibilities to uniformly generate rather big volume of plasma in different gases (including electronegative [1]) at atmospheric and elevated pressures [2]. This property enables the development of plasma processing technology at ambient pressure and temperature without using of special vacuum technique, and can be used in a wide range of applications, e.g. in medicine and surface modification. This type of discharge was realized in non-uniform electric field by applying of high voltage pulse of 2 ns full width at half maximum to cathode with small curvature radius. As a plane anode, stainless steel foil of 200 μm thickness was used. In a runaway electrons preionized diffuse discharge (REP DD), the anode experiences complex action of several factors:

- dense nanosecond discharge plasma with a specific power up to 1 GW/cm^3 in a pulse mode [3];
- a shock wave which was recorded with a calorimeter [4];
- UV, VUV radiation, as well as X-rays from the discharge plasma [5];
- a supershort avalanche electron beam (SAEB) with a wide energy range [4].

In the present study, REP DD was employed for surface treatment of stainless steel in air atmosphere. The aim of the work to study changes in microhardness of near-surface layer, its roughness characteristics, as well as surface element composition which was occurred under the action of REP DD.

2. Experimental setup and methods of measurements

High voltage pulsed generator RADAN-220 provided negative voltage pulses with amplitude up to 280 kV (in the open-circuit regime) and pulse front of 0.5 ns (on a matching load) was used for REP DD formation. At such experimental conditions, the specific input power in air plasma was about 100

MW/cm^3 per pulse. Diffuse discharge was formed between cathode and anode in ambient air of atmospheric pressure. The distance between high voltage tube cathode of diameter 6 mm and plane anode was 16 mm.

Before the experiments, plane samples of 150-μm-thick stainless steel foil was cleaned twice in ultrasonic unit Elmasonic S 10H (Elma-Hans Schmidbauer GmbH & Co.) with distilled water and alcohol at 65 °C. The pulse plasma processing of the steel surface was performed with frequency of 1 Hz.

Microhardness was measured by Berkovich diamond indenter on the system of NanoTest 600 (Micro Materials Ltd., UK) by Oliver-Pharr method [6] at loads of 0.5, 1 and 2 mN. After REP DD treatment, concentration changes of the main chemical elements within the surface layers were studied by Auger electron spectroscopy (AES) with the use of STIL-2 Auger-spectrometer [7]. The surface roughness of treated samples was studied with using of a 3D contactless profilometer Micro Measure 3D Station (STIL).

3. Results and discussion

It has been found that samples modified by REP DD have microhardness almost 2 times higher than untreated ones, and its values increase with increasing of number of pulses (see fig. 1). So microhardness of the initial stainless steel sample was around 5.5 GPa, whereas after the exposure to 3000 discharge pulses it increased ~1.5 times, and the value of the sample treated by 6100 pulses increased ~2 times.

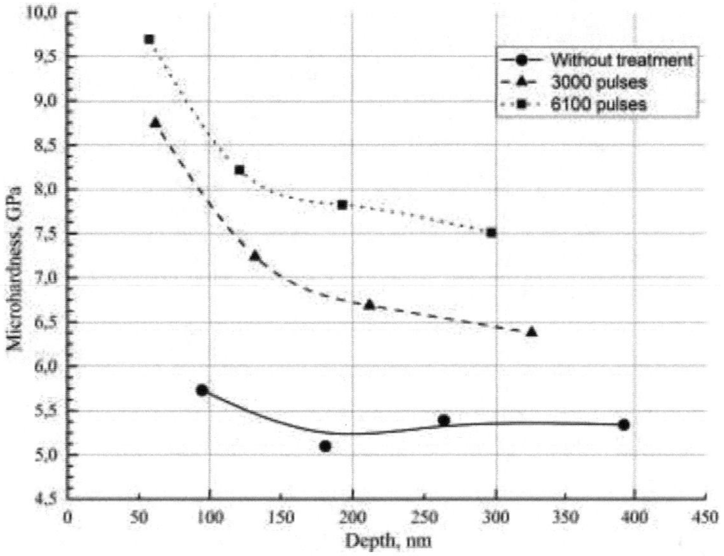

Figure 1. Dependences of surface microhardness versus its depth for initial stainless steel samples, and samples treated by REP DD.

Moreover, study of the sample's surface topology has shown that under the action of REP DD, the planarization of the surface has been occurred. Thus roughness parameter R_a (roughness average, μm) of the initial sample was equal to 0.0315 μm, after treatment by 500 discharge pulses – 0.0234 μm, and for the sample treated with 6100 pulses it decreased in 3 times and was equal to 0.0104 μm. The 2D-profilogram, 3D- topography and statistics data of the geometrical properties of stainless steel surface upon the exposure to 6100 discharge pulses of REP DD are shown on fig. 2.

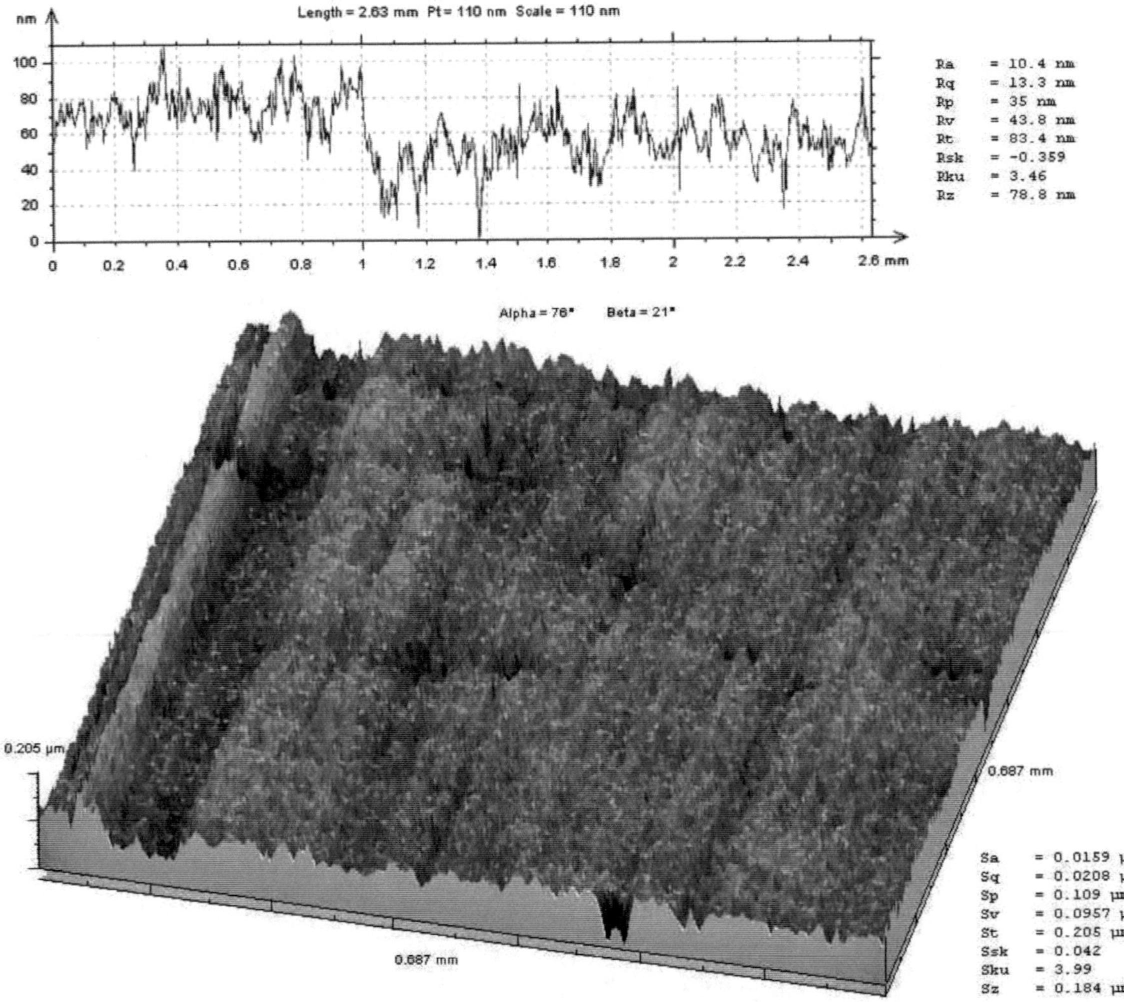

Figure 2. 2D-profilogram, 3D-topography and statistics data of the geometrical properties of stainless steel surface upon the exposure to 6100 discharge pulses of REP DD.

The effect of the diffuse discharge plasma on the examined specimen surface was estimated from the degree of variation in its composition. For this purpose, surface layers of thickness 100–200 nm before and after treatment were analyzed by Auger spectrometry. Fig. 3 shows the carbon and oxygen concentrations in the stainless steel surface layer before and after treatment. It is seen from fig. 3a that the most substantial changes in the carbon concentration of treated samples occurs at a depth of up to 10 nm and decreases with increasing of number of discharge pulses. Thus, after exposure to 1200 pulses the carbon concentration decreases ~2 times, and after exposure to 6100 pulses, it decreases almost 5 times.

Moreover, Auger spectroscopy also showed that under the action of REP DD, the formation of oxide layer on the sample's surface took place. The oxygen concentration and oxide depth in the threated samples increases with increasing of number of discharge pulses. It is seen from fig. 3b that the oxygen concentration and oxide depth for the stainless steel sample treated by 3000 pulses was 10% higher and 1.5 times deeper than in the initial one.

Earlier in [8] it was shown, that in the pulsed periodic mode of voltage pulses with frequency up to 2 kHz and specific input power of 10 MW/cm^3 per pulse, the hardening effect of steel took place. The reason of the hardening effect is the increasing in scalar dislocation density, the formation of two-phase regions of nanosized structure, a multi-stage process proceeds consisting in fracture (dissolution) of initial Fe_3C particles, carbon redistribution in the grain volume, and repeated precipitation of carbide particles. The structural phase transformations revealed in the steel suggest that treatment with a runaway electron preionized diffuse discharge produces thermomechanical action on the material.

The optical technique based on the relative radiation intensity of rotation structure of electronic-vibrational molecular transitions was used for measurement of gas temperature *Tg*. The values of *Tg* was ~ 80°K at condition under study [9], that was not sufficient for modification due to melting of metal surface layers. Electron beam also could not induce it because of its short time duration and low energy. Most likely microhardness was increased due to change in surface layers chemical composition by mixing the oxide and substrate layers after shock waves.

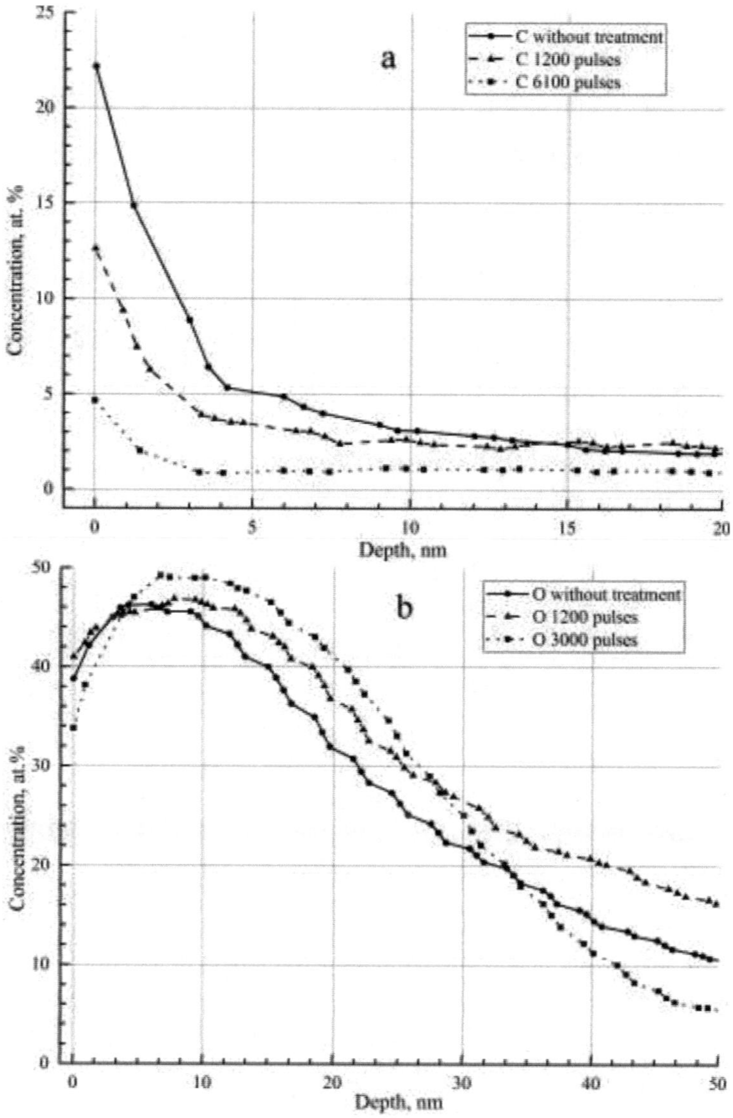

Figure 3. Carbon (a) and oxygen (b) concentration in near-surface layer of stainless steel before treatment and after treatment in air by REP DD with 1200 pulses and 6100 pulses.

4. Conclusion

Thus, experimental results of the action of a runaway electron preionized diffuse discharge show that at atmospheric pressure the planarization of the stainless steel surface, as well as its oxidation and cleaning from carbon took place. It was shown, that mechanical action of shock waves and interaction of near-surface layers with product of plasma chemical reactions produced in diffuse discharge leads to the substructure transformation and may induce hardening effect of stainless steel surface.

Acknowledgement

The authors would like to thank Dr. M.I. Lomaev for the measurements of gas temperature.
The work is performed in the framework of the Russian Science Foundation (the project #14-29-00052).

References

[1] Baksht E Kh, Burachenko A G, Erofeev M V, Lomaev M I, Rybka D V, Sorokin D A and Tarasenko V F 2008 *Laser Phys.* **18** 732
[2] Kostyrya I D, Skakun V S, Tarasenko V F, Tkachev A N and Jakovlenko S I 2004 *Tech. Phys. Lett.* **30** 411
[3] Alekseev S B, Gubanov V P, Kostyrya I D, Orlovskii V M, Skakun V S and Tarasenko V F 2004 *Quantum Electron.* **34** 576
[4] Tarasenko V F, Baksht E K, Burachenko A G, Kostyrya I D, Lomaev M I and Rybka D V 2008 *Plasma Devices Oper.* **16** 267
[5] Baksht E Kh, Lomaev M I, Rybka D V and Tarasenko V F 2006 *Quantum Electron.* **36** 576
[6] Oliver W C and Pharr G M 1992 *J. Mater. Res.* **7** 1564
[7] Shulepov I A 2004 *Development and application of complex of atomic and nuclear-physical methods for studying of modified layers of materials* (PhD thesis, Tomsk)
[8] Shulepov M, Erofeev M, Ivanov Yu, Oskomov K and Tarasenko V 2015 *J. Phys. Sci. Appl.* **5** 33
[9] Tarasenko V F (Editor) 2014 *Runaway electrons preionized diffuse discharges* (New York: Nova Science Publishers, Inc.)

Propagation of the pulsed electron beam of nanosecond duration in gas composition of high pressure

G Kholodnaya, R Sazonov, D Ponomarev, G Remnev
National Research Tomsk Polytechnic University, Lenin Avenue, 30, Tomsk, Russia

E-mail: galina_holodnaya@mail.ru

Abstract. This paper presents the results of the investigation of the propagation of an electron beam in the high-pressure gas compositions (50, 300, and 760 Torr): sulfur hexafluoride and hydrogen, sulfur hexafluoride and nitrogen, sulfur hexafluoride and argon. The experiments have been performed using the TEA-500 laboratory accelerator. The main parameters of the accelerator are as follows: an accelerating voltage of 500 kV; an electron beam current of 10 kA; a pulse width at half maximum of 60 ns; a pulse energy of 200 J; a pulse repetition rate of up to 5 pulses per second, a beam diameter of 5 cm. The pulsed electron beam was injected into a 55 cm metal drift tube. The drift tube is equipped with three reverse-current shunts with simultaneous detecting of signals. The obtained results of the investigation make it possible to conclude that the picture of the processes occurring in the interaction of an electron beam in the high-pressure gas compositions is different from that observed in the propagation of the electron beam in the low-pressure gas compositions (1 Torr).

1. Introduction

For years, in Russia, especially in Tomsk, the particle accelerators of different types have been developed and implemented in technological processes [1-4]. Among a great variety of the accelerators, the DC accelerators and the linear accelerators are widely applied for commercial purposes (through a high efficiency of electrical energy conversion to beam energy, stability of beam parameters, etc.). However, having many advantages the DC accelerators and the linear accelerators have a huge disadvantage, namely, a high cost, which is the result of dc high voltage application, as well as high vacuum diodes (10^{-7} Torr). Using the pulsed accelerators, especially nanosecond ones, enables to sharply decrease the dimensions of high voltage formation schemes through the increase in insulating strength under pulse action. Besides, the nanosecond vacuum diodes are significantly simpler in construction and are capable to operate at relatively low vacuum up to forevacuum ~10^{-2} Torr.

The sources of pulsed energy flows of average power (10^6-10^8W/cm^2) as an instrument in engineering sciences have a high potential of practical application in many fields of economy: mechanical engineering, semiconductor electronics, chemical and biochemical industry, agriculture, medicine, etc. [5-8]. The important factor which constrains the development of particular technological processes in these fields is the necessity to modify the physical model of the interaction of the pulsed electron beams with the matter in a gas phase, the model being supported by the experimental data of the processes which occur in the interaction of such beams not only with model objects in a gas phase, but also with the objects with the complex chemical compound, which are the basic objects in the technological processes.

The physical model complexity is determined by the nonlinear nature of the energy absorption transferred by the beams, by the formation of the charged and excited particles, by the chemical reactions in the interaction zone, by the secondary radiation, and by a set of the phenomena which accompany the interaction of the pulsed electron beams with the matter in a gas phase. A significant part in solving this problem is assigned to the experimental data both regarding the results for forming the model and the model testing and regarding the quantitative description of the processes following the development of the particular technological processes [9-12]. In [13], the investigation results are presented for the electron beam propagation in vacuum and gases (nitrogen, helium, and carbon dioxide). The experiments are performed using the accelerator with the following parameters: an electron energy of 430 keV, a beam current of 10 kA, and a current pulse duration of 15 ns. The beam current was detected using the Faraday cup, when an electron beam was propagated to the drift tube. The authors noted that in the case of the electron beam propagation in nitrogen under a pressure of 50 mTorr, the gas underionization occurs and therefore a partial current neutralization of the beam. The electron beam is dissipated under the action of the electrical repulsion forces. Nevertheless, for nitrogen the current recorded using the Faraday cup is much higher than for the same distances in vacuum. When the pressure in the drift tube reaches 100 mTorr, the effective ionization occurs and the bulk charge neutralization is realized. With increasing the pressure in the drift tube, the determining role in the electron beam propagation is played by the current neutralization for all the types of gases used in this work. In [14], the authors present the results of the experimental and theoretical investigations of the pulsed electron beam in the drift tube filled with the N_2 gas. The pressure range is from 10^{-3} to ~ 10^{3} Topp. The experiments were performed using the Hermes-III accelerator (an electron energy of 19 MeV, a beam current of 700 kA, a duration of 25 ns). The authors demonstrate the results, which indicate that for the given type of the electron beam there are two modes of the effective propagation, which are in their turn related to the nitrogen pressure in the drift tube. The first mode is described with the pressure range from 1 to 100 mTorr (low-pressure region), the second mode is from 1 to 100 Torr (high-pressure region). Beyond the given types of the pressure, the fast-developing instability was observed, which results in the total decomposition of an electron beam. In the first mode of propagation, the beam is significantly ionized, and after the plasma ion generation the bulk ion charge is formed, which will restrict the electron beam. The authors of paper [15] studied the behavior of the intensive relativistic electron beam in its propagation in a high-voltage gas media. The resistive firehose instability occurs under such conditions. The gas media was a mixture of argon with air. The authors showed the possibility to decrease the impact of instability during propagation of the electron beam varying the parameters of the beam (including a beam current, a beam radius, a pulse duration reduction).

Nowadays, there is a lack of experimental results of the electron beam propagation in different gas compositions. The data mainly concern the single-component gas media. The purpose of the current work was to study the efficiency of the pulsed electron beam propagation in the high-pressure gas compositions (50, 300, and 760 Torr): sulfur hexafluoride and hydrogen, sulfur hexafluoride and nitrogen, sulfur hexafluoride and argon. The choice of the specified gas compositions is determined by the wide application of the given gases in the technological processes of many industrial technologies as a work media.

2. Experimental setup

The pulsed electron beam propagation in this work has been studied using the TEA-500 pulsed electron accelerator [16-17]. The electron beam in the TEA-500 accelerator is formed caused by the explosive emission from a surface of the graphite cathode with a diameter of 45 mm. The beam was ejected through an anode window, which was a support grid (with an optical transparency of 95%), and through an aluminum foil of 140 μm thickness. The anode-cathode gap for all the experiments was 13 mm. The stability of the accelerator operation was controlled using a Rogowski coil and a capacitor voltage divider. The dispersion of the values of current and voltage recorded by sensors did not exceed 5 %.

The electron charge ejected behind the anode foil is 390-430 μC. A drift tube equipped with reverse current shunts was designed for performing experimental investigations of the pulsed electron beam propagation in the high-pressure gas compositions (Fig. 1). The drift tube was a metal pipe of 15.1 cm in diameter and 55 cm long.

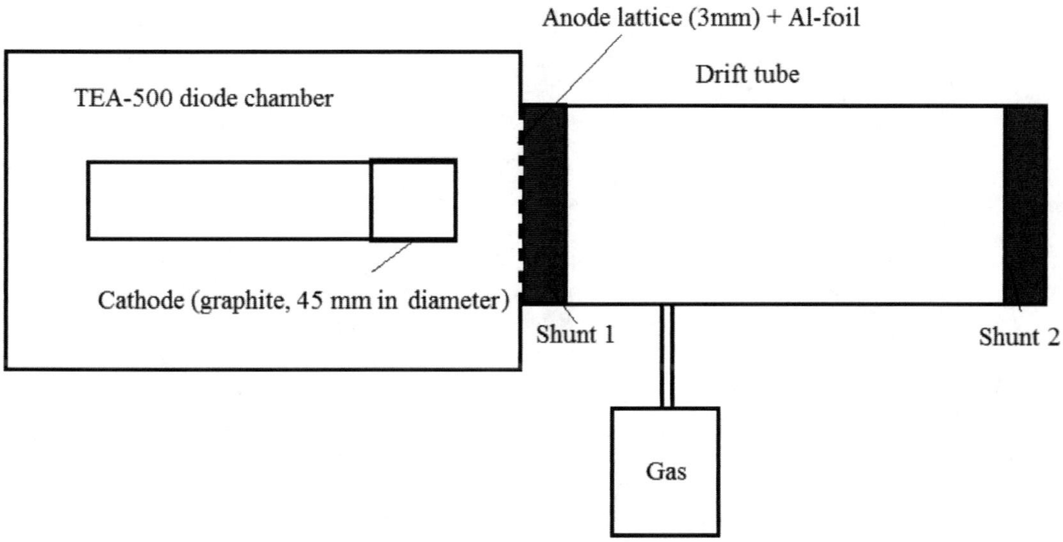

Figure 1. Scheme of the experimental stand to study the pulsed electron beam propagation in gases using reverse current shunts

The experiments were carried out as follows: the drift tube was filled with the gas under study, the electron beam was injected, and the reverse current shunts detected oscillograms of currents at the same time. Using Shunt 1, we registered the total reverse current closed on the walls of the drift tube; Shunt 2 was used to register the electron current reached the back flange of the drift tube. The value for the charge of the electrons fallen on the walls of the drift tube as a quantitative assessment of the efficiency of the pulsed electron beam propagation in high-pressure gas compositions: sulfur hexafluoride and hydrogen, sulfur hexafluoride and nitrogen, sulfur hexafluoride and argon. The experiments were performed with the transportation both in pure argon, nitrogen, hydrogen, and sulfur hexafluoride and in the mixtures of the specified gases with sulfur hexafluoride. The total pressure in the drift tube was fixed as 760 or 300 or 50 Torr, and the relations Ar: SF_6= 1:1, Ar : SF_6= 2:1, and Ar : SF_6= 1:2. The analogical relations are also for the other gas compositions.

In our work, we also used a calorimeter, which was located in the drift tube at the varied distance from the output window of the electron accelerator. The calorimetric investigations are performed for the gas compositions Ar:SF_6=1:1, Ar:SF_6=2:1, Ar:SF_6=1:2 under a pressure of 300 Torr.

3. Results and discussions

Figure 2 presents the oscillograms obtained for the gas compositions under a pressure of 50 Torr. I_1 is the total reverse current closed on the walls of the drift tube, I_2 is the electron current reached the back flange of the drift tube.

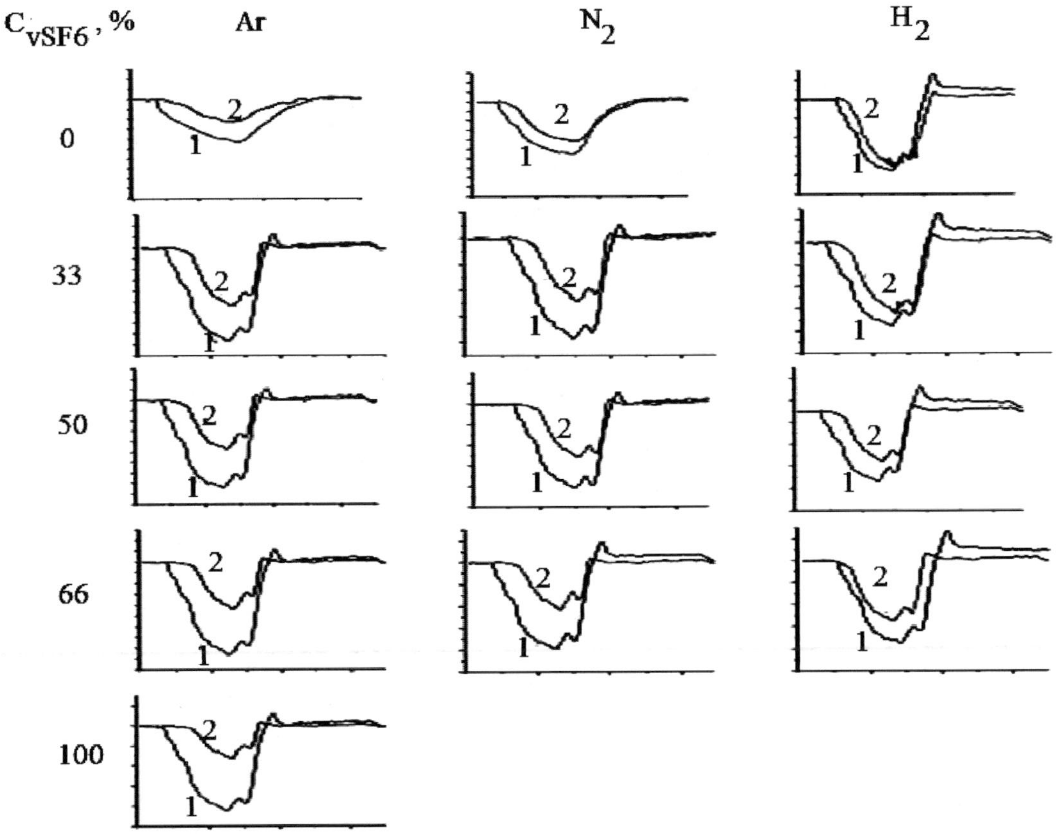

Figure 2. The oscillograms of the total reverse current closed on the walls of the drift tube (1) and the electron current reached the back flange of the drift tube (2).

Figure 2 shows that during the propagation of the electron beam in argon and nitrogen with the total pressure in the drift tube of 50 Torr the amplitudes of the total reverse current closed on the walls of the drift tube are almost similar, but the shapes differ. In nitrogen and argon under a pressure in the drift tube of 50 Torr, the handover of the falling edge of the signal is observed, which is more pronounced for argon. However, such a characteristic of the falling edge is not typical to hydrogen. In hydrogen, the amplitude of the total reverse current closed on the walls of the drift tube and the current of electrons reached the back flange of the drift tube correlate, the minimum scattering of the beam electrons onto the walls of the drift tube occurs. During the propagation of the electron beam in sulfur hexafluoride, the charge of the electrons reached the back flange of the drift tube is 98 μC, and the charge of the electrons fallen on the walls of the drift tube is 293 μC, respectively. For argon, nitrogen, and hydrogen, the charge of the electrons fallen onto the walls of the drift tube is 175 μC, 98 μC, and 43 μC, respectively. When in the drift tube a gas composition $Ar:SF_6$ = 2:1 or $N_2:SF_6$ = 2:1, the value of the charge of electrons reached the back flange of the drift tube and the charge of the electrons fallen on the walls of the drift tube correlate within the limits of experimental error ($Ar:SF_6$: 195 μC and 212 μC, $N_2:SF_6$: 171 μC and 185 μC). In the case when in the drift tube the amount of sulfur hexafluoride is higher, then a large dissipation of an electron beam occurs ($Ar:SF_6$: q=255 μC, $N_2:SF_6$: q=226 μC). In the gas composition when $H_2:SF_6$=1:1, 2:1, 1:2 the values of the charge of electrons reached the back flange of the drift tube q_1 and the charge of the electrons fallen on the walls of the drift tube q_2 are 183 μC /250 μC, 243 μC /98 μC, 203 μC /141 μC, respectively. The main feature typical to the specified pressure and all the ratios of the gas compositions is that the value of the total charge of the beam closed on the walls of the drift tube is always less than the charge of the electrons ejected behind the anode foil. It can be caused by the beam reverse current shorting on the skin-layer of plasma.

For the experiments with a higher pressure of 300, 760 Torr, the results of the experiments are presented in Table 1.

Table 1. The value of the electron charge reached the back flange of the drift tube q_1 and q_2 fallen onto the walls of the drift tube under the total pressure of 300 and 760 Torr (q_1/q_2, μC).

	300 Torr			760 Torr		
C_{vSF6}, %	Ar	N_2	H_2	Ar	N_2	H_2
0	0/163	77/244	55/272	0/163	42/273	76/267
33	0/356	0/329	24/297	0/289	0/313	0/275
50	0/357	0/336	0/322	0/256	0/262	0/257
66	0/350	0/282	0/310	0/246	0/232	0/253
100	0/337	-	-	0/145	-	-

Analyzing the data presented in the table, we can make the following conclusions: despite the increase in the total pressure in the drift tube, the most effective transportation of the electron beam is observed in nitrogen, hydrogen, and gas composition $H_2:SF_6=2:1$ (judging from the value of the charge of electrons reached the back flange of the drift tube). With increase in the pressure in the drift tube for all the gas compositions, the maximum dissipation of the electron beam on the walls of the drift tube is typical. Adding sulfur hexafluoride to argon, nitrogen, and hydrogen causes more qualitative energy absorption of the electron beam. The value of the charge of the electrons fallen on the walls of the drift tube in argon under the considered pressures within the given work is similar within the limits of experimental error and is 170 μC. Under the pressures of 300 and 760 Torr for argon and all the gas compositions with it, the charge of electrons reached the back flange of the drift tube is 0. Here when there is pure argon in the drift tube (300 and 760 Torr), the value of the charge of the electrons fallen on the walls of the drift tube is minimum compared to all other gases and gas compositions presented in the given work. It can be related to the fact that the electron beam is turned off as a result of the formation of the virtual cathode or the beam reverse current is closed on the skin-layer of plasma. To realize if the virtual cathode is formed, we performed the calorimetric study for argon, gas compositions $Ar:SF_6$ for three pressures 50, 300, and 760 Torr. The typical dependence of the ejected beam energy on the distance in the gas mixture of argon and sulfur hexafluoride, 300 Torr, is represented in Figure 3.

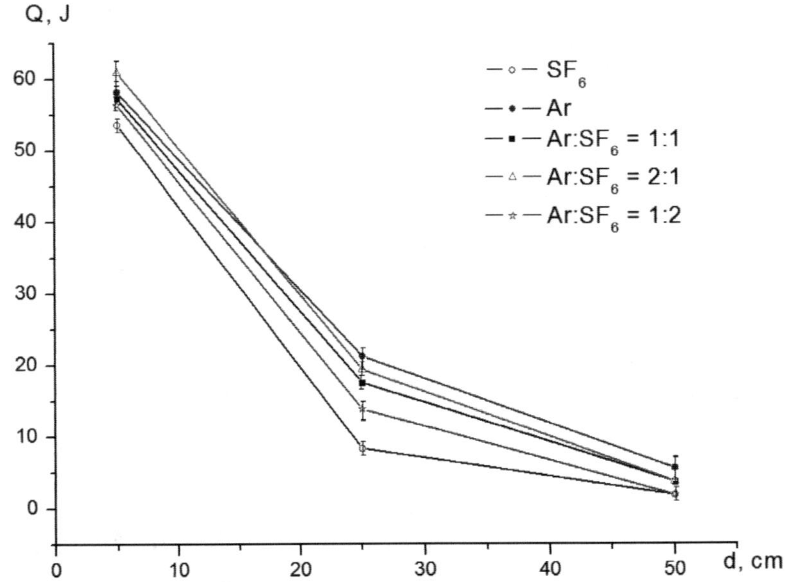

Figure 3. The ejected beam energy vs. the distance in the gas mixture of argon and sulfur hexafluoride, 300 Torr.

Figure 3 shows that the electron beam is propagated throughout the length of the drift tube. The virtual cathode is not formed.

4. Conclusion

Thus, the obtained results of investigations make it possible to conclude that during the propagation of the pulsed electron beam in argon, nitrogen, hydrogen, sulfur hexafluoride, and gas compositions on their basis under the total pressure in the drift tube of 50, 300 or 760 Torr the virtual cathode is not formed. The specific feature of the pulsed electron beam transportation in argon is plasma formation, because the conductivity of plasma is enough for the beam current to close over it. Consequently, argon is more effective as a plasma-forming gas in technological processes regarding the optimization of energy pumping into the gas media.

Acknowledgements

The research was conducted with the financial support of State Task “Science”, Ministry of Education and Science of the Russian Federation. Project number: 11.939.2014/k.

References

[1] Egorov I, Esipov V, Remnev G, Kaikanov M, Lukonin E, Poloskov A 2013 *J. IEEE Trans. Dielect. Electr.Insulation.* **20** 1334

[2] Pushkarev A I, Isakova Y I 2013 *J. Surface and Coatings Technology.* **228** S382.

[3] Devyatkov V N, Koval N N, Schanin P M, Grigoryev V P, Koval T V 2003 *J. Laser and Particle Beams.* **21** 243.

[4] Schanin P M, Koval N N, Devyatkov V N 2001 High-current low-energy plasma electron sources *28th IEEE International Conference on Plasma Science/ 13th IEEE International Pulsed Power Conference (United States, Las Vegas,17- 22 June 2001).* P1B06.

[5] Egorov I S, Kaikanov M I, Lukonin E I, Remnev G E, Stepanov A V 2013 *J. Instr. Exp.Tech.* **56** 568.

[6] Sokovnin S Yu, Il'Ves V G 2012 *J.Ferroelectrics.* **436** 101.

[7] Ponomarev D V, Remnev G E, Sazonov R V, Kholodnaya G E 2013 *J.IEEE Trans. Plasma Science*. **41** 2908.

[8] Sazonov R, Kholodnaya G, Ponomarev D, Remnev G, Razumeko O. 2011 *J. Korean Phys. Society* **59** 3508.

[9] Koval T V and Le X Z 2013 *Russian Physics Journal* **56** 103.

[10] Ozur G E, Grigoryev V P, Karlik K V, Koval T V and Le X Z 2011 *J. Tech. Phys* **81** 100.

[11] Sazonov R V, Kholodnaya G E, Ponomarev D V, Remnev G E 2014 *J. Phys. Plasmas* **21** Article number 072302.

[12] Kholodnaya G, Sazonov R, Ponomarev D, Remnev G, Vikanov A. 2014 *J. Phys: Conference Series* **552** Article number 012017.

[13] Luches A, Nassisi V, Perrone 1979 *J. Appl. Phys.* **50** 2502.

[14] Sanford T W L 1995 *J. Phys. Plasmas* **2** 2539.

[15] Antoniades J, Myers M, Murphy D, Hubbard R, Peyser T, Fermier R, Pechacek R, Santos J, Meger R 1992 *Proceedings of the 9th International Conference on High-Power Particle Beams* **2** 1245.

[16] Remnev G E, Furman E G, Pushkarev A I, Karpuzov S B, Kondrat'ev N A and Goncharov D V 2004 *J. Instr. Exp, Tech* **47** 394.

[17] Isakova Y I, Pushkarev A I, Kholodnaya G E 2011 *J. Instr. Exp, Tech* **54** 183.

Journal of Physics: Conference Series **652** (2015) 012041 doi:10.1088/1742-6596/652/1/012041

On plasma jet formation in vacuum arc with composite cathode

D L Shmelev[1,*], **S A Barengolts**[2], **I V Uimanov**[1], **M M Tsventoukh**[3], **K P Savkin**[4]

[1] Institute of Electrophysics, RAS, 106 Amundsen St., Ekaterinburg, 620016, Russia
[2] Prokhorov General Physics Institute, RAS, 38 Vavilov St., Moscow, 119991, Russia
[3] Lebedev Physical Institute, RAS, 53 Leninsky Ave., Moscow, 119991, Russia
[4] Institute of High Current Electronics SB RAS, 2/3 Akademichesky Ave., Tomsk 634055, Russia

E-mail: shmelev@iep.uran.ru

Abstract. This paper deals with the computer modeling of vacuum arc with composite multicomponent cathode. This arc is typical for certain kind of ion sources, plasma generator and vacuum interrupters. The described hybrid model treats the electrons as an inertialess fluid and ions as macroparticles. The macroparticle dynamic is calculated with the use of particle-in-cell method. Ion-ion Coulomb collision is considered with the use of Monte Carlo method. The model can simulate vacuum arc as a whole including separate cathode plasma jets, mixing zone, and common plasma column. The dependence of ion angular current distribution on the cathode composition reproduced with the help of developed model agrees well with experimental results.

1. Introduction

Research on plasma produced by vacuum arc remains an area of certain scientific interest in plasma physic [1, 2]. This plasma is used for various technical applications [2]. In particular, the vacuum-arc-produced metal plasma is used in ion sources [3-6] to produce intense ion beams for ion implantation, in plasma generator for coating deposition and in high-current interrupters [7, 8]. In order to produce the multielement ion beam or to form the composite metal coating the vacuum-arc-produced plasma from composite multicomponent cathode is used.

Recently, it has been found [3, 4] that ion angular distributions in the plasma jet of vacuum arc are different for the various ion types. Moreover, the angular distribution of ions in the plasma arc with a pure (single component) cathode differs from the angular distribution for the same type of ions in an arc with the composite cathode [3].

The dependence of the angular distribution from the charge of the ion can be explained by the inhomogeneous distribution of electron temperature in the region of strong ionization in the vicinity of the cathode spot. However, the dependence of the angular distribution on the composition of the cathode is difficult to explain as well. It is assumed that the angular distribution of ions in arc plasma is formed under the influence of the mixing of the plasma jets originated from the separate group-spots. However, the simulation of mixing the plasma jets presents some difficulties because this process can not be described in the framework of magnetohydrodynamic approach. In the area of the mixing zone the ion velocity distribution strongly deviates from Maxwell distribution. A suitable choice in this case is to use a hybrid model [8], which describes the electron subsystem as a massless fluid, but ions are described as macroparticles with the help of the particle-in-cell approach. Thus, the motivation of the present work is to try to reconstruct the experimental [3, 4] ion angular distributions with the help of the newly developed hybrid model, which is able to simulate the separate jets of multicomponent plasma from cathode spots, the mixing of the jet and the common plasma column.

2. Model description

A two-dimensional hybrid model was developed to study the formation of the plasma jet originating from the composite cathode. According to the hybrid approach the electron subsystem is described as an inertialess fluid, but ions are described as macroparticles. The developed model is analogous to model [8], but the described model is modified to account for the dynamics and interaction of the ions of various masses and charges. The governing equations in case of the 2D axial symmetrical geometry are the following:

$$\frac{d\vec{r}_i}{dt}=\vec{V}_i; \qquad m_i\frac{d\vec{V}_i}{dt}=e_iE+\frac{e_i}{c}\vec{V}_i\times\vec{B}+\frac{m_en_e}{n_i}\frac{\vec{u}_e-\vec{V}_i}{\tau_{ei}}, \tag{1}$$

$$n_i=\frac{1}{H_c}\sum_{k=1}^{N_c}S(x-x_k); \quad \vec{u}_i=\frac{1}{H_c}\sum_{k=1}^{N_c}\vec{V}_k\,S(x-x_k), \tag{2}$$

$$n_e=\frac{1}{e}\sum_i e_in_i; \qquad \vec{u}_e=\frac{1}{en_e}\left(\sum_i e_in_i\vec{u}_i-\vec{J}\right), \tag{3}$$

$$\frac{\partial B_\theta}{\partial t}+\frac{\partial u_{ez}B_\theta}{\partial z}+\frac{\partial u_{er}B_\theta}{\partial r}=\frac{c^2}{4\pi}\left(\frac{\partial}{\partial r}\left(\frac{1}{r\sigma}\frac{\partial rB_\theta}{\partial r}\right)+\frac{\partial}{\partial z}\left(\frac{1}{\sigma}\frac{\partial B_\theta}{\partial z}\right)\right), \tag{4}$$

$$\vec{J}=\frac{c}{4\pi}\nabla\times\vec{B}, \qquad \vec{E}=-\frac{\vec{\nabla}P_e}{en_e}-\frac{1}{c}\vec{u}_e\times\vec{B}-\frac{m_e}{e}\sum_i\frac{\vec{u}_e-\vec{u}_i}{\tau_{ei}}, \tag{5}$$

where V_i—velocity of ion (of type i) macroparticles, r_i—ion macroparticles radius vector, S—shape function of macroparticle (bilinear interpolation was used), H_c—form factor of cell, N_c—number of macroparticles in cell, n_i—ion density, n_e—electron density, u_i—ion drift velocity, u_e—the electron drift velocity, J—total arc current density, τ_{ei}— electron-ion collision time, σ—plasma (Spitzer) conductivity, B—magnetic field (only θ component exists in a given geometry), E—electric field, P_e—electron pressure. It is assumed in addition that electron temperature T_e remains constant during the calculation for simplicity.

The macroparticles undergo ion-ion Coulomb collisions. These collisions have been considered with the help of Monte Carlo method [9].

Two types of cathode boundary conditions (CBC) are tested to analyze the influence of CBC of certain types on the plasma jet formation. The first type of boundary conditions (CBC-1) is conceptually identical to supersonic CBC used in the computer models of high current vacuum arc [7]. It is assumed that the numerous individual plasma jets from the separate cathode spots mix together in extreme vicinity of the cathode surface to produce one common plasma jet with the averaged laminar ion flux. Practically it means the specification of one solid cathode attachment zone with the effective current density, which is defined as a total arc current divided by the area of the attachment zone. Plasma parameters (velocity, charge composition, temperature) at the boundary are assumed to be identical to the parameters typical for the cathode spot plasma. The plasma density is obtained from the measured specific erosion. In addition, it is assumed that the cathode attachment zone emits ion flux in a cone with the angle of $\pi/2$.

The second type of the tested cathode boundary conditions (CBC-2) is constructed to represent certain distribution of the cathode group-spots over the cathode surface. The arbitrary distributed small

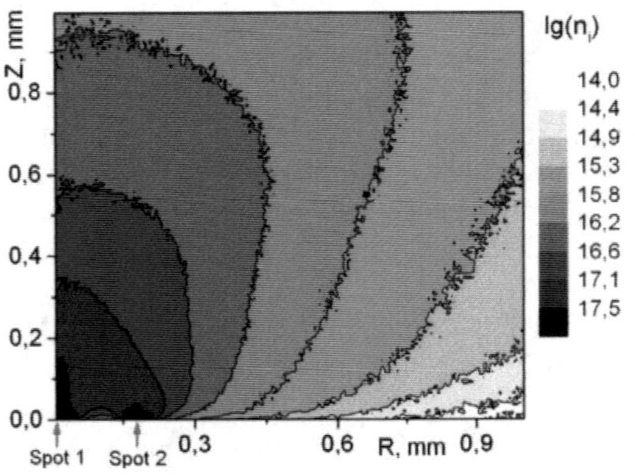

Figure 1. Spatial distribution of total ion density (Pb+Zn, all charge states) n_i (cm^{-3}) for CBC-2. Total current 500 A.

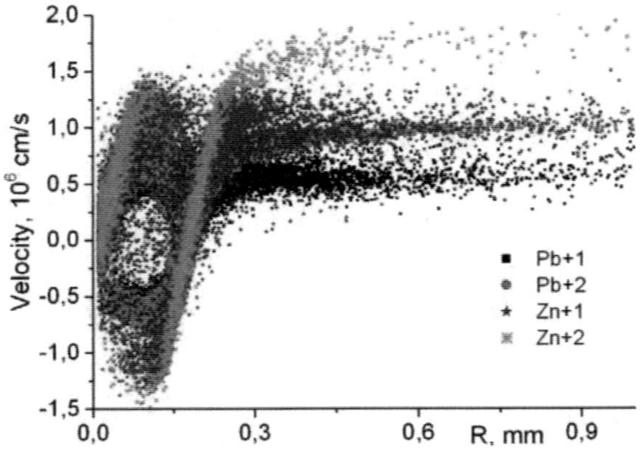

Figure 2. Phase portrait (r, V_r) of ions at the distance of 25 μm from the cathode surface; CBC-2; Total current 500 A.

attachment areas at the cathode (let us call it a quasi-spot, in order not to confuse it with the real group-spot) emit current and ion within the certain angle. The parameters of plasma emitted from the quasi-spots are the same as in the case of CBC-1. The cathode surface between the quasi-spots only absorbs ions.

The cathode attachment areas (quasi-spots), which emit the plasma jet, mimic the cathode group-spot. In the present calculations the simplest two-spot arrangement is used (Figure 1). Actually the spot 1 has a circle shape but the spot 2 has a ring shape (because of axial symmetry). The spot 2 mimics a number of separate group-spots surrounding the spot 1. The lateral dimension of the quasi-spot and the distance between them are visible in Figure 1. Effective current density in the quasi-spots (for total current 500 A) is about of 10^6 A/cm^2. Current densities through the cathode attachments in the case of CBC-1 and in the case of CBC-2 are equal.

The position of the quasi-spot is fixed during the calculations. In principle the hybrid model is a nonstationary one. Hence, it is possible to model the moving spots. But currently we are modeling a "snapshot" of arc for simplicity (like in [8]).

3. Results and discussion

Results discussed in this section are obtained for arc plasma with composite cathode $Zn_{0.4}Pb_{0.6}$. For this cathode the following initial values are used: initial velocity for Zn – 1.15*10^6 cm/s, for Pb – 0.55*10^6 cm/s [3]; charge state composition for Zn – 80% Zn+1, 20% Zn+2, for Pb – 36% Pb+1, 64% Pb+2 [3]; specific erosion for Zn – 54.5 μg/C, for Pb – 172.8 μg/C [6]; ion temperature – 0.5 eV for all types; electron temperature – 2.5 eV [2].

Typical steady state spatial distribution of ion density for CBC-2 is shown in Figure 1. That is the summarised density of all ion components. It is seen that the plasma jets originated from the separate sources (Spot 1 and Spot 2) mix together and produce one common jet at a distance less than 0.2 from the cathode. In general, the mixing takes place at a distance comparable with the size of the cathode attachment.

At the mixing area the kinetic energy of ion counter-streams (originated from different spots, see Figure 2) is converted to chaotic energy by means of ion-ion coulomb collision. Due to this an average effective temperature at the mixing zone reaches 20 eV (for the case shown in Figure 1). After mixing, the jet expands so that the effective ion temperature decreases rapidly. At the distance of 1.5 mm from the cathode the ion temperature decreases to 0.5 eV. The ions of different types heated up at the mixing zone come to local thermodynamic equilibrium (at least in certain volume). Thus, the light ions (Zn in our case) get higher chaotic velocity than the heavy ions. Further, during the common jet expansion the light ions should expand faster than the heavy ions. Particularly, this effect should be

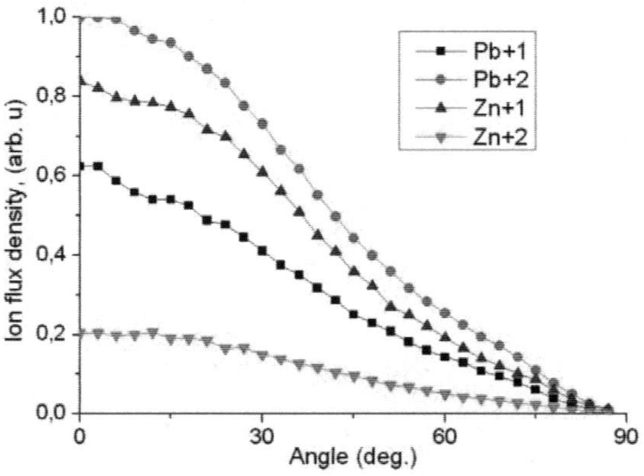

Figure 3. Angular distributions of ion flux density; $Zn_{0.4}Pb_{0.6}$ cathode; CBC-1

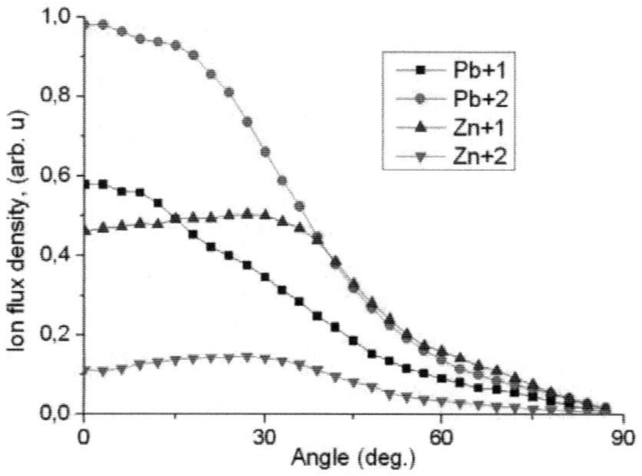

Figure 4. Angular distributions of ion flux density; $Zn_{0.4}Pb_{0.6}$ cathode; CBC-2.

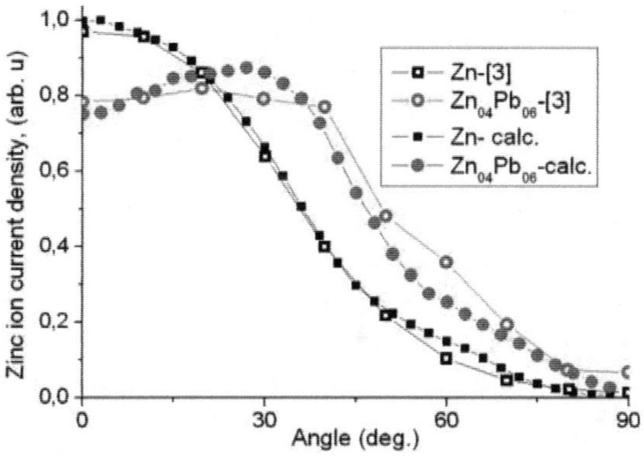

Figure 5. Comparison of Zn ion current angular distribution with experimental results [3] for pure Zn and for composite $Zn_{0.4}Pb_{0.6}$ cathode. CBC-2.

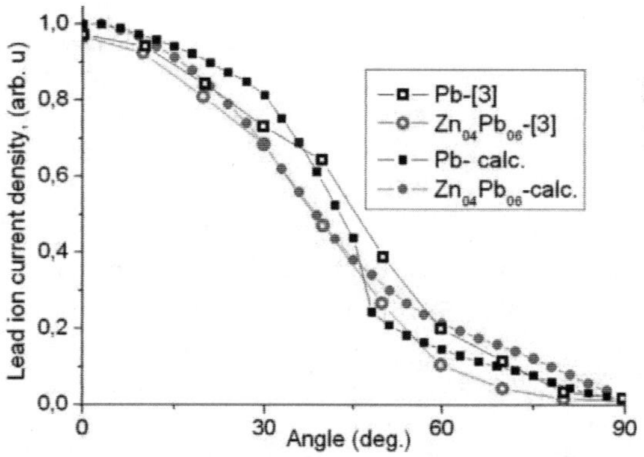

Figure 6. Comparison of Pb ion current angular distribution with experimental results [3] for pure Pb and for composite $Zn_{0.4}Pb_{0.6}$ cathode. CBC-2.

more evident along a line which has a certain angle with the jet axis. The ion density relatively slow decreases along the axis so that the ion-ion interaction will equalize more effectively the velocities of ions of different atomic weights. The heavy ions, compensating the momentum that gone with the light ions should expand more slowly than it was in case of one-component cathode. The calculations made with the help of our model prove the described scenario.

Angular ion flux density distributions shown in Figure 3 and Figure 4 are calculated at the distance of 1 mm from the cathode center (Figure 1). It is seen that in the case of CBC-1 (solid attachment, no jet mixing) the curves of angular distributions of different ions differ in amplitude but have a similar shape. In the case of CBC-2 (Figure 4) the shape of light ions flux distribution is noticeably different from the shape of heavy ion flux distribution. The comparison of calculated angular distributions with ones obtained in [3] is shown in Figure 5 and Figure 6. In accordance with the experimental results the calculated distribution of Zn is wider for composite cathode than the same distribution for one-component cathode. For Pb ions (Figure 6) the agreement with the results of [3] is observed though it is not so clearly as in the previous case. Thus, the jet mixing in front of the cathode attachment affects the ion current angular distribution properly.

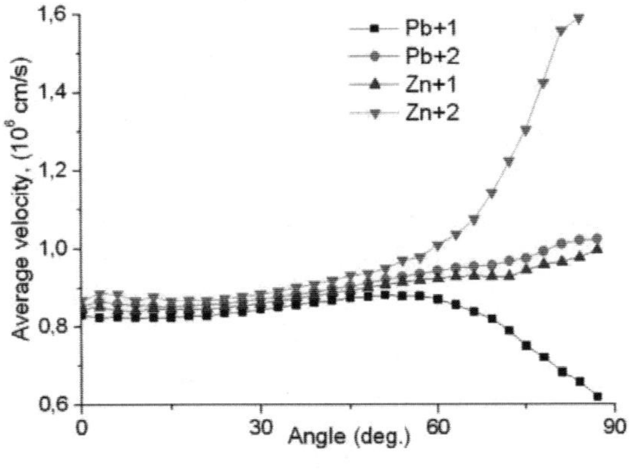

Figure 7. Angular distributions of average ion velocities. $Zn_{0.4}Pb_{0.6}$ cathode; CBC-2.

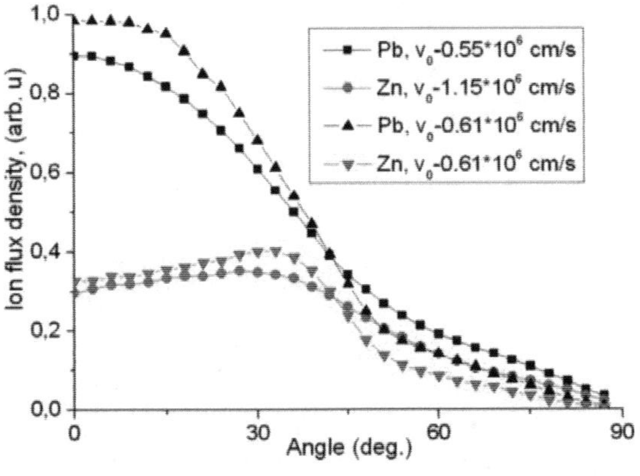

Figure 8. Comparison of angular distributions of ion flux densities for different initial ion velocities. $Zn_{0.4}Pb_{0.6}$ cathode; CBC-2.

Angular distributions of average ion velocity for ions of various types are shown in Figure 7. It is seen that the velocities of different ions are close at the angles less than 45 degrees and differ from each other at the higher angles. The reason for this is the ion-ion coulomb collisions. In the vicinity of the jet axis the ion density decreases slower than at the jet periphery (see Figure 1) so that the ion-ion collisions keep effectiveness at small angles for longer distance. At the jet periphery the ion-ion collisions are ineffective but electron-ion interaction (last term on RHS in (1)) seems to be excessively effective. Probably it indicates certain restriction of the developed model.

The results shown in Figures 4-7 were obtained in assumptions that Zn and Pb ions have different initial velocities ($1.15*10^6$ cm/s and $0.55*10^6$ cm/s respectively [3]). It is roughly means that the cathode spot size is less than the pure element grain so that ions of different elements are produced in different cathode spots at a certain distance from each other. If due to some treatment the mixture of elements is improved then the ions of the different elements can be produced in one cathode spots. In this case the ions of different mass will have the same average velocity due to very strong ion-ion collision in dense cathode spot plasma [10]. In the framework of the model of [11] the calculated average velocity of Zn-Pb plasma was $0.61*10^6$ cm/s for all ion types. It is necessary to clarify that the ions originated from the cathode spots have quite a wide distribution of kinetic energy [12]. This is in agreement with ecton theory [1, 13], which declares a cyclic behaviour of the cathode spot phenomena. At different stages of the cycle the ions emitted from the spot have different drift velocities, but at any stage the ions with different charges have the same drift velocity [14]. From this viewpoint the velocity obtained within the model of [11] can be treated as a velocity averaged over the ecton cycle [1, 13]. The ion flux angular distributions calculated for this case are shown in Figure 8. The comparison with previous results shows only weak dependence on the ion initial velocity. This effect confirms the important role of jet mixing in the formation of common ion flux from composite cathode.

Acknowledgments
The work was supported by Russian Science Foundation; Project No. 14-19-00083.

References

[1] Mesyats G A 1995 *IEEE Trans. Plasma Sci.* **23** 879
[2] Anders A 2008 *Cathodic Arcs: From Fractal Spots to Energetic Condensation* (Springer, New York)

[3] Nikolaev A G, Oks E M, Savkin K P, Yushkov G Yu, Frolova V P and Barengolts S A 2014 *J. Appl. Phys.* **116** 213303
[4] Nikolaev A G, Savkin K P, Yushkov G Yu, and Oks E M 2014 *Rev. Sci. Instrum.* **85** 02B501
[5] Savkin K P, Yushkov Yu G, Nikolaev A G, Oks E M, and Yushkov G Yu 2010 *Rev. Sci. Instrum.* **81** 02A501
[6] Anders A, Oks E M, Yushkov G Yu, Savkin K P, Brown I G, and Nikolaev A G 2005 *IEEE Trans. Plasma Sci.* **33** 1532–1536
[7] Schade E, Shmelev D L 2003 *IEEE Trans. Plasma Sci.* **31** 890–901
[8] Shmelev D L, Uimanov I V 2015 *IEEE Trans. Plasma Sci.* **43** 2261-2266
[9] Bobylev A V, Potapenko I F 2013 *J. Comp. Phys.* **246** 123-144
[10] Shmelev D L, Barengolts S A, Shchitov N N 2014 *Technical Physics Lett.* **40** 783-786
[11] Shmelev D L, Barengolts S A, Savkin K P 2015 *Technical Physics Lett.* **41** 500-503
[12] Zhirkov I, Oks E, and Rosen J 2015 *J. Appl. Phys.* **117** 093301
[13] Mesyats G A 2013 *IEEE Trans. Plasma Sci.* **41** 676-694
[14] Shmelev D L, Barengolts S A, 2013 *IEEE Trans. Plasma Sci.* **41** 1964-1968

Journal of Physics: Conference Series **652** (2015) 012042 doi:10.1088/1742-6596/652/1/012042

Pulse volume discharges in high pressure gases

V A Yamshchikov

Institute for Electrophysics and Electric Power RAS, 18 Dvortzovaya nab., Saint Petersburg, 191186, Russia

E-mail: yamschikov52@mail.ru

Abstract. New approach for suppression of plasma inhomogeneities and instabilities in the volume self-sustained discharge is offered. The physical model is offered and conditions of obtaining extremely homogeneous self-sustained discharge are defined (with full suppression of plasma inhomogeneity and instability). Results of calculations agree with experiments.

1. Introduction

Interest in self-sustained volume discharge (SSVD) was largely provoked by the development of electric-discharge lasers that operate at sub-atmospheric and super atmospheric pressure. A necessary condition for obtaining SSVD is the presence of preionization and the homogeneity of the electric field in the discharge gap (DG). This ensures the homogenization of ionization processes in the active volume [1,2]. However, this condition is not sufficient due to the instability of the discharge. Powerful electrophysical devices require stable operation with a pulse repetition rate of 1 kHz or with the discharge volume of nearly 100 liters. Self-sustained discharge in such cases should be quasi-continuous or it should have a burning duration of 10 μs. For these purposes the fundamental problem is the complete suppression of plasma inhomogeneities in volume self-sustained discharge. Unlike conventional SSVD it can be called a extremely homogeneous self-sustained discharge. In this paper we propose a new approach to the suppression of plasma inhomogeneities and instabilities in SSVD [3,4].

2. Main types of inhomogeneities and instabilities in SSVD

Let's consider the most distinctive types of local inhomogeneities.

A) The stage of discharge formation is a stage of discharge development from the moment of applying voltage pulses to the electrodes up to appearance of the cathode potential drop. At this stage the ionization multiplication of initial electrons and removal of electrons through the anode take place. These processes have wave character mainly. Due to electrons removing from the cathode region their concentration may become essentially less, than concentration of initial electrons n_0. Therefore, even at performance of a condition for overlapping primary avalanches, in this area streamers are able to emerge from not overlapping secondary avalanches. Also ionization waves lead to large-scale cross-section irregularities. Cathode wave moves most quickly in the discharge center, creating more dense plasma also in the center [5].

In [6] there was described the discharge model which considered that the velocity of electrons motion along the field lines is more than velocities in other directions. Therefore the electron reproduction occurs in the form of chains of the avalanches trailing each other. As a result, the plasma column appears non-homogenous and consists of plasma micro strings extended in a direction of a field.

The effect of noted negative phenomena decreases with growth of n_0 and becomes not essential at $n_0 = 10^8$-10^9 cm^{-3}. But even in these conditions, development of cathode instability occurs as follows [7]. At the initial stage of discharge plasma column removes from the cathode, forming electron-poor region. In this region filamental distribution of current occurs. As a result there are local places with raised electron emission on cathode, where cathode spots appear.

B) Discharge burning stage starts from occurrence of cathode layer, which provides effective electron emission due to bombardment of cathode by ions. The inhomogeneities, arisen at the stage of SSVD formation, grow quickly under the action following energy input. Cathode spots turn into unfinished spark channels. As a result, the duration of sustained discharge burning decreases with increasing pump power density [3,4].

Also on the burning stage of SSVD different types of volume instabilities may occur. They are ionizing-overheating instability, ionizing instability, instability caused by stepwise particle ionization. The condition of volume instabilities is usually certain threshold energy density released in the positive column. For example, in mixtures of CO_2-lasers at atmospheric pressure with 50 % content of molecular components threshold value of specific energy input is $w_{th} = 0.4 \div 0.5$ J/cm^3[3,4].

Thus, at the stage of SSVD formation the initial irregularities occur in the form of streamers, nonuniform distribution of plasma density across the discharge, or cathode spots. At burning stage of the discharge these inhomogeneities cause the growth of spark channels, what reduces duration of steady discharge burning. Therefore, if we exclude the formation stage from discharge development and do not exceed threshold value of energy input, it is possible to eliminate the occurrence of inhomogeneities and instabilities considered above and to ignite extremely homogeneous SSVD with maximum possible duration of burning.

3. Critical conditions for ignition of extremely homogeneous self-sustained discharge

Let's consider a mechanism and ignition conditions, when discharge arises right from the burning stage without formation stage. Figure 1 shows general view for distribution of electric field in the discharge gap (DG) filled by initial plasma with electron concentration n_0. After applying to DG the field E_0 a charge separation takes place and near the cathode surface a layer with ion extra charge Δ. The field distribution at x distance from left column edge is given by:

$$E_x = E + 4\pi e n_+(\Delta - x) = E + 4\pi e n_0(\Delta - x) \qquad (0<x<\Delta) \qquad (1)$$

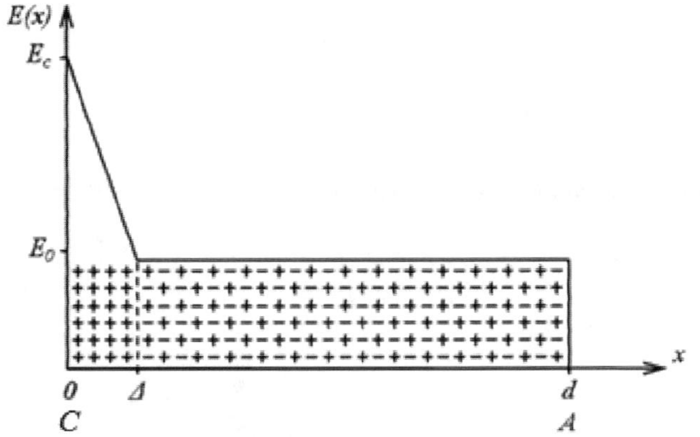

Figure 1. General view for distribution of electric field $E(x)$ and charged particles before ionization processes in discharge gap with width d between cathode (C) and anode (A). E_0 is an external electric field; E_C is a cathode field; $\Delta \leq x \leq d$ is a plasma column formed by preionization; $0 \leq x \leq \Delta$ is a cathode layer.

The burning stage of SSVD begins with the formation of cathode layer and corresponding cathode drop. It means that conditions for cathode layer formation must precede the occurrence of impact ionization in the plasma column. Thus, it is necessary to fulfill a condition

$$\Delta < \alpha^{-1}, \qquad (2)$$

where α is a coefficient of impact ionization in plasma column depending on *E*.

Townsend discharge is a process providing the current in cathode layer. A dependence of cathode layer thickness on discharge current density is falling, which limits the size of the border Δ separating the charges by the value corresponding to the condition of the Townsend breakdown in this layer. The condition for occurrence of Townsend breakdown in the cathode layer has the following form [2,8]:

$$\int_0^\Delta \alpha_C(E_x)dx = \ln\left(\frac{1}{\gamma}+1\right), \tag{3}$$

where $\alpha_C(E_x)$ is a coefficient of impact ionization in cathode layer depending on E_x. If we replace the linear distribution of the field in the cathode layer with a step with value of the field at the cathode E_c and use an empirical formula for $\alpha = pAe^{-Bp/E}$ (*A*, *B*, γ are constants [2,8]), than from the (1), (2), (3) we obtain the connection between n_0 and *E*

$$n_0 > \frac{1}{4\pi e} pAe^{\frac{-Bp}{E}} \frac{C^* E^2}{(Bp - C^* E)}. \tag{4}$$

Figure 2 shows the dependence of n_0 on *E* for ignition border right from the burning stage of SSVD in gas mixture of CO_2:N_2:He = 1:2:3 at p = 760 torr, A = 5.3 (cm·torr)$^{-1}$, B = 135 V/(cm·torr), $\gamma = 10^{-2}$, $C^* = \ln\left[\ln\left(\frac{1}{\gamma}+1\right)\right] = 1.53$. It is calculated by (4). This mixture has value of field equal E_{qs} = 12.5 kV/cm for the phase of quasi-stable discharge burning. Suggesting the field value for ignition of extremely homogenous self-sustained discharge in mixture of CO_2-laser belongs to interval of $E_1 = E_{qs} \le E \le 2E_{qs} = E_2$, from (4) we have $\mathrm{n}_1 = 1.7\times10^9 \le \mathrm{n}_0 \le \mathrm{n}_2 = 5.5\times10^{11}\,cm^{-3}$.

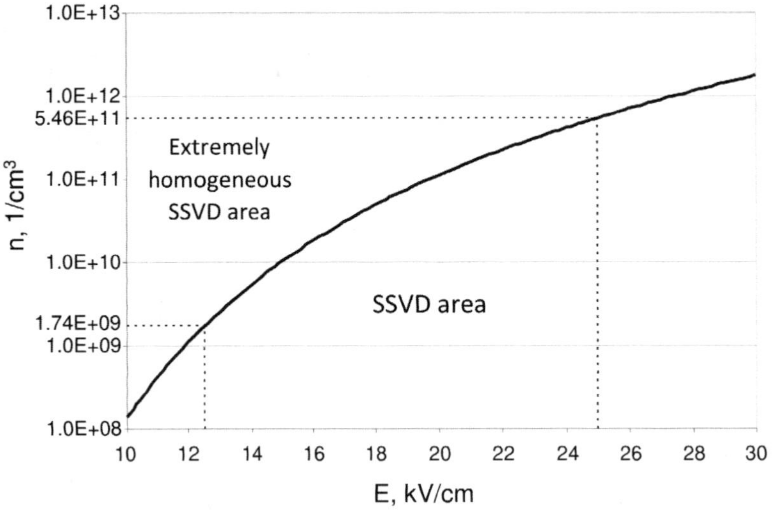

Figure 2. Calculated dependence of boundary concentration of initial electrons n_0 on the field strength *E* in the DG for producing extremely homogeneous SSVD in the mix of CO_2:N_2:He = 1:2:3 at p = 1 atm.

4. Upper limit for n_0

SSVD with the optimal mode of the energy input, when all the electrical energy is dissipated in the plasma over the first half-period of the discharge current, represents the greatest practical interest. This mode is executed under the condition $U = 2U_{qs} = 2E_{qs}d$ with calculated for it value of $n_0 = 5.5\times10^{11}$ cm^{-3}. If we increase n_0 from its minimal value $n_0^{min} = 5.5\times10^{11}$ cm^{-3}, then at some value of n_0^{max} during the initial period of development stage the volume unsustained discharge (VUD) occurs. To minimize the stage of VUD it is necessary to restrict the n_0 of the top. To estimate this value we assume that U_0 is an output voltage of pump generator, Z is a wave impedance of discharge circuit, U_p is a plasma voltage, I is a current in discharge circuit. Therefore, plasma voltage is given by $U_p=U_0-IZ$. The discharge will be self-sustained under the condition $U_p \geq U_{qs}$, where from it is easy to find $n_0^{max} \leqq d(U_0/U_p-1)(e\mu_e SZ)^{-1}$, S is a discharge square. Then, considering that $\mu_e = 7\cdot 10^2$ cm^2(V·s)$^{-1}$; $S = l\cdot h = l\cdot d$, where l is a discharge length, h is a discharge height. l = 100-200 cm, Z = 5-3 Ohm, for the mode of optimal energy input ($U_0 = 2U_{qs}$) we have $n_0^{max} \leqq 10^{13}$ cm^{-3}.

Thus, for producing extremely homogeneous volume self-sustained discharge in the mode of optimal energy input it is necessary that:

$$n_0^{min} < n_0 < n_0^{max},\ U_0 = 2U_{qs},\ w < w_{th}.$$

5. Experimental researches of conditions for initiation of extremely homogeneous SSVD in gas mixes CO2: N2: He

The considered theoretical model corresponds to the experimental results received earlier at initiation of SSVD by a low-current electron beam. The voltage on electron gun and beam current density were change in the following ranges U_e = 160÷220 kV, j_e = 2÷20 mA/cm^2 ($n_0 = 10^7$-10^{13}). The duration of electron current T_e was less than 1 μs. The pressure of gas mixtures was changed from 1 to 2 atm [3,4].

Figure 3 shows photographs of glowing SSVD (w = 0.3 J/(cm^{3}atm), the duration of discharge current was about 4 μs) in gas mixture of CO_2:N_2:He = 1:2:3 at p = 760 torr with different values of plasma concentration n_0 forming by low-current electron beam in a gap with distance between electrodes d = 9 cm (cathode on the right). At relatively low preionization level ($n_0 = 5\times10^7$ cm^{-3}) a typical pattern of a discharge inherent to the exciting systems with an UV irradiation is observed. There were clearly visible inhomogeneities in the form of unfinished spark channels sprouting from the cathode by almost a third length of the electrode gap. A cathode directed channel starts growing from the anode. Even at $n_0 = 2\times10^{10}$ cm^{-3} there is still a large number of channels regularly distributed over the cathode surface. They are likely to originate from cathode spots. And only at $n_0 = 5\times10^{11}$ cm^{-3} the discharge takes on the shape of a completely uniform positive column with an evenly glowing narrow cathode sheath.

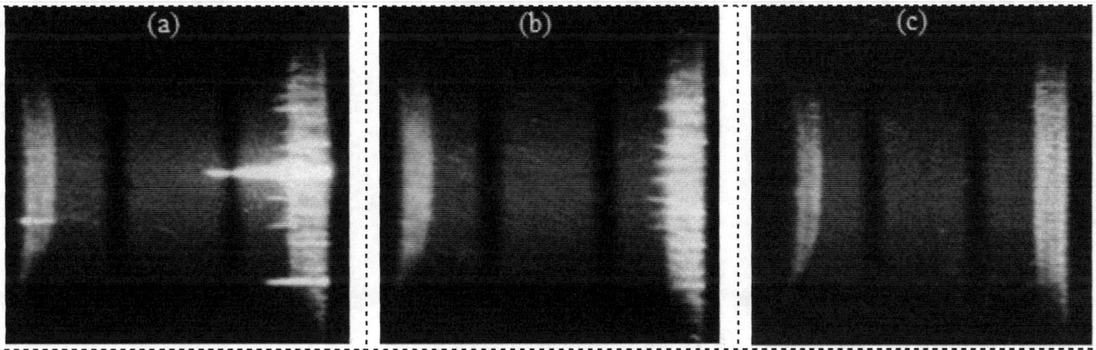

Figure 3. Photographs of SSVD glowing in mixture CO_2:N_2:He = 1:2:3. a) $n_0 = 5\cdot10^7$ cm^{-3}; b) $n_0 = 2\cdot10^{10}$ cm^{-3}; c) $n_0 = 5\cdot10^{11}$ cm^{-3}. Anode is on the left side, cathode is on the right side. (Two dark vertical stripes do not apply to the discharge, they form due to support grid placed in front of discharge chamber window).

An experiment was conducted to determine the duration of the stable burning of extremely homogeneous SSVD in the discharge volume $V \sim 60$ l with d = 18.5 cm and a total energy input of about 10 kJ. The current was elongated by incorporating of inductors into the discharge circuit. When the inductance L = 30 μH the total duration of the discharge current was T = 10 μs for mixture CO_2:N_2:He = 1:2:3 at atmospheric pressure.

6. Conclusions

The new approach for suppression of plasma inhomogeneities and instabilities in the self-sustained volume discharge is offered. It consists in two special aspects. The first aspect is in excluding the formation stage, at which main types of inhomogeneities arise, from development of the discharge. And the second one is not to exceed a threshold on a specific energy input at which the volume instabilities occur.

We offer the physical model and calculate the dependence of boundary concentration of initial electrons n_0 on the intensity of an electric field E demanded for ignition of the self-sustained discharge at burning stage without the stage of discharge formation.

There were defined the conditions of obtaining extremely homogeneous self-sustained volume discharge (with full suppression of plasma inhomogeneities and instabilities) in the mode of an optimal energy input.

In the mixture CO_2:N_2:He =1:2:3 of atmospheric pressure the self-sustained discharge in the volume of 60 litres with burning duration of 10 μs and total energy input of near 10 kJ is received. These values were limited by setup parameters and could be exceeded.

References

[1] Levatter J I and Lin S C 1980 *J. Appl. Phys.* **51** 210-222
[2] Korolev Yu D and Mesyats G A 1998 *Physics of Pulsed Breakdown in Gases* (Ekateriburg: URO-PRESS)
[3] Khomich V Yu and Yamshchikov V A 2011 *Plasma Phys. Rep.* **37** 1182-89
[4] Khomich V Yu and Yamshchikov V A 2014 *Fundamentals of Creating Electric-Discharge Excitation Systems for High-Power CO_2-, N_2- and F_2-lasers* (Moscow: Fizmatlit)
[5] Gadiyak G V, Nasyrov K A, Shveigert V A and Uuemaa O U 1985 Mathematical modeling of gas-discharge lasers *Preprint* №30-85 (Novosibirsk, ITAM Siberian Branch of Academy of Sciences USSR) [in russian]
[6] Osipov V V and Lisenkov V V 1996 *Tech. Phys. Lett.* **22** 74
[7] Osipov V V and Lisenkov V V 2000 *Tech. Phys. Lett.* **70** 27-33
[8] Raizer Yu P 1980 *Fundamentals of the Modern Physics of Gas-Discharge Processes* (Moscow: Mir)

Journal of Physics: Conference Series **652** (2015) 012043 doi:10.1088/1742-6596/652/1/012043

Investigation of electrooptical breakdown threshold in gas mixtures of complex chemical composition

E Y Loktionov, N A Pasechnikov, A V Pavlov, Y S Protasov and V D Telekh

State Lab for Photon Energetics, Bauman Moscow State Technical University, 5-1 2nd Baumanskaya Street, Moscow 105005, Russia

E-mail: stcpe@bmstu.ru

Abstract. The results of combined electrooptical breakdown (DC E~0–13.2 kV/cm + laser λ~213, 266, 355, 532, 1064 nm, $\tau_{0,5}$~18 ns, I_0~10^9–10^{11} W/cm^2) of gases (He, Ne, Ar, Kr) and their binary and triple mixtures at subatmospheric pressure (10^1–10^5 Pa) experimental investigation are presented for the first time. Thresholds of optical, electrical and combined electrooptical breakdown for those were experimentally determined. The possibility of breakdown threshold components decrease at simultaneous optical and electrical impact on the gas, and the possibility of lowering the thresholds of the combined breakdown of doped gases have been investigated.

1. Introduction

Ignition of pulsed breakdown of gases has been investigated for a long time, and spark discharge in DC or low frequency AC electric field is perhaps the most studied and frequently used. Despite this, studies of electrical discharges are ongoing and are aimed primarily on finding ways to improve energy efficiency (e.g. for lighting and lightning protection), also due to combined impact. So, to date, various combinations of DC, AC, RF [1] and microwave electric fields and light fields [2], [3] have been investigated, including combinations with various sources of seed electrons (thermal [4], emission [5], formed by chemical reactions or the preionization gas gap [6]). And in most cases the vector fields for the combined effect were set to collinear (radiation going along the electric field).

Interest in the study of optical and electrooptical breakdown is associated with the ignition of breakdown in hydrocarbon fuels. The advantages of laser ignition of fuel mixtures are that the concentration of harmful compounds (NO_x, CO, SN) in the exhaust is proportional to the temperature in the combustion chamber, which can be reduced when lean mixtures are burnt. The ignition of such mixtures by an electric spark requires a substantial increase of the ignition system power, which reduces spark plug's lifetime. The specific power of the engine is proportional to the degree of compression in the combustion chamber, pressure increase within reasonable limits leads to an increase in the required voltage at the spark ignition, while the laser ignition threshold is reduced. Laser ignition can be effectively used at low pressures in the combustion chamber also, at pulsed periodic ignition of laser plasma (optical plasmatron) in the engines with continuous fuel supply.

The aim of this work is to study the possibility of breakdown thresholds components mutual reduction at simultaneous optical and electrical impact on the gas, and the possibility of breakdown threshold decrease at doping by other gas. Combined impacts could be considered as those with one of

Journal of Physics: Conference Series **652** (2015) 012043 doi:10.1088/1742-6596/652/1/012043

components domination or with their equal effect [7]. At electrical and optical breakdown of gases minimum thresholds occur in the regions of ~10^2-10^3 Pa and ~10^6-10^7 Pa, respectively [8]. In addition to the optimum for each mode of breakdown, there are also regions where combined impact is the most energy efficient. Analysis of mechanisms of electric and optical breakdown suggests the existence of synergistic effects, when formation of seed electrons and their acceleration is combined [9].

2. Experimental setup

ISO 40-KF vacuum cross had been pumped down to 10^{-2} Pa and than filled with gas mixtures (>99.99 %, Linde gas) with pressure varied from 10^5 to 10^1 Pa. (Fig. 1). Radiation of five harmonics (λ~1064 nm, 532 nm, 355 nm, 266 nm and 213 nm) of nanosecond (τ~11-18 ns) Nd:YAG lasers (LS-2147, Lotis TII and LQ929, Solar LS) with maximum pulse energy $E_{1064,max}$~0.91 J, $E_{532,max}$~0.32 J, $E_{355,max}$~0.074 J, $E_{266,max}$~0.046 J and $E_{213,max}$~0.009 J have been used in the experiments, exposure was carried out in a repetitively pulsed mode (10 Hz).

During the experiments radiation was focused by a quartz lens (*F*~150 mm) in a spot diameter of ~ 0.12 mm in the gap (*d*~3–10 mm) between two movable copper electrodes, one of which had a positive potential up to *U*~5 kV (second electrode was grounded), giving electric field strength up to *E*~13.2 kV/cm (current limited 5.25 mA) and radiation limiting power densities I_{1064}~$1.2\cdot10^{11}$ W/cm^2, I_{532}~$4.3\cdot10^{10}$ W/cm^2, I_{355}~$9.2\cdot10^9$ W/cm^2, I_{266}~$6.1\cdot10^9$ W/cm^{2}and I_{213}~$1.2\cdot10^9$ W/cm^2.

Plasma broadband radiation and electric grid shortening were considered as breakdown evidence. For breakdown dynamics experiments Nomarski interferometer with 20 ns exposure time was used.

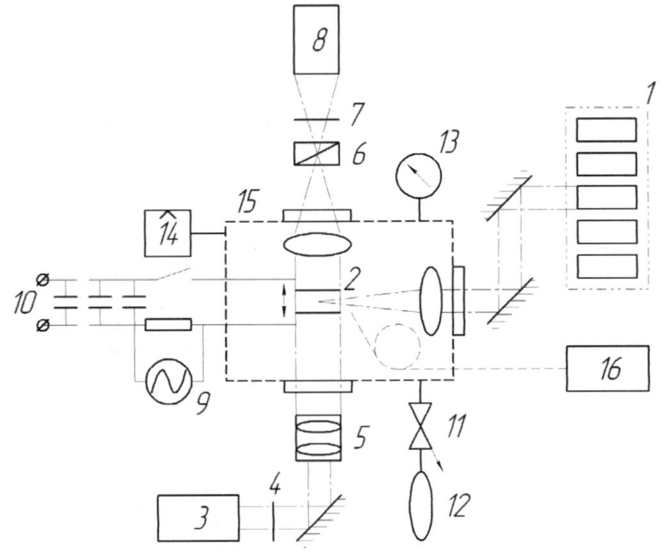

Figure 1. Experimental setup (1 – Nd:YAG laser with harmonic generators; 2 – electrodes; 3 – diode laser; 4 – half-wavelength plate; 5 – telescope; 6 – Wollaston prism; 7 – polarizing filter; 8 – CCD camera with image intensifier; 9 – oscilloscope; 10 – high voltage power unit; 11 – precision leak valve; 12 – gas cylinder; 13 – diaphragm vacuum gauge; 14 – oil-free vacuum unit; 15 – vacuum chamber; 16 – fiber-coupled spectrometer).

3. Results

Optical and electrical thresholds at different pressures were evaluated initially, the results of these reference experiments correspond to published ones [10], and measured the threshold of the combined breakdown: at fixed voltage radiation intensity was increased until spark observation. Threshold value was recorded for breakdown probability of 50% [11].

Electric field influence on combined breakdown threshold becomes more obvious with pressure and wavelength decrease (Fig. 2). This trend might be explained by multi-photon ionization probability increase at short wavelength irradiation, seeding electron recombination decrease during their number doubling period (ca. 20 ps) and electron energy increase due to acceleration by electric field in between of collisions.

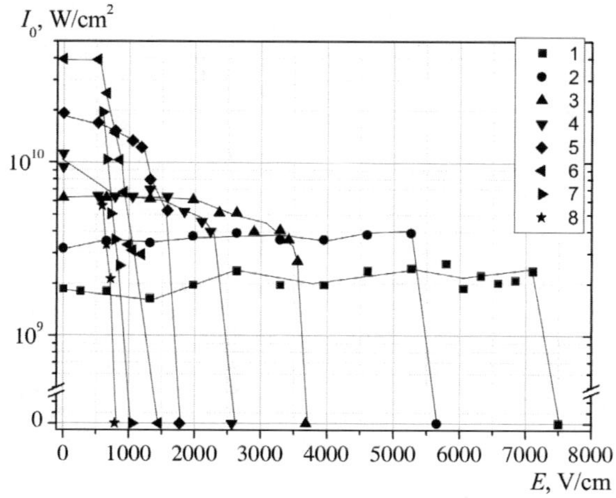

Figure 2. Dependence of electrooptical threshold breakdown Ar at different pressures on the electric field, laser intensity and wavelength λ = 532 nm (1 – p ≈ $1\cdot10^5$ Pa, 2 – p ≈ $4.9\cdot10^4$ Pa, 3 – p ≈ $2.3\cdot10^4$ Pa, 4 – p ≈ $1.1\cdot10^4$ Pa, 5 – p ≈ $5.5\cdot10^3$ Pa, 6 – p ≈ $2.8\cdot10^3$ Pa, 7 – p ≈ $1.4\cdot10^3$ Pa, 8 – p ≈ $6.8\cdot10^2$ Pa).

The next experiments were conducted to determine the required concentration (partial pressure) of doping gas, which would considerably reduce optical breakdown threshold. For this doping was changed as 50 %, 40 %, 10 %, 5 %, 1 %. As can be seen from Fig. 3, if optical breakdown threshold decrease occurred at all, it happened even at small amounts of additives. For Ne – Ar and Ne – Kr mixtures doping rates 1% and 5% of Ar and Kr, respectively, were enough for optical breakdown threshold decrease. No anomalous effect mentioned in [12] for 1% of Ne in Ar (Fig. 3b) was detected. The optical breakdown threshold decrease, as compared to a net background gas was accompanied by an increase of the electrical breakdown threshold, opposite cases were not detected in our experiments. For a mixture of 90% He – 10% Ne ($p \sim 9\cdot10^4$ Pa) we observed an increase of the electrical breakdown threshold (fig. 4a), without reducing the optical one, as pressure decreases, this effect is less pronounced (fig. 4b).

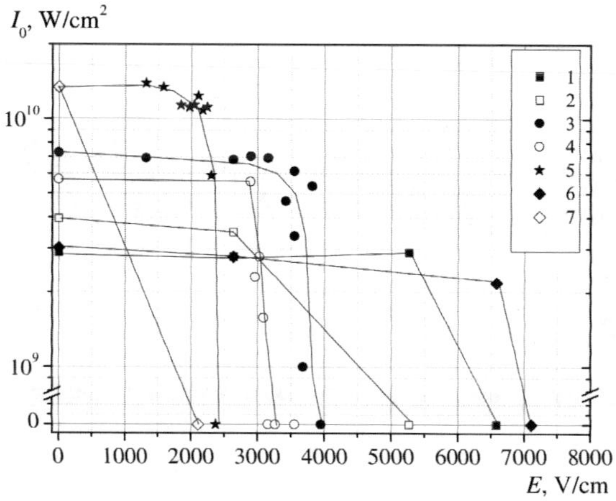

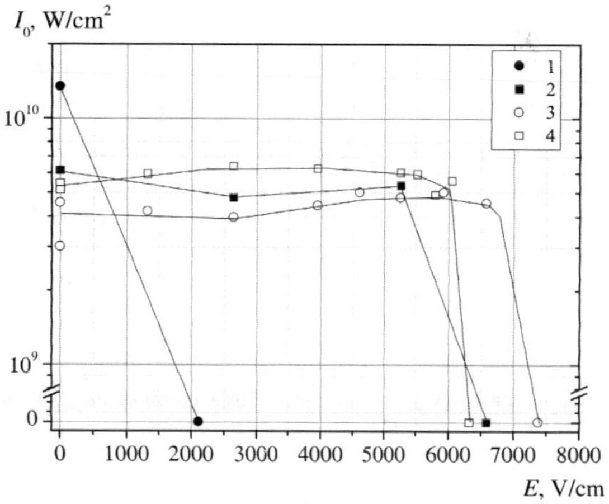

Figure 3. Dependence of electrooptical breakdown threshold of Ne mixtures with Kr (a) and Ar (b) on the electric field and laser intensity (p~$9\cdot10^4$ Pa); a: 1 – 50% Ne – 50% Kr, 2 – 60% Ne – 40% Kr, 3 – 90% Ne – 10% Kr, 4 – 95% Ne – 5% Kr, 5 – 99% Ne – 1% Kr, 6 – 100% Kr, 7 – 100% Ne; b: 1 – 100% Ne, 2 – 100% Ar, 3 – 99% Ne – 1% Ar, 4 – 99% Ar – 1% Ne.

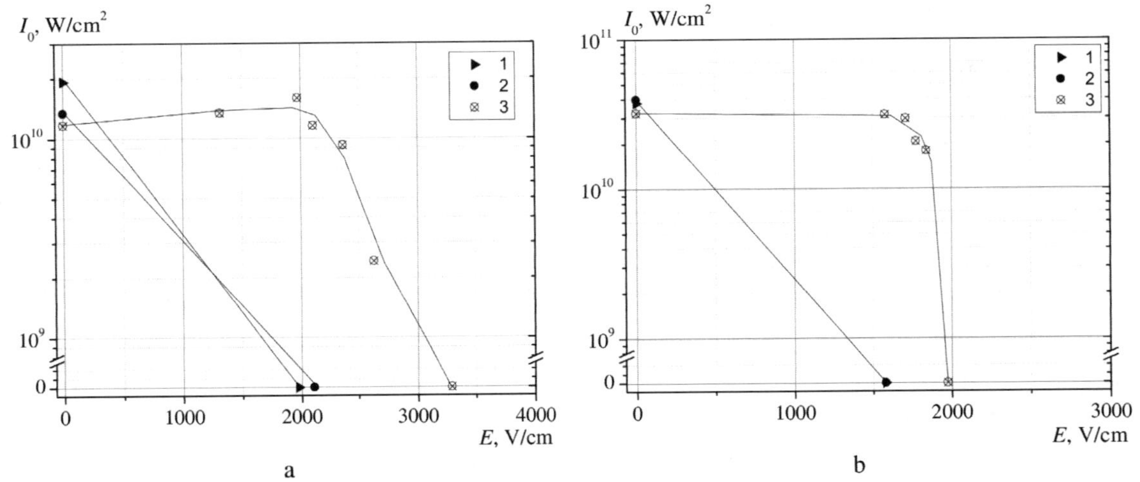

Figure 4. Dependence of electrooptical breakdown threshold for Ne – He mixture (a – p~$9{\cdot}10^4$ Pa, b – p~$3.4{\cdot}10^4$ Pa) on the electric field and intensity of laser radiation (1 – 100% He, 2 – 100% Ne, 3 – 90% He – 10% Ne).

Unlike pure gases and binary mixtures, we could not find in the literature the results of any study of breakdown phenomena in triple mixtures. Performing experiments with gases doped with small amounts of two additives we have found the effect of those additives mutual suppression (negative interference). This phenomena for argon and krypton (both are Penning additives – Fig.2) was observed in the neon (Fig.5).

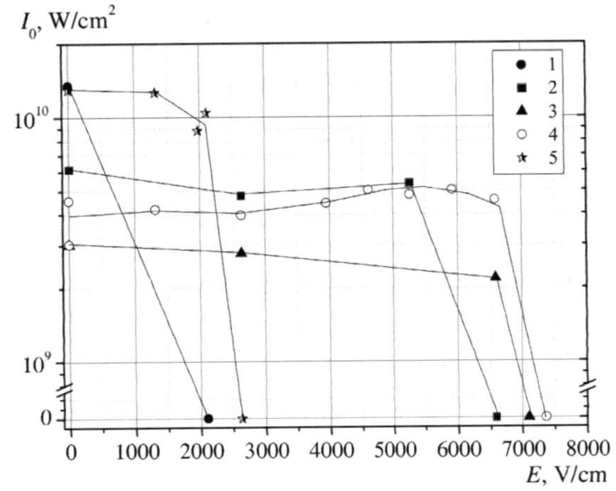

Figure 5. Dependence of electrooptical breakdown threshold for Ne – Ar – Kr mixture (p~$9{\cdot}10^4$ Pa) on the electric field and intensity of laser radiation (1 – 100% Ne, 2 – 100% Ar, 3 – 100% Kr, 4 – 99% Ne – 1% Ar, 5 – 98% Ne – 1% Ar – 1% Kr).

4. Conclusions

We experimentally determined the thresholds of optical, electrical and combined electrooptical breakdown for noble gases and their binary and triple mixtures with different ratio of components. It was found that the decrease of the threshold for optical breakdown in the presence of the electric field is particularly evident at low pressures. For some binary mixtures the observed decrease of the optical breakdown threshold (the threshold of electrical breakdown in relation to the electrical breakdown threshold of the buffer gas increases), while this effect is already apparent at low concentrations of additives ~1-5%. For triple mixtures we found mutual suppression effects (negative interference) of the additives on breakdown thresholds.

Acknowledgements
This work was performed at the facility "Beam-M" (RFMEFI59014X0001), supported by the State Task of the Russian Ministry of Education and Science, and the Russian Foundation for Basic Research (grant 13-08-01391, 15-38-20890).

References
[1] Lisovskiy V A, Kharchenko N D, Yegorenkov V D 2010 *J. Phys.* D: *Appl. Phys.* **43** 425202
[2] Zhou W, Li K, Qian H, Ren Z, Yu Y 2012 *Appl. Optics* **51** B42
[3] Ikeda Y, Tsuruoka R 2012 *Appl. Optics* **51** B183
[4] Varaksin A Yu 2013 *High Temperature* **51** 377
[5] Kostyrya I D *et al* 2004 *Techn. Physics* **49** 987
[6] Hong L, Cheng Z, Zuo D, Wang X, Zai B, Yu L 2007 *Proc. SPIE* **6823** 68230Y-7
[7] Lisovskiy V A, Kharchenko N D, Yegorenkov V D 2008 *J. Phys.* D: *Appl. Phys.* **41** 125207
[8] Raizer Y P 1980 Optical discharges *Soviet Physics Uspekhi* **23** 789
[9] Morsy M H 2012 *Renewable and Sustainable Energy Reviews* **16** 4849
[10] Raizer Y P 1997 *Gas Discharge Physics* Springer: New York
[11] Voronina E I *et al* 2008 *Technical Physics Letters* **35** 1024
[12] Mul'chenko B F and Raizer Y P 1971 *J. Experim. Theor. Phys.* **33** 349

Journal of Physics: Conference Series **652** (2015) 012044 doi:10.1088/1742-6596/652/1/012044

Sterilization of dielectric containers using a fore-vacuum pressure plasma-cathode electron source

D Zolotukhin[1], V Burdovitsin1[1], E Oks[1, 2], A Tyunkov[1], Yu Yushkov[1]
[1] Tomsk State University of Control Systems and Radioelectronics, Tomsk 634050, Russia
[2] Institute of High Current Electronics SB RAS, Tomsk 634055, Russia

E-mail: ZolotukhinDen@gmail.com

Abstract. We describe our work on sterilization of 10 ml glass and 60 ml plastic cylindrical containers using a fore-vacuum pressure, plasma-cathode, electron beam source. Beam plasma is formed inside the vessel by injection of a low-energy electron beam at 3 - 6 keV energy and current of 50 mA, at a working gas (air) pressure of 8 Pa. The gas composition was tracked by a quadrupole gas analyzer type RGA-100. As a test biological object for sterilization we used *E. coli* ATCC 25922 bacteria, the inner surface of each vessel was inoculated with a bacterial suspension. We find a smooth dependence of the degree of sterilization on the total energy density injected into the vessel. The efficacy of sterilization of container inner surfaces using a fore-vacuum pressure, plasma-cathode e-beam source of relatively low energy (a few keV) electrons is thus demonstrated.

1. Introduction

Sterilization is a concern and of great importance in the pharmaceutical, food, chemical industries and in medicine. Sterilization technologies need to provide high bactericidal effect with great reliability and performance. In recent years, in addition to traditional sterilization methods (such as dry and moisture heating, and chemical exposure) low-temperature plasma sterilization techniques have been actively developed. Effective plasma sterilization technologies have been demonstrated for heat- and chemical-sensitive products at atmospheric pressure [1] and under vacuum conditions (at a gas pressure of 1-100 Pa [2]). To date, plasma sterilization of products made of dielectric material (glass, plastic) has been carried out using various kinds of gas plasmas such as barrier discharge [1,3] and microwave plasma [2]. However the generation of plasma by a dielectric barrier discharge and its variations allows sterilizing only flat objects [4]. RF discharge sterilization technology is well-developed. An RF discharge allows formation of relatively dense (10^{10}-10^{11} cm^{-3}) spatially uniform plasma for sterilization of three-dimensional objects [4] and a variety kinds of containers [5]. At the same time, RF-formed plasmas have some disadvantages, such as low efficiency of energy transfer from the power supply to the plasma and the problem of matching the RF generator to the plasma [6]. Some novel possibilities for plasma generation are opened by application of electron beams [7]. Fore-vacuum pressure, plasma-cathode electron beam sources can produce relatively dense plasma not only in metal vacuum chambers but also within dielectric vessels by injection of a low-energy electron beam (2 - 8 keV) through the vessel opening [8]. We describe here our work in developing this approach for sterilizing the inner surfaces of glass and plastic vessels by plasma formation by an electron beam at fore-vacuum pressures.

Journal of Physics: Conference Series **652** (2015) 012044 doi:10.1088/1742-6596/652/1/012044

2. Experimental

A simplified schematic of our experimental setup is shown in figure 1.

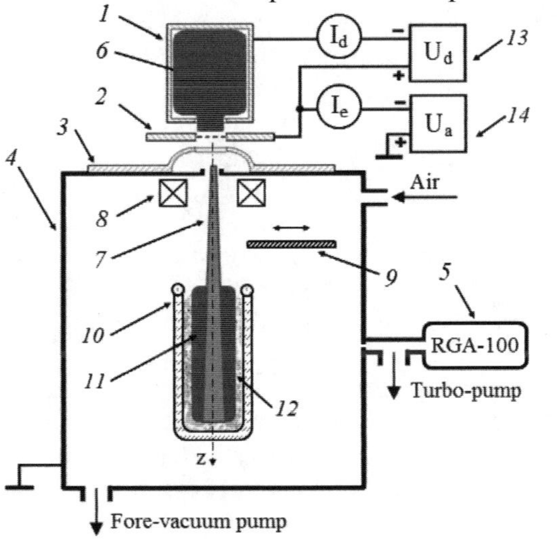

a)

b)

Figure 1. (a) Experimental set-up: 1 – hollow cathode, 2 – anode, 3 – extractor, 4 – vacuum chamber, 5 – residual gas analyzer RGA-100, 6 – discharge plasma, 7 – electron beam, 8 – focusing system, 9 – shutter, 10 – vessel, 11 – beam plasma, 12 – bacteria suspension, 13 – discharge power supply; 14 – acceleration voltage power supply. (b) Photo of the beam plasma during the sterilization process.

We used a steady-state fore-vacuum pressure plasma-cathode electron source [7] (*1-3*) with hollow cathode. In this source, a glow discharge was established between the hollow cathode *1* and anode *2*. The discharge current I_d was 140 - 160 mA with a discharge voltage U_d of 400 - 420 V. The working gas was air at a pressure of 8 Pa. Pressure in the vacuum chamber *4* was regulated by varying the flow rate of air flowing into the chamber, which was pumped only by a mechanical fore-vacuum spiral pump type ISP-1000C. The working gas composition was monitored using a quadrupole residual gas analyzer *5*, an RGA-100 [9, 10]. The analyzer volume was additionally pumped by a turbo-molecular pump to pressures of about 10^{-2} Pa. When an electron-accelerating voltage U_a of from 3 to 6 kV is applied between the anode *2* and the accelerating electrode *3*, a continuous, cylindrical electron beam *7* is extracted from the discharge plasma *6* through the anode grid. The value of electron beam current was measured as the current I_e in the accelerating voltage supply circuit and was set to 50 mA. The e-beam diameter was about 5-10 mm. The magnetic system *8* was intended to focus the e-beam. Above the dielectric vessel *10*, a shutter *9* was mounted. This shutter defines the beginning and end of the e-beam exposure session; in the closed position the shutter prevents penetration of the electron beam into the vessel. The duration τ of each exposure session was 5 s; during experiments, we made from 1 to 18 exposure sessions. When the electron source is operating appropriately, the shutter is opened and the sterilization process proceeds for a predetermined number of sessions. Thus the total exposure time is determined by the number of sessions, each session being of 5 seconds duration. The electron beam penetrates into the cylindrical vessel and produces inside it beam plasma *11*. For our experiments we used two types of vessels – a 10 ml container made of glass, and a 60 ml container made of plastic (polypropylene). The glass vessels had an inner diameter d_1 = 20 mm and height h_1 = 55 mm, and the polypropylene vessels had dimensions d_2 = 36 mm and h_2 = 62 mm. Each vessel, when it was placed in the vacuum chamber, was co-axial with the axis of the electron source. After 5 s of exposure, the shutter closes, and the accelerating voltage is switched off.

In order to assess the sterilizing effect, the inner surface of the vessel (both bottom end and side walls) was inoculated by *E. coli* ATCC 25922 -containing suspension with bacterial concentration of about 1-2 McFarland turbidity standards [11], which is approximately equal to 1×10^9 colony forming units per ml (cfu/ml) of *E. coli*. Preparation of the bacterial suspension for inoculation of the vessel surface and the control Petri dish was performed using following standard methods. An *E. coli* culture was grown on a simple agar plate for 24 h. After 24 h of culture growth, the *E. coli* suspension of 1-2 McFarland turbidity standards was prepared using 0.9% NaCl solution. This suspension was spread on the bottom and sidewalls of the vessel using a sterile swab, and the vessel placed in the vacuum chamber for plasma and e-beam processing. After exposure, the vessel was removed from the chamber and sent for bacteriological examination. To prevent the unwanted contamination, processed vessels were placed in a closed container, and in such "packed form" were delivered to bacteriological lab located on the other side of the city. Swabs were taken from the inner surface (separately from the bottom and from the sidewalls) in 2 ml of 0.9% NaCl solution, and incubated at 37 ^{0}C for 1-2 days in selective Endo medium in a separate Petri dish. Each experimental result was compared with non-exposed seeding of *E. coli* in a control Petri dish with Endo medium. If the sample from the container surface from which the swab was made was not sterile, the *E. coli* culture growth was observed in the corresponding Petri dish sector (Figure 2).

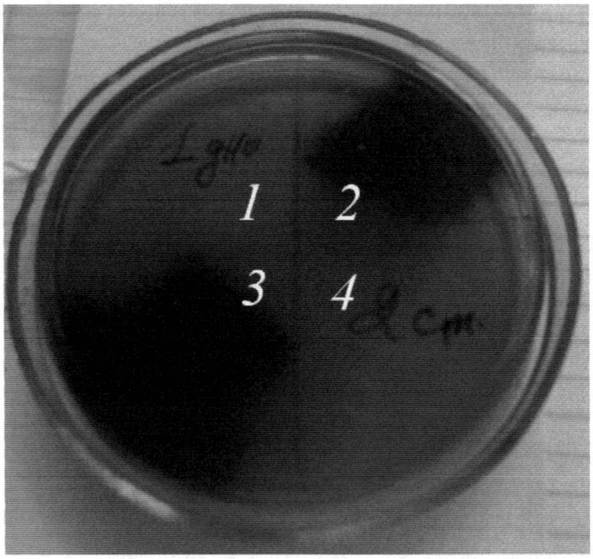

Figure 2. Petri dish containing swabs from sterile (1, 4) and non-sterile (2, 3) sites of the vessels.

As primary experimental exposure parameter we chose the total energy E_Σ that is injected into the container over the complete exposure time, as calculated using

$$E_\Sigma = I_e U_a \cdot n \cdot \tau, \tag{1}$$

where n is the number of exposure sessions, selected so that for different beam energies U_a the value of E_Σ was roughly constant, I_e is the e-beam current, and τ is the duration time of each exposure (5 seconds).

To assess the sterilization effectiveness, we calculated the total energy density D_E (in J/mm^2), which we define as the ratio of the total energy given by equation (1) to the vessel inner surface area S:

Journal of Physics: Conference Series **652** (2015) 012044 doi:10.1088/1742-6596/652/1/012044

$$D_E = \frac{E_\Sigma}{S} \tag{2}$$

The experiments were carried out at a constant gas pressure $p = 8$ Pa and with beam current $I_e = 50$ mA for three different beam energies (3, 4.5 and 6 keV). The sterilization degree was determined as the ratio of the number of sterile samples from the vessel surface to the total number of samples in each series.

3. Results and discussion

The experimental results for sterilization of glass and polypropylene vessels are summarized in table 1 and table 2 respectively.

Table 1. Glass vessel sterilization results

Test No.	U_a (keV)	Number of sessions, n	D_E (J/mm^2)	Result		Sterilization degree (%)
				Bottom end	Sidewalls	
	3	1		Sterile	Non-sterile	
1	4.5	1	0.56	Non-sterile	Sterile	50
	6	1		Sterile	Non-sterile	
	3	4		Sterile	Sterile	
2	4.5	3	1.55	Sterile	Non-sterile	66.7
	6	2		Sterile	Non-sterile	
	3	9		Sterile	Sterile	
3	4.5	6	3.22	Sterile	Sterile	83.3
	6	4		Sterile	Non-sterile	
	3	18		Sterile	Sterile	
4	4.5	12	6.69	Sterile	Sterile	100
	6	8		Sterile	Sterile	

Table 2. Polypropylene vessel sterilization results

Test No.	U_a (keV)	Number of sessions, n	D_E (J/mm^2)	Result		Sterilization degree (%)
				Bottom end	Sidewalls	
	3	1		Non-sterile	Non-sterile	
1	4.5	1	0.25	Non-sterile	Non-sterile	0
	6	1		Non-sterile	Non-sterile	
	3	4		Sterile	Sterile	
2	4.5	3	0.69	Non-sterile	Non-sterile	33.33
	6	2		Non-sterile	Non-sterile	
	3	6		Non-sterile	Sterile	
3	4.5	4	0.99	Sterile	Non-sterile	66.67
	6	3		Sterile	Sterile	
	3	6		Non-sterile	Sterile	
4	4.5	8	1.99	Sterile	Sterile	83.33
	6	12		Sterile	Sterile	
	3	8		Sterile	Sterile	
5	4.5	11	2.68	Sterile	Sterile	100
	6	16		Sterile	Sterile	

As shown in the tables, the sterilization degree increases with the number of exposure sessions for all electron beam energies. With increasing total energy density D_E, the sterilization degree reaches 100% (figure 3). Note that the bottom ends, as well as the sidewalls of the vessels, are completely sterilized. It should also be noted that there were no visible surface deformations of polypropylene vessels at a beam energy of 3 keV for any number of exposure sessions. In all cases the *E. coli* culture seeded in a control non-exposed Petri dish demonstrated marked growth.

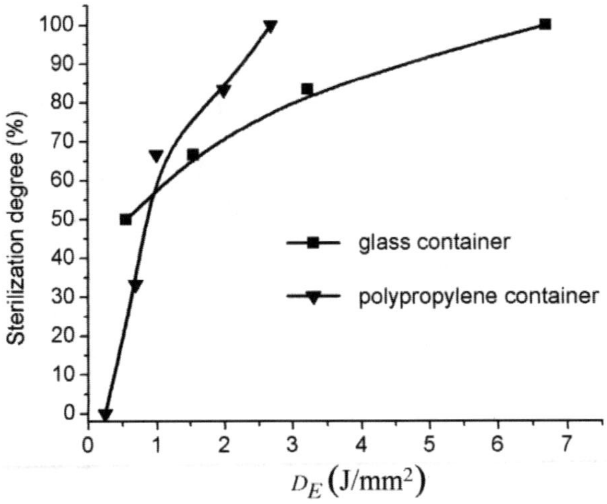

Figure 3. Sterilization degree as a function of total energy density injected into the vessel.

The working gas mass spectrum (figure 4) indicates the presence in the gas composition of a considerable amount of molecular nitrogen and oxygen along with a smaller amount of atomic nitrogen and oxygen, the latter of which determine the bactericidal properties of the plasma, as has been reported [12]. With increasing number of sessions and hence total exposure time of the active plasma components on the microbial cells, the sterilizing effect increases.

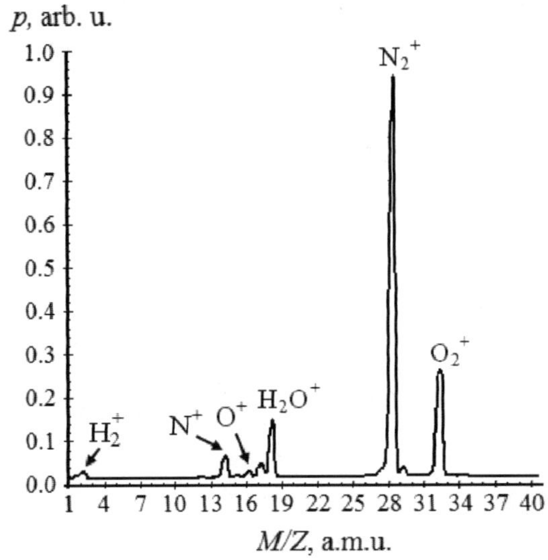

Figure 4. Mass spectrum of working gas obtained using RGA-100, at a gas pressure of 8 Pa.

4. Conclusion

We have demonstrated the sterilization of the inner surfaces of glass and polypropylene vessels by means of a fore-vacuum pressure, plasma-cathode e-beam source of low energy (a few keV) electrons, without using toxic gases or other chemicals and extremely high heating. The degree of sterilization depends on the total energy density injected into the vessel, suggesting the existence of experimental regimes for effective sterilization of glass and heat-sensitive plastic containers. More detail understanding of the mechanisms responsible for successful sterilization by fore-vacuum electron sources requires further experiments, and these issues are the subject of our on-going research.

Acknowledgements

The work was supported by the Foundation for Assistance of Small Innovative Enterprises in Science and Technology (Grant "UMNIK 2014", agreement #3898GU1/2014). Special thanks to Ian G. Brown (Berkeley Lab) for useful discussion and English improvement, and to Mrs. Maria Arkina for bacteriological research.

5. References

[1] Liu Xiaohu, Hong Feng, Guo Ying, Zhang Jing, and Shi Jianjun 2013 *Plasma Science and Technology* **15** 439
[2] Hueso J L, Rico V J, Frias J E, Cotrino J and Gonzales-Elipe A R 2008 *J. Phys. D: Appl. Phys.* **41** 092002
[3] Hu Miao, Guo Yun 2012 *Plasma Science and Technology* **14** 735
[4] Halfmann H, Bibinov N, Wunderlich J and Awakowicz P 2007 *J. Phys. D: Appl. Phys.* **40** 4154
[5] Sakudo N, Ikenaga N, Ikeda F, Nakayama Y, Kishi Y, Yajima Z, Jiro Matsuo, Masataka Kase, Takaaki Aoki, Toshio Seki 2011 Simultaneous sterilization with surface modification of plastic bottle by plasma-based ion implantation *Ion Implantation Technology 2101: 18th International Conference on Ion Implantation Technology IIT 2010 (Kyoto, Japan, 6–11 June 2010) (AIP Conference Proceedings* vol *1321)* pp.266-269
[6] Conrads H, Schmidt M 2000 *Plasma Sources Sci. Technol.* **9** 441
[7] Burdovitsin V A, Oks E M 2008 *Laser and particle beams* **26** 419
[8] Zolotukhin D B, Burdovitsin V A, Oks E M 2015, *Technical Physics* **60** 772
[9] Zhaokui Wang, Yanhui Lou, Kuixun Lin and Xuanying Lin 2006 *International Journal of Mass Spectrometry* **261** 25
[10] Navamathavan R, An Soo Jung, Chang Young Kim, Heon Ju Lee and Chi Kyu Choi 2008 *Journal of the Korean Physical Society* **53** 1468
[11] Koch A L 1994 *Methods for General and Molecular Bacteriology* (Washington, DC: American Society for Microbiology)
[12] Villeger S, Cousty S, Ricard A and Sixou M 2003 *J. Phys. D: Appl. Phys.* **36**: L60

Radiation-induced hydrogen transfer in metals

Yu I Tyurin[1], V A Vlasov [1,2] and A S Dolgov[1]
[1]National Research Tomsk Polytechnic University, 634050, Tomsk, Russia
[2]Tomsk State University of Architecture and Building, 634003, Tomsk, Russia

E-mail: tyurin@tpu.ru

Abstract. The paper presents processes of hydrogen (deuterium) diffusion and release from hydrogen-saturated condensed matters in atomic, molecular and ionized states under the influence of the electron beam and X-ray radiation in the pre-threshold region. The dependence is described between the hydrogen isotope release intensity and the current density and the electron beam energy affecting sample, hydrogen concentration in the material volume and time of radiation exposure to the sample. The energy distribution of the emitted positive ions of hydrogen isotopes is investigated herein. Mechanisms of radiation-induced hydrogen transfer in condensed matters are suggested.

Keywords: hydrogen, condensed matter, radiation-induced transfer, non-equilibrium release, plasma

1. Introduction

The hydrogen behavior in metals is the important scientific and technical problem for a wide range of basic and applied research. The problem of hydrogen permeability and accumulation is interesting due to the need for new construction materials for nuclear, thermonuclear and hydrogen power engineering and solutions of materials science problems [1-7]. It was found that the radiation exposure and control for hydrogen concentration in the bulk of solids provide creation of non-equilibrium thermodynamic systems, the synthesis of which is not possible to achieve using traditional methods [8-13].

Complex experimental and theoretical studies were performed to understand the behavior of hydrogen in metal-hydrogen systems exposed to ionizing radiation (X-rays, electrons with energies below 100 keV).

The obtained results indicate [5,14-18] that hydrogen isotopes taking regular positions in metal, form its own hydrogen subsystem. Energy introduced during the radiation exposure is accumulated by the hydrogen subsystem and hydrogen isotope atoms acquire energy greater than that of the metal matrix for the time period much longer than at electron-phonon relaxation. Hydrogen isotope movement stimulates the diffusion of impurity atoms and leads to restructuring of the metal defect structure. The unique accumulating properties of hydrogen subsystem in metals are described by the following experimental facts:

- Radiation-induced hydrogen yield from metals at room temperature or below;
- Superlinear hydrogen yield rate from metals depending on the concentration of injected hydrogen and radiation intensity;
- A non-equilibrium desorption of H and D atoms and H^+ and D^+ ions from the metal surface saturated by hydrogen at electron beam exposure;

Journal of Physics: Conference Series **652** (2015) 012045 doi:10.1088/1742-6596/652/1/012045

- Removal of hydrogen from the total metal volume in irradiating a local surface area.

2. Experimental

2.1. Methods of mass-spectrometric analysis of desorption from hydrogenated metals and non-equilibrium migration of deuterium through metals. In situ investigations.

The design of the installation is shown in Figure 1. The installation [5,14] (Figure 1) allows performing electrolytic saturation of metal samples with hydrogen (deuterium), monitoring hydrogen (deuterium) migration through metal and hydrogen (deuterium) yield in vacuum. The yield of H, D atoms is stimulated by accelerated electrons (E = 1 - 100 keV, $j_e = 1 - 150$ mA/cm^2) or X-ray radiation ($E_x = 1.0 - 100$ keV, $j_x = 10^{13} - 10^{15}$ photons / (cm^2·s)). Recording of H and D atoms is carried out by time-of-flight mass-spectrometer. The samples are presented by Pd with 0,1 mm thickness; Ti with 0,05 mm thickness; Nb with 0,1 mm thickness. The electron beam is directed from the gun and focused on the saturating sample or converted into the X-ray beam after a tungsten plate passing. This installation allows mass spectrometric measurements of metal deuterium yield rate when exposed to the electron beam during the linear heating.

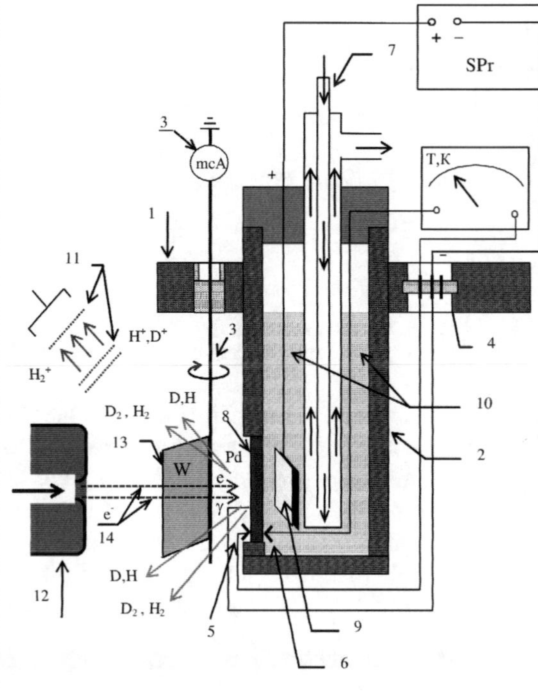

Figure 1. Schematic view of the cell for in situ studying non-equilibrium migration of hydrogen and deuterium in metals: 1 - flange; 2 - electrolysis glass; 3 - measuring device for electron beam and X-ray generation; 4 - vacuum current leads; 5, 6 - thermocouples; 7 - flow refrigerator; 8 - sample; 9 -platinum electrode; 10 - electrolyte; 11 - mass spectrometer; 12 - electron gun; 13 - tungsten foil; 14 - electron beam; 15 - cell power supply

2.2. Method for studying non-equilibrium yields of atomic hydrogen from metals under irradiation

Having a high sensitivity, the mass-spectrometric technique, recoil nucleus method does not answer the fundamental questions: whether hydrogen at the local irradiation exposure yields form the entire surface of the sample or from the point of the beam impact. The hydrogen atom yield from the metals is observed using the phenomena of heterogeneous chemiluminescence [15] and the bleaching effect of some kinds of organic dyes applied to a catalytic substrate [16]. In order to avoid effects associated with X-ray and cathodoluminescent glow, a method of H, D atoms visualization was chosen using the appropriate dyes.

2.3. Deuterium migration and yield under electron and X-ray radiation

Study of migration, diffusion, and hydrogen isotope release from metals under the action of accelerated electrons and X-ray radiation was carried out on stainless steel, niobium and palladium samples. These materials substantially differ in hydrogen isotope solubility. The minimum solubility is in stainless steel, the maximum - in palladium. Fig. 2a shows the yield rates of deuterium ($I_D cm^{-2} s^{-1}$) from stainless steel under the electron beam exposure. In these measurements, the temperature of the outer sample side in the electron beam area is not over 60 ° C, while on the back side – 40 °C. At these temperatures, in absence of the electron beam, the equilibrium deuterium yield is insignificant. The most pronounced effect from radiation-induced deuterium release under the electron beam exposure is in palladium (Figure. 2b).

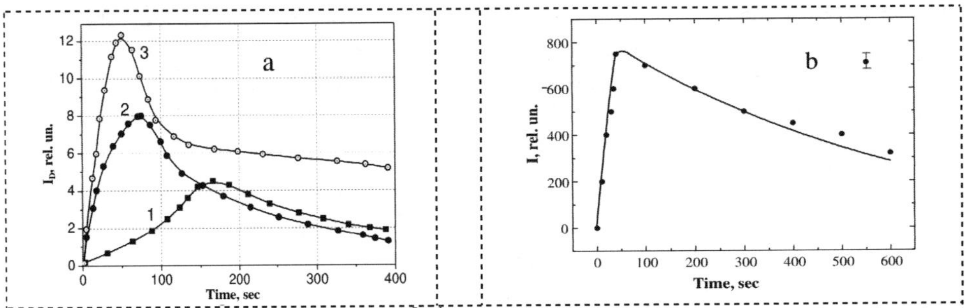

Figure 2. (a) Deuterium ($I_D cm^{-2} s^{-1}$) yield rate from stainless steel under the irradiation. Electron energy E = 20 keV, current density: j, mA/cm^2: 1 – j_1 = 250, 2 – j_2 = 500, 3 – j_3 = 750.
(b) the intensity of deuterium yield from Pd under electron beam (T = 40 °C, E = 20 keV), the current density j = 100 mA/cm^2

The rate of deuterium realese from metals under electron beam superlinearly increases with increasing the beam current and injected deuterium concentration.

The superlinear dependence indicates unequivocally the collective processes in the metal-hydrogen system under ionizing radiation.

The flow chart of experiment on atomic hydrogen release from PdDx is shown in Fig. 3 [17,18]. The electron beam with energy of 20 keV, diameter of 3 mm formed with copper (thickness t = 0.1 cm) and iron diaphragms (t = 0,4 cm) is directed to 15 x 10 mm palladium sample with thickness of 60 μm electrolytically saturated with deuterium. The 20 keV electron run in Pd is 1 μm or 60 times less than the thickness of the sample. Except for the electron beam formation, diaphragms performed the function of attenuation of bremsstrahlung radiation produced by the electron gun. Dyes were placed above and below the sample. In order to observe the intensity distribution of atom release along the surface, the additional copper plate with holes was placed between Pd and dye (Fig. 3). The whole "sandwich" consisting of the top, bottom dyes, palladium and copper plate is compacted to eliminate lateral gaps between them.

Under the influence of the electron beam, dye bleaching is observed (Fig. 4a) indicating the atomic deuterium yield both on the beam side (Fig. 4b) and the reverse side of the sample (Fig. 4c) not directly irradiated by electrons. Thus, electrons that have the run in the metal are much lower than its thickness, can excite the deuterium (hydrogen) subsystem in the whole volume of the sample.

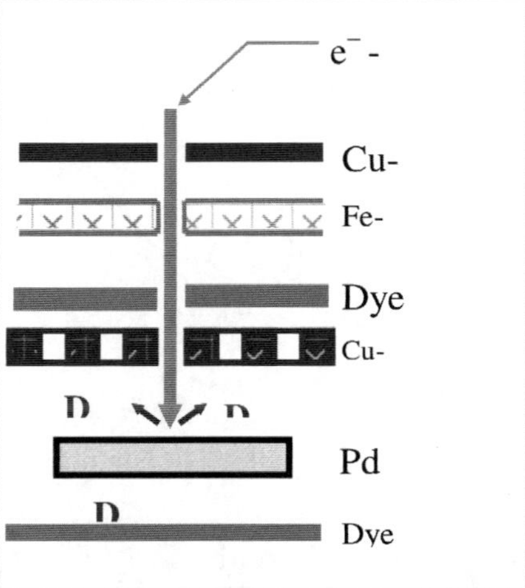

Figure 3. The flow chart of experiment on atomic hydrogen release from PdDx induced by 20 keV electron beam energy

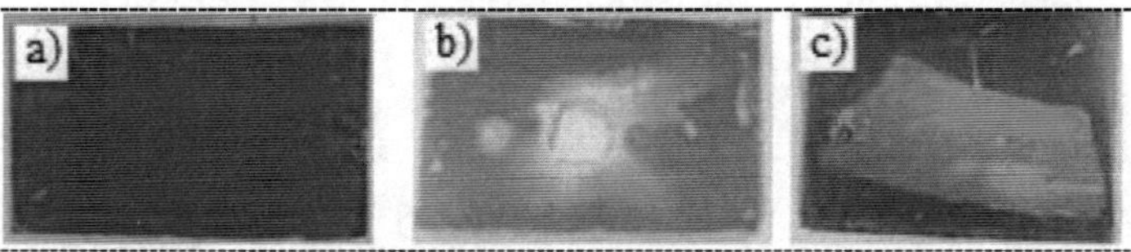

Figure 4. Deuterium PdDx release under 20 keV electron beam energy, a - original dye; b - upper dye; c - lower dye

2.4. Emission of positive ions under Pd irradiation saturated by hydrogen and deuterium

Radiation by metal electrons saturated by hydrogen (deuterium) stimulates the non-equilibrium yield from Pd (D, H), Nb (D, H) of not only H_2, DH, D_2 molecules [16], but also free H, D atoms [5,16,18]. Since even the thermally equilibrium heating of metals with the adsorption layer is accompanied by the emission of positive ions, it is expected that the metal supply with the saturated by hydrogen (deuterium) radiation energy will lead to the excitation of the internal hydrogen atmosphere and can induce the non-equilibrium emission of positive ions from the metal surface. The cell with a sensor was additionally placed in the installation (Fig.5a) to record positive ions (Fig.5b). The electron beam current could vary from 1 to 100 uA, and beam energy from 0.1 to 100 keV.

A significant degree of non-equilibrium in the hydrogenous (deuterium) palladium subsystem ($\hbar\omega/kT \cong 10$ at 313 K) can cause not only the yield of free atoms from Pd surface, but also the emission of positive ions (H +, D +) under electron irradiation.

In irradiating Pd by the electron beam with 50 keV energy and I_e =2–10 uA current the emission of positive ions was detected with the maximum current of 0.2 uA at beam current of 10 mA (0,1 at 6 uA, 0.04 at 4 uA and 0.012 at 2 uA). The emission current of positive ions decreases with the removal of hydrogen (deuterium) from Pd (Fig. 5 (b), curves 4, 5). Curves 2, 5 shown in Figure 5 (b) were obtained at interval corresponding to the passage through the cathode charge ~ 10^{-2} Kl (~ $6{\cdot}10^{16}$ electrons, $\Delta t = 10^4$ s). The increase in the emission of positive ion current correlates with the intensity increase of hydrogen and deuterium bands in mass spectra.

Small emission current of positive ions is observed at irradiating Pd (D, H) by γ-rays. An interesting feature of obtained results is the quadratic increase in the emission current of positive ions from the electron beam current. A good approximation of this relationship is the formula (I^+ and I_e in uA): $I^+ = 2.7{\cdot}10^{-3} I_e^2$.

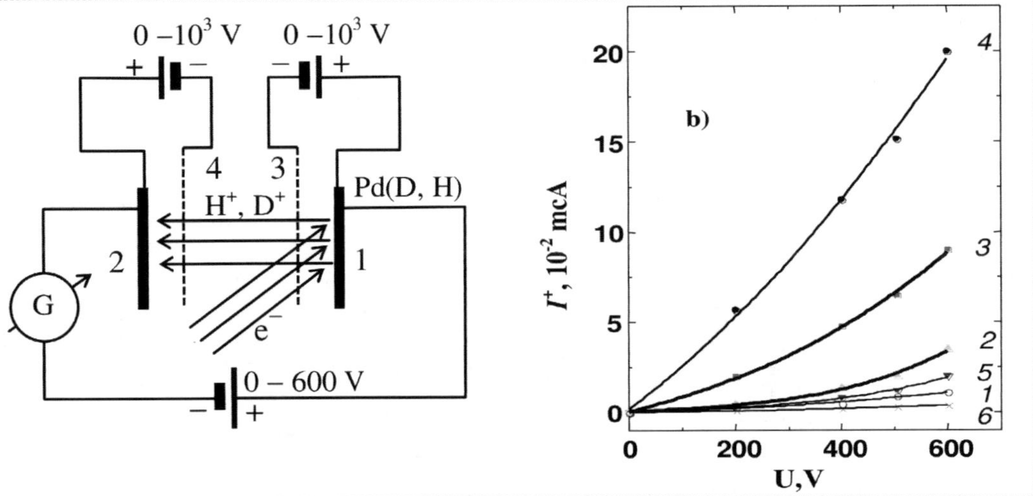

Figure 5. a) a schematic diagram of the cell for rerecording positive ions releasing from the Pd (D, H) under electron beam radiation: 1 - irradiated Pd (D, H) 'cathode'; 2 - anode collecting (protons); 3, 4 - grids locking (electrons). b) current-voltage characteristics of ion Pd (D, H) emitter irradiated by electrons. Electron beam currents: 1 - 2 mkA; 2 - 4 mkA; 3 - 6 mkA; 4 - 9 mkA; 5 - 4 mkA (irradiation at lower D and H concentrations in Pd as compared to (2)); 6 - X-ray irradiation of Pd (D, H).

The nonlinear increase in the emission current of positive ions from the electron beam is typical for processes involving the excited hydrogen atmosphere of metals. Excitation of the hydrogen atmosphere in metal results in the increase of H, D atom amount yielded from the metal onto the surface and the number of positive ions removed from the surface. A combination of these factors leads to a non-linear increase of the emission current of positive ions from the electron beam current.

The dependences between the positive ion current and the energy of accelerated electrons and between the ion distribution and the energy of the electron beam are shown in Figure 6. The positive ion current decreases with decrease of the accelerated electron energy within 6 keV region that is not associated with the decrease of power supplied to Pd (D, H) electron beams. In the experiment, the power supply $I_e E = 0.3$ W is constant - beam current increases with the decrease of the electron beam energy.

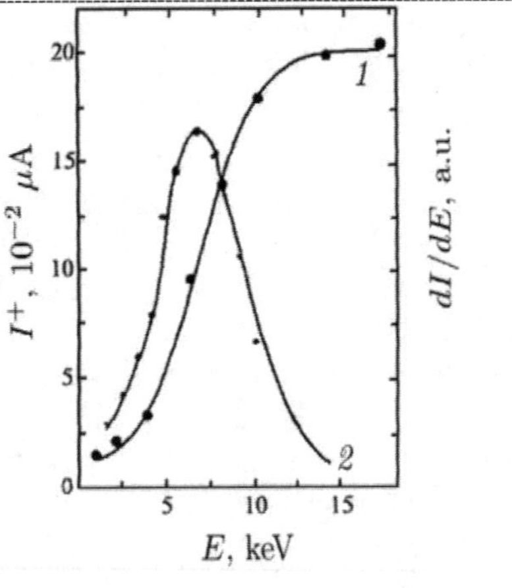

Figure 6. Dependences between the positive ion current and the energy of accelerated electrons (1) and the ion distribution and the energy of the electron beam (2). The power supply of Pd(D, H) ($I_e E = 0.3$ W) is constant.

6. Conclusion

A study of the hydrogen behavior in metals under ionizing radiation shows:

- Hydrogen in metals forms its own subsystem;

- Hydrogen subsystem in metals transforms to the excited state during the exposure to ionizing radiation;

- Migrating hydrogen provides the defect and impurity atom diffusion and organizes the metal structure.

A mechanism of non-equilibrium hydrogen yield and its isotopes from metals is connected with the specific accumulative properties of hydrogen in metals. Given that the degree of Pd saturation with hydrogen is high and corresponds approximately to one H (D) atom per Pd atom, then the energy of plasma oscillations in the hydrogen subsystem Pd:

$$\hbar\omega = \hbar e\sqrt{\frac{\rho_{Pd}N_A}{M_{Pd}m_p\varepsilon_0}} \cong 0.2eV\ , \qquad (1)$$

where ρ_{Pd}, M_{Pd} are the density and molar mass of palladium respectively; m_p is the mass of a proton; N_A is the Avogadro's number; ε_0 is the dielectric constant; e is the elementary charge.

The value of the non-equilibrium oscillation energy can be obtained from a comparison of the effective and equilibrium diffusion coefficients of hydrogen in Pd.

The applied nature of the observed phenomenon is as follows:

- the development of methods of low-temperature hydrogen removal from construction materials and recovering their mechanical properties;
- dosed hydrogen extraction from the hydrogen storages (hydrogen energetics);
- radiation technologies of hydrogen brittleness prevention.

Acknowledgement

The authors gratefully acknowledge the financial support of the project of the Ministry of Education and Science of the Russian Federation and as part of the state task in the field of scientific work on assignments #11.351.2014/K.

References

[1] Alefel'd G and Fekl' I 1981 Hydrogen in Metals, third ed. (Moscow: Mir)
[2] Nechaev Yu S 2008 *Physics-Uspekhi* **178** 709
[3] Ovchinnikov V V 2008 *Physics-Uspekhi* **178** 991
[4] Il'in A A, Kolachev B A and Pol'kin I S 2000 Titanium Alloys: Composition, Structure, Properties (Moscow: Moscow State Aviation Technological University Press)
[5] Tyurin Yu I , Chernov I P, Krening M X and Baumbakh Kh 2000 Radiation-stimulated hydrogen transfer from metals (Tomsk: Tomsk Stats University)
[6] Tyurin Yu I, Larionov V V, Chernov I P and Sklyarova E A 2011 *Technical Physics* **56** 30
[7] Prognimak A M 1995 *International Journal of Hydrogen Energy* **20** 449
[8] Gel'd P V, Ryabov R A and Kode E S 1979 Hydrogen and imperfections in metal structure (Moscow: Metallurgy)
[9] Sirotina R A, Mitrokhin, Zakharova A P and Verbetsky 1993 J. of Alloys and Comp. 202 41
[10] Evard E A , Gabis I E and Voyt A P 2005 *J. of Alloys and Comp.* **404-406** 335
[11] Heller E MB, Vredanberg AM and Boerma D O 2006 *Appl. Surface Sc.* **253** 771
[12] Elias R J, Corso H L and Gervasioni J L 2002 *I. J. of Hydrogen Energy* **27** 91
[13] Gabis I E , Kompaniets T I and Kurdyumov A.A , Surface processes and hydrogen permeability of metals 1987 (Moscow: Pleiades Publishing Ltd.)

[14] Tyurin Yu I, Semenov A M and Nikitenkov N N 2004 *J. Surface Investigation. X-ray, Synchrotron and Neutron Techniques* **5** 90

[15] Tyurin Yu I, Styrov V V, Sigfusson T I, Sivov Y A, Khoruzhy V D, Plotnikova E Y and Chistyakova N V 2012 *Bulletin of the Russian Academy of Sciences: Physics* **76** 687

[16] Nikitenkov N N, Khashkhash A M, Sigfusson T I, Kudryavtseva E N , Sypchenko V S , Tyurin Yu I and Chernov I P 2012 *A. Acta Physica Polonica* **121** 78

[17] Nikitenkov N N, Khashkhash A M, Khoruzhy V D, Tyurin Yu I, Chernov I P and Kudryatseva 2009 *Instruments and Experimental Technicue*s **52** 865

[18] Chernov I P, Rusetsky A S, Krasnov D N, Larionov V V, Sigfusson T I and Tyurin Yu I 2011 *Journal of Engineering Thermophysics* **20** 360

Journal of Physics: Conference Series **652** (2015) 012046 doi:10.1088/1742-6596/652/1/012046

Magnetron deposition of TCO films using ion beam

O Asainov[1], S Umnov[1] and A Chinin[1]
[1]Physical Technical Institute, Tomsk Polytechnic University, Tomsk, Russia

E-mail: usp@tpu.ru

Abstract. Thin films of tin oxide (TO) were deposited on the glass substrates at room temperature using reactive magnetron sputtering at various oxygen partial pressures. After the deposition the films were irradiated with argon ions beam. The change of the optical and electrical properties of the films depending on the irradiation time was studied. Films optical properties in the range of 300-1100 nm were investigated by photometry as well as their structural properties were studied using X-ray diffraction. Diffractometric research showed that the films, deposited on a substrate, have a crystal structure, and after argon ions irradiation they become quasi-crystalline (amorphous). It was found that the transmission increases proportionally with the irradiation time, but the surface resistance -disproportionally.

1. Introduction

The transparent conductive oxide (TCO) films are widely used in various industries. These optically transparent electrodes are for displays, solar panels, photoelectric devices, touch panels, etc. Currently, one of the most widely used TCO substances is indium tin oxide (ITO) because it has high electrical conductivity and optical transparency.

Recently, because of the high cost of indium, the research is conducted to find ITO alternative replacement. This is, in particular, the compounds based on tin oxide (dopant-Sb, F, As, Nb, Ta), zinc (dopant- Sn, Ge, Mo, F, Ti, Zr, Hf, Nb, W, Te), and others. Although, SnO_2 and ZnO impurities increase their conductivity, but it does not reach the values obtained for the ITO films. However, it is not necessary in some practical applications (very large conductivity). So, in the paper [1] pure oxide films (SnO_2) prepared by electron beam evaporation had transmittance of 90% and a surface resistance of 200 KOhm/□ (substrate temperature is 100°C). Originally, pure SnO_2 films are the conductors if they are deposited in a shortage of oxygen, so they are non-stoichiometric.

For the deposition of thin TCO films various methods are applied, including thermal, electron-beam, laser vaporization, cathode, a high-frequency or magnetron sputtering and others. We would like to focus on the film deposition by magnetron sputtering with additional use of ion beam. The use of ion beams in the film coatings deposition leads to almost all films properties change [2-3].

The papers [4-8] present data showing that the use of ion assisting leads to the changes in the optical and electrical properties of the deposited ITO thin films. It means that ion beams allow achieving the optimum relationship of the metal and the oxygen components in the film. So, the transmission of the films treated in that way is at 90%, and the surface resistance of the best samples is

[1] E-mail address: usp@tpu.ru .

Journal of Physics: Conference Series **652** (2015) 012046 doi:10.1088/1742-6596/652/1/012046

20 Ohms/▯. According to the authors, this is due to such factors as the changes in the mobility of the adatoms and crystal lattice damage.

In their works, the authors have used different types of ion sources that had been used at different deposition conditions (the ion beam parameters, the temperature of the substrate). In the papers mentioned above, assisting ion beams were used in the deposition of the doped TCO oxides.

Whereas, the energy formation of oxygen vacancies and tin interstitials in SnO2 is very low and, thus, these defects form readily, explaining often observed high conductivity of pure, but non-stoichiometric, SnO2 [9], so the ion beams could be a good tool to stimulate this process.

In this paper, we studied the influence of the irradiation of the pure tin oxide (without dopant) films with argon ions after its deposition on the optical and electrical properties.

2. Experimental details

Tin oxide films with thickness of about 30 nm were deposited by reactive magnetron sputtering of the target made of tin (99.5 wt.%) on the glass substrates at a room temperature. The glass substrates before the film deposition were purified by a chalk suspension in isopropyl alcohol.

The substrates were placed on a carrier that was able to carry out a reciprocating motion under the magnetron and the ion source in a vacuum chamber. The speed of carrier moving was 3.62 cm / second. The working gas was a mixture of oxygen and argon (partial pressure ratio is O2 – 60%, Ar – 40%). The pressure in the vacuum chamber during sputtering was 0.3 Pa. The gas flow was controlled by an electronic meter and the gas pressure was measured by an ionization and thermocouple gauge. After the deposition, tin oxide films were irradiated with argon ions.

The ions were generated by the ion source with a closed electron drift with 2800 V voltage and a 0.28 a discharge current. The greatest number of argon ions in the beam has energy of 400-600 eV. The chamber pressure in such case was 0.166 Pa. For measuring the surface electrical resistance Rsq. the samples were deposited with aluminum strips, which provided a reliable contact with the film. The thickness of the deposited films was measured using profilometer MicroMeasure 3D Station. To measure the transmission spectrum in the range of 300-1100 nm the spectrophotometer SF-256was used. The structure and morphology of the films was studied using the X-ray diffractometer XRD6000 and atomic-force microscope NT-MDT.

3. Results and discussion

All freshly prepared tin oxide films were transparent dielectrics. This is due to the deposition by reactive magnetron sputtering that was performed in an atmosphere with excess oxygen content. Under these conditions, there was enough oxygen to have all the metal oxidised to a state of complete oxide, so all the oxygen vacancies, creating conductivity, are busy and SnO2 semiconductor film is deposited as a dielectric. After the tin oxide films deposition the samples were irradiated with argon ions. Figure 1 shows the results of measurement of the films transmittance both the non-irradiated and irradiated samples. As it seen from the figure the transmission is strongly dependent on exposure time. So, when the exposure time is less than ~ 4.5 sec. the transmission slightly improves. For longer times ($\geq$ 8.8 sec.) the transmission is increased to 100% in almost the entire range of the wavelengths. It should be said, when the irradiation time is 13 - 22 seconds, there is the increase in transmittance at wavelengths less than 360 nm and a slight increase in the infrared region. The increase in transmission, especially in the short-wavelengths region, of the irradiated films is related to the change of the plasma frequency [10]. This change is a consequence of the electrons concentration increasing.

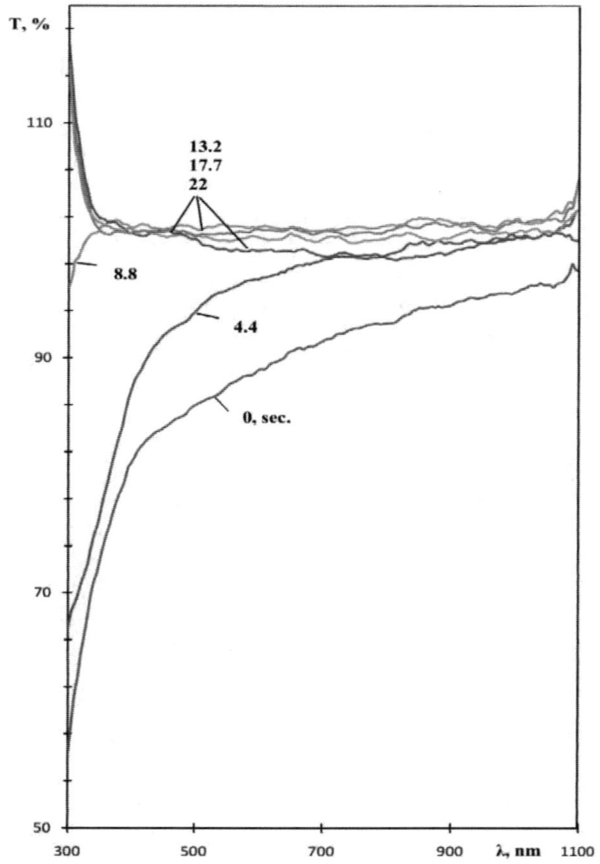

Figure 1. Dependence of the transmission on the irradiation time

However, the nature of the transmission is defined by the frequency of the incident light wave and the plasma frequency.

Table 1 shows the results of the surface resistance measurements, an average transmittance value at various irradiation times. The table shows that the film surface resistance is disproportionately dependent on the exposure time. With exposure time increasing, it initially decreases and, then, increases. The film transmittance almost uniformly increases and quickly reaches 100%.

As the deposition was carried out at excess oxygen content, the films were deposited in the SnO2 stoichiometric form which is a good insulator.

During the Ar ions irradiation there is their collision with the film surface and the power transfer o in the local area to the neighbouring atoms of the crystal lattice, imbedded and adsorbed on the grains boundary. As a result, there are physical and chemical processes that lead to the break or reduction of chemical bonds, causing desorption or chemical reaction that alters the structure of the coating.

The concentration of the electrons increasing is connected both with the break of the Sn-O chemical bonds and desorption of oxygen, which leads to the formation of non-stoichiometric tin oxide. Thus, the film exposure leads to increase in the number of oxygen vacancies, which, in its turn, leads to reduce in the films electric resistance. However, the further irradiation time increase leads to the film resistance increase. This is, possibly, connected with the formation of Sn self-point defects [11, 12]. Whereas, the point defects interact with all conductive electrons, so the electrons, located near the defects, are "bound" and do not participate in conductivity. As a result, the film resistance increases.

Table 1. Change in the surface resistance and the average transmittance of the exposure time

Parameters of the ion sours	Exposure time, sec.	Sheet resistance, MOhms/□	Average transmission (300-1100 nm), %
I=0.78 A	0	∞	88.82
U=2800 V	4.42	39	94.99
	8.84	∞	100.4
	13.26	0.92	100.11
	17.68	0.105	101.33
	22.1	19	101.66

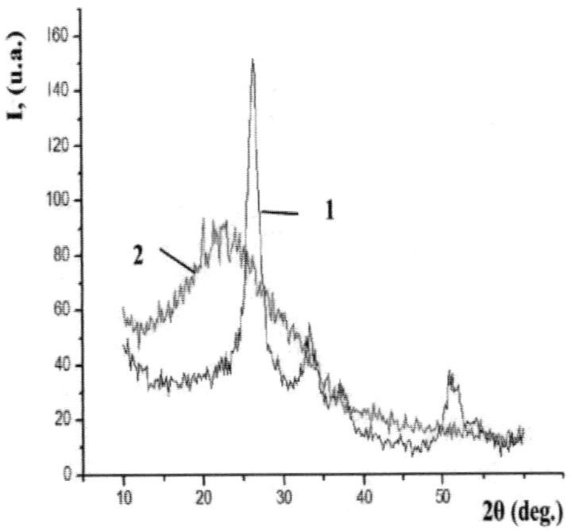

Figure 2. XRD diagram samples SnO_2: 1- before irradiation; 2-after irradiation

Figure 2 shows the results of the study of the deposited films structure. The figures show that the deposited tin oxide films have a crystal structure (tetragonal) with the grain size bottom of ~ 9.8 nm. After argon ions irradiation the structure becomes quasi-crystalline (amorphous - the red curve). This is due to the opening of the crystal lattice chemical bonds and forming new ones, which lead to formation of quasi-crystalline structure.

4. Conclusions

Pure SnO_2 films were deposited at room temperature with reactive magnetron sputtering, followed by Ar ions beam irradiation with energy of 400-600 eV. It allows obtaining the films with 100% transmittance and satisfactory surface resistance of 100 KOhm/□. The surface resistance value can be reduced by optimizing the deposition conditions and ion beam parameters. Thereby, transmission increases proportionally with irradiation time increasing, while the surface resistance - disproportionately. The crystal structure of SnO_2 film after ion irradiation is transformed into a quasi-crystalline.

References

[1] Shamala K S, Murthy L C and Narasimha Rao K 2004 *Bull. Mater. Sci.* **27** 295
[2] McNeil J R, Barron A C, Wilson A R, et al 1984 *Applied Optics* **23** 552
[3] McNeil J R, Al-Jumaily G A, Jungling K C et al 1985 *Applied Optics* **24** 486 [4] Meng L-J, Gao J, dos Santos M P, Wang X et al 2008 *Thin Solid Films* **516** 1365
[5] Yang Y, Huang Q, Metz A W et al 2004 *Advanced Materials* **16** 321
[6] Lee *S H*, Cho *H C*, Kim *H J et al* 2012 *Mol. Cryst. Liq. Cryst.* **564** 185
[7] Bae J W, Kim J S, Yeom G Y 2001 *Nuclear Instruments and Methods in Physics Research B* **178** 311
[8] Gilo M, Dahan R, Croitoru N 1999 *Opt. Eng.* **38(6)** 953
[9] Kılıç C and Zunger A 2002 *Phys. Rev. Lett.* 88, 095501
[10] Ginley D S and Bright C 2000 *MRS Bull.* **25** 15
[11] Umnov S and Asainov O 2015 *IOP Conf. Series: Materials Science and Engineering* **81** 012001
[12] Asainov O and Umnov S 2012 7th International Forum on Strategic Technology, IFOST 2012; Tomsk; Russian Federation, art. # 6357554

Electro-explosive alloying of VT6 alloy surface by boron carbide powder with the subsequent electron-beam treatment

DA Romanov[1], SV Raykov[1], VE Gromov[1] and YuF Ivanov[2,3,4]

[1]Siberian State Industrial University, Novokuznetsk, Russia
[2] Institute of High Current Electronics Siberian Branch of Russian Academy of Science, Tomsk, Russia
[3]National Research Tomsk State University, Tomsk, Russia
[4]National Research Tomsk Polytechnic University, Tomsk, Russia

E-mail: romanov_da@physics.sibsiu.ru

Abstract. The formation of electro-explosive alloying zone with the thickness up to 50 μm has been revealed. It has been shown that it has a gradient structure, characterized by the decrease of carbon and boron concentration with the increase of the distance up to the treatment surface. The subsequent electron-beam treatment of alloying zone leads to flattening of alloying surface relief and is accompanied by the formation of a multilevel structure at the depth up to 30 μm, characterized by the interchange of some layers with a different level of alloying, having structure of a submicro- and nanoscale level.

1. Introduction

The shortcomings of titanium alloys are their low antifriction properties. To increase the metal and alloy wear resistance different surface modification methods (coating formation, surface alloying and mechanical treatment, etc.) are used. One of them is electro-explosive alloying (EEA), in which the effect instrument on the surface are pulsed plasma jets, formed at the discharge of capacitive energy storage through a current-conducting material. The working substance of plasma accelerator serves both for heating of a surface layer of a modified material, and for its alloying [1, 2]. In some works it has been shown that in the capacity of the instrument, that allows avoiding disadvantages of EEA, one can use the sources of pulsed high-intensive low-energy electron beams (electron energy is up to 30 keV, energy density is up to 100 J/cm^2, pulse duration is 50-200 mks) [2]. Pulsed electron beams with such parameters provide remelting and high-speed crystallization of material surface layers with cooling rates up to 106 K/s and give the possibility to form nanoscale surface layers with a low roughness level and the increase of physical-mechanical properties.

The aim of work is the analysis of a structural state of a surface layer of VT6 titanium alloy, formed as a result of electro-explosive alloying with boron carbide powder and the subsequent irradiation with high-intensive pulsed electron beam.

2. Materials and research methods

In the capacity of the research material the alloy based on VT6 titanium has been used. It is used for the production of large-size welded and built-up constructions of aircrafts, for the production of the containers, acting under the internal pressure in large temperature intervals from 196 to 450°C, and a series of other structural components [3]. By the structure type it refers to the class of two-phase alloys, containing α-Ti and β-Ti [4]. The chemical composition of the alloy has corresponded to the State Standard 19807-91 (Table 1).

Table 1. Chemical composition of VT6 alloy VT6 (weight %)

Fe	C	Si	V	N	Ti	Al	Zr	O	H	Impurities
to 0.6	to 0.1	to 0.1	3.5 – 5.3	to 0.05	86.45 – 90.9	5.3 – 6.8	to 0.3	to 0.2	to 0.015	other 0.3

For EEA the authors of the work has used a laboratory electro-explosive unit EVU 60/10 (EEU 60/10) (power-intensity – to 60 kJ; discharge natural frequency – 10 kHz; maximal value of the charge – 5 kV; mean power consumption – 0.55 kW) with typical values of the absorbed power density at the treatment of the material surface ~10^9 W/m2, the pressure in a shock-compressed plasma layer near the irradiation surface ~106–107 Pa, treatment time - 100 μs, the thickness of the alloying zone in its central area 20-50 μm. In the capacity of the burst conductor a titanium foil with the thickness of 20 μm and mass of 0.25 g. has been used. In the area of the explosion the samples of boron carbide powder B4 Cwith the mass of 0.5 g. have been located. The treatment mode has been set by the value of a charging voltage of the capacitive energy storage, passage diameter of the booster nozzle and the distance from its section to the irradiation surface. At the same time the absorbed power density has been 5.5·109 W/m^2.

The subsequent electron-beam treatment (EBT) of the alloying layer has been carried out by a high-intensive pulsed electron beam at the unit SOLO of the Institute of High Current Electronics Siberian Branch of Russian Academy of Science [5]. The irradiation has been implemented at the following work parameters of electron sources: electron energy is 18 keV; energy density of an electron beam is 50 J/cm^2 and 60 J/cm^2; pulse duration of electron beam impact is 100 μs; impulse repetition frequency is 0.3 s^{-1}; the number of irradiation pulses is equal to 10.

The structure research of the modified material has been fulfilled by the methods of scanning electron microscopy. The elemental composition of a surface layer has been analyzed by the methods of X-ray spectral microanalysis.

3. Results and their discussions

The structure research of the modified material has been fulfilled by the methods of scanning electron microscopy. The elemental composition of a surface layer has been analyzed by the methods of X-ray spectral microanalysis [6].

The subsequent EBT leads to the essential transformation of the relief and the distribution of the alloying elements in the alloying zone. Namely, the surface relief is flattened; black-and-white contrast on the image of the alloying surface is replaced mainly by a grey one; that indicates a more even distribution along the surface of the alloying elements, initiated by the treatment.

High-speed melting and the subsequent high-speed self-hardening of a surface layer owing to the heat removal into an integrally cold sample volume leads not only to the ironing of the relief, but to the essential transformation of the material structure (fig. 1). The analysis of the treatment surface has revealed two character types of the structure, formed exclusively as a result of EBT. The first ones are the areas with a needlelike structure. The longitudinal sizes of the needles reach 10 μm, and the cross ones – 1 μm (fig. 2). The needles are located mainly perpendicularly to the irradiation surface, i.e. along the direction of the heat removal. The second ones are rather smooth areas, the element sizes of which reach 100 nm).

The results of the micro-X-ray spectroscopic analysis are the evidence of the fact that the areas, having a brightly evident dark contrast, are formed exclusively by the alloying elements and oxygen. Therefore, one can suppose that they are formed by the particles of the basic powder, which has not been dissolved at the EEA and the subsequent EBT.

Journal of Physics: Conference Series **652** (2015) 012047 doi:10.1088/1742-6596/652/1/012047

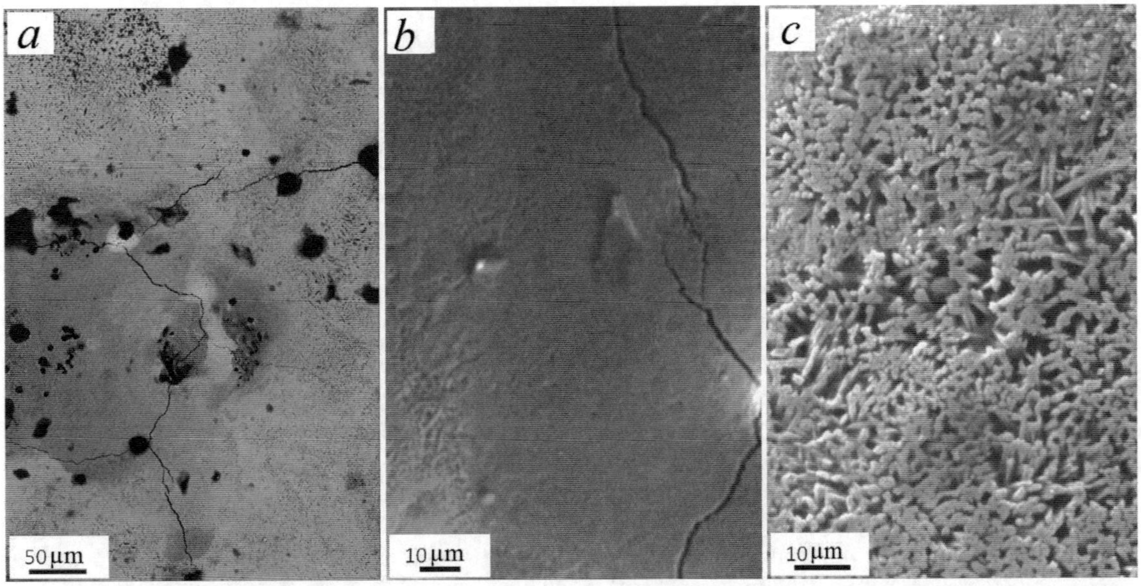

Fig. 1. The structure of the surface area electro-explosive alloying VT6 alloy after electron beam processing a - the central zone of the sample, b, c - areas on the periphery of the sample

Fig. 2. The structure of the surface area electro-explosive alloying VT6 alloy after electron beam processing a - the zone with the dendritic crystallization, b - smooth zone

The increase of energy density of electron beams up to 60 J/cm^2 leads to the formation of the structure of a needlelike type mainly. X-ray spectroscopic microanalysis of the areas with a needlelike structure has revealed the presence of both alloying elements, and the elements of a basic alloy. This fact is an indication of the increase of the degree of the dissolution of baron carbide particles in titanium with the development of energy density of the electron beams, i.e. of the increase of the homogeneity level of the modified surface layer.

The structure transformation of the surface layer volume has been analyzed, when studying the chemically etched cross sections. EBT at the energy density of the electron beams of 50 J/cm^2 does not lead to the formation of a homogeneous structure. The surface layer with the thickness to 30 µm has got a needlelike structure; the substructure of an underlying layer is similar to the substructure, formed in the material at EEA. Therefore, EBT at this treatment mode allows to modify the layer with the thickness, which is not more than 30 µm. The structure of the modified layer has often got a laminated structure.

One can also reveal the layers which are enriched and united by the alloying elements. At the same time the concentration of the alloying elements within the limits of each layer depends poorly on the distance up to the irradiation surface. This fact is the evidence of the formation of a multilayer structure in the material; in this structure the layers with the increased alloying level (more durable layers) interchange with the layers with the reduced alloying level (less solid layers).

Consequently, the irradiation of the alloying layer with high-intensive pulsed electron beam does not lead to the homogenization of the modified layer; as well as a peculiar laminated structure is formed. One can expect that the essential difference in the concentration of the alloying elements in the revealed layers will lead to the difference of their strengthening and tribological properties.

4. Conclusion

Electro-explosive alloying of VT6 titanium alloys with plasma, formed at the explosion of a titanium foil with the powder sample of boron carbide has been performed. The formation of the modified layer with the thickness up to 50 μm with a gradient structure has been revealed. It has four layers with a different degree of a chemical etchability. The distribution of the alloying elements both along the depth and each layer is characterized by the heterogeneity. At the same time the concentration of carbon and boron decreases with the depth.

The subsequent alloying surface irradiation with high-intensity pulsed electron beam of submillisecond action time has been carried out. It has been shown that, electron-beam treatment leads to the ironing of irradiation surface and the redistribution of alloying elements along the surface. The areas are revealed, which elemental composition is close to the composition of boron carbide particles, as well as the areas with a high degree of titanium substrate alloying, having a needlelike structure, and the areas with a low degree of alloying, having a nanodimensional structure. The increase of a surface energy density from 50 to 60 J/cm2 leads to a more even distribution of alloying elements along the surface. Along the zone depth of a combined treatment some layers with a different alloying degree are revealed. The increase of a surface energy density leads to the thickness decrease of layers and the increase of their number. The zone depth of electron-beam impact reaches 30 μm. It has been established that high cooling rates, initiated by the electron-beam treatment, leads to the structure formation of a submicro- and nanoscale level, that allows forecasting high strengthening and tribological properties of a modified material.

Acknowledgement

The research is fulfilled at the financial support of the Grant of the President of the Russian Federation for state support of young Russian scientists and the Candidates of Sciences MK-4166.2015.2 and Doctoral Candidates MD-2920.2015.8, RFBR within the frameworks of scientific projects # 13-02-12009 ofi_m, 15-08-03411, 14-08-00506a, state tasks of the Ministry of Education and Science # 2708 and 3.1496.2014/K for the fulfillment of scientific researches. The research is fulfilled in the centre of «Physical metallurgy» in SibSIU.

References

[1] Bagautdinov A Ya, Budovskikh E A, Ivanov Yu F and Gromov V E 2007 *Physical basis of electro-explosive alloying of metals and alloys* (Novokusnetsk, Publishing House of SibSIU).

[2] Raykov S V, Budovskikh E A, Gromov V E, Ivanov Yu F and Vashchuk E S 2014 *Formation of structural-phase states and properties of titanium alloy surfaces at electro-explosive alloying and the subsequent electron-beam treatment* (Novokusnetsk: Publishing House «Inter-Kuzbass»).

[3] Kolachev B A, Elagin V I and Livanov V A 1999 *Physical metallurgy and heat treatment of non-ferrous metals and alloys* (Moscow, Publishing House MISiS).

[4] Il'in A A, Kolachev B I and Pol'kin I S 2009 *Titanium alloys. Composition, structure,*

properties (Moscow, Publishing House MATI).
[5] Koval' N N and Ivanov Yu F 2008 *Izvestia vuzov. Physics* **5** 60-70.
[6] Brandon D and Kaplan U. 2006 Microstructure of materials. Methods of research and control

Investigation of the stability of the electron source with a multi-aperture plasma emitter generating a large cross-section electron beam

M.S. Vorobyov [1,2], **N.N. Koval**[2,3], **S.A. Sulakshin**[2]. **V.V. Shugurov**[2]

[2]Institute of High Current Electronics SB RAS, 2/3 Akademichesky Ave., Tomsk, 634055, Russian Federation
[3]National Research Tomsk State University, 36 Lenin Ave., Tomsk, 634050, Russian Federation

E-mail: vorobyovms@yandex.ru

Abstract. An experiment was performed to investigate the electric strength of the high-voltage acceleration gap of an electron source with a multi-aperture plasma emitter generating a beam of large cross section (750 × 150 mm^2) extracted into the atmosphere through a thin metal foil. It has been shown that the use of a mask in the plasma emitter which partitions the overall emission region to produce a plurality of small-cross-section beamlets, so that the extracted beam is a superposition of beamlets formed by individual emission units whose plasma boundary is stabilized by a fine metal grid, increases the electric strength of the high-voltage acceleration gap. This is of critical importance in cases where the electron source is operated in a repetitive pulse mode at high average power of the beam. In addition, an increase in the electric strength of the acceleration gap is promoted by that the modernized cathode assemblies of the plasma emitter are arranged normal to the axis along which electrons are extracted into the acceleration gap.

1. Introduction

Increasing the electric strength of the acceleration gap in high-voltage electron sources is one of the main problems that must be solved in designing and using systems of this type [1–3]. It is well known that the use of plasma emitters where electrons are emitted only from certain sites bounded with an emission grid increases and stabilizes the parameters of the generated beam [4–7]. In systems of this type, called multi-aperture electron sources, large- cross-section beams are generated due to the superposition of individual beamlets, each being generated in the its own acceleration unit consisting of two or more electrodes with coaxial holes. The use of multi-aperture plasma cathodes allows additional stabilization of the plasma emission boundary, thereby providing a wide range of the electron beam parameters [4, 5].

It is also well known that the cathode spot operating on the surface of a cathode causes intense sputtering of the cathode material [8]. This not only leads to a considerable consumption of the cathode but also reduces the electric strength of the high-voltage acceleration gap, as the microdroplets of the molten cathode material, reaching, for instance, the emission grid and settling on it, cause a local increase in electric field, the occurrence of a cathode spot on the grid, and, consequently,

[1] To whom any correspondence should be addressed.

electrical breakdown of the acceleration gap. The present study was carried out to investigate the electric strength of the high-voltage acceleration gap in an electron source with a multi-aperture plasma cathode in which the emission plasma is generated by a cathode assembly with a screened cathode spot initiated by electrical breakdown in a low-pressure gas.

2. Experimental procedure

The experiment was carried out on DUET, a modernized wide-aperture pulsed electron source with a grid plasma cathode (Fig. 1) capable of producing an electron beam with the following key parameters: an electron energy of up to 200 keV, emission current of up to 50 A, pulse duration of up to 100 μs, pulse repetition rate of 50 s^{-1}, and beam dimensions of 750 mm × 150 mm with the current density distribution over the beam cross section uniform to within ± 15% of the average current density.

It is well known [9] that the arrangement of the cathode assemblies in a plasma emitter normal to the axis along which electrons are extracted increases the electric strength of the high-voltage acceleration gap. This is probably due to a decrease in the number of the cathode material microdroplets formed as a result of the operation of the cathode spot [8] both on the emission grid and in the region of the acceleration gap.

To reduce the number of cathode material microdroplets falling onto the surface of the emission grid, a cathode assembly has been designed, fabricated, and experimentally investigated where the cathode spot is line-of-sight screened from both the hollow anode and the emission grid. The cathode assembly design and operating principle are described in detail elsewhere.

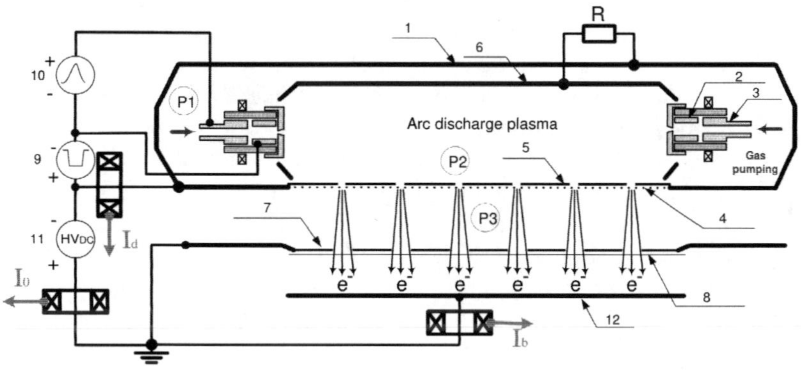

Figure 1. Schematic of a wide-aperture electron source: 1 – hollow anode, 2 – cathode, 3 – igniter electrode, 4 – emission grid, 5 – mask; 6 – screen 7 – support grid of extraction foil window, 8 – extracion foil, 9 - power supply of discharge, 10 – power supply of igniter, 11 – high voltage source, 12 – collector, P1, P2, and P3 – PMI-2 type manometric transducers.

A cathode assembly of this type has a constricting hole providing a pressure difference between the regions of electron beam generation and acceleration. This reduces the gas pressure in the acceleration region and thereby increases the electric strength of the high-voltage acceleration gap. This is especially important when operating the system at high average beam powers under the conditions of desorption of working and residual gases from the high-voltage electrode surfaces. When a metal mask is placed in the emitter, an additional pressure difference between the region of emission plasma generation and the acceleration gap should be attained due to the low geometric transparency of the mask. To check this assumption, three PMI-2 type manometric transducers were placed in the electrode system of the electron source (see Fig. 1): P1 downstream of the plasma emitter, P2 in the plasma emitter, and P3 in the acceleration gap. In the absence of a mask in the emitter, the readings of the transducers were approximately the same. When a mask of total geometric transparency ≈ 13% was placed in the emitter and the working gas pressure in the acceleration gap was 20 mPa, the pressure in the plasma emitter was a factor of 1.5 higher, equal to about 30 mPa. This provided stable

Journal of Physics: Conference Series **652** (2015) 012048 doi:10.1088/1742-6596/652/1/012048

operation of the plasma emitter at a slightly reduced working gas flow rate (from $Q \approx 0.1 \frac{m^3 Pa}{s}$ without a mask to $Q \approx 0.07 \frac{m^3 Pa}{s}$ with a mask). According to preliminary estimates, the stability of operation of the electron source, which is inversely proportional to the number of electrical breakdowns per 1000 beam current pulses, increased in inverse proportion to the geometric transparency of the mask placed in the plasma emitter. Thus, the stability was increased from 2 to 12.

3. Results and discussion

The mechanism of plasma generation in the plasma emitter electrode system is the following: The electrons produced due to the operation of the cathode spot are accelerated in the cathode layer and enter the space of the main discharge hollow anode with an energy roughly corresponding to the discharge voltage. The primary electrons interact with the atoms of the working gas, and the gas is ionized. The area of the anode in this case is much larger than the area of the cathode, and the electrons having higher mobility than the ions produced as a result of ionization of the working gas may arrive at the hollow anode if they have energies high enough to overcome the potential barrier induced by the positive charge of the plasma. Therefore, to maintain a balance of currents, the plasma reacts to the loss of electrons by increasing its potential [3]. This is, among other things, due to the lower potential barrier in the electron extraction region that should be overcome by the electrons when leaving the discharge gap toward the anode. However, the increase in plasma potential relative to the walls of the hollow anode is always limited. For instance, an increased plasma potential can be responsible for the formation of a cathode spot on the hollow anode surface, occurrence of the discharge current high-frequency noise, distortion of the discharge current waveform and amplitude, etc. The cathode spot formed on the hollow anode wall causes a local increase in plasma density and, as a consequence, an uncontrolled increase in discharge current accompanied by an increase in emission current. This, in turn, promotes the breakdown of the high-voltage acceleration gap [12]. It is also well known [12] that the increase in plasma potential can be significantly affected by the high-energy ions entering the plasma emitter from the acceleration gap through the emission grid, whose geometric transparency is always less than unity.

The presence of ions in the acceleration region is evidenced by the autograph of the ion beam left on emission grid 4 (Fig. 2) and on hollow anode 6 (not shown in Fig. 2). This experiment was carried out in the absence of a mask in the emitter. A somewhat smaller autograph of the ion beam on the emission grid and even smaller one on the hollow anode are probably due to the presence of a chamfer at the place of contact of the vacuum chamber with the support grid, causing the ion beam to focus. In this case, the ions can be produced, for instance, by ionization of the gas desorbed from the surfaces of the extraction foil and support grid both by the accelerated beam electrons and by the electrons reflected from the extraction window foil.

In addition, examination of the beam autograph shown in Fig. 2 suggests that each ion beamlet formed in the channel of the extraction window support grid always has a diameter less than the diameter of the holes in the support grid, which, probably, is also due to the focusing of each individual beamlet.

The ions accelerated in the high-voltage acceleration gap and arrived at the emission grid knock out secondary electrons from the grid. The number of these electrons is determined by the cathode material and the energy of the ions [13].

The absence of a mask in the emitter has the result that some portion of the ion flow reaches the emission grid, knocking out secondary electrons in accordance with the coefficient of secondary ion-electron emission. Another portion of the ion flow, determined by the geometric transparency of the emission grid, is injected into the emission plasma and partially reaches the hollow anode of the discharge. This results, first, in an increase in plasma potential and, second, in ejection of secondary electrons from the other electrodes of the plasma emitter, e.g., from hollow anode 6. All this adversely affects the electric strength of the high-voltage acceleration gap and leads to an increase in the potential of the emission plasma, to an uncontrolled increase in the emission current associated with the ion-electron emission, to charging of the dielectric inclusions that are always present on electrodes when a diffusion pump is used, etc. The decrease in electric strength of the high-voltage acceleration gap is accompanied by the loss of one of the main advantages of electron sources with plasma emitters with grid stabilization of the plasma emission boundary, namely the independence of the beam parameters from each other, as in this case, the processes occurring in the acceleration gap affect the processes occurring in the discharge gap.

Figure 2. Ion beam autograph left on the emission grid on the side of the high-voltage acceleration gap in the experiment with no mask in the emitter.

Thus, it is of importance that the paths of the primary electrons emitted from the plasma emitter and accelerated mostly normal to the extraction foil window are substantially different from the paths of the desorbed gas ions accelerated, due to focusing, at an angle to the emission grid. Knowing this, one can significantly reduce the ion flow injected into the emission plasma that increases the plasma potential. This can also be achieved by using a low-geometric-transparency mask in the plasma emitter; in the case under consideration, the transparency was ≈ 13% (see Fig. 1). Assume that all electrons emitted into the acceleration gap are extracted into the atmosphere through the foil with no loss of the beam current at the support grid. At the same time, the interaction of the electron beam with the foil results in the occurrence of ions, as evidenced by the beam autograph shown in Fig. 2. However, in view of the geometric transparency of the emission grid equal to ≈ 50%, the total geometric transparency of the emission electrode makes only 7% (with no mask in the emitter, the geometric transparency of the emission electrode is ≈ 50%) and thus almost the entire ion flow is involved in ion-electron emission. This mechanism of electron beam generation makes it possible to significantly reduce the potential of the emission plasma and restore one of the major advantages of the plasma emitter, namely the possibility to control the electron beam parameters independently from each other.

In an experiment on the extraction of a beam from the acceleration gap into the atmosphere, it was shown that at an accelerating voltage U_0 = 200 kW, the beam current extraction efficiency β, which is the ratio of the beam current extracted into the atmosphere, I_b, to the current in the acceleration gap, I_0 ($\beta = I_b/I_0$), was no greater than 0.75. This indirectly confirms the presence of an ionic component in the acceleration gap, which is estimated to make about 10% of the total current in the acceleration gap.

4. Conclusion

The experiments have shown that the electric strength of the high-voltage acceleration gap in an electron source with a wide-aperture plasma emitter operating based on a low-pressure arc can be increased by using cathode assemblies with the cathode spot screened from the line of sight of the hollow anode and emission grid, which reduces the number of cathode material microdroplets, and by using a low-geometric-transparency mask in the emitter, which provides an additional pressure difference between the plasma generation and electron acceleration regions and reduces the potential of the emission plasma at a higher emission current density due to a reduced effect of the high-energy ion beam on the plasma and redistribution of the discharge current in the plasma emitter.

The presence of a mask in the emitter is a condition for stable operation of the electron source. With no mask, the average power of the beam in the acceleration gap was no greater than 3500 W, whereas with a mask placed in the emitter, it could be increased to 6500 W. Further increase in beam power in the experiment with a mask installed in the emitter was limited by the high-voltage power supply.

Journal of Physics: Conference Series **652** (2015) 012048 doi:10.1088/1742-6596/652/1/012048

Acknowledgement

This work was supported by the Russian Science Foundation (project #14-29-00091).

References

[1] Slivkov I N 1986 Processes at high voltage in vacuum (Moscow, EAI) (in Russian) 256 pp.
[2] Bugaev S P, Kreindel Yu E, Schanin P M 1984 Large section electron beams (Moscow, EAI) (in Russian) 112 pp.
[3] Koval N N, Oks E M, Protasov Yu S, Semashko N N 2009 Emission electronics (Moscow, MGTU) (in Russian)
[4] Kreindel Yu E, Martens V Ya, S'edin V Ya, Gavrintsev S V 1982 J Prib. Tekh. Eksp Issue 4 178
[5] Burdovitsin V A, Oks E M 2008 *Laser and Particle Beams*. **26** **4** 619-635.
[6] Goebel D.M., Schumacher R.W., Watkins R.M. 1992 Proc. Conf. on High Power Particle Beams. Washington, USA., 1992, p.1093.
[7] Kandaurov I, Astrelin V, Avrorov A, Burdakov A, Bykov P, Derevyankin G, et al. 2011 *Fusion Science and Technology* **59** **1** 67.
[8] Mesyats G A 2011 Explosive Electron Emission (Moscow: Izd. Fiz. Mat. Lit.) (in Russian) 280 pp. ISBN 978-5-94052-207-2
[9] Gavrilov N V, Kovalchuk B M, Kreindel Yu E, Tolkachev V S, Schanin P M 1981 Plasma-emitter high-voltage electron source for generation of large-cross-section beams *J Prib. Tekh. Eksp* **24** 3(2) 727
[10] Koval N N, Kreindel Yu E, Tolkachev V S, Schanin P M 1985 *IEEE Trans. Electr. Insul.* **El-20** 735-737.
[11] Zlobina A F, Kazmin G S, Koval N N, Kreindel Yu E 1980 Plasma parameters in the expander of an electron emitter with a constricted arc discharge J Zh. Tekh. Fiz. **50** 6 1203
[12] Devyatkov V N, Koval N N 2014 *Journal of Physics: Conference Series* **552**, 2014, 012014
[13] Gavrilov N V, Kreindel Yu E, Schanin P M 1984 Pulsed electron source based on a non-self-sustained high-voltage glow discharge *J Prib. Tekh. Eksp*. **2** 143

Recovery of the electric strength in a cold cathode thyratron

N V Landl[1,2], Y D Korolev[1,2,3], O B Frants[1,2], V G Geyman[1], A V Bolotov[1]
[1] Institute of High Current Electronics SB RAS, 634055 Tomsk, Russian Federation
[2] Tomsk State University, 634050 Tomsk, Russian Federation
[3] Tomsk Polytechnic University, 634050 Tomsk, Russian Federation

E-mail: landl@lnp.hcei.tsc.ru

Abstract. The paper deals with the investigations of the recovery process of electric strength in a cold cathode thyratron. Method which allows extracting the plasma of the preceding discharge not only from the cathode cavity but also from the main gap of the thyratron is proposed. Method is based on the usage of a low-current nonsteady state discharge in the pause between the pulses.

1. Introduction

The cold cathode thyratron (or the pseudospark switch) is considered as an advanced alternative to ignitrons and vacuum switches in the facilities that require an extremely high current [1–9]. The triggering with a small jitter makes it possible to use the pseudospark switch instead of the classical thyratrons in the electric circuits with a fast current rise. In general, the principle of operation of the pseudospark switch resembles that for the classical thyratron [4]. On the one hand, a usage of the cold cathode is quite definite advantage from the viewpoint of increasing the switching current [3–5]. On the other hand, the problem of increasing the pulse repetition rate in the pseudospark switch is more severe than that in the thyratron with hot cathode [4, 10, 11].

The upper level of the pulse repetition rate in the switch is determined by the conditions when a characteristic recombination time for plasma in the gap becomes comparable with the time interval between the pulses [11–15]. In these conditions, a residual plasma from a preceding discharge remains in the gap to the instant when a successive pulse arrives to the electrodes thereby the pulsed breakdown voltage for the switch decreases.

One of the methods to enhance a limited value of the pulse repetition rate is based on the idea to extract the products of the preceding breakdown from the cathode cavity in the pause between the pulses. For this purpose a so-called blocking electrodes can be applicable [2, 11]. These electrodes are intended to extract the electrons from the cathode cavity and to suppress the prebreakdown electron current in the main gap.

In this paper we investigate the method which allows extracting the plasma of the preceding discharge not only from the cathode cavity but also from the main gap of the switch. This method also allows us to obtain the information on the process of deionization in the main gap.

2. Experimental setup and principle of operation of the electric circuit

To investigate the recovery process of the main gap we used the electric circuit shown in figure 1. The main electrode system of the thyratron is formed by grounded anode *1* and hollow cathode *2*. The main gap distance, the borehole diameters, and the thickness of the flat part of electrodes *1* and *2* are

Journal of Physics: Conference Series **652** (2015) 012049 doi:10.1088/1742-6596/652/1/012049

equal to each other and amount to 4 mm. A thickness of the upper flat part of electrode *3* is 1 mm and an aperture diameter is 12 mm. A height of the cathode cavity h = 16 mm and its diameter D_c = 30 mm. The working gas was xenon.

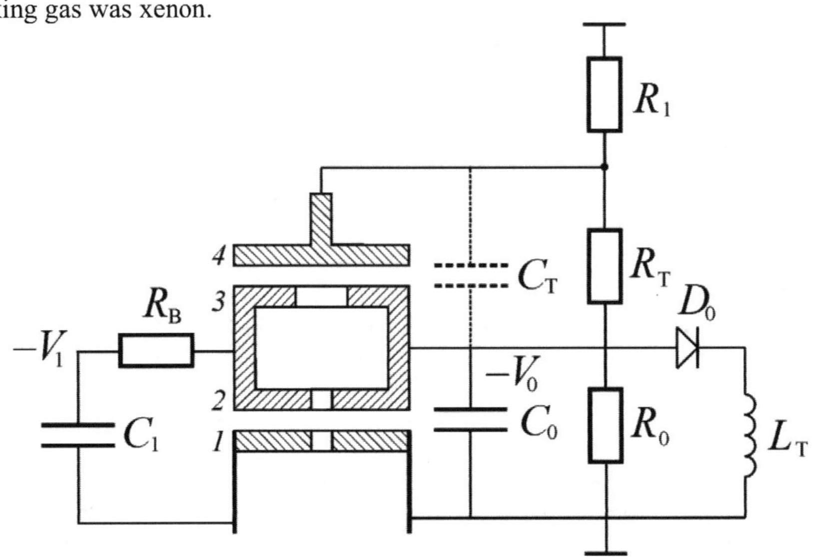

Figure 1. Design of electrode system and electric circuit. C_0 = 1 nF, R_0 = 70 kΩ, R_1 = 100 kΩ, R_T = 100 kΩ or ∞, R_B = (10 – 100) kΩ, C_T is the parasitic capacitance of the gap *3–4*.

The principle of operation of the electric circuit can be described as follows. The main gap *1–2* is powered by means of a low-inductance circuit from a capacitor bank C_0, which is charged to a voltage V_0 from the pulsed transformer L_T during a typical time of about 5 μs and larger. Some fraction of the voltage V_0 is derived to electrode *4* due to the active divider $R_T - R_1$ or due to divider $C_T - R_1$ (when $R_T = \infty$). Electrode *4* can be considered as the blocking electrode [11]. The essence of the blocking effect seems to be as follows. With respect to the main cathode cavity *2*, the potential V_4 of the electrode *4* is positive, and electrons can be extracted from the main cathode cavity *2* to electrode *4* via the aperture in electrode *3*. Then, the process of discharge initiation in the main gap is suppressed that allows operating the switch with an enhanced pulsed breakdown voltage.

Enhancing in efficiency of the blocking effect is achieved when the additional *DC* power supply V_1 is used. A low voltage of about 200 V and less is applied to the main gap. This voltage is not sufficient for appearing a self-sustained discharge [11–13, 16]. However, due to this voltage a certain nonself-sustained current is available in the main gap not only in the time interval during the charging of the capacitor C_0 but also in the pause between the pulses.

In the experiments we measured the voltage V_0 between the electrodes *2* and *1* (i. e., the potential of electrode *3* with respect to the ground) and the potential V_4 of electrode *4* with respect to the ground. In the initial conditions the main capacitor C_0 is charged to the voltage V_1 ($R_0 >> R_B$). When the high-voltage pulse appears at the capacitor C_0 the self-breakdown in the main gap can occur. Typical breakdown current is about 1 kA and period of the current oscillations is 60 ns.

As a result of the breakdown the discharge plasma appears in the main gap and in the main cathode cavity *2* so that the current from power supply V_1 flows in the gap in the pause between the pulses. This current is determined by the voltage V_1, the ballast resistor R_B, and by the properties of the plasma in the main gap. The voltage V_1 and the resistance R_B are selected from the condition of preventing the self-sustained discharge in the gap *1–2* (the voltage V_1 is lower than the discharge burning voltage for a glow type of discharge [12, 16]). Correspondingly, measurements of the potential V_0 during breakdown and in the pause between the pulses offer a possibility to investigate the process of recovering the electric strength for the main gap.

3. Results and discussion

Typical voltage waveforms are shown in figure 2. On the basis of these waveforms we can envision the following interpretation of the processes in the main gap and in the cathode cavity.

In the initial conditions the capacitance C_0 is under a voltage V_1 = 100 V (the time interval before t = 200 μs). When the high voltage pulse is formed by the pulsed transformer L_T the capacitance C_0 is charged to a voltage V_0 = 3 kV for a time of about 5 μs and the self-breakdown of the main gap occurs. The instant of breakdown corresponds to t = 200 μs. Then the capacitor bank C_0 is completely discharged via the main gap. (Note that the voltage pulse of 3 kV is not visible in figure 2 since the pulse duration of 5 μs is much less than the oscilloscope time scale).

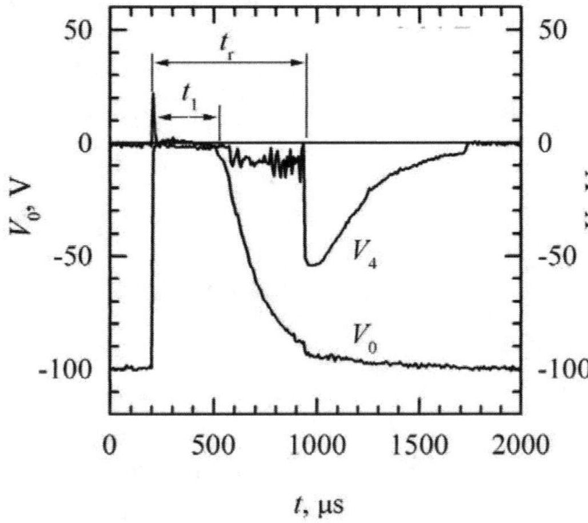

Figure 2. Voltage waveforms at electrode *4* and at the main gap. p = 0.04 Torr, V_0 = 3 kV, V_1 = 100 V, R_B = 10 kΩ, R_0 = 70 kΩ, R_1 = 100 kΩ, $R_T = \infty$.

As a result of the breakdown the gas discharge plasma of a high density is generated in the main gap and in the cavity *2* so that the current from the power supply V_1 starts flowing through this plasma. During the time interval t_1 we can speak of a low-current discharge with extremely low discharge burning voltage. The gas discharge plasma seems to be a short-circuiting bridge for power supplier V_1. Correspondingly, the potentials V_0 and V_4 are close to zero. At this stage the plasma density in the main gap decreases mainly due to volume recombination processes and due ambipolar diffusion to the surface of the electrodes.

After the time interval t_1 the plasma density decreases in such extent that the total resistance of the gap *1–2* becomes comparable with the resistance of R_B and R_0. It means that the capacitor C_0 becomes to charge from the power supplier V_1. However the charging time is still larger than C_0R_B = 100 μs. In other words, the resistance of the main gap gradually increases with time, which leads to increasing the voltage V_0. As applied to this stage we can say that the charge carriers are extracted from the main gap due to the voltage V_0. In succeeding description for the sake of definiteness we will use the term "extracting current" for the current between the electrodes *1* and *2* and the term "blocking current" for the current to electrode *4*.

A special case is the behavior of the voltage V_4 at the stage under discussion. Note that for the conditions in figure 2 the resistor R_T is absent ($R_T = \infty$) and the voltage V_4 is determined by the spurious capacitance C_T, by the resistor R_1, and by the properties of the discharge between the hollow cathode *2* and electrode *4*. If the voltage V_4 was equal to zero, we would definitely say that the residual gas discharge plasma in the gap *2–4* is absent. In these conditions the capacitor C_T would be charged to a voltage V_0 at each instant of time (a characteristic charging time for this capacitor is extremely small, C_TR_1 = 2 μs).

If the gap *2–4* was filled with high-density plasma, than the voltage V_4 would be equal to V_0. For our particular case we observe the chaotic oscillations in the voltage V_4 at the stage ($t_r - t_1$). It means that a low-density plasma is available in the gap *2–4*. A blocking current through this plasma is responsible for the voltage behavior at the resistor R_1.

At the instant t_r the abrupt increase in the voltage V_0 is observed. We interpret it as a complete de-ionization of the main gap (the charge carriers have been extracted from the main gap to the instant t_r). As a result, the charging time for the capacitance C_0 becomes to determine by the resistor R_B which is accompanied by the fast rise in the voltage V_0. Simultaneously, the abrupt increase in the voltage V_4 takes place. According to the above interpretation we can say that the current from power supply V_1 is intercepted by the blocking discharge which burns in the gap *2–4*. The initial current of this discharge is equal to $V_4/R_1 = 0.55$ mA. After the time t_r the charge carriers continue to be extracted from the gap *2–4*. Correspondingly, the voltage drop at the gap *2–4* increases and potential V_4 decreases. To a time $t = 1600$ μs the non-self sustained blocking discharge in the gap *2 – 4* is stopped.

The time interval t_r corresponds to complete recovering of the electric strength of the main gap. After the time t_r a small current is still available in the gap *2–4*. This current can be considered as a blocking current, which is useful from the viewpoint of preventing an occasional breakdown in the main gap when a successive voltage pulse arrives [11].

The summary of experimental data on measuring the time t_r is presented in figure 3. We fixed the value R_B = 10 kΩ and changed the voltage V_1. It should be stressed that the voltage V_1 cannot be increased unlimitedly. With a high voltage V_1 the situation is possible when the voltage drop at the main gap achieves of about 250 V. Then a steady state self-sustained glow discharge appears in the main gap. It is evident that the electric strength of the gap decreases dramatically in this case. With taking this fact into account we normally use the voltage V_1 at a level of 200 V and less.

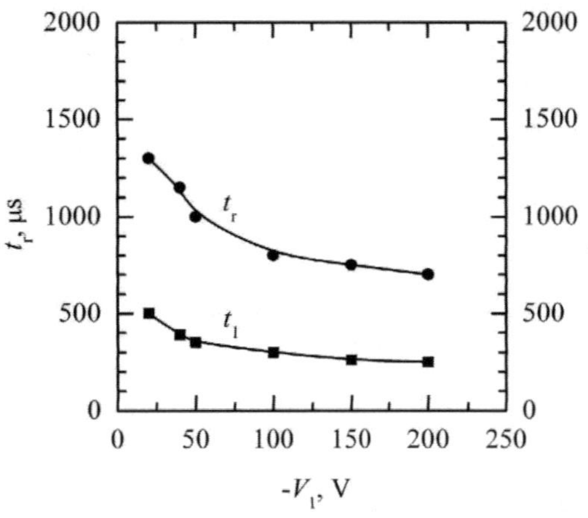

Figure 3. Recovery time t_r and time t_1 versus the voltage of power supplier V_1. R_B = 10 kΩ, R_1 = 100 kΩ, $R_T = \infty$, p = 0.03 Torr.

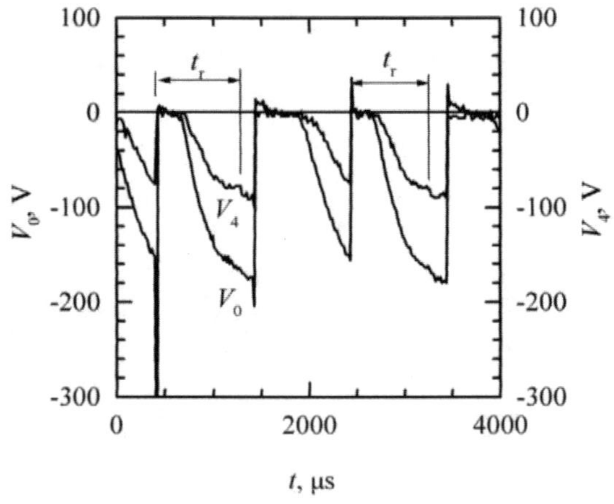

Figure 4. Voltage waveforms at electrode *4* and at the main gap. V_0 = 4 kV, V_1 = 200 V, R_B = 100 kΩ, R_1 = 100 kΩ, R_T = 100 kΩ, p = 0.04 Torr.

Due to an increase in the extracting current the time t_r can be decreased approximately by two times. However, the curve saturates and we obtain a minimal $t_r \approx$ 800 μs. Note that a decrease in the time t_r is mainly due to reduction of the first stage of recovering (the time t_1).

The recovery time t_r is influenced both the extracting current between the electrodes *1* and *2* the blocking current between the electrodes *3* and *4*. As for the latter, we can vary this current by means of variation of the voltage ($V_0 - V_4$) across the gap *3 - 4*. A convenient method to do it is to insert the resistor R_T in the circuit. Then the relation between R_T and R_1 determines the potential difference $(V_0 - V_4)$.

Figure 4 shows voltage waveforms at the electrode *4* and at the main gap for the conditions of figure 2, but a resistor R_T = 100 kΩ is available in the circuit. The time scale is chosen from the consideration to demonstrate four pulses at the main gap, which follow one after another at the times: t_1 = 400 µs, t_2 = 1400 µs, t_3 = 2400 µs, and t_4 = 3400 µs. We can see that in the time interval $(t_2 - t_1)$ the gap is completely recovered its electric strength. Then at an instant t_2 breakdown in the main gap occurs, and breakdown voltage $V_{br} = V_0$ = 4 kV. After that, in the time interval $(t_3 - t_2)$, the recovery time is larger than 1000 µs. As a result, at an instant t_3 we have the breakdown at low voltage ($V_{br} \approx$ 1.5 kV). Then again, in the pause $(t_4 - t_3)$ the time t_r becomes less than 1000 µs, and the breakdown occurs at a high voltage and so on.

A characteristic feature of the above described experiments is that we used the main capacitor bank C_0 = 1 nF. In this case we can obtain a high pulse repetition rate but a value of the main discharge current is rather small for these conditions (less then 500 A). It could be expected that the discharge current in the main gap would influence to its recovery process. However with the electric circuits that we currently have we are not able to provide simultaneously both a large value of C_0 and a high pulse repetition rate.

On the other hand the method of measurements that we have proposed allows us to obtain direct information on recovery time for the main gap. The experiments presented below are the investigation of the recovery process in the main gap for a high value of capacitor bank (C_0 = 35 nF) and correspondingly for a high pulsed discharge current.

Experimental arrangement is shown in figure 5. Here the intermediate capacitor bank C_{ch} = 35 nF is charged by a positive voltage from the power supply V_{ch}. After closing the pseudospark switch *S* the resonant charging of the main capacitor bank C_0 occurs and the voltage pulse with a voltage rise time of 5 µs forms at the main gap. The main gap operates in self-breakdown mode. Due to the power supply V_1 an extracting voltage up to 200 V is attached to the main gap in the pause between the pulses.

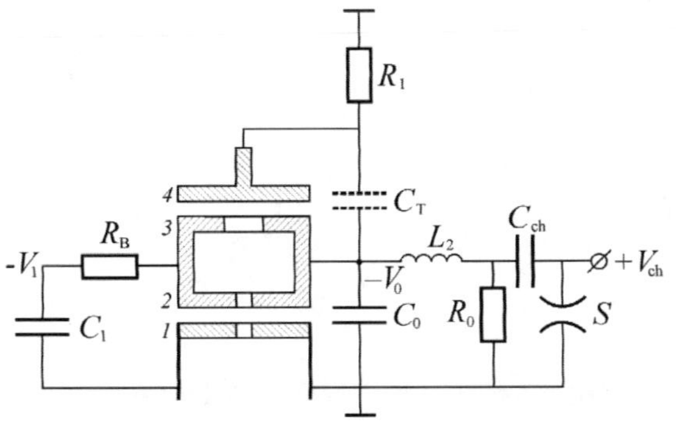

Figure 5. Design of electrode system and electric circuit for investigation of the recovery of the main gap at a high value of C_0. C_T is the parasitic capacitance of the gap *3* – *4*; C_0 = 35 nF, C_{ch} = 35 nF, R_0 = 50 kΩ, R_1 = 100 kΩ, R_B = 1 kΩ.

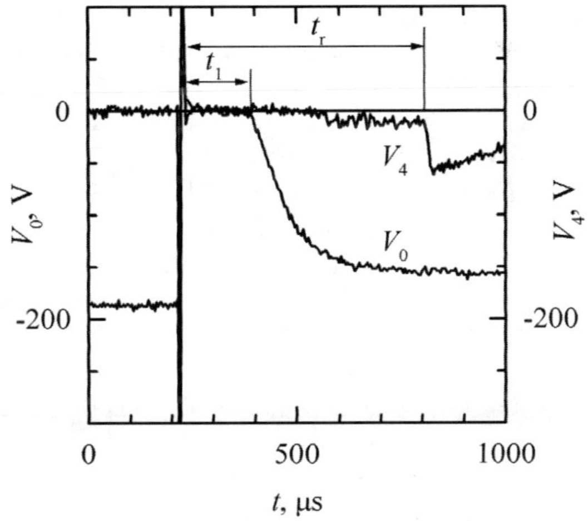

Figure 6.Voltage waveforms for the electric circuit shown in figure 5. V_1 = 200 V, R_B = 1 kΩ, R_1 = 100 kΩ, $R_T = \infty$, V_0 = 4 kV, p = 0.05 Torr.

Typical voltage waveforms are shown in figure 6. The maximum current in this case is about 4 kA. The presented data allows us to estimate the recovery time for the main gap as $t_r \approx$ 600 µs. It means that an essential increase in the discharge current does not lead to increasing the recovery time. The pulse repetition rate for the pseudospark gap in similar regime can be at a level of 1000 Hz.

Acknowledgments
The work was supported by Russian Science Foundation under the Grant # 14-19-00139.

References

[1] Frank K and Christiansen J 1989 The fundamentals of the pseudospark and its applications *IEEE Trans. Plasma Sci.* **17** 748-53

[2] Stetter M, Felsner P, Christiansen J, Frank K, Gortler A, Hintz G, Mehr T, Stark R and Tkotz R 1995 Investigation of the different discharge mechanisms in pseudospark discharges *IEEE Trans. Plasma Sci.* **23** 283–93

[3] Kozyrev A V, Korolev Y D, Rabotkin V G and Shemyakin I A 1993 Processes in the prebreakdown stage of a low-pressure discharge and the mechanism of discharge initiation in pseudospark switches *J. Appl. Phys.* **74** 5366–71.

[4] Bochkov V D, Dyagilev V M, Ushich V G, Frants O B, Korolev Y D, Shemyakin I A and Frank K 2001 Sealed-off pseudospark switches for pulsed power applications (current status and prospects) *IEEE Trans. Plasma Sci.* **29** 802–8

[5] Korolev Y D, Frants O B, Landl N V, Shemyakin I A and Geyman V G 2013 High-current stages in a low-pressure glow discharge with hollow cathode *IEEE Trans. Plasma Sci.* **41** 2087–96

[6] Meena B L, Rai S K, Tyagi M S, Pal U N, Kumar M and Sharma A K 2010 Characterization of high power pseudospark plasma switch (PSS) *J. Phys. Conference Series* **208** 012110

[7] Zhang J, Zhao J P and Zhang Q G 2014 The breakdown characteristics of single-gap pseudospark discharge under nanosecond pulsed voltages *IEEE Trans. Plasma Sci.* **42** 2037–41

[8] Hu J and Rovey J L 2012 Experimental investigation of formation time in single-gap pseudospark discharge *J. Phys. D: Appl. Phys*. **45** 465203

[9] Kondrat'eva N P, Koval N N, Korolev Y D and Schanin P M 1999 A spectroscopic investigation of the near-cathode regions in a low-pressure arc *J. Phys. D: Appl. Phys*. **32** 699–705

[10] Rosier O, Apetz R, Bergmann K, Jonkers J, Wester R, Neff W and Pankert J 2004 Frequency scaling in a hollow-cathode-triggered pinch plasma as radiation source in the extreme ultraviolet *IEEE Trans. Plasma Sci.* **32** 240–6

[11] Bochkov V D, Kolesnikov A V, Korolev Y D, Rabotkin V G, Frants O B and Shemyakin I A 1995 Investigation of the effect of blocking potential on the static breakdown voltage and discharge initiation in the pseudospark switches *IEEE Trans. Plasma Sci.* **23** 341–6

[12] Korolev Y D, Frants O B, Geyman V G, et al 2012 Transient processes during formation of a steady-state glow discharge in air *IEEE Trans. Plasma Sci.*, **40** 2951–60

[13] Korolev Y D and Matveev I B 2006 Non-steady state processes in a plasma pilot for ignition and flame control *IEEE Trans. Plasma Sci.* **34** 2507–13

[14] Korolev Y D, Frants O B, Landl N V, Kasyanov V S, Galanov S I, Sidorova O I, Kim Y, Rosocha L A and Matveev I B 2012 Propane oxidation in a plasma torch of a low-current nonsteady-state plasmatron *IEEE Trans. Plasma Sci.* **40** 535–42

[15] Korolev Y D, Frants O B, Landl N V and Suslov A I 2012 Low-current plasmatron as a source of nitrogen oxide molecules *IEEE Trans. Plasma Sci.* **40** 2837–42

[16] Kozhevnikov V Y, Kozyrev A V and Korolev Y D 2006 Drift model of the cathode region of a glow discharge *Plasma Phys. Reports* **32** 949–59

Low-current hollow-cathode discharge in a trigger unit of a cold cathode thyratron

N V Landl[1,2], **Y D Korolev**[1,2,3], **O B Frants**[1,2], **V G Geyman**[1], **A V Bolotov**[1]
[1] Institute of High Current Electronics SB RAS, 634055 Tomsk, Russian Federation
[2] Tomsk State University, 634050 Tomsk, Russian Federation
[3] Tomsk Polytechnic University, 634050 Tomsk, Russian Federation

E-mail: landl@lnp.hcei.tsc.ru

Abstract. The paper deals with the investigations of the effect of cathode emissivity on the regimes of low-current hollow-cathode glow discharge. It is shown that an increase in the emissivity due to the so-called high-emissivity tablet allows the essential decrease in the discharge initiation voltage and the discharge burning voltage. The model of current sustaining for the hollow-cathode discharge, which takes into account the external emission current has been developed. On basis of the model, the current-voltage characteristics of the discharge have been interpreted.

1. Introduction

Since the end of 1980's, a considerable interest has been generated in a new type of low-pressure switching device with a cold cathode, often named the pseudospark switch or the cold cathode thyratron [1–8]. A range of operating pressures of the switch corresponds to the conditions of the left branch of Paschen's curve when the electron free path for ionization is much in excess of the electrode separation. Under such conditions, for both self-breakdown of the main gap of the switch and for external discharge triggering a considerable prebreakdown electron current is required [2–4]. This current is provided due to a trigger unit that is placed in the cathode cavity of the main gap [3].

Various types of the trigger units are used in the switches [1, 3]. One type of the trigger devices is based on an auxiliary low-current hollow-cathode glow discharge. The conditions of the auxiliary discharge burning significantly determine the rating characteristics of the switch itself. Therefore, the investigations of the auxiliary discharge seem to be of a great importance.

In this paper, the data on the influence of the cathode emissivity on the regimes of the auxiliary glow discharge are presented. The model for current sustaining in hollow-cathode discharge has been developed. The current-voltage characteristics are interpreted with a usage of the model.

2. Experimental setup and summary of experimental data

The experiments were carried out with the sealed-off switches TPI1-10k/50 [3] whose schematic arrangement is shown in figure 1. The ceramic casing of the device has an external diameter of 95 mm. The operating pressure is maintained due to a hydrogen reservoir that is powered by a voltage V_H at a current of about 2 A. The auxiliary glow discharge in the trigger unit is initiated and sustained due to the power supply V_1. The cathode cavity has the following dimensions: cavity diameter D = 3 cm, cavity height h = 2.9 cm. The cavity C_1 communicates with the main cathode cavity C via the axial hole with a diameter D_1 = 1.2 cm. Due to this hole a parasitic current i_2 inevitably flows from the anode A_1 to the cathode cavity *C*. The presence of this parasitic current can lead to decreasing the breakdown voltage of the main gap. In the experiments we measured the separate components of the total discharge current *i*, as figure 1 shows ($i = i_1 + i_2 + i_G$).

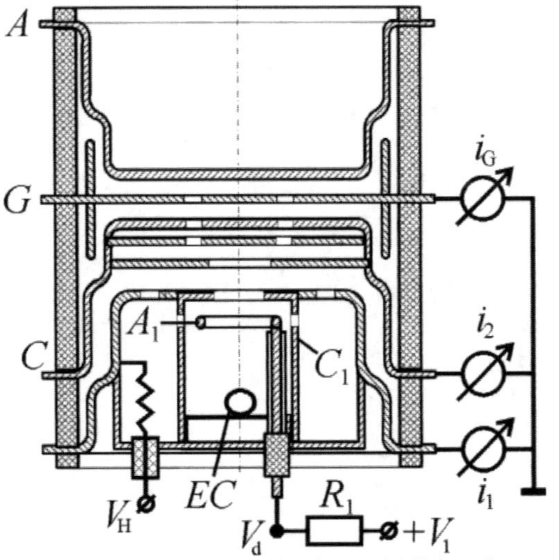

Figure 1. Schematic arrangement of the switch TPI1-10k/50 and the method of measurement of current-voltage characteristics. *A* – anode of the switch, *C* – hollow cathode of the switch, *G* – intermediate gradient electrode, A_1 – ring anode of the auxiliary glow discharge, C_1 – hollow cathode of the auxiliary glow discharge, *EC* – high emissivity cylinder, V_1 – voltage for powering the auxiliary glow discharge, V_H – voltage for powering of a hydrogen reservoir, R_1 = (20 – 65) kΩ – ballast resistor.

To increase the cathode emissivity the so-called high-emissivity tablet (emissivity cylinder) *EC* is used. The tablet is placed in the hollow cathode C_1. It represents a hollow cylinder that is fabricated from the powder materials by means of hot-pressing and sinter technology [3]. The basic component of the cylinder is the powder tungsten (90 %) to which aluminium oxide and cesium carbonate are added. In experiments we also used the switches without high-emissivity without tablet.

As far as the experiments were carried out with the sealed-off switches, a gas pressure in the switch was determined by the hydrogen reservoir that was powered by a voltage V_H. Then from one switch to the other, the same gas pressure can be achieved at different values of V_H. Since the geometry of the main gap was identical, we estimated a gas pressure inside the switches by measuring the static breakdown voltage in the main gap.

Figure 2 shows a typical current-voltage characteristic of the auxiliary glow discharge and a fraction of parasitic current in total discharge current i_2/i for the switch without the high-emissivity cylinder. The characteristic is rather intricate. Discharge is able to burn in two regimes. Regime *I* corresponds to the so-called hampered glow discharge with hollow cathode. In this regime a length of the cathode voltage drop region is comparable with the radius of hollow cathode *D*/2 and a negative glow region is not distinctly observed. In this regime, increasing of the voltage V_1 shifts the operating point over the curve to the right. When we decrease a voltage of power supply V_1, then in a vicinity of point *1*, the discharge is extinguished.

A critical value of the discharge current corresponds to point *3* in the current-voltage characteristic. A further increase in voltage V_1 leads to a sharp transition to the regime *II* (to point *4*). In this mode of operation, the negative glow region appears. The discharge conditions are transformed from the hampered regime to an ordinary glow discharge with hollow cathode. In the regime *II* a hysteresis of current-voltage characteristic takes place. When we decrease a voltage V_1, the operating point is shifted to the left up to point *5*. After that a reverse abrupt transition to the hampered regime *I* occurs.

The same characteristics with hysteresis are typical for a wide range of gas pressure (voltages V_H). The main features of the discharge without high-emissivity tablet are rather high values of the discharge initiation voltage and of the discharge burning voltage V_d. For example, for V_H = 6.2 V and V_H = 5.6 V the discharge initiation voltages are 1230 V and 3500 V correspondently.

A current-voltage characteristic for the case of the enhance cathode emissivity is shown in figure 3. The principal difference from the conditions without tablet is the absence of the abrupt transition from regime *I* to regime *II*. The discharge burning voltage is essentially decreased (up to a value $V_d \approx 130$

V). As for the discharge initiation voltage (the static breakdown voltage in the trigger unit), this value corresponds to 380 V for $V_H = 5$ V and to 280 V for $V_H = 5.6$ V.

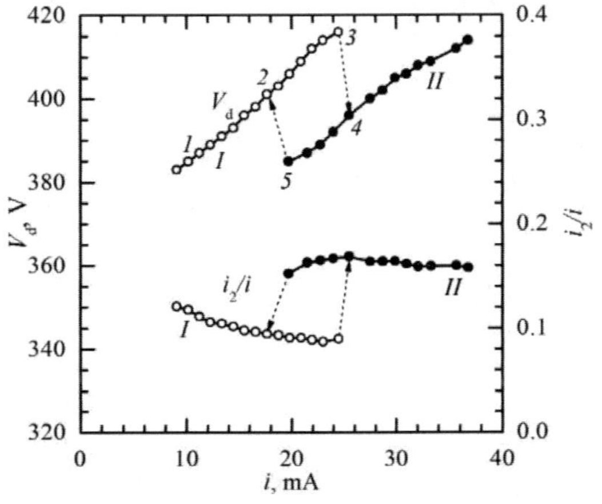

Figure 2. Current-voltage characteristic for the auxiliary glow discharge and a fraction of parasitic current in total discharge current i_2/i for the trigger unit without the emissivity cylinder.
$V_H = 6.2$ V, $R_1 = 25.6$ kΩ.

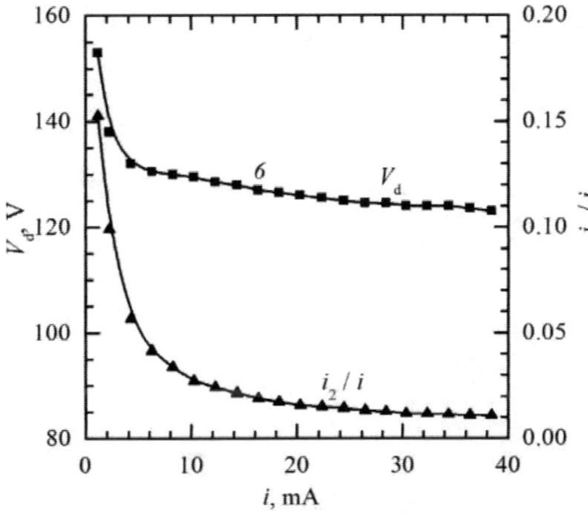

Figure 3. Current-voltage characteristics for the auxiliary glow discharge and a fraction of parasitic current in total discharge current i_2/i for the trigger with the emissivity cylinder.
$V_H = 5.6$ V, $R_1 = 65$ kΩ.

As noted earlier, when the glow discharge burns between the electrodes A_1 and C_1, some fraction of the current inevitably flows from the anode to the main cathode cavity C. In general, this current has to be minimized. For the cases of the cathode with the high emissivity a current to the electrode C does not exceed 4 % of the current i in wide range of the discharge current in the trigger unit, whereas for the cathode with an ordinary emissivity the fraction of current can achieve to 20 %.

3. Description of the model and interpretation of the experimental data

It is convenient to interpret the regimes of the auxiliary glow discharge and to carry out the corresponding estimations with a usage of the model that had been developed earlier for the high-current hollow-cathode pulsed glow discharge [5]. Illustration of the essence of this model is shown in figure 4. The cathode cavity C is filled with the negative glow plasma NG. The electrons emitted from the cathode are accelerated in the cathode layer l_c by the voltage drop V_c and the energy of the fast oscillating electron is spent for sustaining the negative glow plasma.

A steady state plasma density in the hollow cathode is established due to a balance between the process of ionization and the processes of the losses of charged particles. We do not take into account the recombination process. The ions disappear from the negative glow plasma due to their outflow to the cathode. The ions move in a collisionless regime under a potential difference $kT_e/2e$, which is applied to the so-called pre-sheath layer. Since the recombination losses are negligibly small and $l_c \ll D/2$, a length of the pre-sheath layer corresponds to $D/2$.

Total discharge current at the cathode surface i is a sum of the emission current i_{em} and the current of ions to the cathode i_i. The discharge current to anode is provided by the fast electrons from plasma that are able to overcome the potential barrier ΔV near the anode.

In the generally accepted models, the ion emission current at the cathode surface is provided by the classical γ processes ($i_{em} = \gamma i_i$) [9]. In our model we introduce the additional emission current at the cathode which appears due to a source of external emission i_{ext} [5]. It should be stressed that an external emission current is inevitably available in any type of glow discharge. For example, the negative glow plasma has to be considered as the external source of radiation with respect to the cathode surface [10]. Then the component of external emission current is provided due to photoeffect at the cathode.

Beside that, there is a great variety of the discharges in which the other physical reasons for appearing the external current exist. The widespread reason for arising the current i_{ext} is formation of the spark cathode spots as a result of explosive processes at the cathode [2, 5, 11–14]. In the sources of electron and ion beams with plasma cathode [15–17], the additional emission current is often provided due to external injection of electrons in the plasma of hollow cathode or hollow anode. As for the particular case of this experiment, the high-emissivity tablet is responsible for the external emission current.

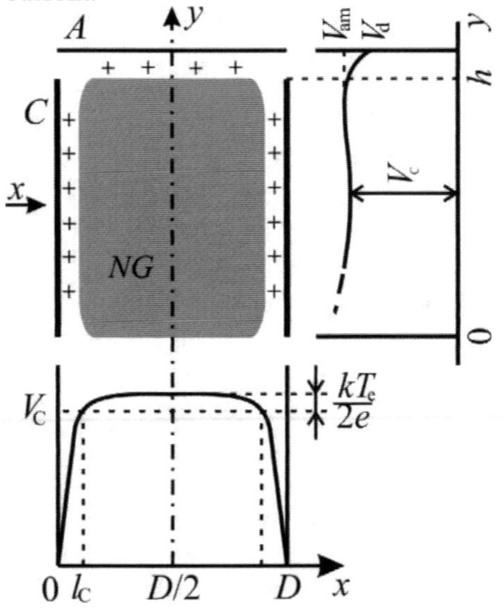

Figure 4. Schematic illustration of the discharge regions and the potential distributions as applied to the model of current sustaining in hollow cathode glow discharge.
NG – plasma of the hollow cathode (negative glow plasma; l_c – cathode voltage drop region; $\Delta V = (V_{am} - V_d)$ – negative potential barrier near the anode; $i = (i_{em}+i_i)$ – total discharge current at the cathode surface that is equal to current at the anode surface and in the external circuit.

Then in terms of the model, the emission current from the cathode can be written as

$$i_{em} = i_{ext} + \gamma i_i = i_i \left(\gamma + \frac{i_{ext}}{i_i} \right). \tag{1}$$

It is convenient to use an external parameter δ as a fraction of the external emission current in the emission current appearing due to ion bombardment of the cathode

$$\delta = \frac{i_{ext}}{\gamma i_i}. \tag{2}$$

Equations (1) and (2) allow us to introduce a generalized coefficient of the secondary processes Γ

$$\Gamma = \gamma + \frac{i_{ext}}{i_i} = \gamma \left(1 + \frac{i_{ext}}{\gamma i_i} \right) = \gamma(1+\delta). \tag{3}$$

After simple manipulations we can obtain the expression for total discharge current

$$i = i_c = i_a = i_{em} + i_i = (1+\Gamma) i_i = \frac{1+\Gamma}{\Gamma} i_{em}. \tag{4}$$

Proceeding from the power balance in the hollow cathode the condition for discharge sustaining can be obtained [5]

$$\Gamma \frac{V_c}{V_2} K_S = 1, \tag{5}$$

where $K_S = (1 - S_D/S_C)$ is a geometrical factor (S_D – anode area, S_C – area of the surface of cathode cavity) and

$$V_2 = V^* + \frac{2kT_e}{e} + \Delta V + \Gamma\left(\frac{3kT_e}{2e} + \Delta V\right). \tag{6}$$

The physical meaning of V_2 is the total cost of ionization under the effect of electron beam and V^* is the energy which the accelerated electron spends for one act of ionization with taking into account only the energy losses for gas excitation and for elastic collisions.

As distinct to [9], in the framework of this model, we can obtain a correlation between the total discharge current and the external emission current as

$$i_c = i_{ext}\left(1 + \frac{1+\gamma}{\gamma\delta}\right) = i_{ext}\left(1 + \frac{1+\gamma}{\Gamma - \gamma}\right) = i_{ext} K. \tag{7}$$

The results of estimations are presented in the Table below. In the calculations we have used the following parameters of the discharge: $eV^* = 55$ eV, $kT_e = 5$ eV, $K_S \approx 1$, $\Delta V \approx 0$, $V_c = V_d$ [5]

Table 1. The results of estimates as applied to data presented in figures 2 and 3. The first line shows the operating points of the current-voltage characteristics.

	2	3	*4*	6
V_d, V	398	416	396	127
i, mA	16.6	24.5	25.5	16.3
Γ	0.166	0.159	0.167	0.544
δ	0.387	0.326	0.394	3.533
$K = i/i_{ext}$	25.11	29.63	24.67	3.64
l_c, cm	0.863	0.732	0.694	0.425

The point *1* is not suitable for estimation since the cathode layer length l_c is comparable with $D/2$. Minimal value of $\Gamma = 0.159$ corresponds to point *3* of the hampered glow discharge. From this point a sharp transition to the regime *II* occurs. This transition is accompanied by a slight increase in generalized coefficient of secondary electron emission Γ. We believe that this increase is associated with the fact that the contribution of external current to emission current slightly increases.

The largest contribution of external current to total discharge current is observed in the case when high emissivity tablet is available in the cathode cavity. These conditions correspond to point *6* of current-voltage characteristic. Discharge burning voltage in this case is much less than for the regime *II* in figure 2, and coefficient Γ is three times higher.

The ratio of the external current to the current due to the ion bombardment is characterized by the parameter $\delta = i_{ext}/\gamma i_i$. To estimate this parameter we have to know the classical coefficient of secondary emission γ. In the calculations we use $\gamma = 0.12$ [5]. Then for the case of the cathode with the high-emissivity tablet we obtain $\delta = 3.53$. It means that the current of external emission is more than three times higher as compared to the current due to ion bombardment of the cathode surface.

For the ordinary glow discharge with a hollow cathode a fraction of external current in total emission current i_{em} is small. Nevertheless, for the cases when emission current is set artificially, the contribution of the external current to the total emission current increases. For example, the balance of the currents at the cathode for the point *6* at figure 3 looks like as follows: $i_i = 10.56$ mA, $i_{ext} = 4.478$ mA, $\gamma i_i = 1.32$ mA. Although the ion current is still higher than the emission current, this excess is not so considerable as for the classical glow discharge. This conclusion is illustrated by the estimation of the coefficient K. For the hampered glow discharge, the typical values of the current enhancement coefficient is higher than 20. For the case of the cathode with the high emissivity, we obtain $K \approx 3.5$.

Acknowledgments
The work was supported by Russian Science Foundation under the Grant # 14-19-00139.

References

[1] Frank K and Christiansen J 1989 The fundamentals of the pseudospark and its applications *IEEE Trans. Plasma Sci.* **17** 748-53
[2] Korolev Y D and Frank K 1999 Discharge formation processes and glow-to-arc transition in pseudospark switch *IEEE Trans. Plasma Sci.* **27** 1525–37
[3] Bochkov V D, Dyagilev V M, Ushich V G, Frants O B, Korolev Y D, Shemyakin I A and Frank K 2001 Sealed-off pseudospark switches for pulsed power applications (current status and prospects) *IEEE Trans. Plasma Sci.* **29** 802–8
[4] Kozyrev A V, Korolev Y D, Rabotkin V G and Shemyakin I A 1993 Processes in the prebreakdown stage of a low-pressure discharge and the mechanism of discharge initiation in pseudospark switches *J. Appl. Phys.* **74** 5366–71.
[5] Korolev Y D, Frants O B, Landl N V, Shemyakin I A and Geyman V G 2013 High-current stages in a low-pressure glow discharge with hollow cathode *IEEE Trans. Plasma Sci.* **41** 2087–96
[6] Bochkov V D, Kolesnikov A V, Korolev Y D, Rabotkin V G, Frants O B and Shemyakin I A 1995 Investigation of the effect of blocking potential on the static breakdown voltage and discharge initiation in the pseudospark switches *IEEE Trans. Plasma Sci.* **23** 341–6
[7] Meena B L, Rai S K, Tyagi M S, Pal U N, Kumar M and Sharma A K 2010 Characterization of high power pseudospark plasma switch (PSS) *J. Phys. Conference Series* **208** 012110
[8] Zhang J, Zhao J P and Zhang Q G 2014 The breakdown characteristics of single-gap pseudospark discharge under nanosecond pulsed voltages *IEEE Trans. Plasma Sci.* **42** 2037–41
[9] Ul'yanov K N 1999 Superdense glow discharge: Theoretical model of the cathode region *High Temperature* **37** 337–47
[10] Kozhevnikov V Y, Kozyrev A V and Korolev Y D 2006 Drift model of the cathode region of a glow discharge *Plasma Phys. Reports* **32** 949–59
[11] Kondrat'eva N P, Koval N N, Korolev Y D et al 1999 A spectroscopic investigation of the near-cathode regions in a low-pressure arc *J. Phys. D: Appl. Phys*. **32** 699–705
[12] Korolev Y D, Frants O B, Geyman V G, et al 2012 Transient processes during formation of a steady-state glow discharge in air *IEEE Trans. Plasma Sci.*, **40** 2951–60
[13] Korolev Y D, Frants O B, Landl N V, Bolotov A V and Nekhoroshev V O 2014 Features of a near-cathode region in a gliding arc discharge in air flow *Plasma Sources Sci. Technol.* **23** 054016
[14] Korolev Y D, Frants O B, Landl N V, Kasyanov V S, Galanov S I, Sidorova O I, Kim Y, Rosocha L A and Matveev I B 2012 Propane oxidation in a plasma torch of a low-current nonsteady-state plasmatron *IEEE Trans. Plasma Sci.* **40** 535–42
[15] Lopatin I V, Schanin P M, Akhmadeev Y H, Kovalsky S S and Koval N N 2012 Self-sustained low pressure glow discharge with a hollow cathode at currents of tens of amperes *Plasma Physics Reports* **38** 583–89
[16] Ryabchikov A I, Arsubov N M, Vasilyiev N A and Dektyarev S V 1991 The Raduga multipurpose ion source for surface modification of constructed materials *Nuclear Instruments Methods in Phys. Research Sect. B* **59** 124–7
[17] Ryabchikov A I, Ryabchikov I A, Stepanov I B and Usov U P 2007 High-frequency short-pulsed metal plasma-immersion ion implantation or deposition using filtered DC vacuum-arc plasma *Surface Coatings Technology* **201** 6523–25

Journal of Physics: Conference Series **652** (2015) 012051 doi:10.1088/1742-6596/652/1/012051

Accuracy and Roughness of TiN Coatings Deposited by Vacuum Arc Plasma

V Budilov[1], K Ramazanov[1], I Yagafarov[1] and R Vafin[2]
[1] Ufa State Aviation Technical University, Ufa - 45000, Russian Federation
[2] Nosov Magnitogorsk State Technical University, Magnitogorsk - 455000, Russian Federation

E-mail: ramazanovkn@gmail.com

Abstract. Nitride coatings were deposited by vacuum arc plasma in an atmosphere of argon and nitrogen using different deposition conditions of part location and configuration, angle between plasma flow and processing surface. The coating thickness, part dimensions and surface roughness were measured before and after coating deposition. The type of part poor shape precision after coating deposition by vacuum arc plasma was defined. An impact of coating deposition by vacuum arc discharge on the part dimensional and form accuracy was revealed. The effect of parts location on dimensional and radial surfaces form accuracy was induced. The effect of coating surface polythickness on part dimensional and form accuracy for parts with different stiffness was defined. The impact of part location area and parts quantity on coating thickness, surface roughness, dimensional and form accuracy of part was revealed. The coating thickness distribution law, parts quality class and accuracy degree after vacuum ion plasma coating deposition were assigned.

1. Introduction

Required coating thickness, surface roughness, form and dimensions accuracy after coating deposition by vacuum arc plasma assurance leads to quality conformance of high precision parts [1-2].
Precision of vacuum arc plasma coating is determined by coating thickness scatter. Influence of deposition conditions on coating thickness was studied earlier [3-7]. Known data is impossible to use during accuracy assessment, as real parts have dimensions and complicated configuration.

2. Experimental technique and equipment

TiN coatings were deposited by vacuum arc plasma on plants with different cathode diameters (80mm and 180 mm).

An effect of cathode material and coating thickness on surface roughness was studied on TiN, ZrN, CrN, CrCN, MoN, TiC, CrC, TiCN coatings. Pieces of heat-resistant steel 13Cr11Ni2W2MoV (∅15 mm, h=10 mm) were finally processed by grinding (Ra = 1.0 - 1.5 μm) and smoothing (Ra = 0.02 μm).

Influence of pre-process machining (turning, air dynamic hardening, grinding, superfinishing) and coating thickness on surface roughness was studied on structural alloyed steel 30CrMnSiNi pieces. Effect of plasma source – coated surface spatial relationship on coating thickness and surface

Journal of Physics: Conference Series **652** (2015) 012051 doi:10.1088/1742-6596/652/1/012051

roughness was studied on flat pieces of size 14 × 20 × 2 mm with different initial surface roughness (Ra 0.63 μm and Ra 0.04 μm).

Effect of simultaneously processed parts quantity on coating thickness scattering was studied on bushing batches. The bushings were located parallel to the plasma flow (Fig. 1). The location areas were 60 and 200 mm in size (radius). The batch quantities were 50 and 100 parts, respectively.

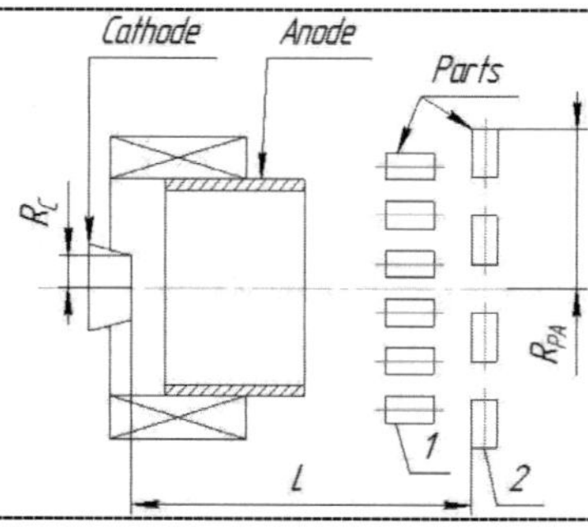

Figure 1. Experimental diagram (1 – parallel part orientation; 2 – perpendicular part orientation).

3. Results

Surface roughness depends on coating thickness and type of preprocess machining, and may decrease (if initial Ra > 1.6 μm) or increase (if initial Ra < 1.6 μm). The surface roughness increases by the coating thickness increasing. Cathode material affect the surface roughness as follows: Ra (Ti) > Ra (Cr) > Ra (Zr) > Ra (Mo);

Table 1. Dependence of TiN and ZrN coating roughness from coating thickness. Coatings were deposited on motionless substrates.

Coating thickness h, μm	8	16	24	32
TiN coating surface roughness Ra, μm (Ra_{init} = 0.02μm)	0.2	0.2	0.23	0.25
TiN coating surface roughness Ra, μm (Ra_{init} = 1.0-1.5μm)	-	2.0	2.0	2.3
ZrN coating surface roughness Ra, μm (Ra_{init} = 0.02μm)	0.11	0.11	0.13	0.14

Surface roughness after deposition by vacuum arc plasma (initial roughness Ra = 0.01–0.63 μm, coating thickness h = 5–15 μm) can be defined from the following expression:

$Ra = C + Ra_{initial}$,

C=0.2 μm for Ti; C=0.15 μm for Cr; C=0.1 μm for Zr; C=0.07 μm for Mo.

Figs 2-5 display surface roughness of TiN coatings. The arithmetic average roughness values of pieces, located on the axis of the particle flux at a distance of 200 mm from cathode are almost the same for samples with different initial roughness and spatial orientation.

Table 2. Dependence of TiN and ZrN coating Roughness thickness on coating thickness. Coatings were deposited on rotated substrates.

Coating	TiN			Zr		
Coating thickness h, μm	5	10	15	5	10	15
Coating surface roughness Ra, μm (Ra_{init} = 1.0-1.5μm)	1.4	1.4	1.7	1.1	1.5	1.6
Coating surface roughness Ra, μm (Ra_{init} = 0.02μm)	0.19	0.19	0.2	0.11	0.11	0.12

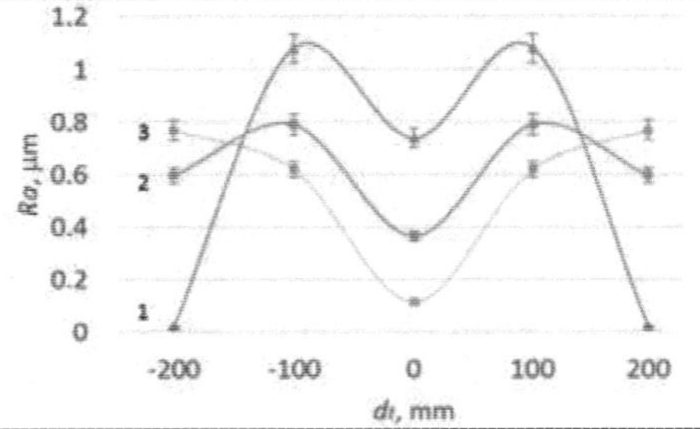

Figure 2. Dependence of surface average roughness *Ra* on the transverse distance d_t from the symmetry axis of plasma flow for different distances *L* to cathode. Initial average roughness Ra=0.04 μm. Cathode diameter 80 mm. 1 – *L* = 200 mm; 2 – *L* = 300 mm; 3 – *L* = 400 mm. Pieces are oriented perpendicular to plasma flow.

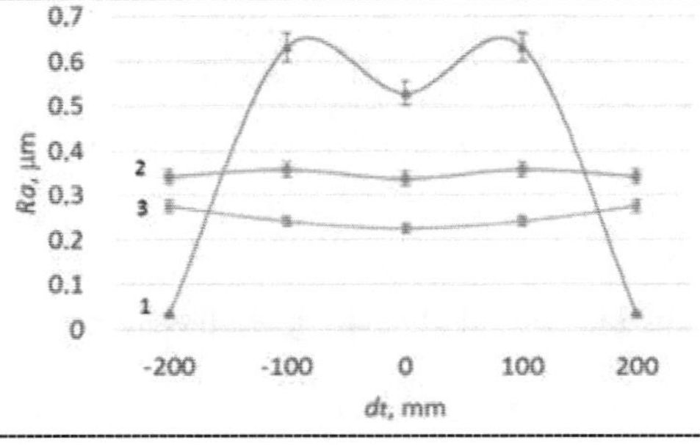

Figure 3. Dependence of surface average roughness Ra on the transverse distance dt from the symmetry axis of plasma flow for different distances *L* to cathode. Initial average roughness Ra=0.04 μm. Cathode diameter 80 mm. 1 – *L* = 200 mm; 2 – *L* = 300 mm; 3 – *L* = 400 mm. Pieces are oriented parallel to plasma flow.

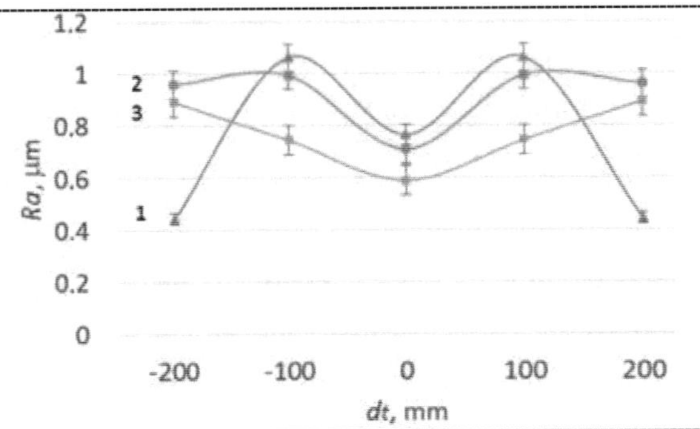

Figure 4. Dependence of surface average roughness *Ra* on the transverse distance d_t from the symmetry axis of plasma flow for different distances *L* to cathode. Initial average roughness Ra=0.63 µm. Cathode diameter 80 mm. 1 – *L* = 200 mm; 2 – *L* = 300 mm; 3 – *L* = 400 mm. Pieces are oriented perpendicular to plasma flow.

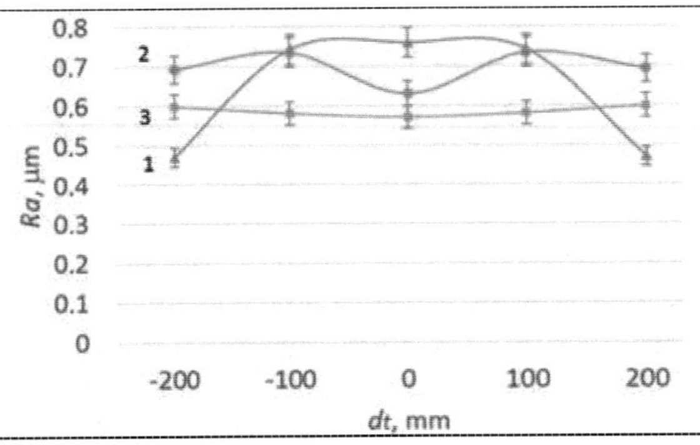

Figure 5. Dependence of surface average roughness *Ra* on the transverse distance d_t from the symmetry axis of plasma flow for different distances *L* to cathode. Initial average roughness Ra=0.63 µm. Cathode diameter 80 mm. 1 – *L* = 200 mm; 2 – *L* = 300 mm; 3 – *L* = 400 mm. Pieces are oriented parallel to plasma flow.

Perpendicular location of coated surface leads to maximum surface average roughness Ra=1.2 µm for both initial surface roughness. Parallel location of coated surface leads to maximum surface average roughness Ra=0.8 µm and depends on initial surface roughness. Maximum roughness corresponds to the deflection angle from the axis of the particle flux 25-30°. The average roughness decreases nonlinearly with a distance from the cathode increasing. At a distance from cathode of 400 mm average roughness Ra = 0.63 µm is achievable. Higher roughness decreases and lower increases during coating deposition by vacuum arc plasma.

Cylindrical surfaces ovality after 4 µm TiN coating deposition by vacuum arc plasma (Fig.6) hasn't changed and depends on pre-process machining. TiN vacuum arc coating of 4 µm thickness deposition on cylindrical surfaces leads to diameter scattering increase from 4 µm up to 7 µm.

Fig. 7 displays the TiN vacuum arc coating thickness distribution on bushing batches. The distribution curve shows correlation with normal distribution law

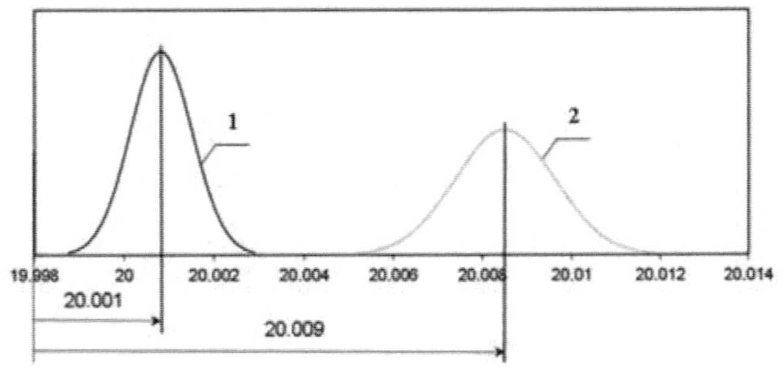

Figure 6. Cylindrical surface diameter distribution curves. 1–after machining; 2–after TiN deposition by vacuum arc plasma. Parts were oriented perpendicular to plasma flow. Cathode diameter 180 mm.

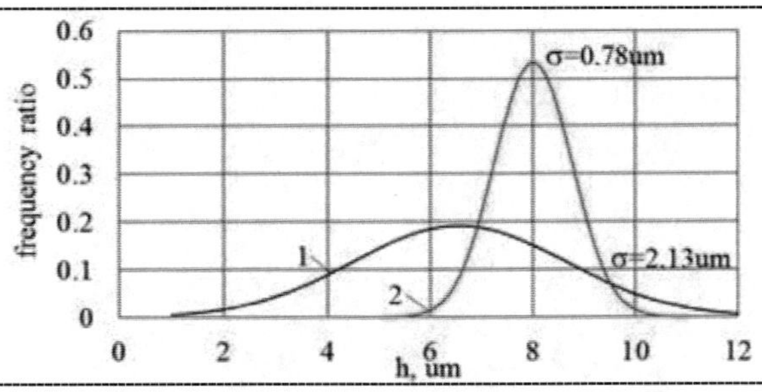

Figure 7. TiN coating thickness on bushings distribution curve. The location area of 200 mm in size (radius). Cathode diameter 180 mm.

The thickness scattering for vacuum arc coating deposition depends on the size (radius) of workpieces location area. For example, reducing the size of workpieces location area from 200 mm to 60 mm leads to coating thickness standard deviation decrease from $\sigma = 2.13$ μm to $\sigma = 0.78$ μm. The coating thickness scatterings were 4 μm and 11 μm for 60 and 200 mm workpieces location area, respectively. The coating thicknesses were 6^{+4} μm and 3^{+11} μm for 60 and 200 mm workpieces location area, respectively.

Effect of coating deposition on form error was studied on thin-walled circular-shaped parts (Fig. 8). Following deposition modes were selected:

a) aflat plasma source spatial orientation;

b) upright plasma source spatial orientation.

Parts were oriented parallel to plasma flow.

As is known increasing of layers quantity leads to parts operating ability increase through decreasing of coating locked-up stresses. So coatings were deposited under two techniques: single-layer TiN coating; multilayer Ti-TiN coating.

Calculated coating thickness, ovality and conicity of thin-walled circular-shaped parts are given in table 3

Measurement results of thin-walled circular-shaped parts after coating deposition are given in table 4.

The experimental data show that coating deposition with aflat plasma source spatial orientation and parts planetary parts rotation leads to ovality tolerance exceeding. Coating deposition with upright plasma source spatial orientation and parts planetary parts rotation fulfil requirements of form accuracy.

Table 3. Calculated ovality and conicity

Process parameters		Coating thickness, µm	Form deviation, µm	
Plasma source spatial orientation	Rotation type		Ovality	Conicity
aflat	about table axis	$1.7^{+5.2}$	8.1	1.7
	planetary	$1.5^{+4.6}$	6.5	1.3
upright	about table axis	$4.6^{+1.9}$	1.8	0.5
	planetary	5.3^{+0}	~0	0.3

Figure 8. Thin-walled circular-shaped parts in vacuum chamber.

Table 4. Results of thin-walled circular-shaped parts form measurement

Process parameters		Coating type	Coating thickness, µm		Ovality, µm	Conicity, µm
Plasma source spatial orientation	Rotation type		min	max		
Aflat	planetary	single-layer	2	7	15.5	10
		multilayer			11	8
Upright	planetary	single-layer	5	6	7	7
		multilayer			7	6

Multilayer coating in comparison with the single-layer leads to essential decrease of form error. That can be explained by decreasing of coating locked-up stresses through structural changes in coating [4]. Experimental data agree with calculated data (taking into account parts initial form error).

According to experimental data the fact of ovality initiation after coating deposition by vacuum arc plasma was established. An ovality originates from polythickness of the coating through the part surface.

Form error decreasing can be achieved by multilayer coating deposition [1] and polithickness decreasing [1,4].

4. Conclusion

- Surface roughness after coating deposition by vacuum arc plasma depends on the manufacturing inheritance (due to the surface roughness after machining), the material evaporation by vacuum arc (due to the microdroplets presence) and coated surface location;
- Vacuum arc plasma coating thickness distribution, as was established, corresponds with normal distribution law. The vacuum arc plasma coating inaccuracy strongly depends on parts location area. The accuracy of coating deposition by vacuum arc plasma depends on part location area radius (Rpa) to cathode radius (Rc) ratio.
- At the ratio $R_{pa}/R_c \leq 1$ plasma flow homogeneity from vacuum arc plasma source is enough to secure required part accuracy and surface quality of high precision parts. The parts location area can be enlarged by cathode radial dimensions increasing or by such parts with plasma source relative movement, which allow to secure deposition conditions equality for different processing surface areas.
- Thin-walled circular-shaped parts form error after deposition by vacuum arc plasma depends on coating polythickness. Deposition of 5 μm coating on 5 accuracy rating parts leads to accuracy deterioration to 7 accuracy rating. Uniform coating deposition does not change accuracy rating at the same time. The ovality after coating deposition is produced from heating and coating locked-up stresses.

Acknowledgments

The study was supported by the grant from the Russian Science Foundation (#15-19-10030)

References

[1] Budilov V V, Kireev R M and Yagafarov I I 2012 *Russian Aeronautics* **55** Issue 2 203–7
[2] Yagafarov I I, Kireev R M and Mukhin V S 2012 *25TH International Symposium on Discharges and Electrical Insulation in Vacuum (ISDEIV 2012)* 552–3
[3] Ali M, Hamzah E, Qazi I A and Toff M R M 2010 *Current Applied Physics* **10** Issue 2 471–4
[4] Budilov V V, Koval N N, Kireev R M and Ramazanov K N 2013 Russian Federation Moscow Mashinostroenie 320
[5] Chokwatvikul Ch, Larpkiattaworn S, Surinphong S, Busabok Ch and Termsuksawad P 2011 *J. of Metals, Materials and Minerals* **21** No.1 115–9
[6] Harris S G, Doyle E D, Wong Y C, Munroe P R, Cairney J M and Long J M 01/2004 **183**(2) 283–94
[7] Straumal B, Gust W, Vershinin N, Dimitriou R and Rabkin E 2000 *Surface & Coatings Technology* **125** 157–60

Application of hollow cathode effect for local ion nitriding of machine parts

V V Budilov[1], K N Ramazanov[1], Yu G Khusainov[1], I V Zolotov[1] and N S Babenko[2]

[1] Department of Machine Building, Ufa State Aviation Technical University, 12 K.Marx Street, Ufa, Russian Federation
[2] Ufa Engine Industrial Association, Ferin str., 2, 450039, Ufa, Russian Federation

E-mail: ramazanovkn@gmail.com

Abstract. Influence of hollow cathode effect (HCE) on local ion nitriding process of 16Kh3NVFMB-Sh steel was studied. Microstructures of nitrided layer, phase composition, microhardness profiles through the diffusion layer were obtained and investigated. Influence of ion nitriding duration on nitrided case depth and wear resistance was studied. Technology of ion nitriding with application to HCE for gear part was developed.

1. Introduction

It's known [1] that the most frequent cause of machine parts working in hard conditions (friction and cyclic loads) destruction is pitting and microscopic cracks formation. Methods of surface hardening, in particular ion nitriding, allow improving the reliability and durability of these parts [2].

In most cases, only local areas of machine parts surface, such as contact area of the gear teeth, are under intense wear conditions [3]. Thus, usually there is no need to treat entire surface of the part; it is enough to nitride only operating area.

In works of Tomsk scientists of ISPMS SB RAS it was experimentally proved that application of hollow cathode effect (HCE) in ion nitriding process allows significantly accelerating diffusion process compared to conventional treatment in glow discharge.

The aim of this work is to determine influence of HCE during local ion nitriding on microstructure, microhardness of diffusion layer, phase composition of 16Kh3NVFMB-Sh steel samples. In addition, technology of ion nitriding with application of HCE of gear part was developed.

2. Methods of study

Experiments were carried out on the multifunctional setup for thermal and thermochemical treatment in vacuum ELU-5M. Samples of thermal pretreated (quenching at 930°C and tempering at 600°C) steel 16Kh3NVFMB-Sh were sputtered in argon at pressure of 10 Pa during 15 min, and then ion nitrided in atmosphere of nitrogen, argon and acetylene (Ar 70%, N_2 25%, C_2H_2 5%) at pressure of 60 Pa during 12 hours. Part of the sample was covered by mesh screen (figure 1). HCE occurred in the cavity formed between the screen and surface of the sample. Process temperature was not higher than 550°C

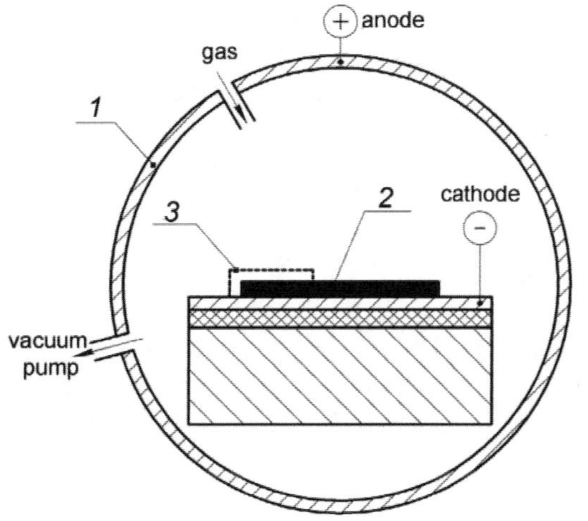

Figure 1. Scheme of experimental setup: 1 – vacuum chamber, 2 – sample, 3 – technological screen

Zeiss Axiotech 25HD microscope was used for optical microscopy investigations. Samples were etched in solution of 5% HNO3 and 95% C2H5OH. Microhardness measured on Struers Duramin michohardness tester. X-ray analysis was performed with help of DRON-4 difractometer.

3. Results and discussing

Presence of nitride elements (Cr, W, Ni, Mo, V) in 16KhNVFMB-Sh steel allows using it for improving wear resistance of machine parts, such as gears, shafts, etc.

As a result of quenching and tempering, formation of ferrite-cementite solution occurs due to diffusion redistribution of carbon atoms. Carbides of iron and other alloying elements appear as dark grain boundaries [4]. Surface microhardness before ion nitriding is 420±20 $HV_{0.1}$.

Photos of nitrided sample microstructure are presented in figure 2.

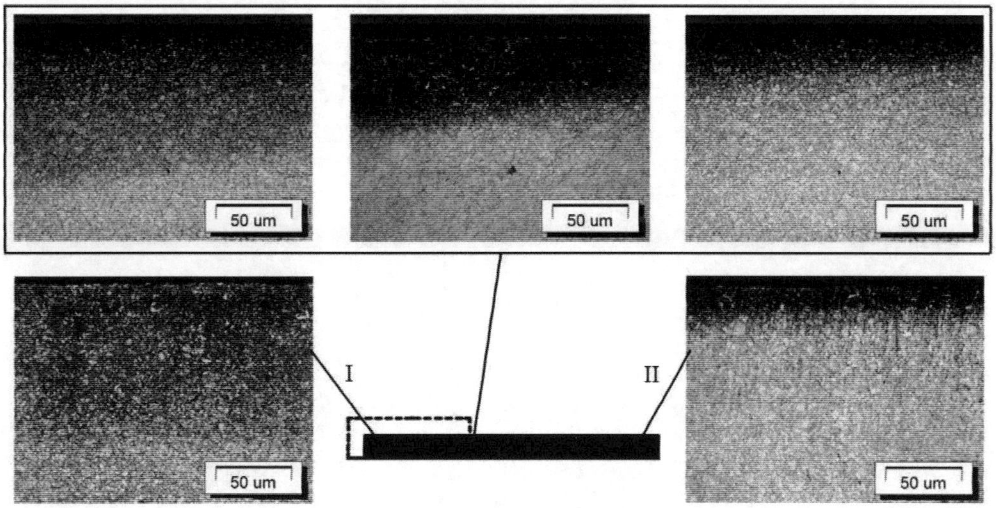

Figure 2. Microstructure of ion nitrided sample I – affected by HCE area, II – unaffected by HCE area

Investigation of microstructure of nitrided sample shows presence of two zones: zone affected by HCE (covered surface area) and unaffected zone. At covered by screen area thicker dark diffusion layer can be observed. Transition from diffusion layer to core material is smooth, that is one of major requirements to microstructure of nitrided layer [5].

Microhardness profiles allow estimating nitrided case depth.

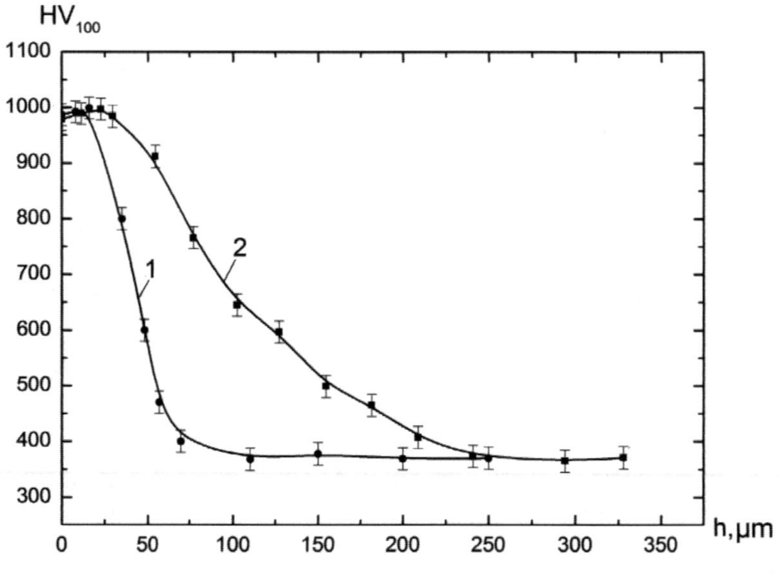

Figure 3. Microhardness profile after local ion nitriding with HCE (1 – covered by screen area, 2 – uncovered area).

Analysis of presented in figure 3 profiles shows that covered area has up to 2 times thicker diffusion layer compared to uncovered area. Transition to core material is also smooth.

X-ray analysis was performed to investigate phase composition of steel surface after nitriding. Obtained diffraction patterns presented in figure 4.

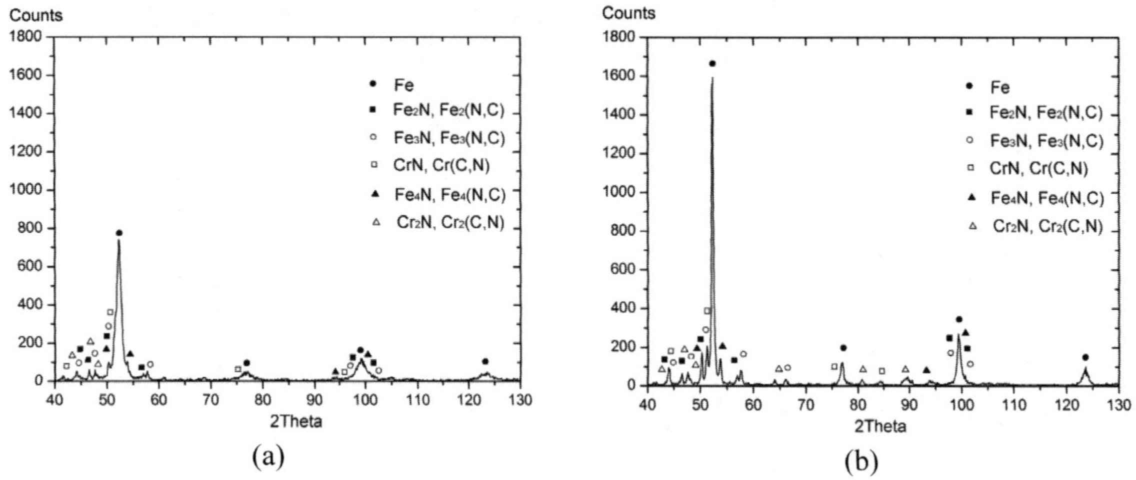

Figure 4. Difraction patterns of surface of the nitrided sample (a) – affected by HCE area, (b) – uncovered area.

The study of obtained diffraction patterns shows the presence of ε-phases Fe2-3(N), Fe2-3(N,C), γ-phases Fe_4N, $Fe_4(N,C)$, and also phases of chromium nitrides and carbonitrides CrN, Cr(C,N).

To study rate of hardened layer growing during ion nitriding with HCE, influence of process time on diffusion case depth was investigated (figure 5). It was established that saturation process during nitriding with HCE obeys the inverse parabolic law, like any diffusion process. Therefore, saturation process slows down over process time.

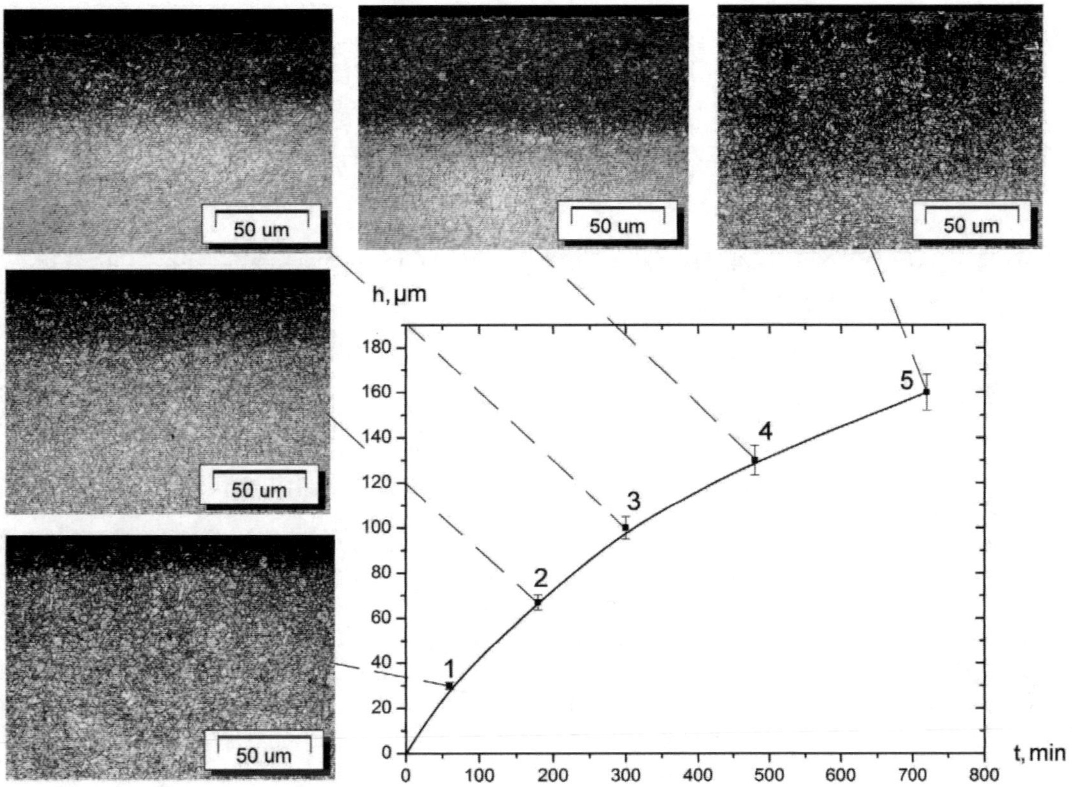

Figure 5. Kinetics of hardened layer growth thickness during ion nitriding with HCE.

Tribological properties were studied by standard method "ball on disk". The dependences of friction coefficient on wear time for different surface areas of the sample presented in figure 6.

Study of the obtained results shows that nitrided with HCE surface bedded after 7 minutes, whereas bedding stage of nitrided without HCE surface ends after 4 minutes. Average friction coefficient of both nitrided with and without HCE areas is 0.6.

As a result of tribological tests it was established that application of HCE leads to decreasing the wear of materials up to 3 times in comparison with the case of conventional ion nitriding in glow discharge. Thus, ion nitriding with HCE is effective way for increasing of wear resistance of 16Kh3NVFMB-Sh steel.

Based on obtained results, technology of ion nitriding of gear part was developed. This process allows obtaining a uniform diffusion layer of 180 μm thickness without nitride network on sharp corners.

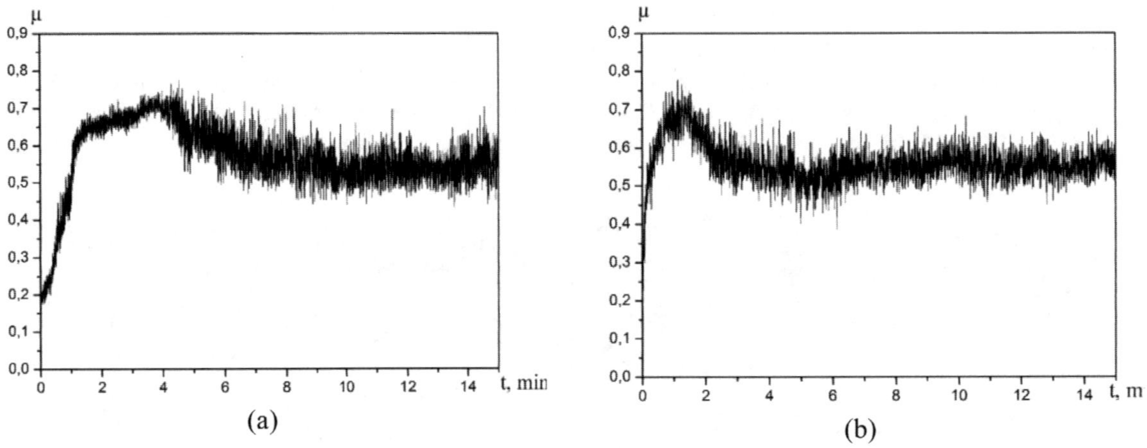

Figure 7. Dependences of friction coefficient on wear duration of the sample (a) – nitriding with HCE, (b) – nitriding without HCE.

4. Conclusions

Two zones form after ion nitriding with HCE: zone affected by HCE (covered surface area) and unaffected zone. Transition from diffusion layer to core material is smooth.

Covered by technological screen area has up to 2 times thicker diffusion layer compared to uncovered area after 12 h of ion nitriding with HCE.

ε -phases $Fe_{2-3}(N)$, $Fe_{2-3}(N,C)$, γ -phases Fe_4N, $Fe_4(N,C)$, and also phases of chromium nitrides and carbonitrides CrN, Cr(C,N) are form on surface of 16Kh3NVFMB-Sh steel after ion nitriding with HCE.

Application of HCE during ion nitriding allows significantly increasing wear resistance for 16Kh3NVFMB-Sh steel surface.

Technology of ion nitriding of gear part with HCE was developed. This process allows obtaining a uniform diffusion layer of 180 μm thickness without nitride network on sharp corners.

References

[1] Kozlovskiy I S 1970 *Thermo-chemicaltreatment of gears* (Moscow: Mashinostroenie)

[2] Arzamasov B N, Batrukhin A G, Eliseyev Yu V, Panayoti T A 1999 *Ion thermo-chemical treatment of alloys* (Moscow: MGTU im. Baumana)

[3] Gerasimov S A , Kuksenkova L I , Lapteva V G 2012 *Structure and wear resistance of design steels and alloys* (Moscow: MGTU im. Baumana)

[4] Budilov V V , Ramazanov K N , Khusainov Yu G, Zolotov I V 2014 Perspectives of using of hollow cathode effect for local ion nitriding of parts from steel 16Kh3VFMB-Sh *Vestnik UGATU* **62** 32-36

[5] Budilov V V, Ramazanov K N, Khusainov Yu G, Zolotov I V 2014 Using of hollow cathode effect for local ion nitriding of design steel 16Kh3VFMB-Sh *Uprochnyayuhcie tekhnologii i pokrytiya* **12** 27-30

Journal of Physics: Conference Series **652** (2015) 012053 doi:10.1088/1742-6596/652/1/012053

Deposition of Functional Coatings Based on Intermetallic Systems TiAl on the Steel Surface by Vacuum Arc Plasma

V Budilov[1], **E Vardanyan**[1], **K Ramazanov**[1]
[1] Department of Machine Building, Ufa State Aviation Technical University, 12 K.Marx Street, Ufa, Russian Federation

E-mail: vardanyaned@gmail.com

Abstract. Laws governing the formation of intermetallic phase by sequential deposition of nano-sized layers coatings from vacuum arc plasma were studied. Mathematical modeling process of deposition by vacuum arc plasma was performed. In order to identify the structural and phase composition of coatings and to explain their physical and chemical behaviour XRD studies were carried out. Production tests of the hardened punching tools were performed.

1. Introduction

Development of new coating materials with advanced physical and chemical properties as well as technologies of their deposition are considered the potential areas of industrial development. Such materials include composite intermetallide-based multicomponent materials with unique properties and ability to maintain their ordered structure up to melting temperatures. Ti-Al system is considered to be rather promising in the field of creating layered metallic and intermetallic composites, first and foremost due to low density and advanced physical and mechanical properties.

Multiple methods are used to obtain intermetallide-based coatings within TiAl system: self-propagating high-temperature synthesis (SHS) [1], ion implantation [2], deposition of composite cathode coatings, high-current pulse electron beam irradiation of Ti-based Al coatings [3], alternate deposition of Ti and Al layers (from 5 to 500 nm) with further thermal treatment (700-1200K). However, the existing methods are technologically complex and require multiple stages of their implementation.

Main intermetallitic compounds of Ti-Al system are formed in the range of temperatures varying from 850 to 1200 °C. However, thin-film coatings are characterized by temperature variations of phase transitions depending on their thickness. Under certain conditions coatings are subject to formation and stabilization of phases, which are not observed in bulk samples [4].

The description of a method to obtain intermetallide-based coatings within TiAl system during simultaneous deposition of vacuum arc discharge plasma is presented in work [5]. The work also includes research results of layer-by-layer deposition of multilayered coatings with the thickness of layers ranging from 1 to 0.1 μm, according to which intermetallides are not formed in coatings. Nanostructured multilayered coatings with the thickness of layers less than 0.1 μm were not studied. Thus, the purpose of this work is to study regularities during the formation of coating intermetallic phases during layer-by-layer deposition of nano-sized layers from vacuum arc discharge plasma.

2. Mathematical modeling

In the course of deposition of intermetallide-based coatings within TiAl system the treated samples pass 4 areas during rotations of a table:

1. Area where only aluminum ion flux gets to the surface;
2. Area where aluminum and titanium ion fluxes get to the surface;
3. Area where only titanium ion flux gets to the surface;
4. Area where the treated surface is located in the shadow zone.

In order to predict phase composition of coatings it is advisable to use the mathematical model of deposition process from vacuum arc discharge plasma. To identify stoichiometric composition of coatings there is a need to: calculate the density of Al and Ti ions fluxes; calculate thicknesses of nano-sized layers of Ti, Al and a transition layer per single rotation of a table. All calculations for stoichiometric composition and thickness of layers were carried out using Budilov and Yagafarov's model thus taking into account various patterns of electric-arc sources.

The following formula [5] was used to calculate the density of ion fluxes:

$$j_i = \frac{\mu_p \cdot I_a \cdot \bar{z} \cdot e}{2 \cdot \pi \cdot m_i \cdot R_\kappa^2} \times \left[1 + \frac{R_c^2 - l^2 - b^2}{\left(\left(R_c^2 - l^2 - b^2\right)^2 - 4 \cdot R_c^2 \cdot b^2 \right)^{1/2}} \right] \tag{1}$$

where i – Ti or Al, μp– coefficient of cathode erosion, Ia – arc current at the cathode, Rc – cathode radius, l – distance from the face of the cathode; b – relative displacement of cathode axis.

The thickness of coatings was defined according to the formula:

$$h = V_k \cdot t \tag{2}$$

The following algorithm was used for calculations of layers thickness during deposition from two electric-arc sources with single component Ti and Al cathodes and rotation of a table round its axis: measurement of the amount of targeted points depending on the table rotation rate and definition of its geometrics (l, b, deposition angle); calculation of density of Ti and Al ions fluxes and their growth rate in every point; definition of the total thickness per one turn of the table. Calculation results of Ti, Al and transition layers thickness at various rotation rates of a table are shown in Figure 1.

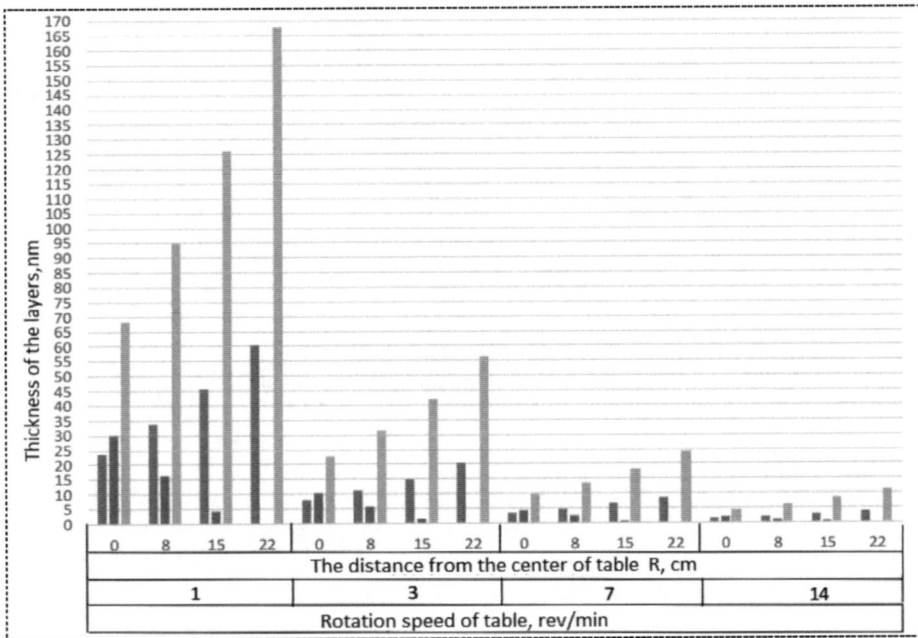

Figure 1. Calculation results of layers thickness per one rotation

Calculations show that with the increase in the rotation rate the thickness of coating decreases from 228 to 8 nm per one rotation, whereas the thickness of Ti layers changes from 60 nm per 1 rpm and at the distance of 22 cm from the center to 2.5 nm per 14 rpm and position in the center of the table. Al thickness varies within the range of 168 nm to 5 nm at the same modes.

3. Experimental methods

Deposition of intermetallide-based coatings of Ti-Al system was carried out using NNV-6.6-I1 unit. Technological modes were adjusted within the following limits: pressure in a chamber P = 10-1 - 10-2 Pa; arc current I = 40 - 120 A; treatment time 60 - 75 min. Deposition scheme of intermetallide-based coatings of Ti-Al system and layout of samples is shown in Figure 2.

During rotation of the table at the rate of 1 rpm multilayered coating with subsequent Ti and Al layers is formed on the surface of a sample. During deposition approximately 120 layers are formed in the coating within 60 min. The thickness of layers decreases with the increase in the rotation rate of the table, while their quantity increases. Thus, the number of layers amounts to 360 at the rotation rate of 3 rpm, 840 - at 7 rpm, and 1680 layers at 14 rpm.

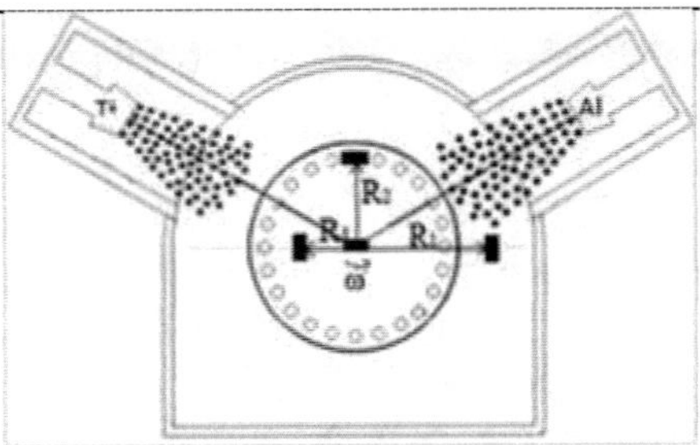

Figure 2. Samples layout scheme

In order to determine structural and phase composition of coatings and explain their physical and chemical behavior X-ray diffraction study was carried out. The X-ray diffraction analysis of coatings was performed using DRON-3 diffractometer in Co K-alpha radiation at 30kV voltage and 20 mA current. Angular spacing of 2-theta scanning = 30-130 ^{0}.

4. Results and discussion

After depositing nanostructured coatings using the developed method the following intermetallic phases of Ti-Al system were identified on XRD patterns of P6M5 steel samples surface: TiAl, $TiAl_3$. The depth of a layer involved in the formation of a diffraction pattern makes ~ 7 μm, and the thickness of coatings varies from 3 to 5 μm. Thus, the analyzed depth during X-ray diffraction analysis of samples exceeds the thickness of coatings. Therefore, XRD patterns have iron peaks, and at quantitative analysis the percentage of iron will depend on the thickness of coatings.

Results of quantitative XRD analysis depending on spatial arrangement and number of layers are presented in Table 1.

Table 1. Results of quantitative analysis

Number of layers	Distance from the center of a table R, cm.																			
	22					15					8					0				
	Fe	Ti	Al	TiAl	$TiAl_3$	Fe	Ti	Al	TiAl	$TiAl_3$	Fe	Ti	Al	TiAl	$TiAl_3$	Fe	Ti	Al	TiAl	$TiAl_3$
120	27	43	3	13		32	34	13	18	3	34	28	0	23	14	26	31	4	17	22
360	22	-	11	-	10	30	32	13	14	11	27	35	-	14	13	28	36	2	17	18
840	13	50	16	12	10	40	35	3	9	13	24	49	0	11	16	35	48	-	4.8	6.2
1680	14	63	0	9	13	17	53	0	11	19	18	60	0	9	13	31	48	0	12	9

Quantitative XRD analysis made it possible to develop dependency diagrams of the percentage of Ti-Al system intermetallide content in a coating at simultaneous deposition from vacuum arc discharge plasma with two unicomponent Ti and Al cathodes in inert-gas Ar on the number of layers and spatial arrangement. The diagrams are shown in Figure 3.

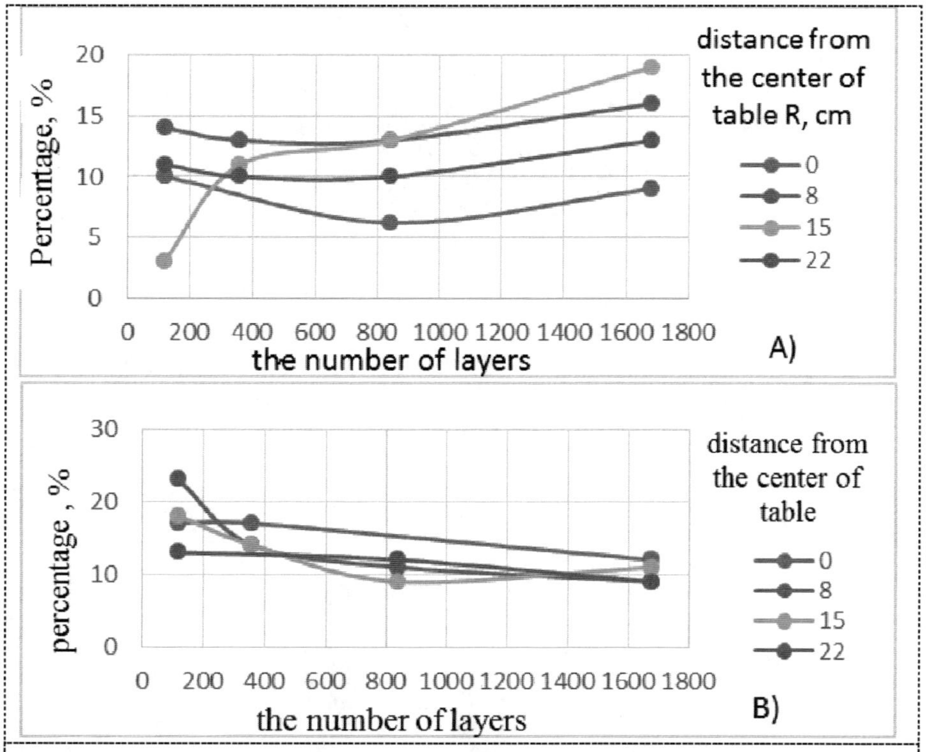

Figure 3. Dependence of the percentage of Ti-Al system intermetallide content at layer-by-layer deposition on the number of layers and spatial arrangement: A) $TiAl_3$; B) TiAl.

The results of quantitative phase XRD analysis show that at various treatment modes (rotation rate, number of layers, thickness of layers) the percentage of Ti increases from 30 to 50% with the increase of the number of layers and decrease in their thickness, thus the percentage of Al content decreases from 4-6% when the number of layers equals 120 to 0 when their number amounts to 1680. The percentage of TiAl intermetallide also decreases from 17 to 12% with the increase in the number of layers, while the amount of $TiAl_3$ phases increases from 5 to 20%. Thus, with the increase of the number of alternating layers and decrease of their thickness almost all aluminum, being part of the coating, reacts with titanium and hence forms intermetallides of $TiAl_3$+TiAl phases. At bigger thickness of >40 nm and a number of ~ 120 layers, not all aluminum is able to react with titanium. In order to increase the percentage of intermetallide content in a coating there is a need to increase the density of aluminum ion fluxes and to deposit a coating at high rotation rate of a table to ensure the increase in the number of layers.

The following combination of layers is suggested as a hardening multilayered composite coating to improve performance properties of punching tools: Ti-$TiAl_3$-(TiAl)N. The thickness of multilayered wear-resistant intermetallide-based coating within Ti-Al system amounts to 4-4.5 μm.

Production test results (Fig. 4) demonstrated the increase in the resistance of punching tools with Ti-$TiAl_3$-(TiAl) N coating by 6-7 times in comparison with the technology applied at BelZAN plant, and by 1.3-1.5 times in comparison with other coatings.

Figure 4. Punching tools with intermetallide-based coating within Ti-Al system

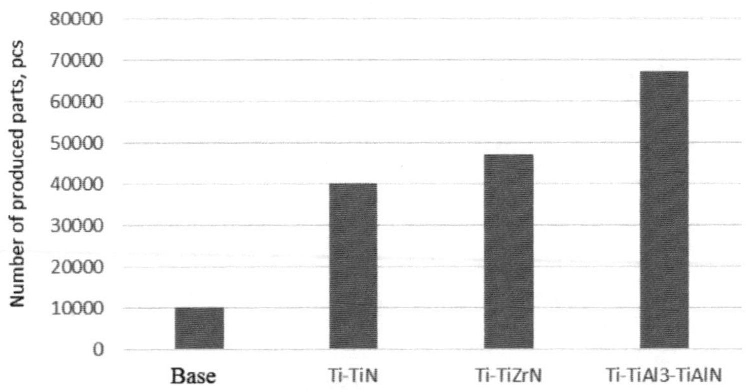

Figure 5. Average results of production tests

5. Conclusions

The results of the undertaken study show that with the increase in the number of alternating layers up to 1680 and decrease of their thickness up to 3-5 μm the percentage of $TiAl_3$ phase increases from 5 to 20%. Moreover, production tests of punching tools with Ti-$TiAl_3$-(TiAl)N composite coating demonstrated the increase in the resistance by 6-7 times in comparison with the technology applied at the plant.

Acknowledgement

The authors are grateful to the Russian Science Foundation for support of this work through the project No. 15-19-00144.

References

[1] Zhirkov I, Petruhins A, Dahlqvist M 2014 *Journal of Applied Physics*. **115** 123-127

[2] Xian-Xiu Mei, Jian-Qiang Fu, Xiao-Na Li, Rotshtein V P, Koval N N, Teng-Cai Ma 2014 *Rare Metals* **33** **2** 155-160

[3] Sun Y, Kulkarni K, Sachdev AK, Lavernia EJ 2014 *Metallurgical and Materials Transactions A* **45** **6** 2750-2758

[4] Budilov V et al 2004 *Mat. Sci.and Eng.A* **375-377** 656-660

[5] Vardanyan E L, Budilov V V, Kireev R M, Ramazanov K N 2014 *26th Int. Symp. on discharges and electrical insulation in vacuum*, Mumbai, 2014, pp. 541 – 543.

Journal of Physics: Conference Series **652** (2015) 012054 doi:10.1088/1742-6596/652/1/012054

Local ion nitriding process with hollow cathode effect computer modelling

K N Ramazanov, Yu G Khusainov, I V Zolotov

Department of Machine Building, Ufa State Aviation Technical University, 12 K.Marx Street, Ufa, Russian Federation

E-mail: ramazanovkn@gmail.com

Abstract. Modelling of glow discharge plasma with hollow cathode effect (HCE) was carried out. The computer model that allows predicting the temperature distribution during local ion nitriding with HCE and depth of the diffusion zone after the treatment was developed. The influence on the HCE on the temperature distribution for the gear was studied.

1. Introduction

It was proved in [1] that occurrence of HCE during ion nitriding raises the efficiency of the process (diffusion rate) and the ionization rate near the treated surface due to raising of generation of charged particles by oscillating electrons. The usage of a HCE is also a promising approach for local hardening treatment. It was experimentally proved [2] that HCE is created with special screens has following features: possibility of local surface treatment; high saturation speed; creation of diffusion layers with given phase composition and structure; high class of surface finish; possibility of ion nitriding and carbonitriding of surface-passivating materials without additional depassivating treatment; significant reduction of total process time since times of heating and charge cooling are reduced and intermediary technological operations for surface activation are removed; high cost effectiveness of the process since electric energy utilization ratio is increased and saturating gases consumption is reduced, which leads to an ecologically friendly process.

HCE has a complicated dependence on various energies, technological and geometrical parameters, thus systemic experimental diagnosis is difficult. The computer simulation hence becomes an important method to reveal the physics of treatment process.

Nitriding process with application of HCE becomes more complicated. In case of local ion nitriding with HCE the heating rate of the surface covered by technological mesh screen is greater than heating rate of remaining surface of the part. Therefore, there is irregularly temperature distribution during the nitriding process.

It is known that the temperature is the critical factor affecting the nitrogen diffusion into steel and diffusion zone formation of the nitrided case. Nowadays, modeling heating and diffusion processes during nitriding is an important task. Investigating and understanding processes occurring in the near-surface material of the part would give the possibility of using optimal process parameters and increasing nitriding efficiency.

In this paper we present our results of glow discharge plasma with HCE and thermal and diffusion processes during ion nitriding with HCE modelling.

Journal of Physics: Conference Series **652** (2015) 012054 doi:10.1088/1742-6596/652/1/012054

2. Methods of study

Calculations were performed with a help of differential equations solver software. Theoretically model of ion nitriding process is based on heat equation and Fick's equation [7].

At the initial time the concentration of nitrogen in steel is equal to zero, the temperature is assumed to be 298 K:

$$c_0 = 0; T_0 = 293 K . \tag{1}$$

The third type boundary condition is used in the model of local ion nitriding process [8]:

$$(-D\nabla c)n = k_m(c_b - c), \tag{2}$$

where D is the diffusion coefficient, c is the concentration of nitrogen, n is boundary's normal vector, k_m is the mass transport coefficient, c_b is the bulk concentration of nitrogen.

$$n(-h\nabla T) = q_0, \tag{3}$$

where h is the heat transfer coefficient, T denotes to temperature, q_0 is the thermal flux through the boundary:

$$q_0 = j_i U, \tag{4}$$

where j_i is the ion current density, U – discharge voltage.

Thermal radiation to environment:

$$-n(-h\nabla T) = \varepsilon\sigma(T_{amb}^4 - T^4), \tag{5}$$

where T_{amb} is environment temperature, ε is emissivity coefficient, σ is the Stefan-Boltzmann's constant.

Current density during ion nitriding is about 1-3 mA/cm^2. HCE leads to increasing of current density up to 3 times comparing to conventional glow discharge and can reach 10 mA/cm^2 [5].

3. Results and discussing

HCE also affects ion concentration near cathode surface. We used results of plasma simulation to estimate this parameter. In Figure 1 showed calculated distribution of ion concentration in cathode 5 mm thickness cavity compared with experimental curve obtained by probe measurements.

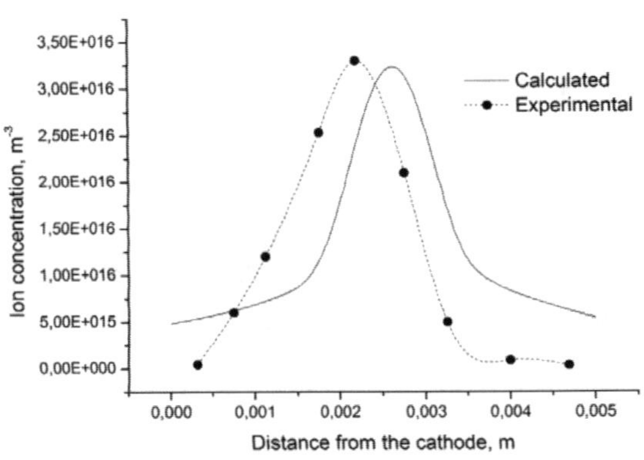

Figure 1. Ion distribution in cathode cavity.

Both curves have a similar shape and order, so obtained simulation data is in good agreement with experimental results. The distribution reaches maximum of about 3.5×10^{16} m^{-3}, which 1.5-2 times grater comparing with conventional glow discharge. It was proved that increasing of near concentration near treated surface leads to accelerating saturation process.

The temperature distribution was obtained as the result of model calculation. Temperature of the gear surface after 8 h of local ion nitriding with HCE is presented at Figure 2. Teethes temperature is about 550°C.

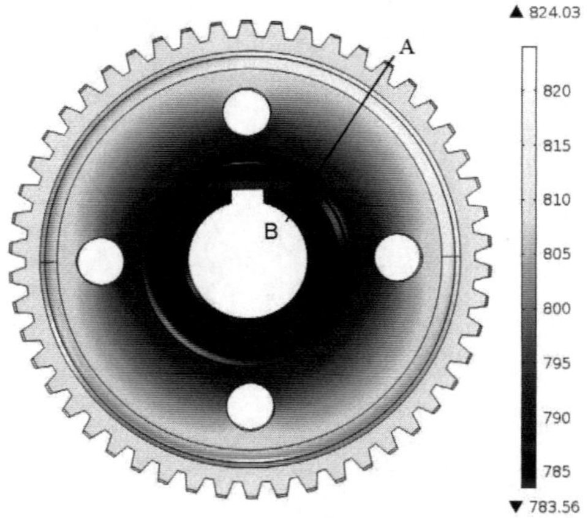

Figure 2. Calculated temperature distribution of the gear surface during nitriding process.

Calculations for gear with diameter of 200 mm show the maximum of temperature difference of 35° C (between area affected by HCE and uncovered area). Temperature distribution depends on size and shape of the part to be treated.

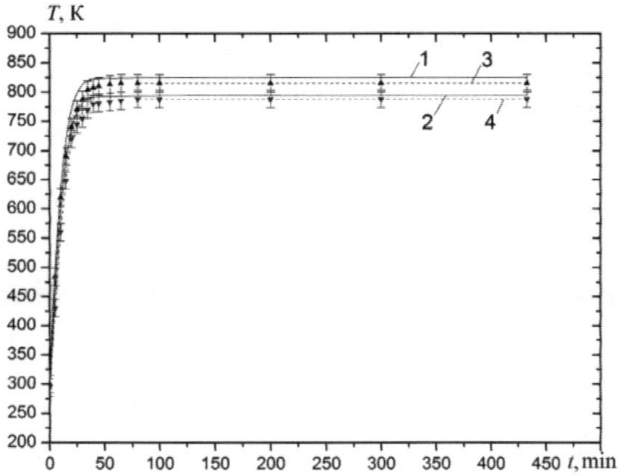

Figure 3. Temperature curves of A and B points of gear surface. 1 – calculated curve, point A, 2 – calculated curve, point B, 3 – experimental curve, point A, 4 – experimental curve, point B.

Figure 3 shows heating temperature curves of two points of part surface (A and B, figure 2), obtained by modelling and experimental measurements. Analysis of this curves shows that after reaching the maximum value of 550°C, temperature remains constant through all the process. It can be linked to a balance between incoming heat flux and thermal radiation. Time of reaching the process temperature is about 40 min (heating rate is 0.2°C per sec).

Calculated and experimental temperature curves (point A) of part cooling stage of the nitriding process are presented at figure 4.

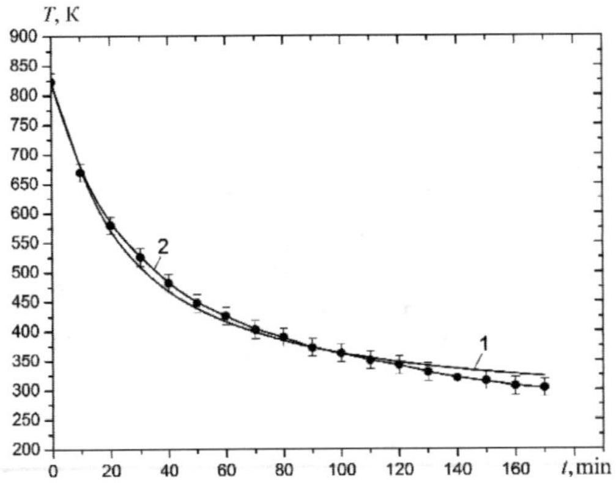

Figure 4. Temperature curves at cooling stage, point A. 1 – calculated curve, 2 – experimental curve.

Presented at Figure 4 curves has a practical application and allow to set the time of load cooling after nitriding with HCE of parts with different dimensions and shapes. As we can see, part cools to the temperature of 250°C (523 K) during about 30 min.

Distribution of nitrogen concentration in surface layer of material was obtained to estimate hardened case depth (figure 5). Modelling shows that nitriding with HCE during 8 hours leads to formation of 200 μm hardened diffusion layer.

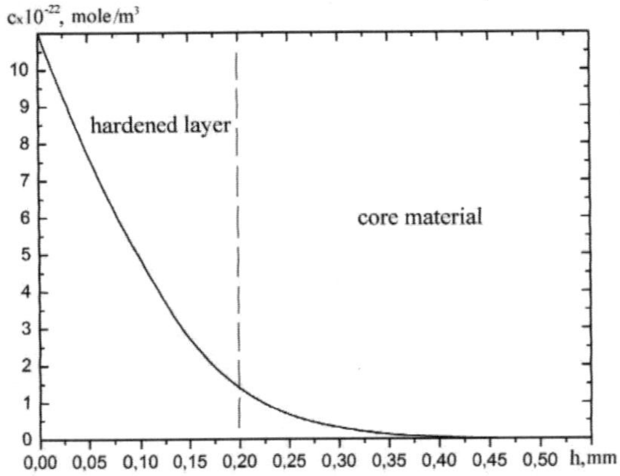

Figure 5. Calculated distribution of nitrogen concentration through the diffusion zone of nitrided case, T=550°C, t=8 h.

4. Conclusions

Increased ion concentration forms near treated surface in ion nitriding with HCE process. It leads to intensification of diffusion saturation process.

The computer model that allows predicting the temperature distribution in gear during local ion nitriding with hollow cathode effect (HCE) and depth of the diffusion zone after the treatment was developed.

The gear with diameter of 200 mm heats up to process temperature 550°C during about 40 min, after that the surface temperature remains constant to the end of the process. It is linked to reaching of balance between heat including and thermal radiation to environment. Maximum of temperature difference on gear surface is 35°C because of uneven heating rate of different surface areas. .

It is established that ion nitriding with HCE is an effective method of surface hardening of steels. Diffusion layer thickness of nitrided with HCE steels is about 0.2 mm.

References

[1] Akhmadeev Yu Kh, Goncharenko I M, Ivanov Yu F, Koval N N, Shcanin P M 2005 *Rus.Pis'ma v ZhTF* **31** 24–30

[2] Kreindel Yu E, Lemeshev N M, Slosman A I 1990 HCE in glow discharge nitriding *Electron.Obrab.Mater.* **6** 38–47

[3] Kogan Ya D, Kolachev B A, Levinskiy Yu V 1987 *Constant interaction of metals with gases* (Moscow: Metallurgiya)

[4] Isachenko V P 1975 *Heat transfer. Textbook for high schools* (Moscow: Energiya)

[5] Budilov V V, Agzamov R D, Ramazanov K N 2007 Technology ion nitriding in glow discharge with a hollow cathode *Metal science and heat treatment* **49** 358–361

Improving the operating properties of parts of titanium alloys by surface hardening in high density plasma of glow discharge

K N Ramazanov, I V Zolotov, Yu G Khusainov and R F Khusnutdinov

Department of Machine Building, Ufa State Aviation Technical University, 12 K.Marx Street, Ufa, Russian Federation

E-mail: ramazanovkn@gmail.com

Abstract. The computer model that allows predicting the temperature distribution during local ion nitriding with hollow cathode effect (HCE) and depth of the diffusion zone after the treatment of part "lever of downhole profilometer" was developed. The influence of the HCE on the temperature distribution for the part was studied.

1. Introduction

Ion nitriding in glow discharge is one of the most common methods machine parts surface hardening [1,2]. Usually parts are subjected to local wear and therefore their operating properties are determined by hardening of work surfaces.

Nitriding of titanium alloys is more difficult due to formation of solid nitride layer that noticeably slows diffusion process. Using hollow cathode effect (HCE) allows intense sputtering nitriding layer by argon ions during the process [3,4]. But it leads to uneven heating of part to be treated. It is known that the temperature is the critical factor affecting the nitrogen diffusion into steel and diffusion zone formation of the nitrided case [5,6]. To ensure effective saturation, treatment is usually carried out at temperature about 700°C. Nowadays, modeling heating and diffusion processes during nitriding is an important task. Investigating and understanding processes occurring in the near-surface material of the part would give the possibility of using optimal process parameters and increasing nitriding efficiency.

In this paper the results of computer modeling of ion nitriding process with HCE of part "lever of downhole profilometer" are presented.

2. Methods of study

Calculations were performed with a help of differential equations solver software. Theoretically model of ion nitriding process is based on heat equation and Fick's equation [7].

The part to be modelled is lever of downhole profilometer. Titanium alloy VT6 is used for this part. Lever is subjected to intense local wear due to friction on downhole walls. Thus, only operating area of part surface is affected by HCE.

To determine the influence of ion nitriding with HCE on depth of hardened layer, samples of VT6 alloy were nitrided. Sputtering was performed in argon during 15 min at pressure of 10 Pa of argon, temperature of the surface was not larger than T=250°C. Ion nitriding was carried in solution of argon, nitrogen (Ar 85%, N_2 15%). Samples were nitrided in plasma of glow discharge at the temperature T=700±10°C during 8 h.

Olympus GX-51 microscope was used to obtain the microstructure photos. Measurements of microhardness were carried out on microhardness tester Micromet-5101.

3. Results and discussing

Temperature distribution was obtained as the result of model calculation. Temperature of the lever surface after 8h of local ion nitriding with HCE is presented at figure 1. Temperature of operating surface is about 700°C.

Figure 1. Calculated temperature distribution of the lever surface during nitriding process.

Calculations show maximum of temperature difference between affected and unaffected by HCE zones of about 33°C.

Heating temperature curves in two points of part surface are presented in figure 2 (point *1* and point *2*, figure 1), obtained by modelling.

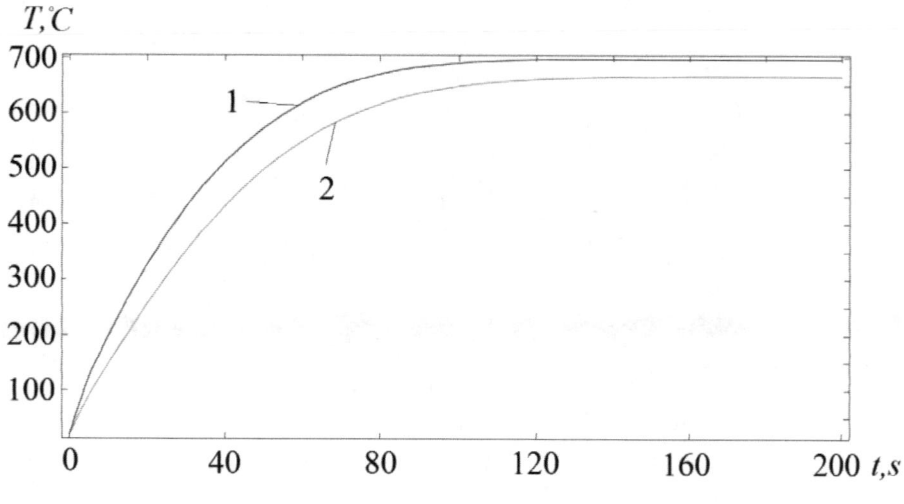

Figure 2. Temperature curves of points *1* and *2* of lever surface.

Presented graph (figure 2) shows that time of reaching process temperature is about 2 minutes.

To study diffusion processes and estimate case depth of nitride layer after treatment, distribution of nitrogen concentration through surface layer was investigated (figure 3).

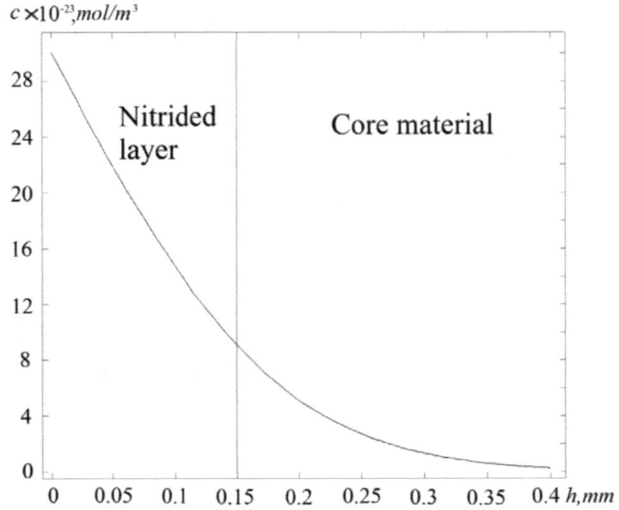

Figure 3. Calculated distribution of nitrogen concentration through the diffusion zone of nitrided case after 8 hours of treatment.

Modelling shows that distribution of nitrogen concentration in surface layer of material was obtained to estimate hardened case depth (figure 3). Modelling shows that nitriding with HCE during 8 hours leads to formation of 130-160 μm hardened diffusion layer.

Figure 4 shows microstructure of nitrided case of alloy VT6 after ion nitriding with HCE.

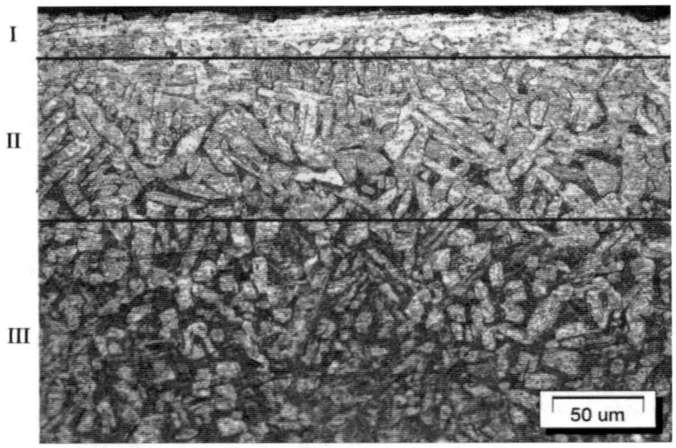

Figure 4. Microstructure of nitrided case of alloy VT6 after ion nitriding with HCE (I – nitride layer, II – diffusion zone, III – core material).

20 μm depth solid nitride layer (I) and diffusion zone (II) are observed. Microhardness profile was obtained to determine the depth of hardened case (figure 5).

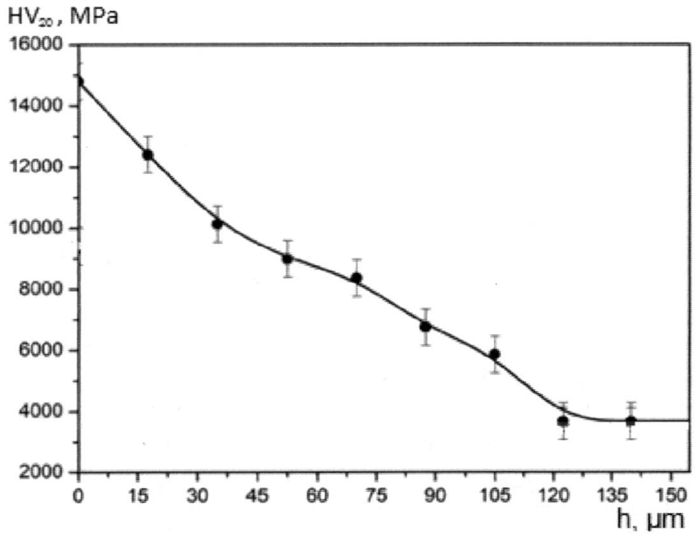

Figure 5. Microhardness profile of the nitrided case.

Hardened case depth is about 120 µm. Thereby, a good agreement between modelling and experimental data were obtained.

4. Conclusions

Presented computer model allows predicting the temperature distribution in Lever of downhole profilometer during local ion nitriding with HCE and depth of the diffusion zone after the treatment.

Lever heats up to process temperature 700°C during about 2 minutes, after that the surface temperature remains constant to the end of the process. Maximum of temperature difference on lever surface is 33°C because of uneven heating rate of different surface areas.

It is established that case depth after ion nitriding with HCE of VT6 alloy is about 120 µm.

References

[1] Zinchenko V M 2001 *Surface engineering gears methods of chemical-thermal treatment* (Moscow: Publishing House of the MSTU N.E. Bauman)

[2] Kozlovski I S 1970 *Chemical heat treatment of gears* (Moscow: Mashinostroenie)

[3] Budilov V V, Agzamov R D, Ramazanov K N 2007 Technology ion nitriding in glow discharge with a hollow cathode *Metal science and heat treatment* **49** 358–361

[4] Budilov V V , Ramazanov K N , Khusainov Yu G, Zolotov I V 2014 Perspectives of using of hollow cathode effect for local ion nitriding of parts from steel 16Kh3VFMB-Sh *Vestnik UGATU* **62** 32–36

[5] Gerasimov S A , Kuksenkova L I , Lapteva V G 2012 *Structure and wear resistance of design steels and alloys* (Moscow: MGTU im. Baumana)

[6] Lakhtin Yu M, Kogan Ya D 1976 *Nitriding steel* (Moscow: Mashinostroenie)

[7] Kogan Ya D, Kolachev B A, Levinskiy Yu V 1987 *Constant interaction of metals with gases* (Moscow: Metallurgiya)

[8] Isachenko V P 1975 *Heat transfer. Textbook for high schools* (Moscow: Energiya)

Journal of Physics: Conference Series **652** (2015) 012056 doi:10.1088/1742-6596/652/1/012056

Modelling of underground geomechanical characteristics for electrophysical conversion of oil shale

A A Bukharkin*, **I A Koryashov, S M Martemyanov, A A Ivanov**
High Voltage Department, Institute of High Technology Physics, Tomsk Polytechnic University, Tomsk 634050, Russia

E-mail: amplexor@ya.ru

Abstract. Oil shale energy extraction is an urgent issue for modern science and technique. With the help of electrical discharge phenomena it is possible to create a new efficient technology for underground conversion of oil shale to shale gas and oil. This method is based on Joule heat in the rock volume. During the laboratory experiments the problem has arisen, when the significant part of a shale fragment is being heated, but the further heating is impossible due to specimen cracking. It leads to disruption in current flow and heat exchange. Evidently, in the underground conditions these failure processes will not proceed. Cement, clay and glass fiber/epoxy resin armature have been used for modelling of geomechanical underground conditions. Experiments have shown that the use of a reinforcing jacket makes it possible to convert a full rock fragment. Also, a thermal field extends radially from the centre of a tree-type structure, and it has an elliptic cross section shape. It is explained by the oil shale anisotropy connected with a rock laminar structure. Therefore, heat propagation is faster along the layers than across ones.

1. Introduction

The attention to the development of oil shale energy potential after the «shale revolution» has not decreased until today [1, 2]. However, shale processing *in-situ* remains an urgent task, in connection with unprofitability of conventional development methods for these inferior solid fossil fuels. A great number of underground methods for rock processing has been offered, but neither of them has achieved the stage of industrial application, except formation hydrolytic fracturing (hydrofrac). The issue of hydrofrac application in the world practice is still opened, as this technology is low-efficient and provokes extremely negative ecological effects. According to the data of theoretical calculations as well as preliminary successful experiments it follows that electrophysical phenomena, occurring in oil shale under the electromagnetic field action, are enables to create a high-efficient and environmentally safety technology for underground processing of these fuels. Electrophysical impact on oil shale *in-situ* leads to the formation of an electroconductive channel directly in the rock strata. With the help of this channel it is possible to carry out heating and pyrolytic conversion of a shale organic constituent (called *kerogen*) into flammable gas and synthetic oil. During preliminary experiments it has been established that the adherence to the geomechanical conditions of bedding is very important for obtaining an adequate model of thermochemical, thermo- and elctrophysical processes.

2. Electrophysical conversion

Electrophysical oil shale conversion includes three relative stages. From the point of the required equipment and physics of running processes, ideally, it is necessary to divide them according to the rock conductivity.

2.1. Low conductivity

Initially, oil shale has high electrical resistance ($10^9 \div 10^{12}$ Ohm/cm), consequently, the current, flowing through the rock, is negligibly small, even under the high field stress. However, when voltage exceeds a specified threshold value, partial discharges (PDs) occur in the material volume. Primarily, PDs are the breakdowns of micro- and macrovoids, and they occur because of different dielectric permittivity of the inclusions and enclosing material [3]. Material fracturing and tree discharge structure forming, called dendrites, occur as a result of PDs plasma impact. The quantity of energy, which is required for the thermal destruction of kerogen, is smaller than for the oil shale mineral constituent. Therefore, the organic component is, mainly, subjected to the thermal decomposition under the PDs impact. Material fracturing by PDs plasma leads to treeing, i.e. the growth of dendrites. In some carbonaceous materials, including oil shale, treeing is accompanied by carbonization, i.e. by the increase of carbon local concentration. A conductive carbon layer is formed on tree walls due to carbonization [4,5].

2.2. Medium conductivity

The rock breakdown occurs, when the discharge structure electrodes are connected. The resistance of an interelectrode gap decreases by $3 \div 5$ degrees due to carbonization. The diameter of a formed channel does not exceed 300 μm. The channel resistance remains relatively high because of its small cross-section, so, current can flow through shale only in the case of high voltage application. Also, because of a small channel volume, the concentration of the input energy takes place there. Hence, a significant heat quantity is locally released even at small energy, which is input in the specimen. This process leads to the abundant emission of conversion gases and the formation of a dust-like suspension of solid particles. The educed aerosol is ionized under the external electromagnetic field and high temperature, and a plasma channel is formed. Due to plasma action the interelectrode gap conductivity and carbonization of the peri-discharge area increase.

2.3. High conductivity

The specimen resistance decreases to the value from one to first hundreds Ohm, when a carbonized channel cross-section is extended significantly. As a result, the difference of potentials between electrodes is reduced, and the discharge plasma dies out. The further heating occurs due to Joule heat in the rock volume and leads to the extension of the carbonization area. The subsequent carbonization provokes a greater decrease in resistance and the conversion of a substantial volume of the rock organic constituent.

3. Methods and equipment

3.1. Sample and medium preparation

A rock fragment with the shape of a parallelepiped and dimensions 145x70x85 mm has been prepared for the study of a full conversion cycle of oil shale. The obtained specimen is put in the reinforcing cage. The bars of glass fiber potted by epoxy resin are used as the armature, as far as this material has a high mechanical strength and electrical resistance. The use of the non-conductive armature is determined by the application of high voltage to the specimen at conversion initiation. The use of the conventional metal armature will lead to the significant distortion of the electromagnetic field applied

Journal of Physics: Conference Series **652** (2015) 012056 doi:10.1088/1742-6596/652/1/012056

to shale, which concentrates on the conducting elements. This process will cause the redistribution of stresses and uncertainty of the obtained data.

The cage with the specimen is infused in concrete. Preliminary, the sample is wrapped with a thin polyethylene film for prevention of water sorbing from concrete. Rock conductivity increases significantly, when it absorbs water, even in small quantities, what influences on the experimental integrity. Two closed holes with the diameter of 12 mm and the depth of 30 mm are drilled at the distance in 100 mm between them in the obtained block for carbon electrodes installation. Carbon is used, because it is resistant to destruction under the high temperature action.

A hermetically sealed chamber with the volume of ~ 60 L and maximum operating pressure up to 3 atm is used to model the fluids pressure in the underground conditions. Nitrogen is used as an inert medium, which simulates fluids and prevents combustion. Before filling the chamber with nitrogen it has been vacuumed, as pyrolysis must progress without any oxidizer.

3.2. Experiment

By the means of the insulators, installed in the wall of the chamber, the electromagnetic field is applied to the electrodes and imposed on the shale fragment. With the help of an adjusting transformer the voltage is increased up to the breakdown value, which is ~ 2 kV. After forming a stable conductive enough channel a high-voltage transformer is switched off, and a high-current adjusting transformer, supplying the variation of the output voltage from 0 to 220 V and the maximum current of 264 A, is directly connected to the specimen. Then, the sample is heated with the constant power of 1500 W during 4 hours. After that the oil shale fragment infused in the concrete cage is sawed up on four parts along longitudinal and cross vertical symmetry axes.

4. Results and discussion

The photographs of longitudinal and cross sections of the oil shale fragment after the conversion by the electrophysical method are shown in Figure 1 b,c,d. Figure 1a shows the heated rock fragment, which is not infused in the concrete cage. During thermochemical reactions the phase transitions of the solid matter into gas occur, it leads to pressure increase within the limits of some volume inside the specimen and occurrence of local mechanical stresses. As a result of a high degree of oil shale heterogeneity and a significant difference in the properties of various fragments, including a thermal linear expansion coefficient, the distribution of mechanical load occurs unevenly. Therefore, due to heating, thermal shocks and excess pressure of conversion gases as well as additional strength reduction during shale carbonization the specimen destruction occurs, primarily, in the form of stratification and splitting. Rock carbonization occurs due to Joule heat at current flowing throw the material. Specimen stratification and separation of rock fragments from the bulk material volume lead to the failure of current flowing paths and, consequently, to the redistribution of heat locations. Also, gas interlayers are formed between partly separated shale fragments, so, heat transfer is disrupted and a thermal field is redistributed. Heat transfer becomes negligibly small at a particular moment, and the further conversion is impossible. Obviously, in the underground conditions the shale strata are surrounded by the enclosing strata, what excludes rock destruction according to the described mechanism. Therefore, for obtaining the adequate data about the thermal field distribution, current flowing paths and conversion of big specimens it is necessary to build the geomechanical conditions of oil shale bedding in the model.

Considering a longitudinal section (Fig. 1 d), it can be noticed that carbonization has occurred from the surface of the shale fragment to the depth slightly bigger than the bottom of electrodes holes. This fact is associated with the contact area expansion. At extension of the carbonized volume the current is distributed evenly by the volume of carbonization, flowing on the path of the least resistance. Therefore, a major part of Joule heat is released directly in the interelectrode area, and carbonization occurs there. After the thermal decomposition of the rock between the electrodes the rest shale part is carbonized, primarily, as a result of heat transfer and only a small current part flows through this volume. Further thermal destruction is possible due to increasing the power input in the specimen and, consequently, temperature and/or increasing the time of electromagnetic field impact on rock. However, the increase of hold-up time does not lead to full rock fragment conversion, because with the extension of a carbonized volume the specific energy input will decrease as well as an absolute temperature value. This process will lead to the thermodynamical equilibrium and termination of thermal field distribution.

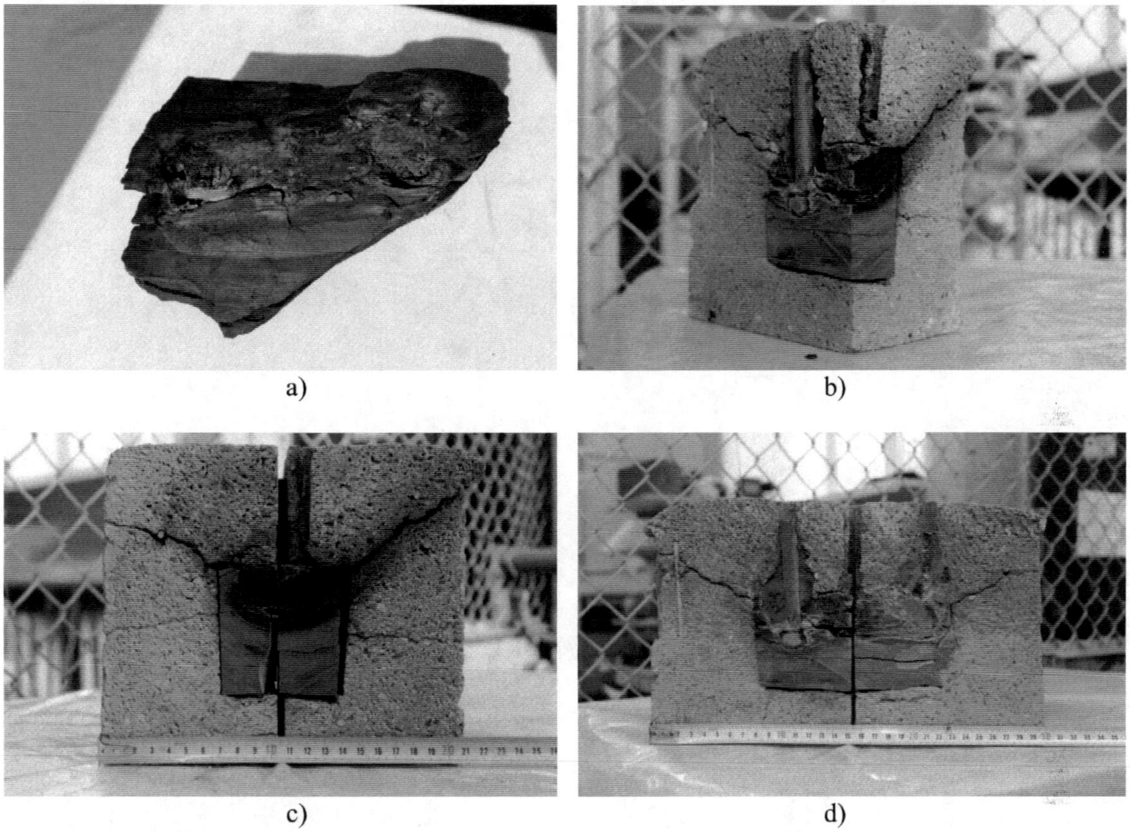

Figure 1. Photographs of the oil shale fragment infused in concrete after the electrophysical conversion; a) a specimen without a concrete cage, b) a lateral section, c) a longitudinal section, d) the quadrant of the sample.

Figure 1 c demonstrates that the cross section of the carbonized area has an elliptical shape with the position of major semiaxis along the layers and a symmetry center on a half-depth of electrodes penetration. Initially, the conductive channel is a dendrite, connecting electrodes. Due to Joule heat, heating, carbonization and channel extension gradually occur in the newly formed structure. The thermal field tends to equilibrium distribution, but the final form of the carbonized area has the elliptical shape as a result of greater thermal conductivity along the layers of oil shale. Directly after its formation the dendrite, connecting the electrodes, has a stochastically branching structure. Channel flattening occurs, because the system tends to equilibrium.

5. Conclusion

The adherence to the geomechanical conditions in the laboratory model is very important for obtaining adequate data about thermal field distribution, which appears to be the key moment in the underground pyrolytic conversion of solid fuels by the electrophysical method.

Journal of Physics: Conference Series **652** (2015) 012056 doi:10.1088/1742-6596/652/1/012056

If the oil shale fragment is infused in the solid cage, which sustains thermal and mechanical overloads, at modeling underground rock heating, it can be seen that a thermal field front is an ellipsoid. The length and position for one of axes is determined by the depth of electrodes penetration and the distance between them, rest two axes are located along and across the layers, semimajor and semiminor, respectively.

At supplying the constant power, inputting in shale, with regard to the extension of heat emission volume, in the particular moment a thermal equilibrium occurs and the further heating is possible only under the energy increase.

References

[1] Hughes J D 2013 *Nature* **494** 307-308
[2] Smil V 2015 *IEEE Spectrum* **52** 28
[3] Boggs S A, Densley J R 2000 *IEEE Electrical Insulation Magazine* **16** 13-18
[4] Dodd S J, Chalashkanov N M, Fothergill J C 2010 *IEEE International Conference on Solid Dielectrics* **1** 1-4
[5] Vaughan A S, Hosier I L, Dodd S J, Sutton S J 2006 *J. Phys. D: Appl. Phys.* **39** 962-978

Journal of Physics: Conference Series **652** (2015) 012057 doi:10.1088/1742-6596/652/1/012057

Thermal Stability of Ti-C-Ni-Cr and Ti-C-Ni-Cr-Al-Si Nanocomposite Coatings

A V Andreev[1], I Y Litovchenko[2, 3], A D Korotaev[1, 2, 3], D P Borisov[1, 3]
[1]Tomsk State University, Tomsk, 634050, RF
[2]Siberian Physical-Technical Institute, Tomsk, 634050, RF
[3]Institute of Strength Physics and Materials Science SB RAS, Tomsk, 634021, RF

E-mail: alexardas@mail.ru

Abstract. The results of the microstructure, mechanical and tribological properties investigations of Ti-C-Ni-Cr and Ti-C-Ni-Cr-Al-Si nanocomposite coatings based on amorphous carbon in the initial state and after the annealing to 900 °C are presented. The effect of annealing temperature on the microstructure, tribological and mechanical properties of Ti-C-Ni-Cr and Ti-C-Ni-Cr-Al-Si nanostructured coatings is discussed.

1. Introduction

Recently, wide attention is focused on the composite materials because of the possibility of creation a protective surface layer having a combination of such properties as high hardness, wear resistance, thermal stability and low coefficient of friction [1].

A promising area of research in this field is development of the coatings based on amorphous carbon (a-C), which are used as the solid lubricant, and nanoscale carbide crystalline particles and (or) nitride phases. These coatings are of the type DNG / AM, (dispersed nanograins in amorphous matrix) and have low internal stresses. These stresses are associated with the decreasing of volume fraction of sp^3 bonds (diamond-like) and increasing proportion of sp^2 bonds (graphite-like) as a result of doping amorphous carbon, for example, by the elements such as Cr, Si, Al, Ti [3].

Coatings of this type can be used to improve the strength properties and reducing the friction coefficient of some manufactures in friction units.

In the present paper the thermal stability, namely, the influence of annealing temperature on the structure-phase state, tribological and mechanical properties of Ti-C-Ni-Cr and Ti-C-Ni-Cr-Al-Si nanostructured coatings have been investigated.

2. Materials and methods

The Ti-C-Ni-Cr coatings were synthesized in the plasma magnetron-arc facility "Legend" [4] with three graphite cathodes (C_1, C_2, C_3) and one composite titanium-carbide cathode using nichrome as a binder. To form a Ti-C-Ni-Cr-Al-Si coating, one of the graphite cathodes was replaced by an AlCrSi cathode. Deposition parameters are given in table 1. The substrates were made from VT1-0 titanium alloy, WCo hard alloy (tungsten-cobalt) and molybdenum.

Investigation of samples by transmission electron microscopy was performed on microscope Philips CM-12 at an accelerating voltage of 120 kV. Microhardness of the resulting coatings was measured by the Vickers method using a special attachment to a Neophot-21optical microscope.

Journal of Physics: Conference Series **652** (2015) 012057 doi:10.1088/1742-6596/652/1/012057

Tribological tests have been carried out in a high-temperature tribometer in a ball-on-disk mode with a normal load of 2 N on the spherical indenter. On the studied coatings anneals at temperatures up to 900 °C under a high vacuum of 2×10^{-5} Torr have been made.

Table 1. Regimes of coatings synthesis.

Coating	P (kW)				U_s (V)	T (°C)	t (min)	P_{Ar+N} (Torr)
Ti-C-Ni-Cr	TiCNiCr	C_1	C_2	C_3	50	450	60	$4 \cdot 10^{-3}$
	1.2-1.3	0.9	0.9	0.9				
Ti-C-Ni-Cr-Al-Si	TiCNiCr	AlCrSi	C_1	C_2	50	450	60	$4 \cdot 10^{-3}$
	1.2-1.3	0.4	0.9	0.9				

3. Results and discussion

Transmission electron microscopy (TEM) investigations reveal the presence in the Ti-C-Ni-Cr coating of nanocrystalline (TiC and NiCr) phases and the amorphous phase (a-C). The diffraction pattern contains a diffuse halo corresponding to the amorphous phase, and the diffraction rings corresponding to the crystalline phase TiC with a small crystallite size and brighter reflections corresponding to relatively large NiCr particles (figure 1 b). It is assumed that chromium in this coating is partially dissolved in the amorphous phase (up to 20 at. %). In the diffraction patterns, these particles (the particles conglomerates) give some brighter reflexes (figure 2 b).

Titanium carbide particles are dispersed in an amorphous matrix sufficiently homogeneous. As areas of high-density particles, where the distance between the particles is comparable with their sizes, as areas, in which the distance between the particles significantly exceeding these sizes, are observed. In the diffraction pattern (figure 1 b) rings of the phase TiC are broadened, that indicates about the small size of the crystallites. Estimates of sizes by the half-width of the diffraction peak component 1/25 Å-1 for reflex g = [111] TiC gives the values of ≈ 2.5 nm, which are consistent with the analysis data of dark-field images.

Investigations of coatings cross-sections have revealed features in the distribution of nanoscale particles. At a distance of more than 0.5 microns from the substrate, a structure of coating is homogeneous. It is represented by an amorphous phase (a - C) with uniformly distributed particles of TiC and NiCr (<10 nm) [5]. At a distance of less than 0.5 microns from the substrate a particles (or conglomerates) NiCr sizes up to 30 nm are observed.

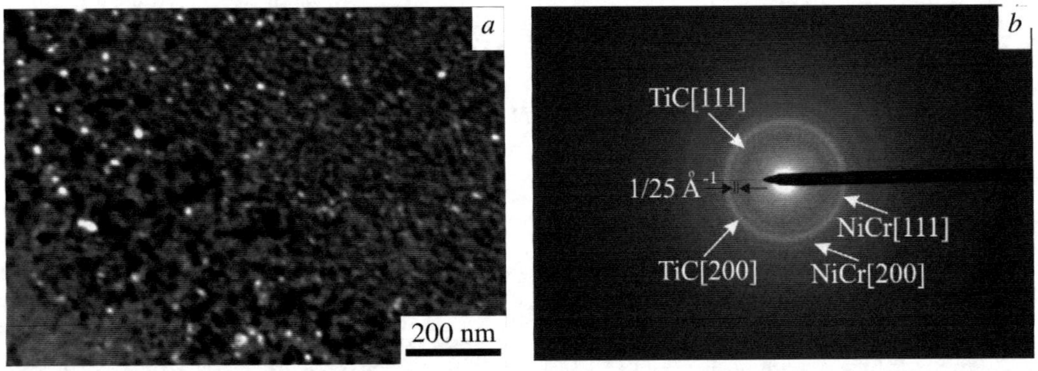

Figure 1. Dark-field image (*a*) and microdiffraction pattern (*b*) of Ti-C-Ni-Cr coating after deposition.

Investigation of the microstructure of the coatings Ti-C-Ni-Cr after the annealing up to T = 900 °C, 1h by transmission electron microscopy revealed that sizes of the nanocrystalline phases (TiC and NiCr) are reserved (figure 2). The formation of new phases did not found.

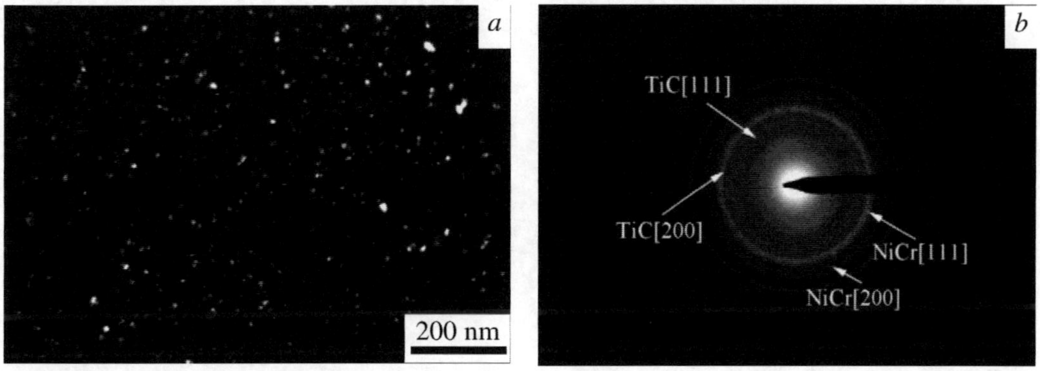

Figure 2. Dark-field image (*a*) and microdiffraction pattern (*b*) of Ti-C-Ni-Cr coating after the annealing at 900 °C.

Ti-C-Ni-Cr-Al-Si coatings after deposition have a microstructure similar to coating Ti-C-Ni-Cr, but different elements content. In Ti-C-Ni-Cr-Al-Si coating the carbon content is significantly less. It should be noted that in crystalline nanoparticles Al and Si did not found. Therefore these elements are dissolved in the amorphous carbon matrix.

Investigations of the microstructure of the Ti-C-Ni-Cr-Al-Si coatings after the annealing up to T = 900 °C, 1h by the methods of TEM also revealed the stability of amorphous and nanocrystalline phases (TiC and NiCr).

The values of the coefficient of friction for Ti-Ni-Cr-C coatings after deposition are in the range $\mu \approx 0.14 - 0.16$, for titanium alloy $\mu \approx 0.5 - 0.6$. These values of the coefficient of friction are retained at the annealing up to 700 °C.

Increasing the annealing temperature up to 900 °C leads to higher values ($\mu \approx 0.4 - 0.5$, figure 3 b) of friction coefficient of Ti-C-Ni-Cr coating, which is comparable with the coefficient of friction of the titanium alloy without coating ($\mu \approx 0.5 - 0.7$). The increasing of the coefficient of friction is caused the formation of oxide phases on the surface of the coating that was confirmed by X-ray analysis.

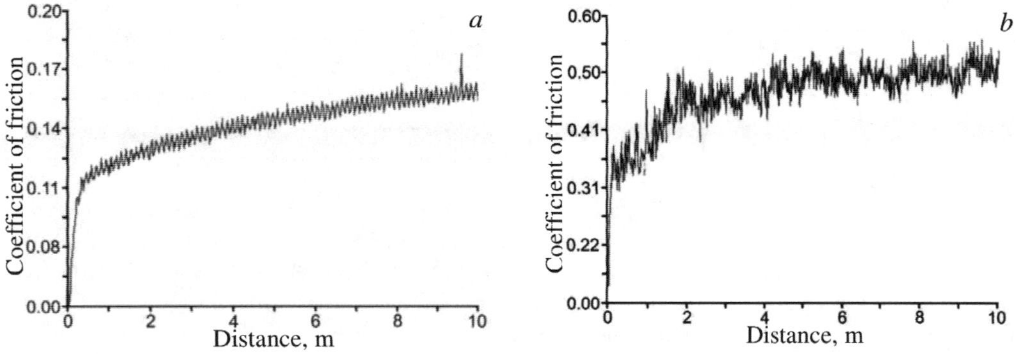

Figure 3. Dependence of the coefficient of friction on the distance traveled by the indenter on the Ti-C-Ni-Cr coating. (a) – after deposition, 1h, (b) – after the annealing 900 °C, 1h.

From the analysis of dependency of the friction coefficient from the distance traveled by the indenter at different annealing temperatures for Ti-C-Ni-Cr-Al-Si coating, it follows that the value of this coefficient after the annealing at T = 900 °C is practically identical with the initial value ($\mu \approx$ 0.34). However, the nature of this dependence is varied considerably. The coefficient of friction achieves this value almost from the beginning of the test (figure 4 b). The changing of the nature of this dependence may indicate on degradation of the coating structure as well as the possible formation of new phases, which increase the coefficient of friction.

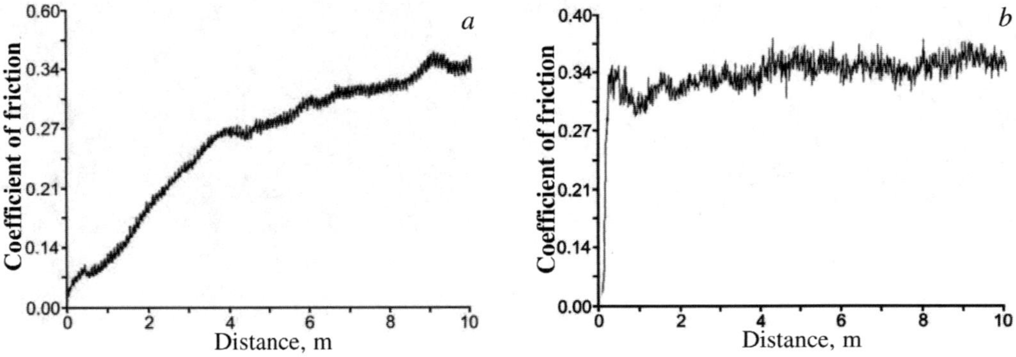

Figure 4. Dependence of the coefficient of friction on the distance traveled by the indenter on the Ti-C-Ni-Cr-Al-Si coating. (a) – after deposition, 1h, (b) – after the annealing 900 °C, 1h.

Investigations of mechanical properties of the samples of titanium alloy with Ti-C-Ni-Cr coating showed that the microhardness of the composition retains its value up to T = 900 °C. Moreover, there is some increase in hardness from ≈ 14 GPa in the initial state to ≈ 18 GPa after the annealing. These values of microhardness are in ≈ 5 - 7 times higher than that of titanium alloy (2 GPa).

It should be noted that at increasing of load the penetration depth of the indenter reaches a value close to the value of the coating thickness. Thus, is actually determined the microhardness of composition "substrate + coating".

Ti-C-Ni-Cr-Al-Si coatings on samples of titanium alloy already after deposition have a somewhat lower hardness (10.7 GPa). The annealing of these coatings causes a reduction of microhardness values up to ≈ 9.4 GPa (900 °C). At present the physical causes of such behavior microhardness do not clear.

Table 2. The microhardness H_μ of the studied coatings before and after the annealing.

Coating	Load, g	H_μ before annealing, GPa	H_μ after annealing 900 °C, GPa
Ti-C-Ni-Cr	20	13.9±0.9	18.3±0.9
Ti-C-Ni-Cr-Al-Si	20	10.7±1.5	9.4±1.1

4. Conclusion

Multielement nanocomposite coatings Ti-C-Ni-Cr and Ti-C-Ni-Cr-Al-Si, having a sufficiently high (11 - 18 GPa) hardness, low (μ <0.2) the coefficient of friction and high thermal stability of the microstructure and properties up to 700 ° C, were obtained. After the annealing at 900 °C the coefficient of friction increases, which can be connected to form an oxide film on the coating.

Coating structure is represented by the amorphous matrix with dissolved Al, Si, Cr, Ni metal elements and nanosized particles of crystalline phases (TiC and NiCr).

The high values of microhardness are caused the structure of amorphous carbon and nanosized particles. The tribological tests have shown that the Ti-C-Ni-Cr coatings have low values ($\mu < 0.2$) of the coefficient of friction, which is necessary for antifriction applications. These values are ensured by the amorphous carbon matrix that plays the role of a solid lubricant. The resulting combination of properties demonstrates the possibility of applying the considered coatings on the manufactures of titanium alloys.

Acknowledgements
This study has been performed with the support of RFBR (Grant No. 13-02-98020 r_sibir_a) and Tomsk State University Competitiveness Improvement Program.

References

[1] Teer D G New solid lubricant coatigs 2001 *Wear* **V. 251** 1068 – 1074

[2] Musil J Hard nanocomposite coatings: Thermal stability, oxidation resistance and toughness 2012 *Surf. Coat. Techn.* **V. 207** 50 – 65

[3] Musil J Physical and mechanical properties of hard nanocomposite films prepared by reactive magnetron sputtering 2006 *Springer* 407 – 463

[4] Borisov D P, Detistov K N, Zenin A A, Korotaev A D, Kuznetsov V M, Tyumentsev A N, Chulkov E V The development of effective vacuum plasma methods of surface modification and obtaining new materials at TSU 2011 *Russian Physics Journal* **V. 54 No 9-2** 19 – 26

[5] Andreev A V, Korotaev A D, Litovchenko I Y, Tyumentsev A N, Borisov D P Microstructure and tribological properties of nanocomposite coatings based on amorphous carbon 2015 *Physical Mesomechanics* **V. 18 No. 1** 73 – 83

The application of an assisting gas plasma generator for low-temperature magnetron sputtering of Ti-C-Mo-S antifriction coatings on titanium alloys

A I Potekaev[1], V M Savostikov[1], A N Tabachenko[1], E F Dudarev[1], E A Melnikova[1] and I A Shulepov[2]

[1]National Research Tomsk State University, Siberian Physical-Technical Institute, 1 Novosobornaya Sq., 634050, Tomsk, Russia

[2]National Research Tomsk Polytechnic University, Physical-Technical Institute, 30 Lenin Street, 634050, Tomsk, Russia

E-mail: svm.53@mail.ru

Abstract. The positive effect of assisting influence of high-density gas plasma formed by an independent plasma generator PINK on mechanical and tribological characteristics of Ti-C-Mo-S magnetron coating on titanium alloys at lowered to 350°C temperature of coating regardless of alloy structural condition was revealed by methods of calotest, nanorecognition, scratch testing and frictional material tests. The coating formed by means of a combined magnetron plasma method reduces titanium alloys friction coefficient in multiple times and increases wear resistance by two orders of magnitude. At the same time the mechanical properties of ultra-fine-grained titanium alloys obtained by nanostructuring do not deteriorate.

1. Introduction

Until recently methods of oxidation, nitriding and borating from gas and solid phase with alloy heating to high temperatures (800°C and higher) and long-term aging [1] have been commonly used to increase wear resistance of titanium alloys and decrease their adhesive and frictional interaction with a counter body in friction pair. However, the conventional high-temperature methods of surface modification mentioned above are inappropriate for alloys in nanostructured condition because of high probability of grain growth and loss of bulk stress-stain properties achieved by nanostructuring. At present the research of alternative methods of titanium alloys surface modification in particular by microarc oxidation, electro-explosive alloying with electron beam post irradiation or ion-beam surface treatment [2] is carried out. Methods of ion-plasma sputtering are also among modern methods of metals and alloys surface modification. Obviously, in order to obtain the adhesive strength of coating to substrate and the entire reaction behavior of ion plasma synthesis of necessary coating compounds it is essential to provide the sufficient diffuse atomic mobility of a deposited coating, which is usually achieved by substrate heating to high temperatures. Owing to the high heating of nanostructured alloys is inadmissible, the application of the assisting influence on a deposited coating by means of high-energy ions generated by an independent source is an advanced method to increase the diffusive atomic mobility of deposited coatings. The research carried out in the synthesis of wear resistant steel substrate coatings [3, 4] is an example of this technological method implementation. In the present work the technological method mentioned above and its potentialities in composite anti-friction Ti-C-Mo-S sputteringonVT6 titanium alloy in nanostructured condition are being investigated.

2. Materials and routine of experiment

The mechanical, physical and tribological investigations of properties of Ti-C-Mo-S coating deposited by the conventional method of magnetron sputtering and the combined method of magnetron sputtering with simultaneous assisting influence of gas-discharged plasma ions, formed by an independent plasma generator (magnetron-plasma method), were examined on VT6 ultra-fine-grained titanium alloy (hereinafter VT6ufg) with the compound of 5.5-7.0% Al, 4.2-6.0% V, Ti- the rest of it [5]. In order to compare the results, some identical investigations were carried out for VT6 and VT14 alloys in coarse-grained condition (VT6c и VT14c, respectively). VT14 alloy composition is the following: 4.5-6.3% Al, 0.9-1.9% V, 2.5-3.8% Mo, Ti - the rest of it [5]. Both alloys examined are alpha-beta titanium alloys. So as to form in VT6 titanium alloy blanks an ultra-fine-grained structure which decreases bulk strength, the method of severe plastic deformation (abc-pressing) with simultaneous decrease of blanks heating temperature in accordance with a stated procedure was used. After this treatment an average grain size was 0.7 μm. The strength properties of the initial VT6**c** alloy and VT6ufg alloy both after ion-plasma doping and/or sputtering were measured at indoor temperature with tensile test of a flat shoulder specimen [6]. The specimen-grid holders in the form of discs with 30 mm diameter and 4 mm thickness were made to carry out the comparative mechanical and tribological tests of initial alloys and deposited alloys. The deposition of Ti-C-Mo-S coating on titanium alloy substrates was produced by means of "Composit-3" [7] setup with a duomagnetron with two Ti-C-Mo-S sputtering targets and a PINK gas plasma generator [8] inside the working chamber. The targets were produced by self-propagating high-temperature synthesis (SHS) method with simultaneous compression of ignited alloy of a set composition from powder components. During the ion plasma spraying, argon was used as plasma-forming gas. Magnetron discharge capacity on each sputtering target was approximately 1500 W. Substrate accelerating potential was 200 V. Discharge current of an assisting plasma generator was regulated so that the substrate temperature did not exceed 350°C. In all cases the method of ion plasma predoping of substrate by Ti, C, Mo, S atoms [9] was applied. The deposition time was 60 minutes. The research of mechanical and tribological properties was performed by means of calotest for measuring coating thickness, hardness measurement by nanorecognition on «NHT-S-AX-000X» device on 0,01 N loading, scratch testing on «Micro Scratch Tester» (CSEM) device with the registration of acoustic emission (AE) signal, friction coefficient testing during frictional material tests together with a counter-body, made of identical titanium alloy in the form of a cone with corner radius R=1mm or with a tempered steel rotating sphere HRC63. The tribological study was realized by means of «THT-S-AX000» (CSM) tribometer in accordance with the following order: a static indentor – rotating disk (pin-on-disc). In all cases the indentation load was 10N. The linear velocity of a counter body slippage was equal to 2 cm/s. The appraisal depreciation of surface coating of initial specimens and deposited ones was realized by «Micro Measure 3D Station» (STIL) by means of the calculation of averaged cross-section wear tack after the fixed friction testing cycle (3000 rotations). Comparative tribological tests were carried out on the same specimen, one side of which was in the initial condition and another was tested with Ti-C-Mo-S deposited coating. The optic microscopy of coating failure on calotest depressions and scratch track was conducted. The scratch test was carried out at the load leading to an explicit signal of acoustic emission.

3. Experimental investigations results and their consideration

The evaluation of coating thickness by means of calotest showed the following. At the same deposition time in 60 minutes, thickness of coatings deposited with simultaneous assisting influence of gas-discharge plasma formed by an independent plasma generator is less than those deposited by the conventional method. The coating thickness on VT6ufg alloy deposited by the combined magnetron plasma method is 1.2 μm whereas by the conventional method the coating thickness is 2.5 μm (figure 1). On VT14c alloy these quantities are 1.7 μm and 2.9 μm, respectively. Quite evidently, it is connected with an additional amplification of spraying intensity deposited material atoms by ions of high-dense gas-discharge plasma argon. Besides, it should be mentioned that in contrast to the magnetron plasma coating failure (comparison figure 1a and figure 1b) the deposited coating formed by the conventional method of magnetron sputtering at moderate (about 350°C) substrate temperature delaminates into large scales around an indenter under calotest circular and load-bearing influence. The same pattern is typical for coatings deposited on VT14c alloy substrate as well.

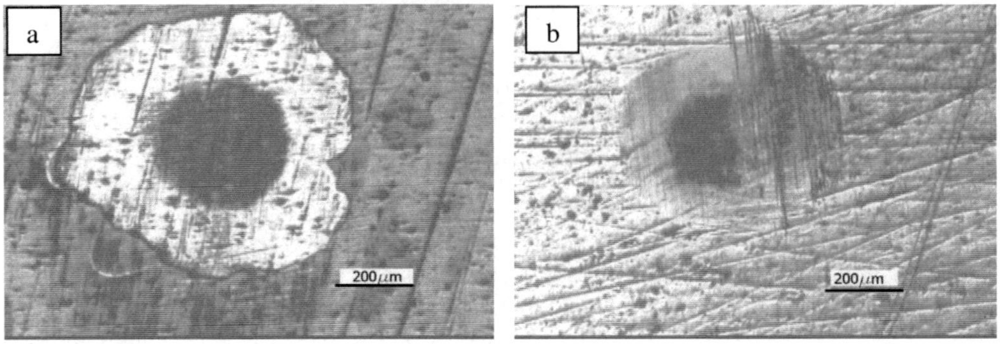

Figure 1. The images of calottes sections of Ti-C-Mo-S coating deposited on a substrate from VT6ufg nanostructured alloy by the conventional magnetron sputtering method (a) and by the magnetron plasma method with an assisting influence of gas discharge plasma from an independent source PINK (b) with indication of thickness index.

In figures 2 and 3 the typical scratch test results with emphasized signals of acoustic emission (AE) and distinctive images of scratch tracks in places of significant initial signals of AE are shown.

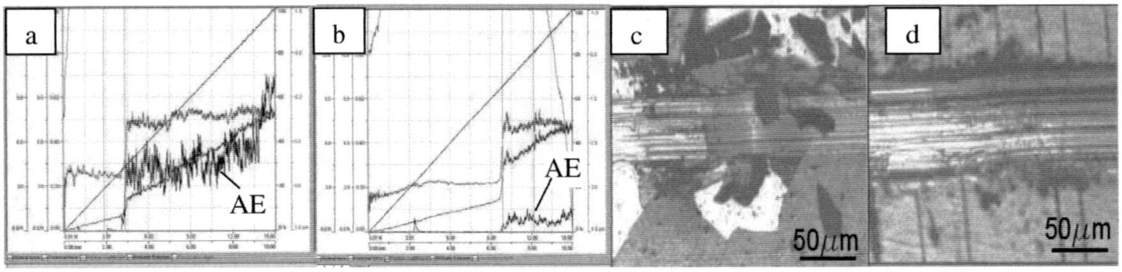

Figure 2. Load dependence of AE signal for Ti-C-Mo-S coating, deposited on nanostructured VT6ufg titanium alloy by means of the conventional method (a) and the combined magnetron plasma method (b) with images of coating failure peripherally a scratch track AE initial signal (c, d)

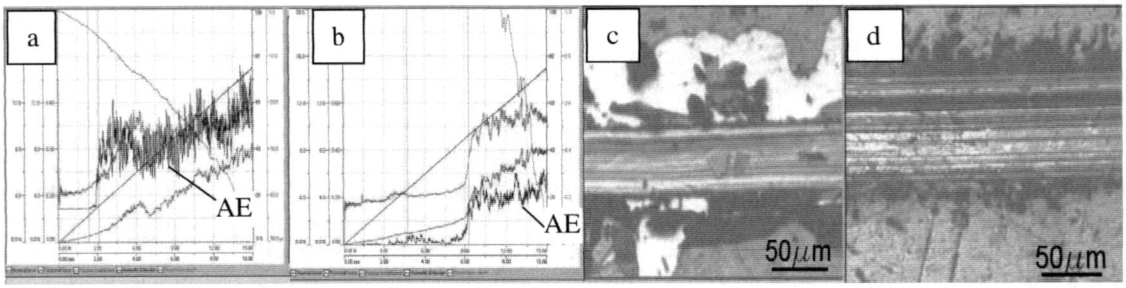

Figure 3. Load dependence of AE signal for Ti-C-Mo-S coating, deposited on nanostructured VT14c titanium alloy by means of the conventional method (a) and the combined magnetron plasma method (b) with images of coating failure peripherally a scratch track AE initial signal (c, d).

Journal of Physics: Conference Series **652** (2015) 012058 doi:10.1088/1742-6596/652/1/012058

During the testing of coatings deposited by the conventional magnetron sputtering, the AE signal initiation is observed at loads ≈3÷4 N for substrates both from VT6ufg and VT14c alloys (figure 2a and figure 3a). For both titanium alloys AE signal at maximal load of 15N is ≈60%. Load dependence of AE signal changes considerably during testing of coatings deposited by the combined magnetron plasma method. The AE signal initiation is registered at bigger for both alloys load (9-10N). The pick/maximum value of signal at load of 15N is 20-30% for substrate from VT14c alloy and not more than does not exceed 10% for substrate from VT6ufg alloy (figure 2b and figure 3b).In accordance with AE curve behavior, peripherally a scratch track there is considerable difference in coating failure inflicted by two mentioned methods. During the conventional method of magnetron sputteringVT6ufg substrate coating fails and delaminates into large scales at load of about 4.5 N (figure 2c). It is accompanied by AE signal increase. In case of the combined magnetron plasma method AE signal increase is observed at load of ≈10N and coating failure occurs with fewer amounts of particles (figure 2d). The similar pattern in coating failure and AE signal generation is observed during comparative scratch tests of coatings on coarse-grained VT14c titanium alloy deposited by means of two mentioned methods (figure 3c and figure 3d).

Hardness tests also demonstrated differences in two methods of sputtering. An averaged value of hardness for VT6ufg alloy substrate coating deposited by the conventional magnetron method is 6.5 GPa while by the combined magnetron plasma method (with assisting influence of gas-discharged plasma) is 7.1GPa. The maximum hardness value of coating deposited by the magnetron plasma method, in comparison with the conventional magnetron sputtering, was also registered on VT14c alloy (8.6 GPa comparatively with 7.9 GPa). From the total analysis of the above results we can conclude that bombardment of deposited material by gas ions of gas-discharged plasma (with higher plasma density than at magnetron discharge) from an independent PINK source notably increases a diffusion process. In its turn, it increases adhesive strength of the coating to the substrate, compound synthesis probability (solid carbide, in particular), has effect on structure and hardness of a deposited coating.

The comparative friction test also showed the advantage of Ti-C-Mo-S magnetron plasma sputtering in comparison with the conventional magnetron sputtering in friction pair both with a counter-body from identical titanium alloy and with a counter-body from SH-15 steel hardened to HRC63 (figure 4).

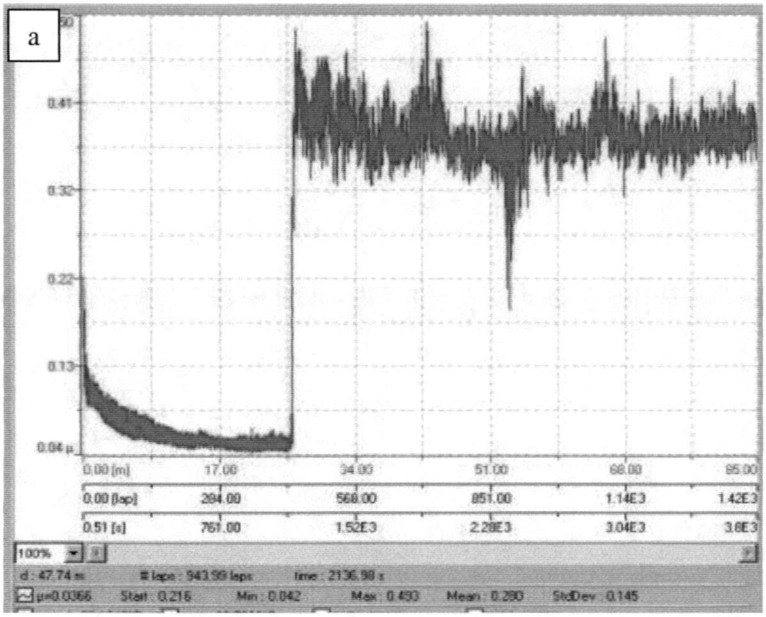

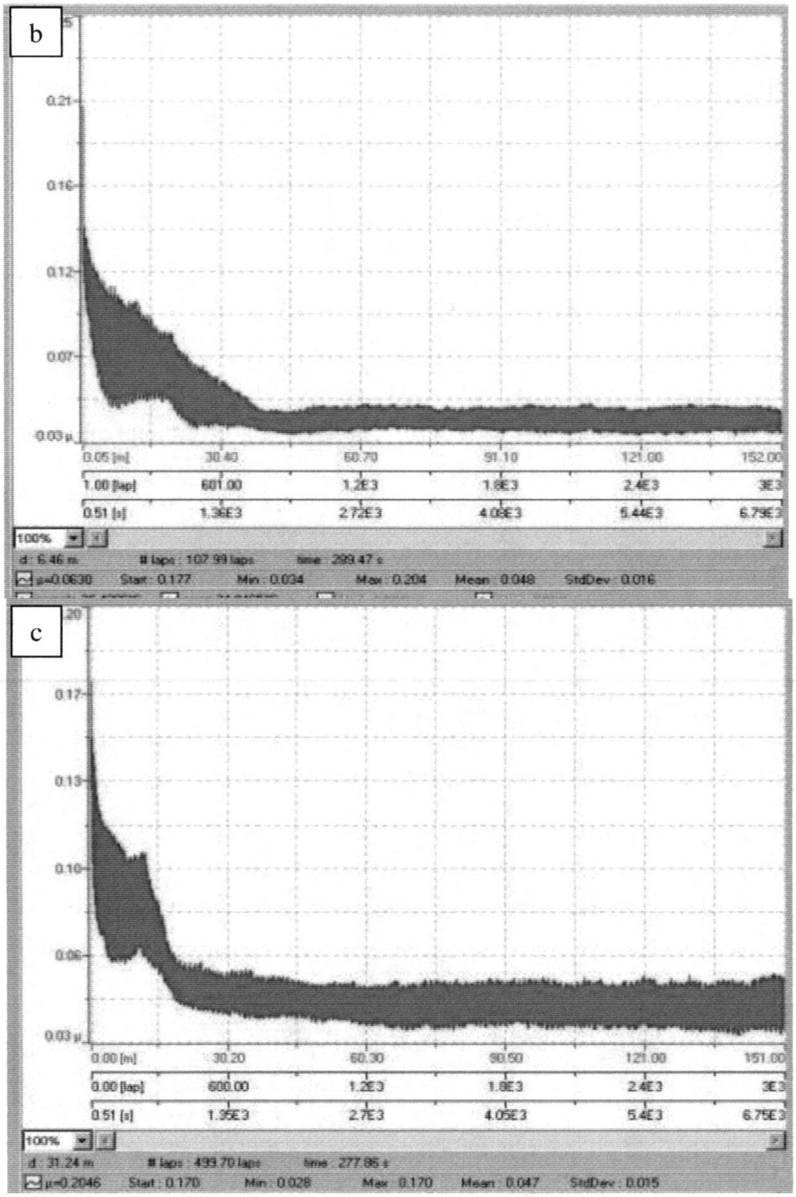

Figure 4. The dependence of friction coefficient μ from turning number of substrate-disc from VT6ufg alloy (friction path) for a coating deposited by the conventional magnetron sputtering method (a) and by the combined magnetron plasma method with an assisting influence of a gas plasma generator PINK (b, c). Counter body material is VT6ufg alloy with modified surface and SH-15 hardened steel (c).

The friction coefficient of coating deposited by conventional techniques after 400 revolutions of the disk increases sharply to 0.4 (figure 4a). Such kind of «a break» or coating delamination is connected with its weak adhesion (minor adhesive strength) with a substrate. It is correlated with the mechanical investigation results mentioned above. During the magnetron plasma sputtering method the minimal friction coefficient (about 0.045) remains while measuring (3000 rotations) when at friction test a counter-body from identical titanium alloy as well as a counter-body from hardened steel are used (figure 4b and figure 4c). The obtained results of friction tests bring out clearly a profound effect of

Journal of Physics: Conference Series **652** (2015) 012058 doi:10.1088/1742-6596/652/1/012058

assisting gas discharge plasma formed by an independent source PINK on strength and wear resistance of Ti-C-Mo-S coating. As it has been reported before [6], μ friction coefficient of titanium alloys decreases in multiple times while coating wear resistance increases by a factor of 10^2.

Control research of the effect of magnetron plasma coating deposited at temperature of 350°C on strength and ductility of an initial coarse-grained and nanostructured (ultra-fine-grained) VT6 titanium alloy were carried out. Control research did not reveal the decrease of the following characteristics (table 1).

Table 1. The effect of Ti-C-Mo-S magnetron plasma coating on ultimate strength (σ_B) and percentage extension before failure (δ) of VT6 alloy in coarse-grained and ultra-fine-grained condition.

Alloy condition	Strength and ductility initial indexes		Strength and ductility indexes after coating magnetron plasma sputtering	
	σ_B, MPa	δ, %	σ_B, MPa	δ, %
coarse-grained	900±10	10.5±0.5	With coating 920±10	With coating 9.5±0.4
ultra-fine-grained	1040±10	10.5±0.5	With coating 1070±10	With coating 13.0±0.3

4. Conclusion

Additional (assisting) influence of high-density gas-discharged plasma formed by an independent PINK source on substrate and deposited material has the potential to increase mechanical and tribological properties of Ti-C-Mo-S deposited magnetron coating at lower temperature of a substrate and to use this combined method of coating synthesis for nanostructured alloys on the basis of titanium, in particular.

References

[1] Tarasov A V 2003 *Metallurgy of titanium* (Moscow: Akademkniga)

[2] Sharkeev Yu P, Kukareko V A, Legostaeva E V and Bely A V 2010 *Russian Physics Journal* **10** 1053–59

[3] Savostikov V M, Borisov D P, Kuzmichenko V M and Pinshin Yu P 2007 *Russian Physics Journal* **9** Addition 424–27

[4] Savostikov V M, Borisov D P, Sergeev S M and Moshkov V Yu 2007 *Russian Physics Journal* **9** Addition 390–93

[5] Gulyaev A P 1977 *Physical metallurgy* (Moscow: Metallurgia)

[6] Potekaev A I, Tabachenko A N, Savostikov V M, Dudarev E F, Bakach G P and Skosirsky A B 2014 *Russian Physics Journal* **2** 222-29

[7] Savostikov V M, Potekaev A I and Tabachenko A N 2011 *Russian Physics Journal* **7** 756–64

[8] Borisov D P, Koval N N and Shanin P M 1994 *Russian Physics Journal* **3** 295–99

[9] Savostikov V M, Potekaev A I, Tabachenko A N, Shulepov I A, Kuzmichenko V M and Didenko A 2012 *Russian Physics Journal* **11** 1232–40

Experience of Pseudospark Switch Operation in Pulse Power Applications

N V Voitenko[1], A S Yudin[1], N S Kuznetsova[1] and V D Bochkov[2]
[1]Tomsk Polytechnic University, 30 Lenina ave., Tomsk, 634050, Russia
[2]Pulsed Technologies Ltd., 5 Yablochkova str., Ryazan, 390023, Russia

E-mail: tevn@hvd.tpu.ru

Abstract. The paper demonstrates the results of TDI1-200k/25SN-P pseudospark switch (PSS) developed by Russian company "Pulsed Technologies Ltd" application. PSS was used in pulsed power unit intended for electric-discharge fracture of rocks and concrete blocks and splitting off from monolith. The pulsed power unit has a pulse current generator with the capacity of 560 μF, stored energy of up to 63 kJ, operating voltage of up to15 kV, current pulse amplitude of up to 200 kA and pulse duration more than 200 μsec. The study also shows the current waveforms determined in the short-circuit experiment of the pulse current generator and in the experiments of the electric-discharge fragmentation of concrete at the charging voltage of 13 kV. PSS was operated in ringing single-pulse mode with the exceedance of more than two maximum permissible parameters: current pulse amplitude, current pulse duration and maximum pulse energy. Internal electrode erosion of PSS is shown and possible reasons of asymmetric current feed are discussed.

1. Introduction

The recent decades demonstrate increasingly growing interest to the development of new environmentally friendly technologies, which is due to severe complication of environmental problems. In most cases, the green technologies are preferred, even though they are less effective than conventional technologies. Such technologies are especially in demand, if their usage reduces the amount of manual labor and operating staff risks. Electric-discharge technologies are considered to be a promising option for their application to various processes in mining and building industry. Therefore, many large laboratories in the developed countries are working on the research and implementation of electric-discharge technologies for fragmentation of large rocks, splitting off blocks from monoliths of natural and artificial origin, digging or expanding tunnels, electric-discharge drilling in hard rocks, crushing and grinding ore, etc. [1-5].

The development of electric-discharge technologies and their successful application in industry depend on the availability of durable, reliable and affordable equipment and its components. A high-voltage switch is one of the basic component in the majority of electric-discharge facilities. The high-voltage switch shall meet a number of strict requirements that are determined by operating parameters of the equipment. The progressive types of switches are TDI and TPI thyratrons with cold cathode, also known as pseudospark switches operating on the left-hand branch of the Paschen's curve in a wide range of anode voltages between 1 kV and 150 kV and the switched current of up to 200 kA [6–9]. The article describes the experience of pseudospark TDI-type switch operation that was developed by the Russian innovation-industrial company "Pulsed Technologies Ltd".

2. Experimental

The specialists of the Department of High Voltage Engineering and Electrophysics of Tomsk Polytechnic University have designed and built a mobile unit for high-voltage and high-current studies. The main application of the mobile unit is electric-blast splitting off from monolith and fragmentation of rock and concrete lumps. The unit is based on a high current pulse generator (HCPG) with the operating voltage of up to 15 kV and the stored energy in the capacitor bank of up to 63 kJ, the current in load can be increased up to 200 kA [10]. In the initial design of the HCPG a custom high-pressure trigatron switch was used. The switch had flat graphite electrodes kept in argon-krypton gas mixture, which enabled the switching currents of up to 500 kA. Despite a considerable operating current, the switch had relatively low triggering stability and required constant pressure monitoring, which affected to productivity of rock destruction.

In order to increase the efficiency of power transmission and reduce resistive losses in discharge circuit, the high-voltage trigatron switch was replaced by TDI1-200k/25SN-P thyratron. PSS TDI1-200k/25SN-P is designed for pulse switching of capacitive energy storage in submicrosecond and microsecond duration range. In the operating state, the switch is filled with a buffer deuterium gas at the pressure of 20–60 Pa. This modification of thyratron has a hollow anode and it is intended for ringing mode operation. The operating and maximum permissible electrical parameters of PSS are shown in Table 1.

Table 1. Electrical parameters of TDI1-200k/25SN-P switch

Parameter description	Optimum value	Maximum permissible value	Actual operating value
Anode voltage, kV	3–20	3–25	5–13
Amplitude of anode current, kA	150	200	150–250
Pulse repetition rate, Hz	0.5	300	≤0.5
Switched energy, kJ	10	40	45–47
Duration of anode current pulses, μsec	0.1–100	0.1–1000	250–500
Amplitude of reverse anode current, kA	up to 90% of Ia	up to 98% of Ia	up to 90% of Ia

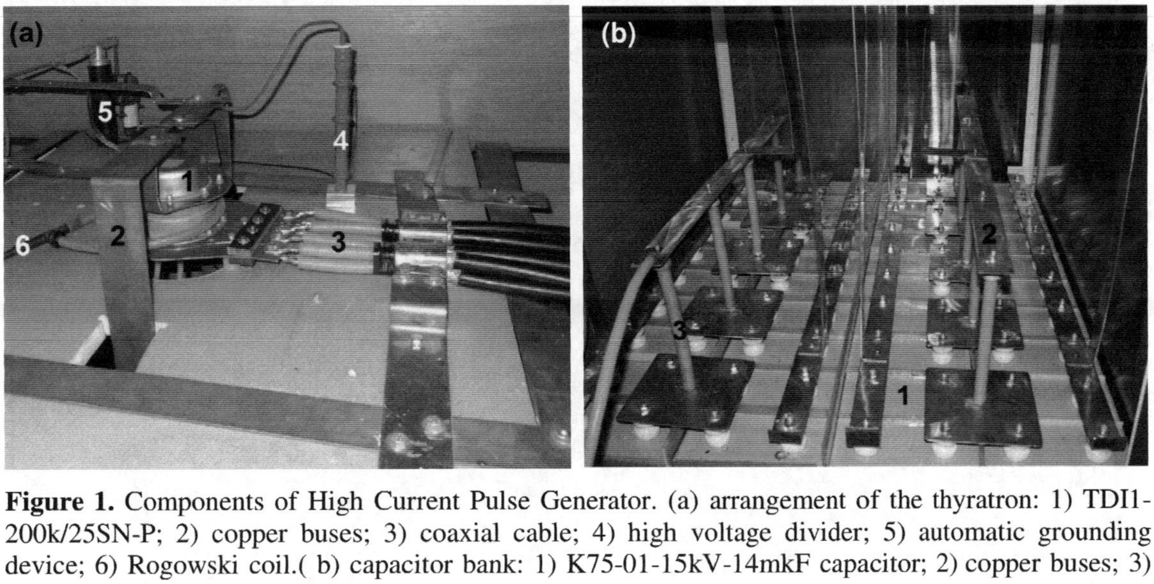

Figure 1. Components of High Current Pulse Generator. (a) arrangement of the thyratron: 1) TDI1-200k/25SN-P; 2) copper buses; 3) coaxial cable; 4) high voltage divider; 5) automatic grounding device; 6) Rogowski coil.(b) capacitor bank: 1) K75-01-15kV-14mkF capacitor; 2) copper buses; 3) fuse.

The High Current Pulse Generator includes a capacitor bank that consists of ten modules. The modules were connected by copper buses in such manner that the current path length from each module to the high voltage switch was practically equal. For reliable operation during the service life of PSS, the manufacturer recommends to provide balanced and symmetrical current feed to the flange of the switch. If the current distribution is asymmetric, then this promotes electrode erosion, which leads to early failure of the device [11]. The arrangement of pseudospark switches and capacitor bank is shown in Fig. 1.

The stored energy was transferred to load by RK50-17-17 coaxial cable. To reduce the equivalent inductance and equivalent resistance of cable the parallel connection of eight cables was used. All cables were divided into two bundles with 4 cables in each. The busses were connected in parallel to reduce the resistance and inductance of the circuit. The symmetrical electric current feed was implemented by connecting two equal sections of capacitor bank from opposite sides to the high-voltage switch (Fig. 2a). For the purpose of balanced current division, in the lower copper flange an incision was made, which divided the flange into several current feed elements (Fig. 2b).

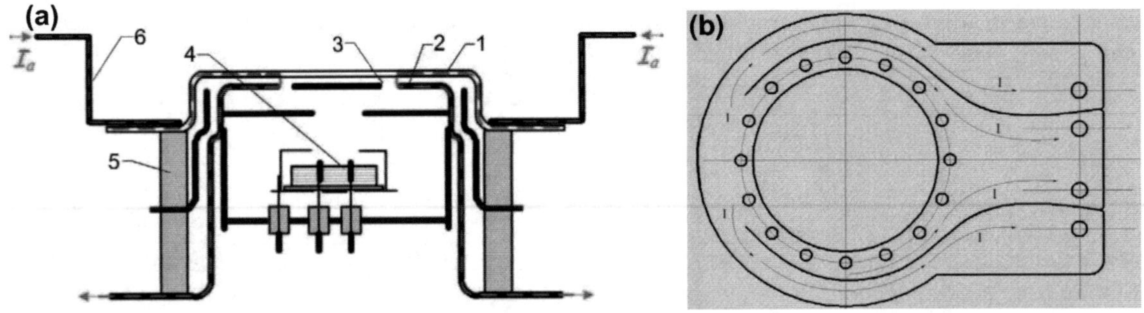

Figure 2. Connection of the pseudospark switch with bus bars. (a) PSS structure: 1) anode; 2) cathode; 3) injection holes; 4) trigger unit; 5) insulator; 6) copper buses. (b) copper flange for current feed.

3. Equipment Testing

The amplitude of current pulse is the most important parameter of electric discharge. Its value directly depends on the resistance and inductance of discharge circuit that are determined by the design parameters of the HCPG. The inductance and resistance of the electrode system can vary according to the configuration and length of the electrode. The values of inductance and resistance were determined in short-circuit experiment by the damping decrement of the current waveform at the charging voltage of 4 kV (Fig. 3a); the operating current was determined in the experiment of concrete monolith splitting at the charging voltage of 13 kV and the electrode length of 50 cm (Fig. 3b).

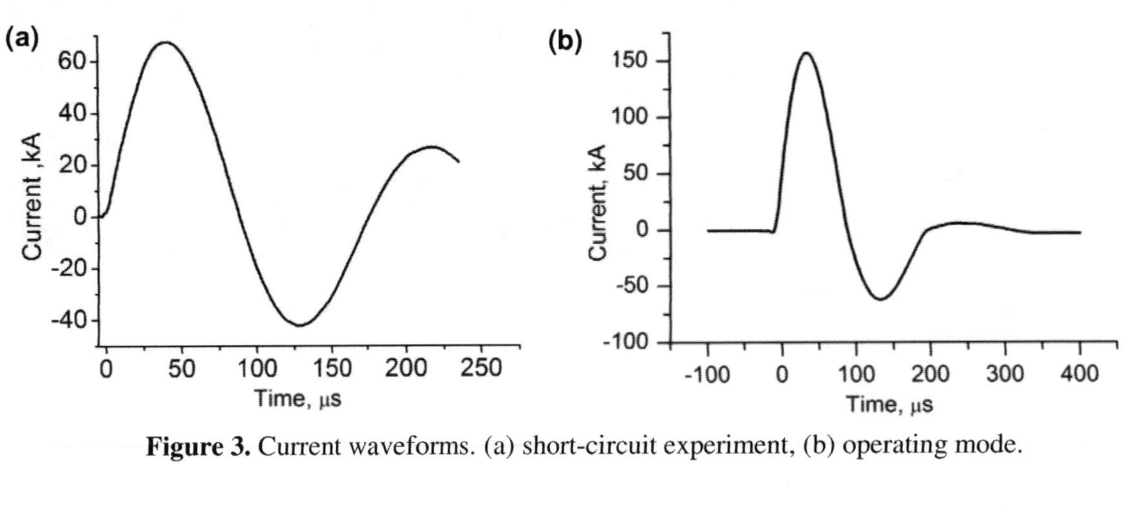

Figure 3. Current waveforms. (a) short-circuit experiment, (b) operating mode.

After a short running life, less than 50 pulses, the pseudospark switch failed to start, while there was no external visible damage. Maintenance work did not lead to the restoration of PSS normal operation. The manufacturer guarantees the satisfactory operation of the PSS and the running life of at least $5 \cdot 10^4$ coulomb of total switched charge, when the thyratron operates in modes without exceeding the maximum permissible parameters. According to Table 1, the actual values of electrical parameters significantly exceed the maximum permissible values. In the experiments of electric-discharge blast of rock and concrete, the PSS worked in the mode with the exceedance of two or more electrical parameters, which probably caused the early failure of the device.

4. PSS Internal Structure Analysis

To detect the internal damage and determine the cause of the failure, PSS was disassembled. The surface condition of the electrodes is shown in Fig. 4. Both electrodes demonstrate the surface erosion of the central region. Fig. 4a shows the cathode; Fig. 4b depicts the anode. In our opinion and according to the information from the manufacturer, this is a normal effect [11].

Figure 4. Internal structure of TDI1-200k/25SN-P

Although the energy of a single pulse was 47 kJ, which is 17.5% higher than the maximum permissible value, and the duration of current pulse was about 200–300 μsec, the overheating or deep surface melting was not observed. "Linking" of the discharge to a certain injection hole (outlined by dashed lines in Fig. 4a) also did not occur. Thus, the excess of three electrical operating parameters, current amplitude, switched energy and pulse duration in one-pulse mode could not be the reason for the early failure of the thyratron; however, this can seriously reduce the service life of the device.

Turned out, the actual reason of the thyratron failure was an asymmetric current feed. A significant damage of the wall (Fig. 4a and 4c), the gradient electrode (Fig. 4d) was identified, which were caused by the effect of electric arc. Localization of the discharge on periphery of the electrodes near the lateral surface led to its thinning and the depressurization of PSS. These processes were conditioned by the Lorentz force that pushes the discharge channel out from the central part to the periphery. Thus, the actions taken to implement the balanced current feed and uniform current collector occurred to be ineffective. Hence, this testifies the demand of more buses connected circumferentially to the thyratron's flange as shown in [11]. In future work, we will use the coaxial inclusion of PSS with the application of external conductive screens.

5. Conclusion

We have replaced the high-voltage trigatron switch that was implemented in the mobile electric-discharge blasting unit for splitting and breaking rocks and concrete blocks by TDI1-200k/25SN-P pseudospark switch. The main parameters of the discharge circuit were determined in short-circuit experiment by the damping decrement of the current curve. Due to the operation of the PSS in the mode with the exceedance of two or more electrical parameters, the lifetime of the switch reduced. As a result of the exceeding current pulse amplitude, the current pulse duration and maximum pulse energy, the electrodes of thyratron were subjected to increased erosion. However it was not the reason of such early failure of the thyratron. Damage of the switch occurred because of the concentration of discharge in the periphery of electrodes due to the asymmetric and unbalanced current feed, which ultimately lead to the depressurization of the thyratron.

Acknowledgements

This work was supported by the Federal Target Program, project No. 14.575.21.0059 (RFMEFI57514X0059).

7. References

[1] Inoue H, Lisitsyn I, Akiyama H, Nishizawa I, Drilling of hard rocks by pulsed power 2000 *IEEE ELECTR. INSUL. M.* **3** pp.19-25

[2] Silva C C M, Stellin A, Hennies W T, Costa E G Electrohydraulic Rock Blasting: An Alternative for Mining in Urban Areas 2002 *INT. J. MINING ENG.* pp.261-269

[3] Hofmann J, Th Weise H G G Pulsed Power Technologies for Commercial Material Reduction and Crushing Applications 1997 *Digest of 11th IEEE Pulsed Power Conf.* (Baltimore) pp. 203-207

[4] Bluhm H, Frey W, Giese H, Hoppe P, et al Application of pulsed power HV discharges to material fragmentation and recycling 2000 *IEEE T. DIELECT. EL. IN.* **7** pp 625-636

[5] Kuznetsova N, Lopatin V, Burkin V, et al Theoretical and experimental investigation of non-conducting materials 2011 *Pros. 17th IEEE Int. Pulsed power Conf.* (Chicago) pp 267–273

[6] Bochkov V D, Dyagilev VM, Korolev Yu D and Ushich V G High-power low-gas-pressure current switches 1998 *Instruments and Experimental Techniques* **41** No.5 pp.676-680

[7] Gribkov V A, Scholz M , Bochkov V D, Dubrovsky A V, Miklaszewski R, Karpinski L, Lee S, Lee P 2004 *J. Phys. D: Appl. Phys.* **37** pp. 2107-2111

[8] Frank K, Petzenhauser I, Blell U Multi-gap Pseudospark Switches for High Voltage Applications 2007 *IEEE T. DIELECT. EL. IN.* **14 (4)** pp. 968-975

[9] Bochkov V D, Djagilev V M, et al Sealed-off pseudospark switches for pulsed power applications (Current status and prospects) 2001 *IEEE T. Plasma Sci.* **29** No.5 pp. 802-808

[10] Yudin A, Kuznetsova N, Lopatin V and Voitenko N 2014 *J. Phys.: Conf. Series*, **552** 012028

[11] Slough J, Pihl C, Bochkov V D , Bochkov D V, Panov P V, Gnedin I N Prospective pulsed power applications of pseudospark switches 2009 *IEEE Pulsed Power Conference* (Washington) pp. 255-259

Journal of Physics: Conference Series **652** (2015) 012060 doi:10.1088/1742-6596/652/1/012060

Application of Microsecond Voltage Pulses for Water Disinfection by Diaphragm Electric Discharge

S V Kakaurov[1,4], I F Suvorov[1], A S Yudin[2], T L Solovyova[3], N S Kuznetsova[1]
[1] Transbaikal State University, Chita, 672039, Russian Federation.
[2] Tomsk Polytechnic University, Tomsk, 634050, Russian Federation.
[3] Chita State Medical Academy, Chita, 672000, Russian Federation.

[4] E-mail: blackhawkserg@yandex.ru

Abstract. The paper presents the dependence of copper and silver ions formation on the duration of voltage pulses of diaphragm electric discharge and on the pH of treated liquid medium. Knowing it allows one to create an automatic control system to control bactericidal agent's parameters obtained in diaphragm electric discharge reactor. The current-voltage characteristic of the reactor with a horizontal to the diaphragm membrane water flow powered from the author`s custom pulse voltage source is also presented. The results of studies of the power consumption of diaphragm electric discharge depending on temperature of the treated liquid medium are given.

1. Introduction

Electric discharge water processing technologies both in industry and in everyday life are increasingly used. This is clearly linked with the improvement of both the technologies themselves and the components and circuitry of the power supplies that are an integral part of any modern electric installation equipment. There are many ways of water treatment by different kinds of electrical discharges: arc, spark, corona, glow, barrier, multi-channel, edge, diaphragm discharges and their subspecies. Each discharge has its own advantages and disadvantages. For example, arc, spark, corona, glow and edge discharges generate a significant amount of H_2O_2 but have high specific electricity consumption for water disinfection [1-5]. Barrier discharge allows one to generate large amount of ozone at relatively low specific energy consumption for water disinfection. But it requires fairly complex and expensive technological equipment [6]. Diaphragm electric discharge (DED) has a number of features due to which it can be used for water disinfection. However, the lack of knowledge of the DED poses a number of questions to researchers, without answering them it is impossible to create a competitive technology. Particularly the question remains unsolved in terms of specific power consumption which is 0.35 kWh/m^3 that reduces the market attractiveness of this water disinfection method [7].

The DED water treatment technology described in [8] consists of the special switching power supply (SPS) and the reactor [9] divided by a diaphragm membrane into two chambers, each of them being equipped with metal electrodes and nozzles. The SPS of the reactor converts the sinusoidal voltage mains 380/220 V in to bipolar pulses with a pulse amplitude of up to 3 kV and a duration of from 0.8 to 16 microseconds. Method of water disinfection by the DED technology is complex and depends on many conditions. The uniqueness of water treatment by the DED is that water is disinfected due to the complex effect of hydrogen peroxide H_2O_2, copper ions Cu^{2+} and Ag^+ silver ions, atomic oxygen O and ozone O_3. The yield of bactericidal agents is affected by a number of factors: the amplitude of the voltage pulses, their duration; conductivity [10], pH and temperature of the treated water; the spatial orientation of the diaphragm membrane.

2. Experiment

Analysis of current-voltage characteristics (CVC) of DED was conducted in the static mode of the reactor operation with water temperature of 30 °C, its pH of 7.4 and conductivity of 0.45 mS/cm. The listed physical quantities were determined by water parameters tester Hanna HI 9813-6. The duration of the voltage pulses was 7 μs, duty cycle was 2.02 and a slew rate was of about 200 V/s. The parameters of the electric signals were recorded by oscilloscope Tektronix TDS 2022B. The voltage was measured by a resistive divider connected to the electrodes of the reactor. The current was calculated as the ratio of measured voltage dropped on resistance connected in series with the reactor and the active nominal value of its resistance. The experiments were carried out with the diaphragm membrane of 5 mm thickness with 72 holes and 1.2 mm of holes diameter. The volume of the reactor chamber is about 1300 cm^3.

Analysis of the dependence of the reactor power consumption of DED on water temperature was carried out under the same conditions as the analysis of DED CVC described in previous para only with constant voltage pulse amplitude of 2150 V and with changing water temperatures from 2 to 100 °C during the experiment.

The experiments of determination of the most effective pulse duration were carried out under the similar conditions as the analysis of DED CVC wherein the amplitude of the voltage pulses was set of 1970 V and the water flow mode was implemented. The model solution of distilled water and NaCl was used as the test fluid. The initial temperature of the test fluid was 20 °C and the final temperature at the end of the experiment was 21 °C. The flow rate was 4 liter/min. At each test point 50 W·h was spent that was defined by the electricity meter with accuracy class of 1.0.

The dependence of copper and silver ions formation on pH of the treated liquid medium was defined in identical conditions as the analysis of DED CVC with constant voltage pulse amplitude of 2150 V and varying pH from 2.5 to 9.5. It was done with a help of liquid conditioners for aquariums «TetraAqua» «pH Minus» and «pH Plus». The model solution was produced from distilled water, liquid conditioners and NaCl to set the conductivity at the level of 0.45 mS/cm.

3. Results and discussion

The pulses amplitude and their duration force on the penetrating ability of the acting voltage which generates a discharge in the vapor bubbles formed in the holes of the diaphragm membrane. Bubbles occur in the holes of the diaphragm membrane due to the Joule heating of water. The electric current density in the holes of the membrane in hundreds of times greater than outside. As a result, the CVC of the DED reactor powered by the SPS with bipolar pulses are highly nonlinear and has a crescent-shaped character (figure 1). Most of all it depends on steam-air mixture formation than on water resistance. In the range of voltages pulse amplitudes from 0 to 525 V the current increases in proportion to the voltage since the discharge under this voltage is absent.

Between the points of curve from 525 V to 755 V the current decrease is observed. Such effect of negative differential resistance is caused by the beginning of the DED formation. The gas-vapor mixture starts to occur in the holes of membrane and thus the resistance of the reactor increases. After the point of pulses amplitude of 755 V the CVC monotonically increases and one can see that after 1.7 kV the current starts rise faster than the voltage because of intensive formation of steam-gas mixture.

The temperature in the holes of the diaphragm membrane rises up to 5000 °K [5] which leads to nonlinear increase of conductivity.

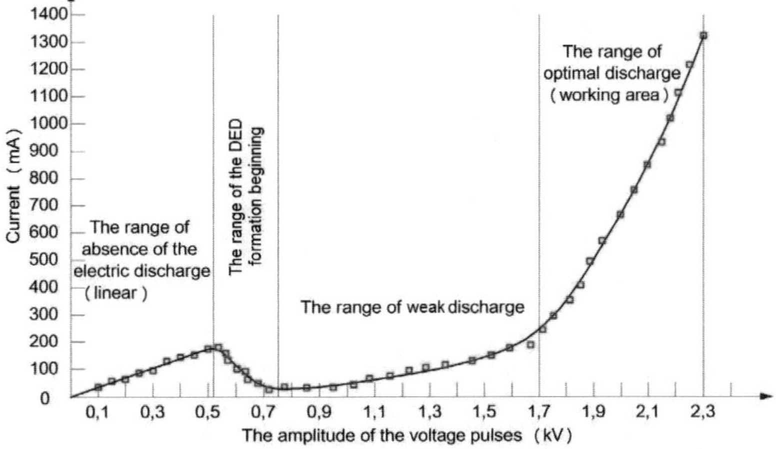

Figure 1. Volt-ampere characteristic of the diaphragm electric discharge.

The CVC of the DED has the same shape of the graph at the temperatures from 2 to 60 °C which confirms the assumption about its dependence mainly on the formation of the gas mixture and not on water resistance which depends from temperature nonlinearly. At higher liquid temperature from 60 to 100 °C more intense evaporation in the holes of the diaphragm membrane takes place and as a result, the power consumption level falls sharply without changing visual intensity of the discharge (figure 2). Based on these results the reactor with a horizontal to diaphragm membrane flow of water was designed [9] instead of the reactor where the water passes through the holes in the diaphragm membrane [7, 11]. In this case water in the upper part of the chamber does not flow and heated almost to the boiling point. Water in the lower part of the chamber moves continuously and heated slightly. As the result one has higher yield of bactericidal agents with the same power consumption [8, 9, 12].

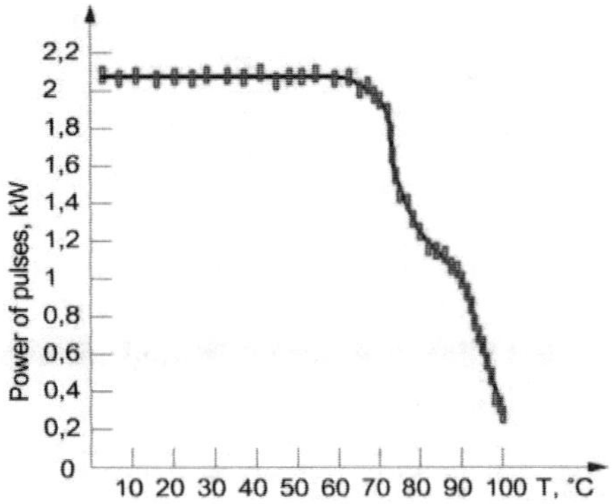

Figure 2. Dependence of the power consumption of the DED reactor from the temperature of the treated water

The yield of bactericidal agents depends on the duration of the voltage pulse. The amount of formed hydrogen peroxide H_2O_2, copper ions Cu^{2+} and silver ions Ag^{+} is changed with the pulse duration changing during the experiments. These dependences are shown in figure 3. The optimal pulse duration with the highest bactericidal agents yield is determined within the range from 4 μs to 8 μs. When the pulse duration is less than 4 μs the poor bactericidal agents yield is observed. This is due to the insufficient time of voltage exposure which is required for the development of the discharge. Bactericidal agents yield decline is also observed in the cases of pulse duration of more than 8 μs. The reason of the bactericidal agents yield lowering is the decrease of the pulse repetition frequency.

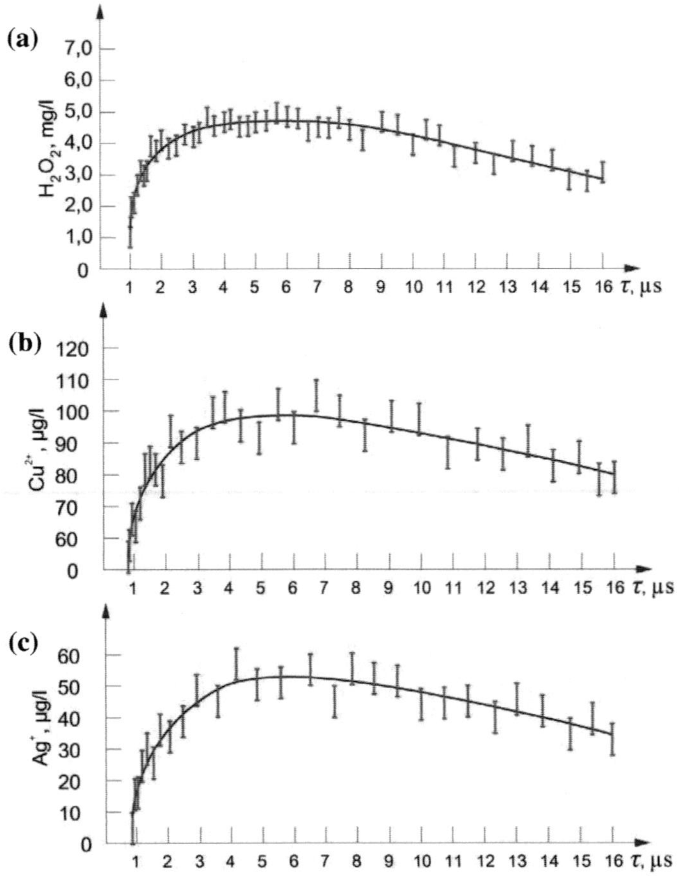

Figure 3. Dependence of the yield of hydrogen peroxide and metal ions from the duration of the voltage pulses. (a) hydrogen peroxide H_2O_2. (b) copper ions Cu^{2+}. (c) silver ions Ag^{+}.

The efficiency of the process of water disinfection is also significantly depends on pH of the treated liquids [11, 13]. However, it is difficult to determine what pH factor of liquid is better for the combined use of silver and copper electrodes: acid, alkaline or neutral. In the case of the copper electrodes using only the yield of copper ions Cu^{2+} increases with acid medium increasing (figure 4a). But on the contrary if the silver electrodes are used only, the yield of silver ions Ag^{+} increases with alkaline medium increasing (figure 4b). That is why in acid medium it is better to use copper electrodes and in the alkaline medium – silver ones.

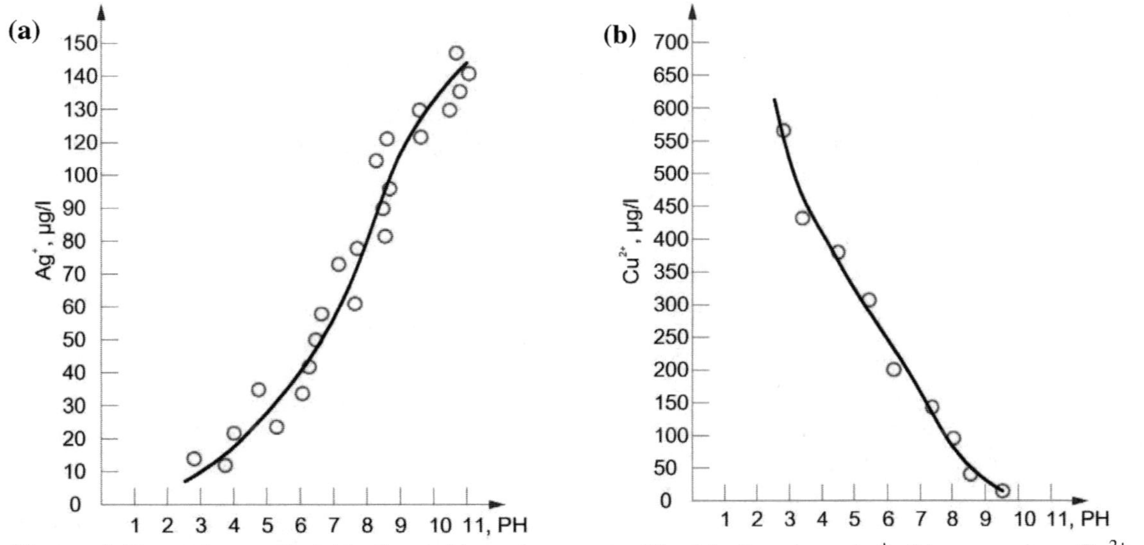

Figure 4. Dependence of yield of metal ions from water Ph. (a) silver ions Ag^+. (b) copper ions Cu^{2+}

4. Conclusions

The knowledge of the optimal pulse parameters for water disinfection by DED will help to create a more perfect and economical system of water disinfection since it is possible to reduce the consumption of electric energy in water treatment process by DED significantly. The investigated range of current pulse duration parameters is not completed yet. It is not clear what results of water disinfection in the pulse duration of less than 0.8 µs and more than 16 µs could be obtained. Dependency of bactericidal agents yield from pulse ratio (used SPS has no capability to regulate pulse ratio) is unknown. In the investigated range of pulse durations from 0.8 µs to 16 µs it is possible to allocate the range with durations from 4 µs to 8 µs. In this range there is a maximum yield of bactericidal agents with a deviation not exceeding the error of methods of determining the amount of these agents. It was found that the most productive amplitude of the voltage pulses is 2.3 kV when the thickness of the diaphragm membrane is 5 mm and the diameter of the holes in it is 1.2 mm. At a higher voltage than 2.3 kV the yield of hydrogen peroxide decreases and the method of water treatment by DED become ineffective. The dependence of the power consumption of the DED reactor on the temperature of the treated water was determined, and based on it a new type of reactor was constructed. This dependence indicates that the electric discharge is more intensive and productive at higher temperatures of liquid media. The new reactor design allows raising the yield of bactericidal agents and decreasing the power consumption.

References

[1] Tverskoj V S and Tverskoj A V 2006 Method of surface treatment: patent № 2286866 Russian Federation

[2] Piskarev I M, Ivanova I P and Trofimova S V 2013 *J. High Energy Chemistry* **47** **5** 247-250

[3] Ivannikov A A, Lelevkin V M and Tokarev A V 2002 Application of glow discharge for disinfection of water *Proceedings of the 3rd International Symposium on theoretical and applied plasma chemistry Ivanovo* pp. 463-469

[4] Hljustova A V, Maksimov A I and Panova D S 2013 The Influence of an electric discharge and oxidizing agents in aqueous solutions of a mixture of two organic dyes *Electronic processing of materials* **49** **4** 6-11

[5] Kuznecov V V 2011 Development of mathematical models of radiation processes in plasma accelerators torch capillary discharge *J. Physico-chemical kinetics in gas discharge* **11 1** 1-6

[6] Kornev Ja I 2005 Water treatment by pulsed discharge in water - air flow: thesis of candidate of technical Sciences. Tomsk Polytechnic University, Tomsk, 162 p

[7] Lapshakova K A 2009 Disinfection of domestic wastewater in small settlements diaphragm electric discharge: thesis of candidate of technical Sciences. Irkutsk State Technical University, Irkutsk, 115 p

[8] Gajsin A F, Suvorov I F, Kakaurov S V and Judin A S 2013 Electrical discharges with liquid electrodes *Bulletin of the Kazan Technological University* **16 23** 199-201

[9] Suvorov I F, Kakaurov S V, Judin A S and Rahlecova T V 2014 The device for water purification and disinfection: patent № 137284 Russian Federation

[10] Kakaurov S V 2013 Features of copper and silver ions formation for water disinfection by diaphragm electric discharge *Bulletin of the Transbaikal State University* **11** 22-27

[11] Judin A S 2010 Development of reactor and system for automatic process control of wastewater disinfection diaphragm electric discharge: thesis of candidate of technical Sciences. Irkutsk State Technical University, Irkutsk 121 p

[12] Son E E, Suvorov I F, Kakaurov S V, Gaisin A F, Samitova G T, Solovyeva T L, Judin A S and Rakhletsova T V 2014 Electrical Discharges with Liquid Electrodes Used in Water Decontamination *J. High Temperature* **52 4** 490–496

[13] Kul'skij L A 1987 *Silver water* (Kiev: Naukova Dumka) 136 p

Journal of Physics: Conference Series **652** (2015) 012061 doi:10.1088/1742-6596/652/1/012061

Enhancement of emission currents in plasma electron sources based on a low-pressure arc discharge

T V Koval[1,a], **V N Devyatkov**[2,b] **and Nguyen Bao Hung**[1,c]

[1] National Research Tomsk Polytechnic University, 30 Lenin Ave., Tomsk, 634050, Russian Federation
[2] Institute of High Current Electronics SB RAS, 2/3 Akademichesky Ave., Tomsk, 634055, Russian Federation

E-mail: [a] tvkoval@mail.ru, [b] vlad@opee.hcei.tsc.ru, [c] baohung.ng@gmail.com

Abstract. The paper reports on a theoretical and experimental study of the discharge plasma generation with an enhanced electron emission current in a plasma electron source based on a low-pressure arc discharge with a grid-stabilized plasma emission boundary. The source operates at a pressure in the working chamber of p = 0.02–0.05 Pa (Ar), accelerating voltage of up to U_a = 10 kV, and longitudinal magnetic field for electron beam transport of up to B_z = 0.1 T. The experiments show that in the mode of electron emission from the plasma, the voltage U_d between the cathode and grid electrode changes its sign. The numerical simulation demonstrates that the plasma potential and voltage U_d depend on the electric field penetrating from the acceleration gap into the discharge region through the grid meshes, and on the discharge current, gas pressure, geometric transparency of the grid, and gas kind. It is shown that the main mechanisms responsible for the increase in the discharge current and electron emission current from the plasma are associated with secondary ion-electron emission from the emission electrode and with positive feedback between the region of cathode plasma generation and the channel of electron beam transport.

1. Introduction

Plasma electron sources with a grid-stabilized plasma emission boundary provide stable generation of emission plasma and electron beams with a high current density of j_b = 1–10 A/cm^2 [1–4]. In these sources, the electrons are emitted from a partially open plasma surface in the central part of the grid meshes and through a potential barrier at their edges. The electron emission from the plasma can not only change its potential but also affect the other parameters of the plasma, resulting in an unstable discharge and high-frequency oscillations [5, 6], in a change of the plasma density and attendant growth of the discharge current. The processes occurring in the region of the emission grid electrode should be taken into account in designing of both plasma emitters and their power supplies. Understanding these processes will allow one to define more accurately the requirements on the parameters of power supplies and to determine the methods that enhance the operation stability of plasma cathodes with a grid-stabilized plasma emission boundary and increase their limiting parameters.

The paper studies the effect of a positive feedback between the discharge region and plasma anode on the generation of plasma and electron beam, discharge current, and electron beam current. The dependence of the plasma potential and voltage between the cathode and emission electrode on the discharge current and gas pressure is investigated. The mode in which electron extraction from the

Journal of Physics: Conference Series **652** (2015) 012061 doi:10.1088/1742-6596/652/1/012061

plasma cathode provides much higher amplitudes of the beam and discharge currents compared to those with no electron extraction is considered.

2. Plasma electron source

The experimental electron source based on a grid plasma cathode is described elsewhere [7]. A simplified schematic of the source is shown on figure 1. The discharge system of the plasma cathode is partially placed in a divergent magnetic field of solenoid *7*; in the emission grid region, $B_g = 25–35$ mT. The electron beam is extracted through an emission window of diameter 60 mm covered with a stainless steel grid shaped to a mesh size of 0.3×0.3 mm and is transported in drift tube *4* of diameter 80 mm to collector *5* spaced by 250 mm from grid electrode *3*. The diameter of anode *2* is 80 mm, and its length is 70 mm. Of significance in the electron source is the interrelation between the processes occurring in the plasma anode, which is formed in the beam drift space during ionization of the working gas, and the processes developing in the plasma cathode due to secondary ion-electron emission from the surface of grid electrode *3* [7–9].

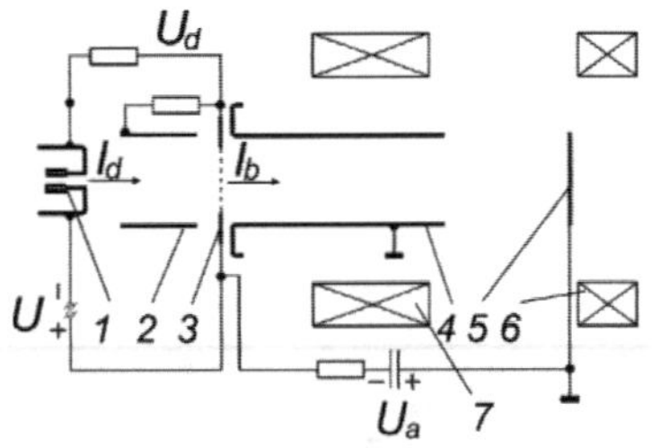

Figure 1. Electrodes of the plasma-cathode electron source: *1* – cathode, *2* – anode, *3* – grid electrode, *4* – drift tube, *5* – collector, *6*, *7* – solenoids.

Figure 2 presents typical waveforms of the discharge current I_d and voltage U_d between electrodes *1* and *3* in the mode with no electron emission (accelerating voltage $U_a = 0$). In this mode, almost the whole discharge current takes the path through grid electrode *3*. The voltage U_d measured in the discharge gap at $p = 0.07$ Pa varies in the range from 20 to 70 V for the discharge current I_d ranging from 80 to 280 A. Typical waveforms of the discharge current I_d, beam current I_b, and voltage U_d in the mode with electron emission at $U_a = 10$ kV are presented in figure 3. Figure 4 shows the voltage U_d depending on the gas pressure (in the plasma anode) for the modes with no electron emission (curves *3*, *4*) and with electron emission (curves *1*, *2*) at a discharge current of 150 and 250 A.

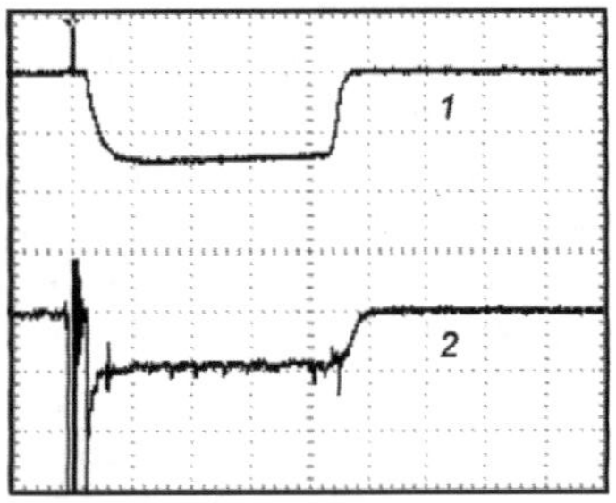

Figure 2. Typical waveforms of the discharge current I_d (*1*) and voltage U_d (*2*) at $B_g = 25$ mT, $p = 0.1$ Pa, $U_a = 0$.
Scale: 100 A/div, 50 µs/div, 40 V/div.

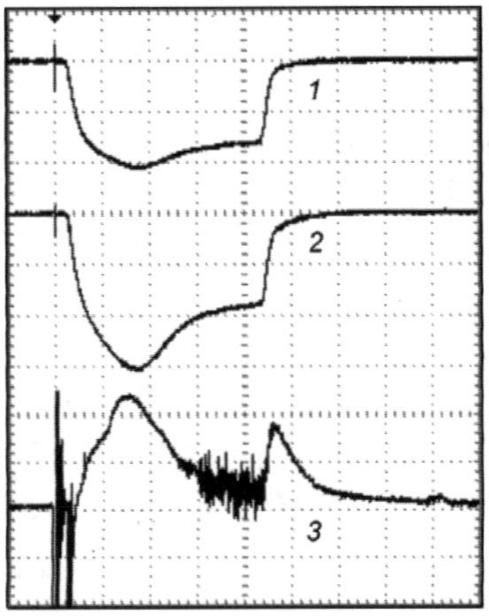

Figure 3. Typical waveforms of the currents I_d (*1*), I_b (*2*), and voltage U_d (*3*) at $B_g = 25$ mT, $p = 0.1$ Pa, $U_a = 10$ kV.
Scale: 100 A/div, 50 µs/div, 40 V/div.

3. Discharge plasma characteristics

The characteristics of the discharge plasma in the electrode system of the plasma cathode were studied using a numerical drift-diffusion model. Figure 5 illustrates the mode with no electron emission, showing the plasma density distribution in an inhomogeneous magnetic field at a gas pressure of 0.1 Pa (argon) in the discharge region; the magnetic field at the grid edge is 25 mT. It is seen that the plasma density distribution is nonuniform, and this is responsible for the radially nonuniform distribution of the beam current density.

Figure 6 shows the plasma density n_e and plasma potential φ depending on the gas pressure in the central region of the emission grid at a discharge current of 150 and 250 A. The plasma temperature T_e at a current of 150 A decreased from 18 to 10 eV with increasing the pressure from 0.05 to 0.2 Pa. The calculations take into account that the pressure in the plasma cathode is 1.5–2 times higher than that downstream of its emission grid in the beam transport region.

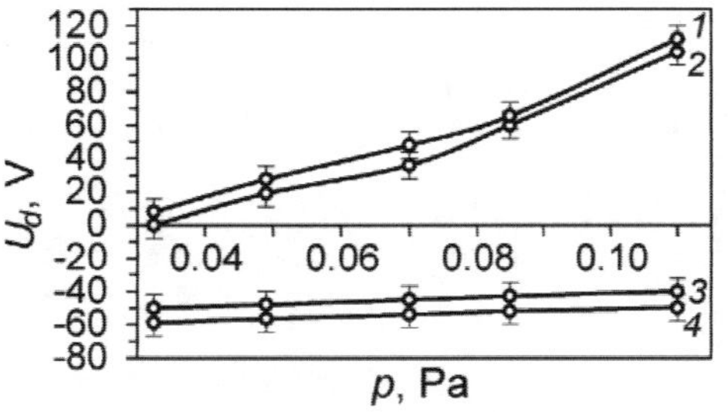

Figure 4. Voltage U_d vs the pressure p in the emission grid region: $U_a = 10$ kV (*1, 2*); $U_a = 0$ (*3, 4*); $I_d = 250$ A (*1, 4*); $I_d = 150$ A (*2, 3*); $B_g = 25$ mT.

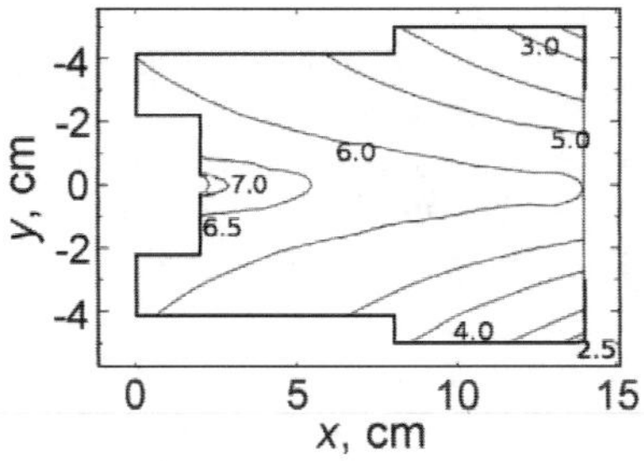

Figure 5. Plasma density distribution in the electrode system (10^{11}cm^{-3}) of the plasma cathode.

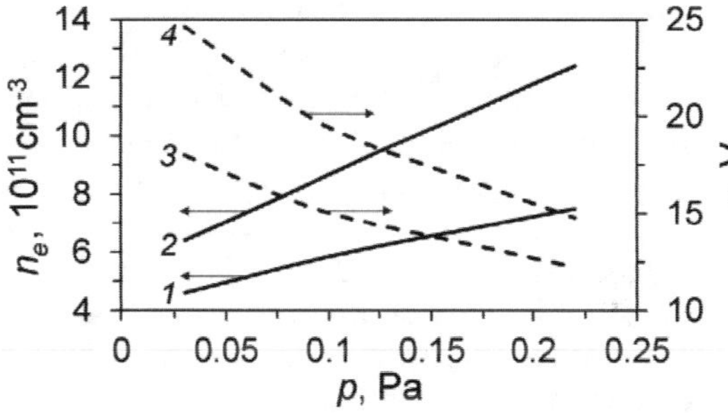

Figure 6. Plasma density n_e (*1, 2*) and plasma potential φ (*3, 4*) vs the gas pressure (argon) at $I_d = 150$ A (*1, 3*) and $I_d = 250$ A (*2, 4*)

4. Mathematical model

Let us consider the physical mechanisms responsible for the dependences of the voltage U_d on the gas pressure p and discharge current I_d (figure 4), taking into account the secondary processes at the cathode, in the discharge region of the hollow cathode, and at the emission electrode. The ion current in the plasma anode and the electron current to the hollow cathode are negligible and ignored. The discharge currents at the cathode, to the emission electrode, and in the acceleration gap are written as

$$I_d = I_{em} + I_{\sigma c} + I_i, \tag{1}$$

$$I_d = I_{eb} + j_{ch} S_e (1 - l_i / r_0)^2 + I_i, \tag{2}$$

$$I_b = I_{em} + I_{\sigma a} + I_{ia} \tag{3}$$

Here $S_e = \pi r_0^2$ is the area of a grid mesh; $S_e(1-l_i/r_0)^2$ is the electron emission area; I_{eb} is the current of fast electrons that not involved in ionization; j_{ch} is the chaotic electron current density; $I_{\sigma c,\ a}$ is the secondary electron current; I_i is the ion current transported from the acceleration gap to the discharge region through the meshes of the emission electrode; I_{ia} is the ion current in the plasma anode; l_i is the layer width separating the plasma from the grid: $l_i = (4\pi n_{ic})^{-1/2}\varphi^{3/4}(ekTe)^{-1/4}$; n_{ic} is the discharge plasma density with regard to its local increase due to bombardment of ions arrived from the acceleration gap [4]; φ is the potential fall across the layer l_i. The plasma density is determined for the equilibrium state from the equation of balance between the rate of charged particle loss and the ionization rate [10, 11]. The system of equations (1)–(3) contains a feedback through the currents in the discharge region and plasma anode.

In the mode with no electron emission, the potential difference between electrodes *1* and *3* (figure 1) is $U_d = -|U_c| + \varphi$, where U_c is the voltage in the near-cathode layer of the space charge, and the plasma potential is $e\varphi/kT_e = \ln[j_{ch}S_a/I_d]$, where S_a is the emission electrode area. In the mode with electron emission, $\varphi = DU$, where

$$\varphi = mc^2\left[\frac{4\sqrt{2\pi}I_d}{NI_A}\right]^{2/3}\left[\left(\frac{j_{ch}S_e}{I_d}\right)^{1/2} - 1\right]^{3/4}, \quad DU = r_0c_1\frac{U_a}{d(t)}\exp\left(-c_2\frac{\rho - z}{r_0}\right) \qquad (4)$$

The potential φ is determined from the condition of current continuity in the plasma cathode in the mode of emission from the open plasma surface. The potential U characterizes the acceleration gap field penetrating into the grid meshes to a certain depth z. The acceleration gap width is determined according to the 3/2 power law: $d(t) = (\sqrt{2}/9\pi)^{1/2}\left[U_a^{3/2}/j_b(t)\right]^{1/2}$; $I_A = 17$ kA; D is the dielectric constant of the emission electrode; N is the ratio between the grid area and the area of a grid mesh; for the grid with a mesh size of 0.3 × 0.3 mm and wire thickness of $\rho = 0.15$ mm, the coefficients are $c_1 = 0.3$, $c_2 = 0.44$; $z = 0$ is the grid plane from the side of the discharge plasma.

The stationary position of the electron-emitting plasma surface z_{pl} adjacent to the region of the electric field is determined by the condition of equality between the gas-kinetic pressure of the plasma $8\pi kT_en_i$ and pressure of the electrostatic field E^2. A change of the extracting voltage or plasma density causes displacement of the boundary of the emitting plasma surface z_{pl}.

A decrease in the layer thickness l_i, i.e., an increase in the emission surface area $S_e(1-l_i/r_0)^2$, breaks the condition necessary for electron acceleration: maximum electron emission current (2) in the acceleration gap may not exceed discharge current (1). The discharge current is cut off due to the processes occurring inside the discharge region. As the discharge current is switched to the acceleration gap, the voltage across the gap drops steeply and the plasma potential decreases. In experiments reported elsewhere [7], the current cutoffs were followed by reignition of the discharge with recovery of its amplitude. This is possible when the gas-kinetic pressure of the discharge plasma reaches the pressure of the accelerating electric field.

The coefficient of electron extraction from the plasma cathode is determined as the ratio of beam current (3) in the acceleration gap to the discharge current $\alpha = I_b/I_d$:

$$\alpha = 1 + (1+\gamma_a)\frac{I_{ia}}{I_d} = 1 + (1+\gamma_a)F(p,\gamma_a), \qquad (5)$$

where the function F is obtained by solving the current continuity and ion balance equations [10], γ_a is the ion-electron emission coefficient. The extraction coefficient α larger than unity is mainly due to the processes in the plasma anode that provide amplification of the beam current through ion-electron emission from the surface of the emission electrode. The coefficient α depends heavily on the gas pressure p, guide magnetic field, accelerating voltage, and also on the material and surface smoothness of the emission grid electrode.

Journal of Physics: Conference Series **652** (2015) 012061 doi:10.1088/1742-6596/652/1/012061

5. Numerical estimates

The voltage between cathode *1* and grid electrode *3* (figure 1) in the mode of electrons emission is $U_d = -|U_c| + U_f + \varphi$, where U_f is the potential of the accelerating field between the face of the drift tube and emission electrode (figure 1). Figure 7 shows calculated dependences of the voltage U_d on the gas pressure (argon) in the mode with no electron emission (*3*, *4*) and with electron emission (*1*, *2*). For the mode with no electron emission (curves *3*, *4*), the dependences were calculated using the values of the plasma density and temperature in figure 6.

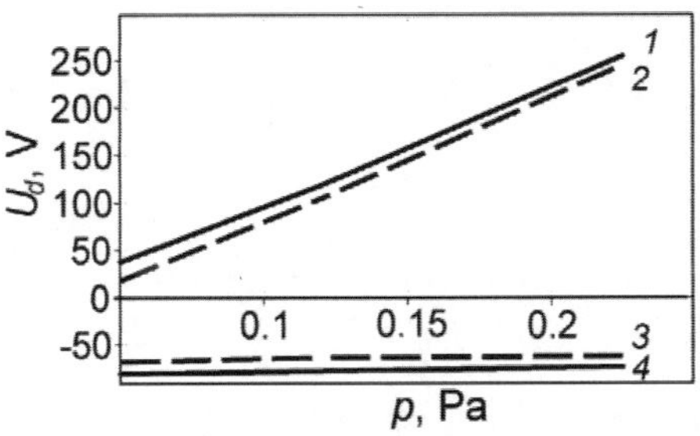

Figure 7. Calculated dependences of the voltage U_d on the pressure p (in the emission grid region): $U_a = 10$ kV (*1*, *2*), $U_a = 0$ (*3*, *4*), $I_d = 250$ A (*1*, *4*), $I_d = 150$ A (*2*, *3*).

The dependences of the voltage U_d for the mode with electron emission (figure 7, curves *1*, *2*) were calculated by formula (4). The polarity reversal of the voltage U_d in this mode (figure 4 and figure 7) owes to the influence of the accelerating field on the formation of the positive charge layer in the region of emission holes. The rather strong pressure dependence of U_d is due to the dependence of the plasma potential φ on the plasma density that increases almost linearly with increasing gas pressure.

Increasing the discharge current I_d from 150 to 250 A causes the voltage U_d to increase almost steadily over the entire pressure range (figure 4 and figure 7). From this, it follows that the emission current varies with I_d mainly due to fast electrons that take the path through the grid meshes to the acceleration gap.

The pulse shape of the beam current is defined in many respects by the processes in the plasma anode. Moreover, the rate of gas ionization during the rise time of the discharge current is higher due to the difference in the rates of supply of fast electrons, generation and loss of plasma particles. The time dependence of the voltage U_d (figure 3), as is seen from relation (4), is defined by the pulse shape of the beam current. In the experiment considered, the discharge and emission currents grow simultaneously due to the positive feedback (figure 3). At the peaks, the relative change of the discharge and emission currents is 1.4–1.5, and as the pulses are flattened, it is 1.2.

As has been shown [12], the pulse shape and amplitude of the beam current I_b depend heavily on the grid material and plane part of the emission electrode, gas kind, and operating mode of the plasma anode: for plasma electrons, the rate of residual gas ionization in the drift space increases in the mode of a reflex discharge [13]. In experiments on the SOLO electron source with a plasma cathode and grid stabilization, the total emission current (with a rectangular pulse shape) was increased up to 2 times without any apparent variation in the discharge current at a magnetic field of 3.6–30 mT and gas (Ar) pressure of 0.015–0.04 Pa. In the experiments reported in the present paper, the discharge current reveals an increase at $p = 0.1$ Pa (figure 3). Increasing the pressure and the magnetic field increases the effect of the feedback on the discharge current due to the ion current from the acceleration gap and to the ions arrived from the discharge region of the cathode to the potential well at the grid surface.

6. Conclusion

Thus, we considered the theoretical model that describes the dependence of the grid electrode potential with respect to the plasma on the parameters of the system with enhanced emission currents. The experimental and calculated dependences show a qualitative agreement, suggesting that the electrons are emitted from the partially open plasma surface in the central part of the emission holes.

The polarity reversal of the voltage U_d in the mode of electron emission owes to the effect of the accelerating field on the formation of the positive charge layer in the region of the emission holes. The voltage U_d depends heavily on the gas pressure and hence on the discharge plasma density.

The processes in the plasma anode that are associated with the rates of bulk gas ionization by the electron beam and ion-electron emission at the emission electrode define the amplitude and pulse shape of the beam current and discharge current. The positive feedback can provide a simultaneous increase in the discharge and emission currents.

Acknowledgement

The experimental part of the work was supported by the Russian Science Foundation (project No. 14-29-00091).

References

[1] Kreindel Yu E 1977 *Plasma Electron Sources* (Moscow, Atomizdat) (in Russian)

[2] Zharinov A V, Kovalenko Yu A, Roganov I S, Tyuryukanov P M 1986 Plasma electron emitter with grid stabilization *Zh. Tekh. Fiz.* **56** Issue 1 66

[3] Koval N N, Kreindel Yu E, Schanin P M 1983 Generation of pulsed electron beams with uniform high current density distribution in systems with a grid plasma emitter *Zh. Tekh. Fiz.* **53** Issue 9 1846

[4] Koval N N, Oks E M, Protasov Yu S, Semashko N N 2009 *Emission electronics* (Moscow, MGTU)

[5] Gavrilov N V, Emlin D R, Kamenetskikh A S 2008 High-efficiency emission from a grid-stabilized plasma cathode *Zh. Tekh. Fiz.* **78**, Issue 10 59

[6] Gavrilov N V, Kamenetskikh A S 2013 Self-oscillating mode of electron beam generation in a source with a grid plasma emitter *Zh. Tekh. Fiz.* **83** Issue 10 32

[7] Devyatkov V N, Koval N N 2014 Effect of electron extraction from a grid plasma cathode on the generation of emission plasma *Journal of Physics: Conference Series* **552** 0102014.

[8] Koval N N, Grigoryev S V, Devyatkov V N, Teresov A D, Schanin P M 2009 Effect of Intensified Emission During the Generation of a Submillisecond Low-Energy Electron Beam in a Plasma-Cathode Diode *IEEE Trans. Plasma Sci.* **37** 1890

[9] Devyatkov V N, Koval N N and Schanin P M 2001 Generation of high-current low-energy electron beams in systems with plasma emitters *Rus. Phys. J.* **44** 9 937

[10] Koval V, Lopatin V, Nguen Bao Hung and Ogorodnikov Alexander S. 2015 Low Pressure Discharge Characteristics in a Large Sized Hollow Cathode. Radiation and Nuclear Techniques in Material Science *Adv. Mater. Res.* **1084** 196

[11] Korolev Y D, Frants O B, Landl N V, Shemyakin I A, Geyman V G 2013 High-Current Stages in a Low-Pressure Glow Discharge With Hollow Cathode *IEEE Trans. Plasma Sci.* **41**. 8. 2087

[12] Grigoryev S V, Devyatkov V N, Koval N N, Teresov A D 2010 Enhanced Emission during Submillisecond Low Energy Electron Beam Generation in a Diode with Grid Stabilized Plasma Cathode and Open Anode Plasma Boundary *Tech. Phys. Lett.*, **36** 2 158

[13] Grigoryev S V, Moskvin P V, Teresov A D 2012 Plasma emitter based on an asymmetric reflex discharge for a submillisecond electron beam *Proc. IV Intern. Kreindel Workshop on Plasma Emission Electronics (Ulan-Ude: Russia), June 25–30,* 112

Journal of Physics: Conference Series **652** (2015) 012062 doi:10.1088/1742-6596/652/1/012062

The vacuum system for technological unit development and design

A M Zhukeshov, A T Gabdullina, A U Amrenova, Sh G Giniyatova, A Kaibar, A Sundetov, K Fermakhan
Science Research Institute of Experimental and Theoretical Physics, Kazakh National University named al-Farabi, Tole bi 96 a, Almaty, 050012, Kazakhstan

e-mail: azhukeshov@gmail.com

Abstract. The paper shows results of development of plasma technological unit on the basis of accelerator of vacuum arc and automated system. During the previous years, the authors investigated the operation of pulsed plasma accelerator and developed unique technologies for hardening of materials. Principles of plasma formation in pulsed plasma accelerator were put into basis of the developed unit. Operation of the pulsed arc accelerator was investigated at different parameters of the charge. The developed vacuum system is designed for production of hi-tech plasma units in high technologies in fields of nanomaterials, mechanical and power engineering and production with high added value. Unlike integrated solutions, the system is a module one to allow its low cost, high reliability and simple maintenance. The problems of use of robots are discussed to modernize the technological process.

1. Introduction

One of actual problems of the FIID state program is replacing of import of technologies by development of home-made hi-tech units. Among various methods of vacuum-plasma surface modification, technologies of magnetron and arc treatment can be considered as the most proved ones [1]. However, they have some shortcomings, such as small treatment area, high consuming current, heating of cathodes, etc. Recently in the world, pulsed systems of plasma treatment are actively developed which promise more effective results [2]. The treatment technique is based on short power exposure of plasma flows on surface layers of material [3]. At erosion of the electrode, definite operation mode of the unit produces plasma which content corresponds to material of the electrode and allows to use this process for deposition of metal onto various materials. Some types of new composite material coatings (multi-cathodes) cannot be obtained in other ways [4]. Due to their unique abilities, pulsed technologies allow to produce new materials of specified properties and have sensible advantages over other methods [5].

During the previous years, the project performers produced various vacuum units and developed unique technologies of pulsed vacuum-arc deposition and plasma treatment. In 2009-2010, a technology was developed for production of new materials on the basis of promising method – pulsed plasma treatment of surfaces with plasma flows obtained at electrodynamic accelerators. These technologies were tested and showed high efficiency for hardening of materials and production of coatings [6-8].

Journal of Physics: Conference Series **652** (2015) 012062 doi:10.1088/1742-6596/652/1/012062

The use of robots and other automation equipment in the plasma process can significantly limit or eliminate participation of the person. In many cases, it provides a significant economic effect. In addition to economic and holds social effect is the removal of a person from danger to his health and life of technological operations [9]. Furthermore, there are operations that cannot be executed by person (operation under vacuum). However, in our case application of the robot is due to operational needs. At work in vacuum for quality coatings production it is necessary to rotate the samples, since the plasma flow is typically propagated in the one direction. Planetary mechanism is applicable only for flat samples, but that is not enough for 3D-surfaces . Therefore, it is necessary to use a robot (hand-manipulator) to capture the details and its free movement in three axes. Furthermore, the plasma flow is not always even and homogeneous in density, so the distance of exposure should be chosen depending on the work mode of installation. The task of the robot is to obtain information about the current situation and the implementation of moving parts. Thus, the creation of intellectual and robotic systems is closely connected with the peculiarities of obtaining and maintaining a high vacuum, which requires developers to choose necessary innovated technical solutions.

2. Experimental device and measured methods

The vacuum installation (Figure 1) developed with technological device– vacuum arc accelerator (VAA). On this installation we experimentally researched the arc plasma flow and carried out automation process. At first, the peculiarities of using microcontrollers in vacuum systems were found out [6]. Taking into account these requirements, a control circuit was designed for the vacuum arc accelerator.

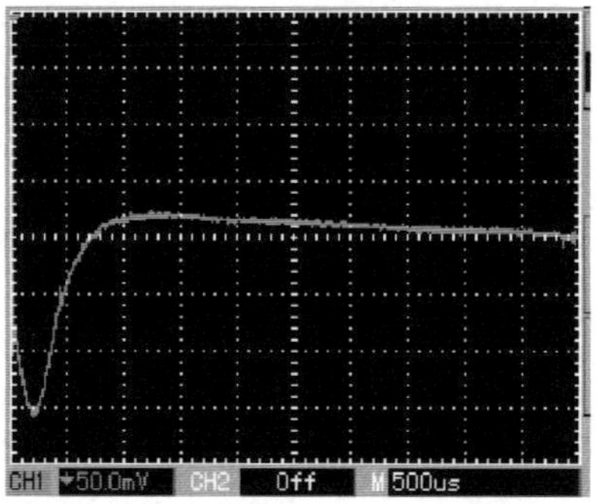

Figure 1 The vacuum-pulsed arc accelerator. **Figure 2** Vacuum arc discharge current map.

The unit is designed for production of refractory material coatings by arc spray deposition as well as experiments on development of dual technology using pulsed flows of plasma as the operating medium.

3. Experimental results and discussion

Pattern of plasma formation in pulsed arc discharge is a way to understand technological features of the unit operation. Investigations with use of current transformers showed that discharge current (Figure 2) is an asymmetric descending signal with the half-period order of 1 ms and consists of two oscillations of different polarities. The same signal of aperiodic type with attenuation was observed in power pulsed coaxial accelerator (further – coaxial) CPA-30. Thus, in principal, operation of pulsed arc accelerator does not differ from operation of classic coaxial, since the both of them use coaxial anode and cathode. However, whereas a coaxial with the long electrode accelerates plasma along the electrode, the VAA constricts it on the surface of the cathode and throws it into the inter-electrode space due to its own magnetic field. In this sense, plasma acceleration mechanism of these accelerators is the same – it is electrodynamic force. Just as the coaxial, the VAA forms plasma flow with radially symmetric distribution of current, i.e. there is a fountaining pinch, in essence. Investigations showed that the arc accelerator demonstrates small amplitude of second half-period of current and there is a series of harmonic oscillations of current with gradual attenuation in the coaxial. Obviously, the difference is due to the fact that the small voltage amplitude in the VAA accelerator sharply limits the current at inverse polarity of capacitors. Besides, it is obvious that the VAA releases ignition energy in the first half-period of current oscillation, thus, plasma formation in the second half-period is complicated.

Further on, the plasma flow density in the inter-electrode space was calorimetrically measured. The Figure 3 shows the results on measurement of plasma flow energy density distribution over the cross section of the accelerator. One can see that the energy density dramatically drops with the distance from the system axis increasing. The scheme of plasma flow formation in the VAA accelerator can be explained on the basis of the plasma flow density measurements: been emitted by the cathode, the electrons move to the anode along the curved trajectories but their drift under the magnetic field shifts the current lines outside the anode. At the same time, ions emitted by the cathode and produced by volume ionization of the gas are drawn out by electric field produced due to formation of space charge region.

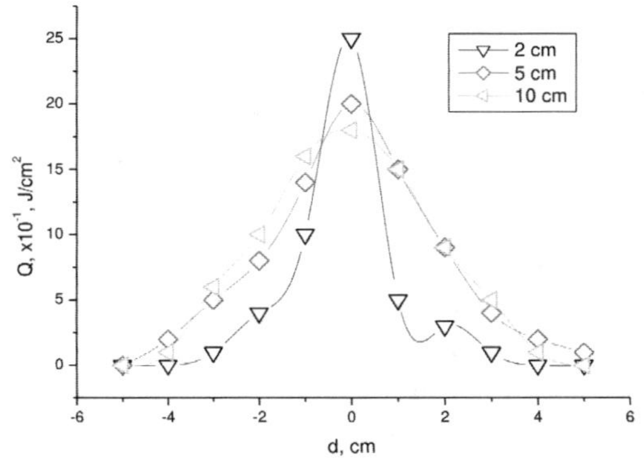

Figure 3 Flow density distribution over the cross section of the accelerator.

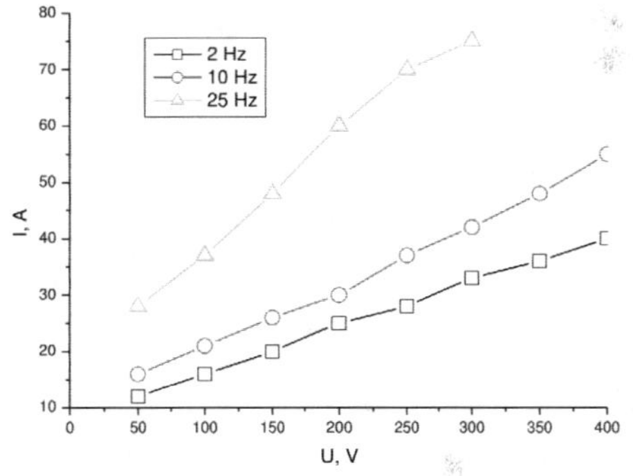

Figure 4 The dependence of discharge current on anode voltage at different frequency.

The carried out experiments defined current-voltage characteristics of VAA, i.e. anode current indications from anode voltage. A series of the relations was obtained at different frequencies of pulse sequence. The Figure 4 shows the data. The measurement was carried out in the voltage range of 50 - 400 V and capacity of the anode collector 500 mcF. The tests showed sustainable operation of VAA at anode voltage higher than 50 V and vacuum of $2 \cdot 10^{-3}$mm Hg. Operation of VAA was tested at frequencies of 2, 10 and 25 Hz with saturation at 50 Hz. After working the technology out, taking into account the features of different processes, the following scheme was suggested for automation of VAA unit (Figure 5).

In the scheme, microcontroller gets analog signal from vacuum sensor, calculates the vacuum level and begins pumping process sending digital code to executive mechanisms – valves, flow controller and inlet unit. When the vacuum level is achieved, the technological process timer starts on and the VAA accelerator gets powered. If emergency, unit shut down signal is sent through the digital gate of MC.

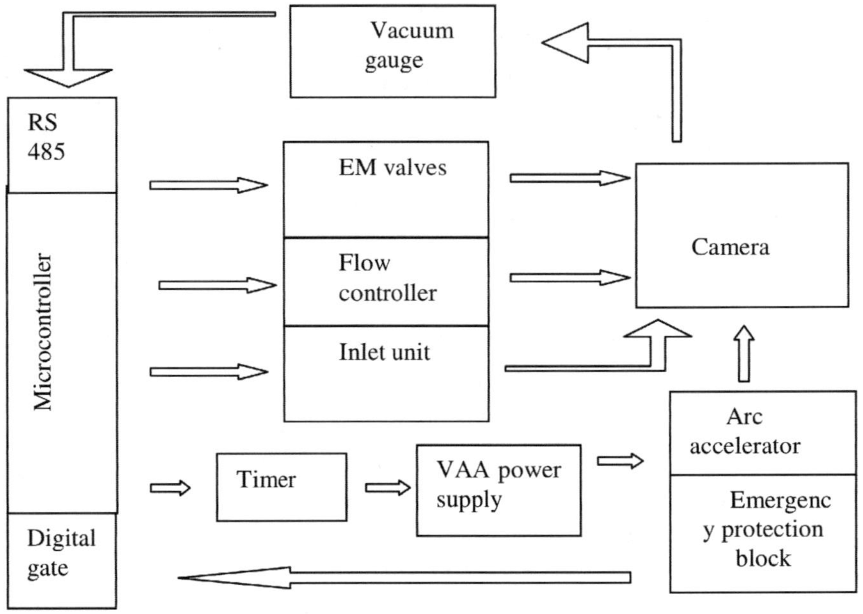

Figure 5 Block scheme of automation of vacuum unit.

Let us consider the concept of the control unit of the executive mechanisms (EM valves, vacuum pumps, inlet valves) in the Figure 5. To match microcontroller outlets and RES-22 electromagnetic relays were chosen. Those relays operate at voltage of 20-24 V and are powered directly by emitter followers of output transistors of digital gates of the microcontroller, thus improving compatibility. In its turn, the microcontroller is connected to Pad via Bluetooth adapter. The manual control is performed by mnemonic menu on the screen of the Pad. Further, work was performed on automation of the vacuum gauge pressure indicators. Vacuum sensor analog signal with the voltage of 0÷10 V was sent to the microcontroller ADC gate and followed by data program linearization according to the formula from the sensor manual. As a result, the Pad screen shows current value of pressure in the vacuum camera as a degree of ten.

4. Conclusions

Thus, there can be made the following conclusions on operation of the vacuum arc unit: the discharge current of vacuum arc accelerator is a double pulse of the current representing the sharp maximum in the first half-period with duration order of 1 ms and the second flatter maximum that can be explained by the VAA ignition feature; the plasma develops over the fanlike trajectory away from the central electrode and the maximal density of the energy is accumulated along the system axis in the form of Gaussian. Along with this, the diameter of the exhibited sample should not exceed the half diameter of the anode to effectively use the plasma energy. The tests showed sustainable operation of VAA at the anode voltage over 50 V and operating currents of 50-70 A with the vacuum up to $2 \cdot 10^{-3}$mm Hg. However, operation of VAA at the frequencies over 50 Hz is undesirable because of sharp increase of the current.

References

[1] Lieberman M A, Lichtenberg A G 1994 *John Wiley & Sons Inc., New York* 450
[2] Piekoszewski J et al. 2000 *Nucleonica.* **45** (3)193-197

[3] Tereshin V I et al. 2002 *Rev. Sci. Instrum.* **73** (2) 1- 3
[4] Yan P X, Yang S Z, Li B. and Chen X S. 1996 *J. Vac. Sci. Technol A.* **14** (1) 115-117
[5] Chebotarev V V, Garkusha I E, Langner J et al. 1999 *Problems of atomic science and technology.Series: Plasma physics*. **3**(3) 273-275
[6] Zhukeshov A.M., Gabdullina A.T., Amrenova A.U e. a. 2013 *Mat. Sci. Appl.* **4**. 35-41
[7] Zhukeshov A M, Amrenova A U, Gabdullina A T 2013 *Int. J. Mat. Sci. Appl.* **3** (2) 115-119
[8] Zhukeshov A M, Amrenova A U, Gabdullina A T, et al. 2013 *American J. Phys. App* **1** 5-9
[9] Bobrow J E, Dubowsky S, and Gibson J S 1985 *Int. J. Robotic Research*. **4**(3) 3-17.

Using the low-temperature plasma in cement production

N A Sazonova[1], N K Skripnikova[2]
[1] Angarsk State Technical Academy, 60 Tchaykovsky st., Angarsk, 665835, RU
[2] Tomsk State University of Architecture and Building, 2 Solyanaya sq., Tomsk, 634003, RU

E-mail: n.a.sazonova@mail.ru

Abstract. The calculation of the raw-material mixtures and mineralogical composition of the cement clinkers which are synthezed on their base taking into account the disbalanced crystallization of the melting and glassing under conditions of the low-temperature plasma was performed. The difference of the actual values from the calculated ones is 0.69–3.73 %. The composition which is characterized as the saturation coefficient 0,88; the silicate module – 3.34, the alumina module – 2.52 in melting of which the alite in amount 78.7 %; $3CaO{\cdot}SiO_2$ – 4 %; $3CaO{\cdot}Al_2O_3$ – 9.8 %; $12CaO{\cdot}7Al_2O_3$ –2.9 %; CaO_{free} – 1 % formed using the lime-stone from the quarry «Pereval» in the town of Slyudyanka and the clay from the deposit «Maximovski» in Irkutsk Region is considered as the optimal one. The structure of the melted clinker is represented as the metastable minerals of alite in the lamellar form with the dimensions up to (3–20)×(80–400) μm and the ratio of length to width 26.6–133. The elongated crystal form may stipulate the high cement activity based on the melted clinkers, which is 82.7–84.2 MPa. Validing the revealed high activity of the viscous substance was confirmed by the results of the scanning electronic microscopy, X-ray phase analysis, with using of which the quantitative and qualitative analyses of the clinker minerals having the deformed crystalic lattice; were performed the morphology of the minerals in the clinker and cement stone, received on its ground, was studied.

1. Introduction

Using the low-temperature plasma (LTP) is known to be one of the progressive methods of physico-chemical process intensification allowing to use liquid-phasic reactions instead of hard-phasic ones in the age of high-speed technologies. Using the high-concentrated heat streams in the production of cement is stipulated by developing the effect of «thermal stroke», which may provide a high reaction ability of raw-material mixture, intensify the decarbonization processes, promote the increase in the amount of the liquid phase up to 90–100 % as well as the decrease in viscosity of (the liquid phase) melting by 27.38–83.93 times relatively to the traditional technologies [1], the complete assimilation of the mineral-forming oxides and, as the consequence, the high clinker quality. Along with this, the plasmic technology allows to solve a number of the actual problems, correlated with the decrease in quality of the raw-material components, use the technogenic wastes; the production of the low-tonnage sets of special cement types, which is considered as the unprofitable task based on the objective technological situation for the traditional technologies. However, it is necessary to take into account that the processes under conditions of the low-temperature plasma are considered as the liquid-phasic ones and characterized by the disbalance which may promote the appearance of the additive amount of the glass, CaO_{fr} and C_3A, $C_{12}A_7$. The disbalanced crystallization is revealed in the absence of the reactions between the crystals and the remaining melting [2]: as well as glassing the remaining liquid phase. This influences on the mineralogical composition of the synthesed cement clinker and, as the consequence, on the calculation of the component composition of the mixture. In this connection the represented work is devoted to the calculation features of the optimal charge composition taking into account the disbalanced crystallization and glassing of the liquid phase in plasma-chemical synthesis of cement clinker and studying the features of the received viscous substance based on this aspect.

2. Technological regimes and study methods

The definite succession of preparing the mixture consisting in pre-dosing the components, the following grinding up to the fraction not more than 160 mcm and homogenization is considered as the distinctive feature of the technological schema in cement production based on the plasma-chemical technology. Moreover, the granulation or briquetting the raw-material mixture which is necessary to remove its blowing in exposure to the high-concentrated heating streams, is considered as the necessary technological stage. Thermal processing the prepared granulated mixture is performed using the experimental installation – the plasma-chemical reactor [3], the temperature in which is 3000 °C. Cooling the samples was performed using the air. The time period of isothermal self-control under conditions of the low-temperature plasma may change from 10 to 12 minutes. Then the melted synthesable cement clinker was grinded in the planetary ball miller PQ-N2 using 3 % gypsum. The instrument PSCH-12 was used to determine the specific surface of the viscous substance.

Studying the raw-materials and the synthesable cement clinker under conditions of LTP was performed based on the chemical analysis method accordingly to the claims of GOST 5382–91 «Cement and materials for cement production. Methods of chemical analyses». Moreover, the roentgen-phasic analysis (RPhA), the scanning electronic microscopy (SEM) were performed. The diffractograms were received using the X-Ray instrument Shimadzu XRD-7000 X-ray Diffractometer. The survey was performed on the copper Cu-K_α radiation with the step 0.05°, in the angle range 3–80°. The qualitative identification of the cement clinker phases and the cement stone was performed based on the program Match, the quantitative – Reflex based on the method by Rietveld [4]. The data base ICDD PDF-2 was used for the qualitative phasic analysis. SEM was performed using the equipment JEOL JIB-Z4500 and Phenom Pro.

3. Raw-material components and mixtures

In the studies the marmoreal lime-stone from the quarry «Pereval» in the town of Slyudyanka and the clay from the deposit «Maximovski» in Irkutsk Region were used. The chemical composition of the rocks studied is represented in table 1. The content of the main oxides in the raw-materials is traditional one [5], which allows to use the lime-stone as the carbonate and the clay – as the alumosilicate components for the synthesis of the cement clinker.

Table 1. Chemical compositions of primary raw-materials.

Component	The content of the main oxides (mass. %)						Sum
	SiO_2	Al_2O_3	Fe_2O_3	CaO	MgO	L.O.I.	(mass. %)
Marmoreal lime-stone	6.92	2.56	0.94	48.96	0.66	39.97	100.00
Clay	62.86	15.42	8.07	1.93	2.38	9.35	100.00

The calculation of the mixtures was performed based on the formulas [6]. The saturation coefficient (SC) of the raw-material mixtures may change from 0.8 to 1.0 (relatively, the mixtures #1 and #21) with the step 0.1. The prognosed mineralo-composition of the cement clinker was determined taking into account the features of the disbalanced crystallization of the melting in dependence on placing the figurative points of the compositions on the concentration triangle (figure 1).

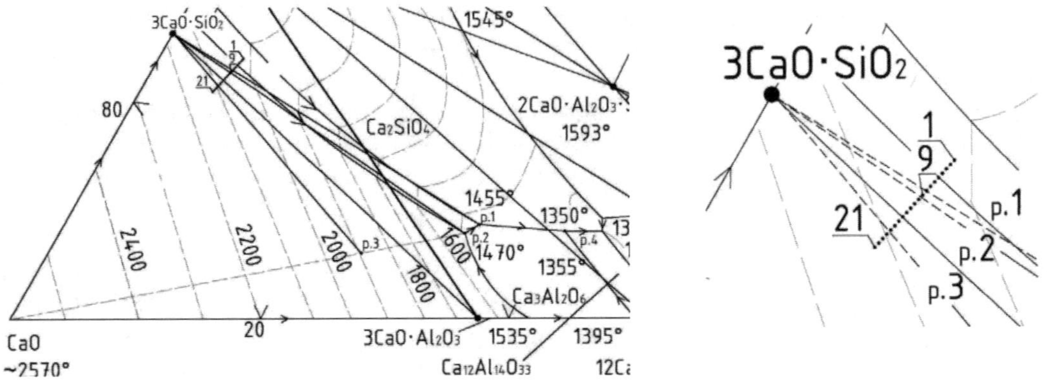

Figure 1. Part of state Diagram CaO-SiO_2-Al_2O_3. Placing the figurative points of the raw-material mixtures #1–21.

Three main aspects are considered: over the line C_3S – p.1; between C_3S – p.1 and C_3S – p.2; under C_3S – p.2. Re-calculation of their composition for the system: CaO, SiO_2 и Al_2O_3 was performed to determine the coordination of the figurative points on the state diagram. With this aim, the translated coefficients, introduction of which was grounded according to the rule the rapid-melted oxides may decrease in the temperature of material melting based on their equivalent weights [7], were used.

4. Results discussion

The calculation results have shown that 79.1–82.9 % of the marmoreal lime-stone and 17.1–20.9 % of the clay are needed for the synthesis of cement clinker under conditions of LTP. Moreover, the module characteristics of the mixture may change in the following ranges: saturation coefficient from 0.8 to 1.0; the silicate module (n) – 3.31–3.38; the alumina module (p) – 2.51–2.55. In the disbalanced crystallization in the cement clinker the alite content may exceed the tradition values 0.9–24.8 %; belite less 23.74–2.74 %; C_3A – 2.47–0.36 % (figure 2).

It was found that the synthesis of the maximal alite amount (more that 60 %) under conditions of LTP is possible with using all the calculated mixtures – #1–21; in the balanced crystallization only based on #10–21. The increase in the phases in different crystallizations may depend on the coordination of the figurative points of the clinkers on the state diagram (figure 1): in their placing over the connected line 3CaO·SiO_2 – p.1 (the compositions #1–8) – the continuous exceeding of the content of tricalcium silicate (24.8 %) relatively to the balanced conditions was observed. It is connected with that the reactions of C_3S with the melting in the p.1 may practically absent and the resting liquid consisting of C_2S, $C_{12}A_7$, C_3A may be crystallized independently. The saturation coefficient is, moreover, higher than SC in the balanced conditions by 1.527 times. As the consequence, the alite amount in the melted clinker may exceed the mineral content in the clinkers of the analogous composition but crystallized with balance.

The composition #9 placed between the lines C_3S – p.1 and C_3S – p.2 may have the increased by 23.88 % amount of alite in connection with that the crystallization before point 1 may go through the line 2–1 without p.2. Calcium aluminates may by formed not only from the liquid (p.1), the crystallization of C_3A is possible up to this point (along the line 2–1 in changing the composition of the liquid phase). The phasic composition of the products of disbalanced crystallization is represented as C_3S, C_2S, C_3A, $C_{12}A_7$. The clinkers under the line C_3S – p.2 may have both the increased by 1.77–20.99 % (the compositions #10–17) and the decreased by 0.84–8.47 % (mixtures #18–21) alite amounts. This is connected with that the crystallization may go through p.2 – the invariant point, in which the reactions between the calcium oxide and the melting may be realized with the temperature 1470 °C. Under conditions of LTP the reactions in p.2 doesn't realize and the resting melting may be crystallized independently along the line p.2–p.1–p.4 (the point of the field intersection of the primary crystallization C_2S–C_3A–$C_{12}A_7$). As the consequence, such compounds as C_3S, C_2S, C_3A, $C_{12}A_7$, CaO_{fr} may present in the disbalanced crystallization. Moreover, the absence of balance in p.1 may increase in the saturation coefficient and in p.2 – decrease. The line C_3S – p.3 is the border direct one, representing the geometric point place which is satisfactory to the balance condition of the saturation coefficients in disbalanced and balanced crystallization. In this connection, the compounds over this line may have the increased amount of alite, under it – decreased one. From the point of view of the disbalanced crystallization, the composition of clinker in which the free calcium oxide may absent and have the maximal alite amount is considered as the optimal one. This mixture is revealed based on the point coordinates on the line C_3S – p.2 or near it. The composition #9 (SC=0.88; n=3.34; p=2.52) in which the alite content 82.43 % may be considered as the actual one (figure 2, d). The composition after melting under conditions of the high-concentrated heat streams was objected to the physic-chemical studies.

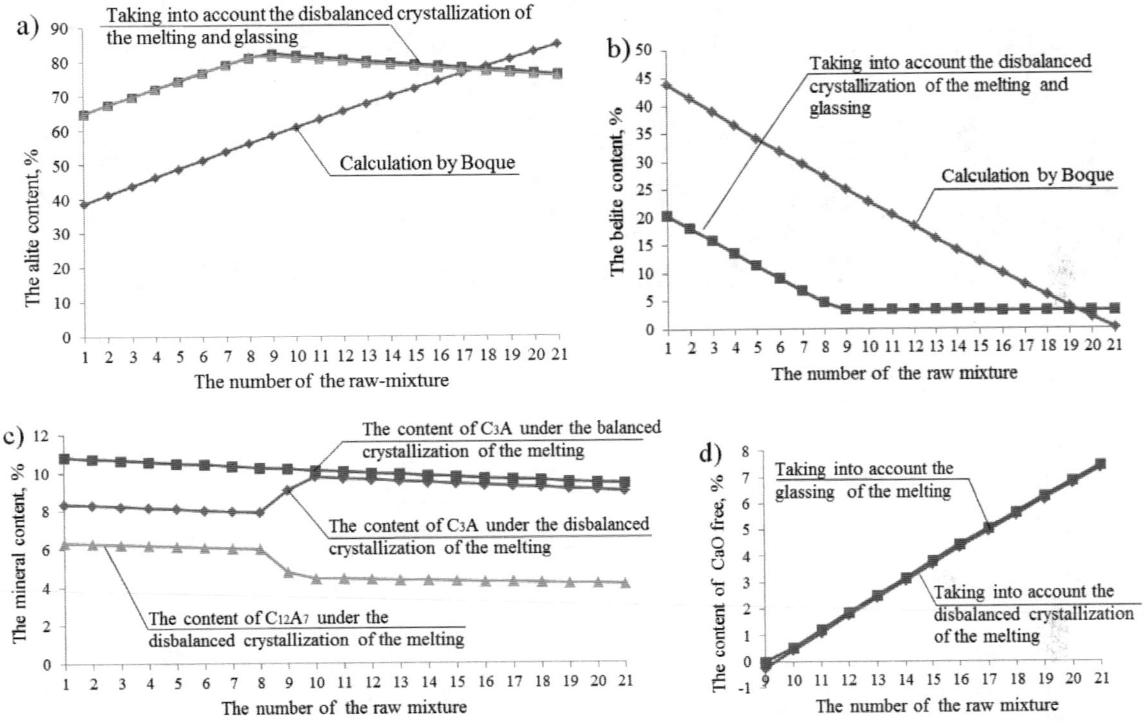

Figure 2. Calculated content in melted cement clinker: a) alite; b) belite; c) $3CaO{\cdot}Al_2O_3$ and $12CaO{\cdot}7Al_2O_3$; d) free calcium oxide (CaO_{fr}).

It was found that the mineralogical composition of the melted clinker #9 is presented by the alite phases in the amount 78.7 %; belite – 4 %; $3CaO{\cdot}Al_2O_3$ – 9.8 %; $12CaO{\cdot}7Al_2O_3$ – 2.9 %; CaO_{fr} – 1 %, which is declined from the calculated value (0.69–3.73 %). While the difference between the actual values of the mineralogical compositions and the received data (calculation by Boque) is 0.42–20.99 %. This may testify the significance of the calculation performed taking into account the disbalanced crystallization of the melting and glassing. The morphology of the clinker minerals formed under conditions of the low-temperature plasma is interesting. The instaneous reactions of clinker formation together with the high temperature gradient in cooling the melting may stipulate the synthesis of the fine-disperse matrix sample model [8] which is monadoblastic one. The structure of the melted clinker (figure 3) is presented by the metastable minerals of alite in the lamellar form with the dimensions up to (3–20)×(80–400) μm and the ratio of their length to width 26.6–133. The elongated crystal form may stipulate the higher cement activity based on the melted clinkers because the more rapid developing sides (in the crystal growth process) are less stable in hydration. The most minerals are fine-dispersive ones: up to 85 % of alite may have the width less than 4 mcm and up to 65 % of belite may have the identical dimensions (figure 4). Along with this, some structure elements of the given compound are more deformed relatively to C_3S (figure 5): alite has the formulae $54CaO\cdot 16SiO_2\cdot Al_2O_3\cdot MgO$ and it is formed in substation of two ions Si^{4+} for two ions Al^{3+} and introduction of Mg^{2+} into interknitting of the lattice. This may increase in the hydraulic activity of the viscous substance received based on the melted clinker [9, 10]. The intensive heat release which exceeds the traditional one by 1.3–4.5 times in the process of cement hydration with the synthesis based on plasma-chemical technology is observed.

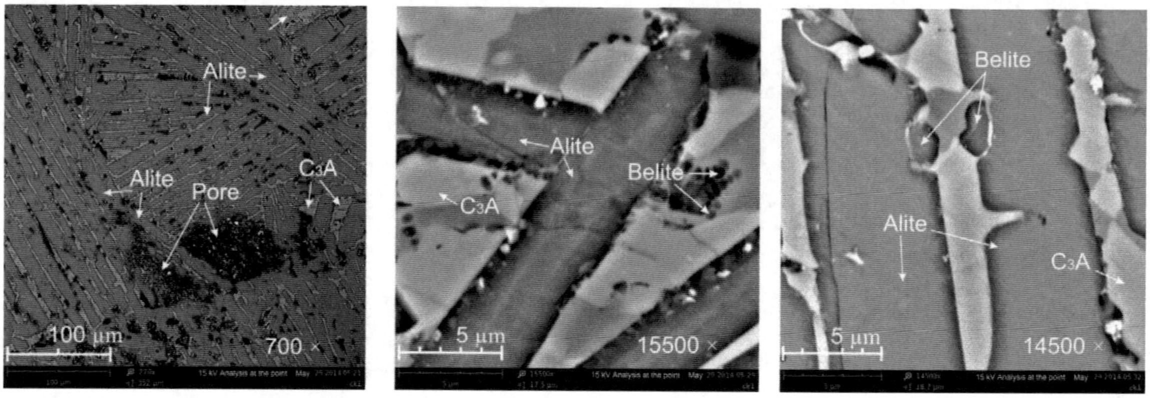

Figure 3.Microstructure of cement clinkers synthezed under conditions of low-temperature plasma based on the mixture #9.

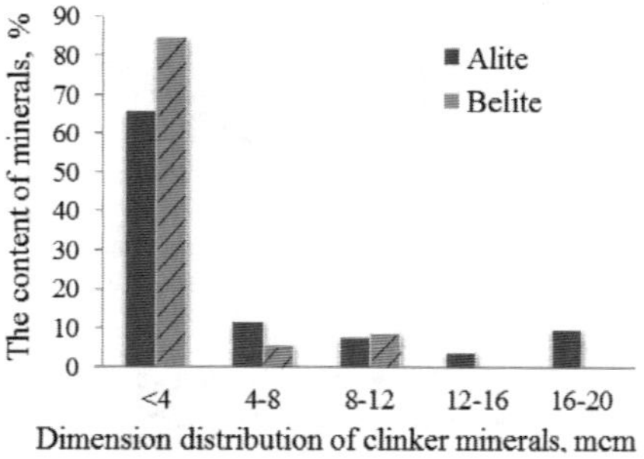

Figure 4. Dimension distribution of clinker minerals.

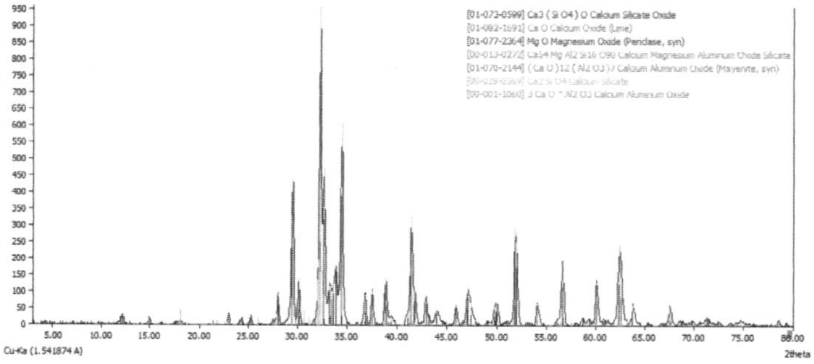

Figure 5. Diffractogram of cement clinker produced under conditions of low-temperature plasma based on the mixture #9.

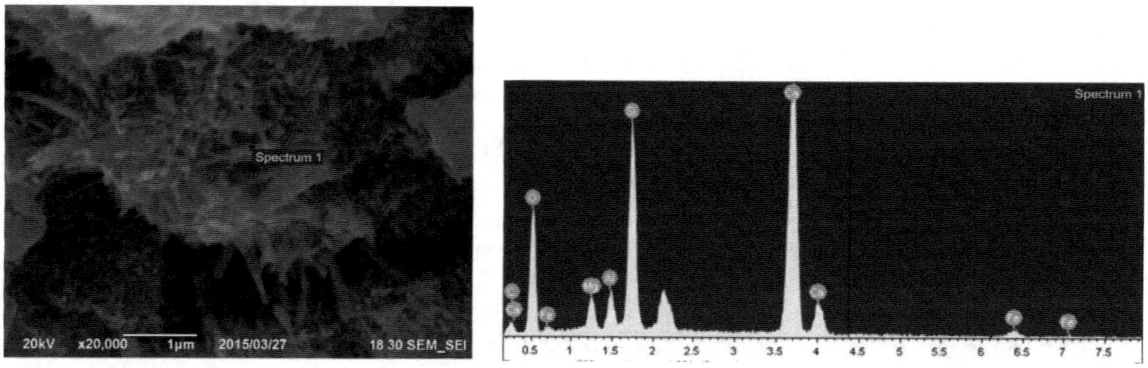

Figure 6. Scanning electronic microscopy and spectral analysis of cement stone in age of 7 days.

The cement gel, portlandite, and the additive nidi of the new-formation growth having the colomn-like habitus are clearly revealed in the structure of the cement stone produced based on the melted clinker. The presented hydrosilicates of calcium (figure 6) are non-traditional ones for the cement systems. The appearance of the analogous compounds is realized in the introduction of nano-crystalic modificators. However, the absence of the additives in the system may testify the type influence of the viscous substance used – the melted cement clinker, in hydration of which the new formations are formed which may cover the available pores and microcracks. This allows to increase in the pressing strength of the samples up to 82.7–84.2 MPa.

5. Conclusions

The study results have shown that in disbalanced crystallization and glassing of the cement clinker synthesed under conditions of the low-temperature plasma (LTP) based on the raw-material mixtures using 79.1–82.9 % of the lime-stone from the quarry «Pereval» in the town of Slyudyanka and 17.1–20.9 % of the clay from the deposit «Maximovski», the sample synthesis which based on the mineral content of C_3S may be considered as the alite (#1–21) and the traditional (#1–9) clinkers based on the rock used, is possible. The composition #9 which is characterized by SC=0.88; n=3.34; p=2.52; and has no CaO_{fr} and represented as alite, the calculated amount of which is 82.43 %, is considered as the optimal composition. The actual mineralogical composition of the melted clinker of #9 is represented as the following compounds: $3CaO{\cdot}SiO_2$ in the amount 78.7 %; $2CaO{\cdot}SiO_2$ – 4 %; $3CaO{\cdot}Al_2O_3$ – 9.8 %; $12CaO{\cdot}7Al_2O_3$ –2.9 %; CaO_{fr} – 1%, these values decline from the calculated ones 0.69–3.73 %. This may confirm the significance of the calculations performed. The structure of the melted clinker is monadoblastic and represented as the metastable minerals of alite with the lamellar form and dimensions up to (3–20)×(80–400) μm and their ratio of length to width 26.6–133. Alite and belite in the amounts 85 and 65 %, relatively, may have the width less than 4 mcm. Hydration of the viscous materials shows the high intensity which is followed by the heat release which exceeds the traditional one by 1.3–4.5 times. The structure of the cement stone is represented by the portlandite, the tobermorito-like gel and the new formations having the colomn-like habitus, they are non-traditional for the cement systems. This allows to increase in the pressing strength of the samples up to 82.7–84.2 MPa.

References

[1] Sazonova N 2011 *Plasma chemical synthesis of cement clinker* (RU, Tomsk: TPU) p 150
[2] Bobrov B 1968 *Processed clinker produced in electric arc furnaces, and cements based on them* (Russia, Moscow: publishing house of the construction) p 240
[3] Volokitin G, Skripnikova N, Pozdnjakova N 2008 High-temperature method of manufacturing the cement clinker using low-temperature plasma and electric warm up (Joule heating) *Vestnik of TSUAB* **4** pp 106–112
[4] Saout G, Kocaba V, Scrivener K 2011 Applicator of the Rietveld method to the analysis of anhydrous cement *Cem. Concr. Res.* vol 41 **8** pp 133–148
[5] Luginina I 2004 *Chemistry and chemical technology of inorganic cementing materials* (Russia, Belgorod: Belgorod State Technological University BSTU named after V.G. Shukhov) volume I p 204
[6] Boqye R 1965 *The Chemistry of Portland Cement* (New York) p 326
[7] Kukolev G 1974 *Book of problems in physical chemistry of silicates* (Russia, Moscow: Publishing house of the construction) p 236
[8] Skripnikova N, Sazonova N 2013 Features of nanostructured matrix model of cement clinker under plasmochemical synthesis *Vestnik of ISTU* **8** pp 33–37
[9] Yamaguchi G, Uchikawa H, Takagi Sh and Koike H 1962 *Yogyo Kyokai shi* **70** pp 147–152
[10] Sazonova N, Skripnikova N, Lutsenko A, Novikova L 2015 Cement clinker structure during plasma-chemical synthesis and its influence on cement properties *IOP Conf. Series: Materials Science and Engineering* **71** 012018

High-current discharge initiation in a vacuum diode

A A Zherlitcyn and B M Kovalchuk

Institute of High Current Electronics, 2/3 Akademichesky ave., 634055, Tomsk, Russia

andzh@oit.hcei.tsc.ru

Abstract. The research results of forming a low-resistance high-current discharge in the vacuum gap by means of forced initiation of the cathode spot by auxiliary discharge are presented. The discharge with current of more than 30 kA at the current rise time of ~100 ns is formed in a gap of 2.5 cm. In a next phase of the current interruption the energy of 180 J is dissipated in the diode gap.

1. Introduction

One way to generate an electron beam is to use the effect of current interruption in the high-current vacuum discharge. For the discharge ignition at voltage below the breakdown voltage and at the residual gas pressure ~10^{-5} torr in the gaps of 1–10 cm it is necessary to originate the cathode spots and to fill the discharge gap with plasma. Surface breakdown [1, 2], auxiliary discharge with pulsed gas puffing [3], laser ablation [4, 5] are used for plasma generation and high-current discharge ignition in the vacuum diodes.

The high-current discharge in a vacuum can also provide metal vapor flow from the cathode spot of the auxiliary discharge [6, 7]. In [7] vacuum low-resistance discharge and subsequent current cutoff were done in the gap of 12 cm. Current amplitude was about 8 kA at the current rise time of ~1 μs.

In the present paper the cathode flare of auxiliary vacuum discharge is tested as a plasma source for e-beam diode. In addition to the implementation of a low-resistance discharge phase with current of more than 30 kA in the diode, the task was effective transfer energy to electron beam in the phase of the current interruption.

2. Experimental arrangement

Electrical scheme and schematic design of the discharge system are shown in Figure 1. Electrical scheme consists of two discharge circuits based on capacitive storages C_1 and C_2, triggered with time delay t_d.

The first circuit supplies the auxiliary discharge. The discharge is ignited between the rod-shaped electrode (1) having a diameter of 1.5 mm and the cylindrical electrode (2) with an inner diameter of 12 mm. Electrode (1) material is copper, electrode (2) material is stainless steel. The discharge gap between (1) and (2) is d_1=3 mm. A negative polarity voltage pulse is applied to the electrode (1) through coaxial line (3) from the capacitor C_1=8 nF, switching by spark gap S_1. Capacitor C_1 charge voltage is 80 kV. The circuit damping resistance is R=10 Ω.

The second circuit supplies the main discharge in the diode gap d_2=25 mm. The circuit is based on capacitive storage C_2=80 nF with a charging voltage of 80 kV. Negative polarity pulse is applied to the electrode (2) from C_1, switching by S_2 with a time delay t_d relative to S_1, whereby electrode (2) becomes the cathode relative to the grounded plate electrode (4). Short circuit with a ferromagnetic core (5) is connected in parallel with the gap d_2. Short circuit allows to have a ground potential at the coaxial (3) input. Core (5) limits the current in the short circuit during the C_2 discharge.

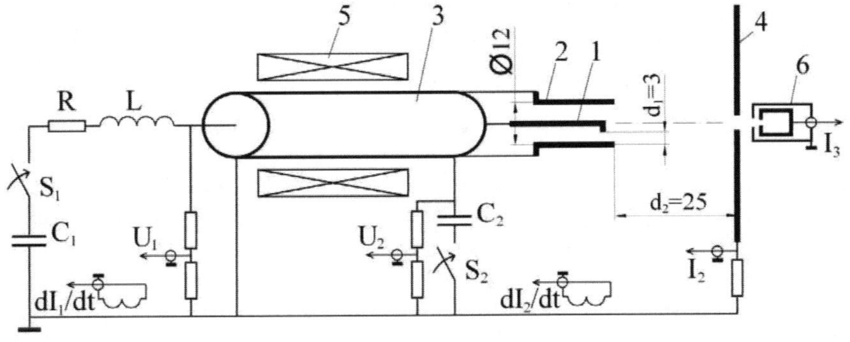

Figure 1. Schematic arrangement: 1– cathode of the auxiliary discharge; 2– anode of the auxiliary discharge and cathode of the main discharge; 3– coaxial line; 4– anode; 5– ferromagnetic core; 6– Faraday cup.

Voltages U_1, U_2 are measured by resistive dividers. The derivative of the auxiliary discharge current dI_1/dt is measured by inductive shunt. The current I_2 in the gap d_2 and its derivative dI_2/dt are measured by resistive shunts and inductive sensor, respectively. In addition, the current I_3 from the auxiliary discharge toward the electrode (4) is recorded by Faraday cup (6). Faraday cup collector is made of graphite. The charged particles flow on the collector is collimated by hole with diameter of 0.6 mm.

The discharge chamber is pumped out by turbomolecular pump to a residual pressure of ~10^{-5} torr.

3. Experimental results

3.1. The auxiliary discharge

Figure 2 shows the waveforms of the total current I_1 and auxiliary discharge voltage across the gap d_1. The voltage is calculated from measured voltage U_1 with taking into account inductive drop on the section between the voltage divider and the end of the electrode (1). The inductance of this section was determined in short-circuit test. The discharge current I_2 is shown in Figure 3. The current density at the anode (4) calculated from Faraday cup current I_3 is shown in Figure 4.

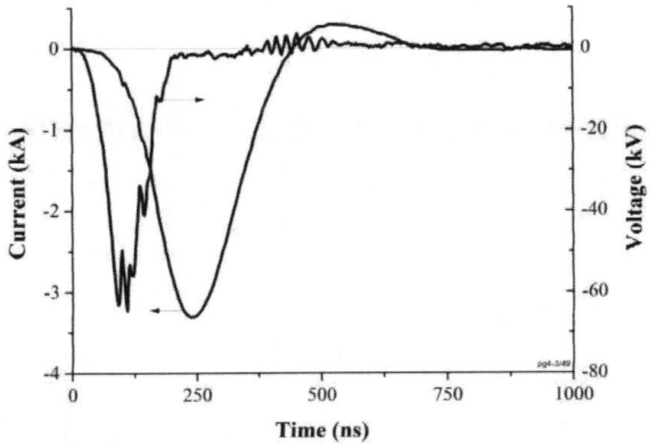

Figure 2. Total current and voltage of the auxiliary discharge.

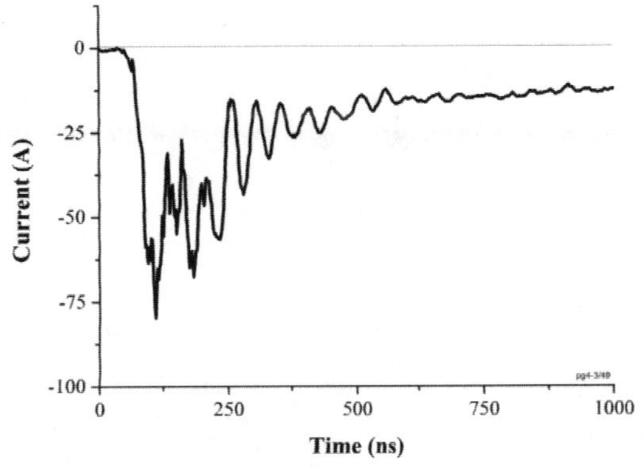

Figure 3. Current to the anode (4).

Journal of Physics: Conference Series **652** (2015) 012064 doi:10.1088/1742-6596/652/1/012064

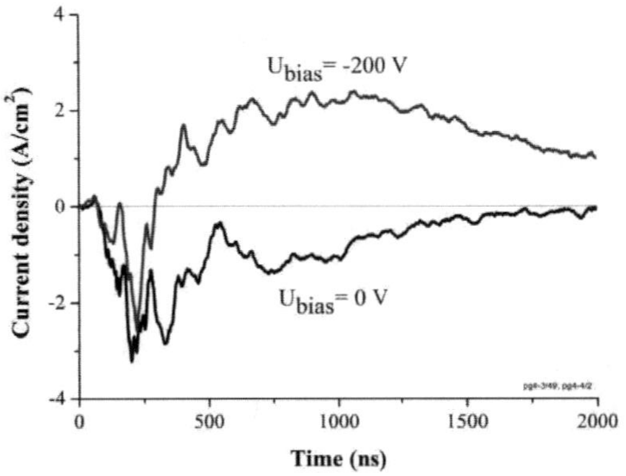

Figure 4. Current density at the electrode (4) from the Faraday cup signal.

The main current of the discharge C_1 flows to the electrode (2). Current amplitude is 3.2 kA, a half period is 400 ns. Part of the current ~50 A also flows to the electrode (4). The current density on the axis measured by Faraday cup at zero bias voltage U_{bias}=0 is above 2 A/cm^2. The current density satisfies the Child-Langmuir law only for a high-voltage discharge phase, when the voltage on the electrode (1) is above 10 kV. Then we have to talk about of ion compensation of the electron current space charge. Faraday cup measurement with a negative bias also suggests the presence of ion flow in the direction of the electrode (4). Negative bias U_{bias}=200 V does not allow cutoff the electronic component of the Faraday cup current in the high-voltage discharge phase and the signal from the Faraday cup is negative. However, signal becomes positive after 250 ns, indicating that the ion flux reaching the electrode (4). Time of appearance of a positive signal corresponds to the ion velocity ~10^7 cm/s. For the copper ions, this velocity corresponds to the energy of more than 1 keV. Ions from the cathode spot toward the anode (4) allow to form a high-current low-resistance discharge in a vacuum gap d_2.

Note, the flow of fast ions from the cathode spot toward the anode is observed in vacuum discharges systematically. The hydrodynamic mechanisms [8] and "potential hump" assumption [7, 9, 10] are invoked for the explanation of the ion acceleration toward the anode.

3.2. Main discharge

Total inductance of the main discharge circuit ~120 nH and circuit wave impedance ~1.2 Ohm were defined in a short-circuit test. In addition, the inductance between voltage divider U_2 and the end of the electrode (2) was defined. This inductance was used to calculate the voltage in the gap d_2 from measured voltage U_2.

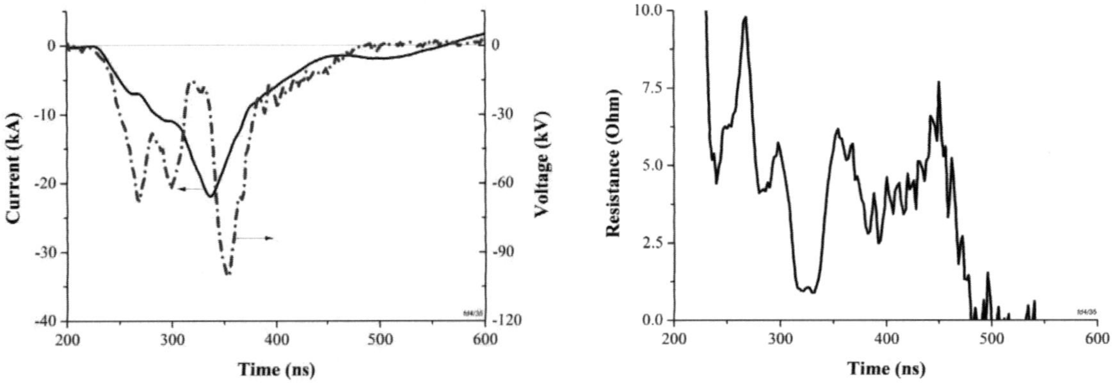

Figure 5. Current, voltage and resistance of the gap d_2 at t_d=200 ns.

The oscillograms of the capacitive storage C_2 discharge at t_d=200 ns and 900 ns are shown in Figures 5 and 6. Time delay t_d=200 ns is not enough to fill the gap d_2 with ions from the auxiliary discharge and to form the low-resistance discharge phase. Discharge is characterized by high resistance of ~5 Ω and a high voltage drop of ~10 kV (Fig. 5). However, the discharge current is noticeably higher than the electron beam limit current in the vacuum. The discharge current amplitude is 20 kA. Several current cutoffs are observed during the pulse. Energy of 80 J is dissipated in the diode gap d_2.

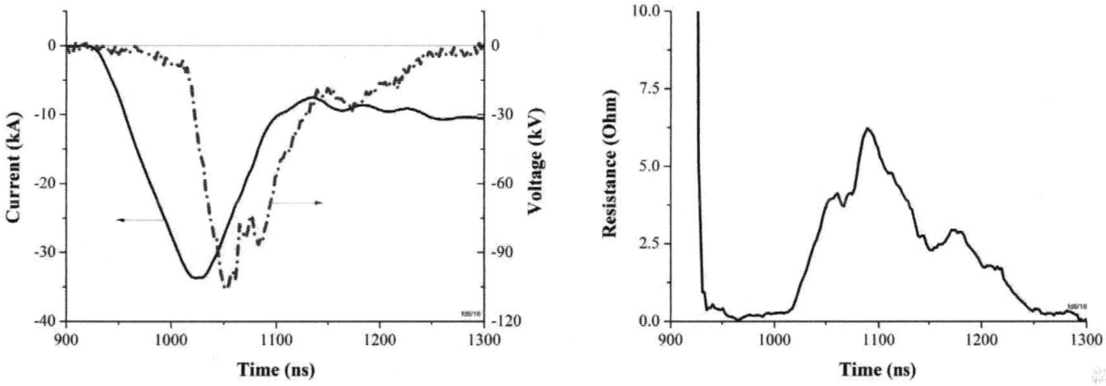

Figure 6. Current, voltage and resistance of the gap d_2 at t_d=900 ns.

Time delay t_d=900 ns is already enough to form discharge with resistance much less than the impedance of the supply circuit and with current amplitude more than 30 kA (Fig. 6). In the current interruption phase the gap resistance increases up to 6 Ω with resistance rise rate of about 0.1 Ω/ns, the voltage pulse and the high-energy electron beam are generated. At stored energy in the capacitor C_2 of 256 J, energy of 180 J is dissipated in the gap d_2 with a peak power of 3 GW.

4. Conclusion

The experimental results confirmed the possibility of forming a low-resistance high-current discharge in the vacuum gap of a few centimeters by means of forced initiation of the cathode spot by auxiliary discharge. The discharge current of more than 30 kA with a rise time of ~100 ns in a gap of 2.5 cm is implemented due to the evaporation of the cathode material. It is used the effect of current interruption in a vacuum discharge for voltage pulse and electron beam generation. Energy of 180 J, which is 70% of stored energy in the primary capacitive storage, is dissipated in the diode gap during current interruption phase.

The absence of dielectric in the ion source allows to avoid life limitation due to evaporation and destruction of the dielectric, as well as to eliminate contamination of the discharge chamber by vaporized dielectric.

Acknowledgments
This work was supported by RFBR grant №15-08-00741-a.

References

[1] Mendel C W, Zagar D M, Mills G S, Humphries S and Goldstein S A 1980 *Rev. Sci. Instrum.* **51** 1641
[2] Renk T J 1989 *J. Appl. Phys.* **65** 2652
[3] Ananjin P S, Karpov V P, Krasik Ya E, Lisitzin I V, Petrov A V and Tolmacheva V G 1992 *IEEE Trans. on Plasma Sci.* **20** 537
[4] Fukuzawa T, Ihara S, Maeda S and Akiyama H 1992 *IEEE Trans. Plasma Sci.* **20** 447
[5] Zherlitcyn A A, Kovalchuk B M, Orlovskii V M and Pedin N N 2012 *Russian. Izv. Vuzov,*

Fizika 55 448
[6] Koval N N, Korolev Yu D, Ponomarev V B, Rabotkin V G, Shemyakin I A and Shyanin P M 1989 *Sov. Fizika Plasmy* **15** 747
[7] Bolotov A V, Kozyrev A V, Kolesnikov A V, Korolev Yu D, Rabotkin V G and Shemyakin I A 1991 *Russian. J. Tech. Fiz.* **61** 40
[8] Mesyats G A and Barengolts S A 2002 *Russian. Usp. Fiz. Nauk* **173** 1113
[9] Plyutto A A, Ryzhkov V N and Kapin A T 1965 *Sov. Phys.-JETP* **20** 328
[10] Bolotov A, Kozyrev A and Korolev Yu 1995 *IEEE Trans. Plasma Sci.* **23** 884

Journal of Physics: Conference Series **652** (2015) 012065 doi:10.1088/1742-6596/652/1/012065

Investigation of Metal Puff Z pinch Based on Multichannel Vacuum Arcs

A G Rousskikh[1], V I Oreshkin[1], A S Zhigalin[1], S A Chaikovsky[1], R B Baksht[2] and K N Mitrofanov[3]

[1]Institute of High Current Electronics RAS, 634055 Tomsk, Russia
[2]Tel Aviv University, 69978 Tel Aviv, Israel
[3]State Research Center of Russian Federation Troitsk Institute for Innovation and Fusion Research, 142190, Troitsk, Russia.

E-mail: *russ@ovpe2.hcei.tsc.ru*

Abstract. The performance of a metal double puff Z-pinch system has been studied experimentally. In this type of system, the outer and inner cylindrical shells were produced by ten plasma guns. Each gun initiates a vacuum arc operating between aluminum electrodes. The net current of the guns was 80 kA. The arc-produced plasma shells were compressed by using a 450-kA, 450-ns driver, and as a result, a plasma column 0.2 cm in diameter was formed. The power of the Al K-line radiation emitted by the plasma for 7 ns was 800 MW/cm.

1. Introduction

The processes that occur in imploding Z-pinches are currently studied by using both terawatt machines and university facilities capable of producing pulse powers of tens of gigawatts [1-5]. The Z-pinch experiments carried out on terawatt setups are aimed at optimizing the conditions for soft X-ray production or at realizing controlled thermonuclear fusion [1,6,7]. These experiments are very time-consuming and expensive. At the same time, as validly mentioned by the authors of Ref. [2], the small university setups, allowing low-cost experimentation, are well suited for testing both new diagnostics and new schemes for realization of proper initial conditions in imploding plasma shells in Z-pinch experiments. In the paper we demonstrate to use for small setups an annular plasma jet (plasma metal puff) which is generated by a vacuum arc discharge (VAD).

2. Experimental results

The diagram (Fig.1) presents a sketch of a metal puff Z-pinch. IMRI-5 [8-10] linear-transformer-driver (3.2 μF, 70 kV) was a driver for the Z-pinch. IMRI 5 has the current of 450 kA and the rise-time of 450 ns. The IMRI 5 current flows through the metal plasma cylinder from the high- voltage electrode to the ground electrode. Ten plasma guns are arranged around a central electrode. The central electrode is the common anode for all plasma guns. The ten arc discharges are powered by a common capacitor bank of capacitance 20 μF charged to a voltage $U_{gun} = 20 - 25$ kV. To sync the IMRI-5 generator to the capacitor bank, a delay generator is used. The time delay between the operation of IMRI-5 and the onset of current in the guns, t_{del}, was varied from 1 to 11 μs. The arcs burn between the rod cathodes and the anode inside a small cavity and the arc plasma spread about the narrow slot. The cathodes and the anode are manufactured from the same metal. Generated plasma flows through the slot between two cylinders and it forms an annular shell. We worked with Al and Mg single and double puffs. We used a Mo plate as a ground electrode that permits to carry out the experiment without disassembling the vacuum chamber after every shot. In this experiment, the net gun current, the pinch current (IMRI-5 current), and the characteristics of the Mg K-line radiation emitted by the pinch were measured. To measure the K-line radiation power, a Diamond Radiation Detector (DRD) filtered with 18-μm thick polypropylene film was used.

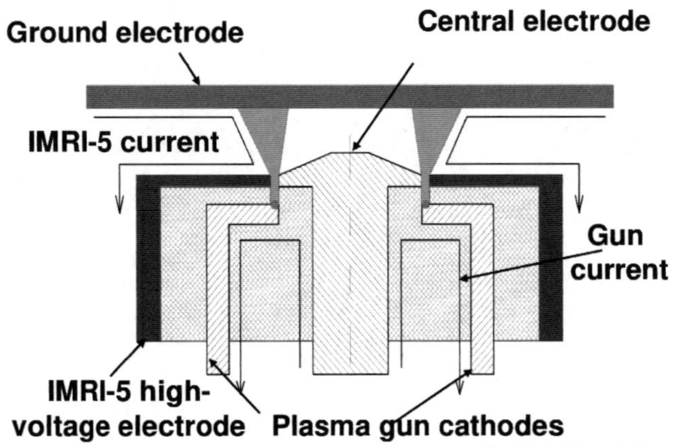

Figure 1. Schematic diagram of the metal puff Z-pinch system. Mo plate with a 10-mm dia hole at the centre is the ground electrode.

The performance of a single metal puff was presented in Ref. [8]. The performance of a double metal puff is described in the paper. Double metal puff consists of two plasma metal shells- outer and inner. For double metal puff we used the same hardware as for single metal puff. But at this case the gun current has another path. First, the gun current flows along the gun electrodes; after that it flows trough the arcs that produce both the outer shell and inner shell. Thus, the arcs of the inner and outer shells connect in series.

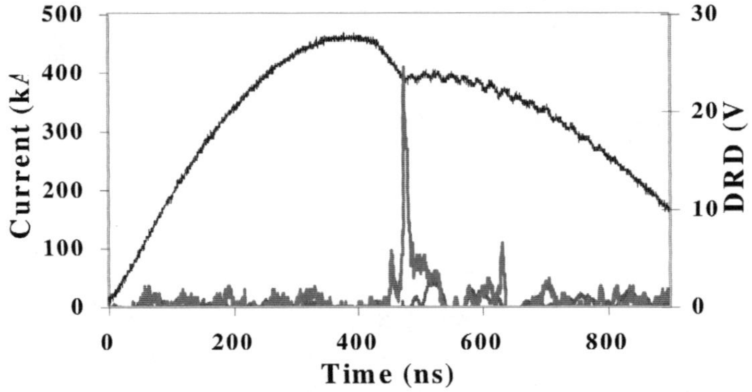

Figure 2. Waveforms of the current and DRD signal for a double metal puff.

Fig.2 shows the current and DRD waveforms for a double Al metal puff. For this shot D_{out} and D_{in} = 40 mm and 8 mm respectively. The current waveform has a pronounced dip and it permits us to estimate the energy delivered to load. For this shot K shell energy and power are 30 J and 8 GW respectively, FWHM = 7 ns, total input energy is 980 J. Pinhole picture showed the stable plasma column with diameter of 3 mm. Hence, the implosion ratio is equal to 12 approximately.

Fig.3 shows the dependences of K-shell power and energy vs t_{del} for Al double puffs. We carried out the experiments with two different inner shell diameters: 10 and 15 mm. The outer diameter was equal to 40 mm for both sets. Both the power and energy are more for the inner diameter of 10 mm. We had the similar implosion time for both modes: something about 500 ns at the time delay of 4 μs. But the K-shell pulse duration is different: 7 ns for diameter of 10 mm and 15 – 20 ns for diameter of 15 mm.

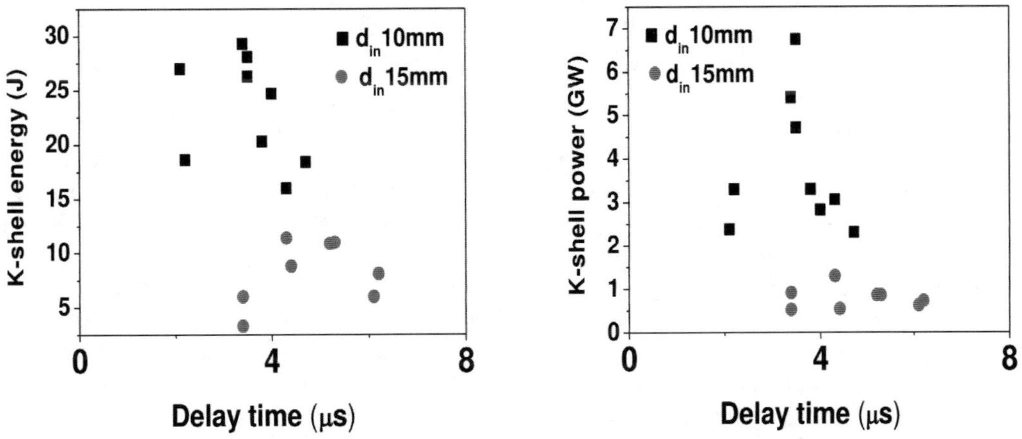

Figure 3. The K-shell energy and power versus the delay time t_{del} between the onset of the current I_{gun} and the onset of the IMRI-5 generator current I_{pinch}.

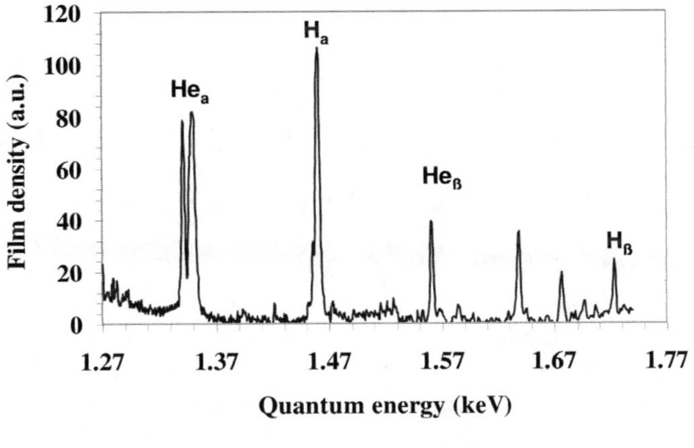

Figure 4. Mg double puff spectral measurements.

We carried out also the experiments with Mg double puff Z-pinches. In whole, the Mg double puff data are similar to those for Al. For the Mg double puff shot K shell energy and power are 80 J and 7 GW respectively, FWHM = 17 ns. In several shots with Mg double puffs, spectral measurements were performed and the line intensity ratio has been found for the hydrogenlike and heliumlike magnesium ions, $I_{H\alpha}/I_{He\alpha}$ (see Fig.4). Knowing this ratio, we can determine the electron temperature Te and the ion concentration Ni for the hottest part of the imploded plasma column. We have calculated the values of Te and Ni with the help of a collision-radiation equilibrium [11] using the values of $I_{H\alpha}/I_{He\alpha} = 2$ and K-line radiation power PK. Thus, we have obtained $T_e \approx 900$ eV and $Ni \approx 1 \cdot 10^{18}$ cm^{-3}.

3. Discussion and conclusion

In this paragraph we compare our results with predictions of the Mosher, Krishnan and Qi model [12]. The authors of Ref.4 plotted a dependence K-shell yield Y_k as a function of ratio kinetic energy to the break energy W_{break}. In our experiment with double metal puff (450 kA, 450 ns) Al K-shell yield is 28 J/cm; kinetic energy estimated according to Ref.[1] at tenfold compression

$$W_{kin} = \frac{\mu}{4 \cdot \pi} \cdot I^2 \cdot \ln \frac{R_0}{R_{fin}} = 2300 \cdot I^2_{pinch} \quad (1)$$

where R_0 and R_{fin} are the original and final radiuses of the Z-pinch. For Al I_{pinch} is equal to 450 kA; W_{break} is equal to 6.4 kJ and $W_{kin}/W_{break} = 0.465/6.4 \approx 0.072$. Thus, for our experimental data $Y_{k(max)}/W_{kin} = 0.06$. According to the model estimates of Ref.[4], $Y_{k(max)}/W_{kin} = 0.05$. Hence, our experimental data are a little above the estimates of the Ref.[12].

It is possible to see that our data are described by the same dependencies that the common Z-pinch loads as the wire-arrays and the gas puffs. Metal puff loads offer the possibility to produce the initial plasma shell for materials which are now not available in the form of thin wires, such as Be and Bi. A metal puff load is a completely ionized plasma shell, where the processes that may result in RT instability in the imploding shell are depressed. Also note that a metal puff experiment can be performed without disassembling the vacuum chamber after every shot, and this reduces the experiment duration. Thus, the metal puffs open new possibility for Z-pinch study.

4. Acknowledgments

The authors wish to thank Prof. G.Yu. Yushkov for his interest and advice. This work was supported in part by the Russian Foundation for Basic Research (Grants # 14-02-00382-a).

5. References

[1] Ryutov D D, Derzon M S and Matzen M K 2000 *Rev. of Modern Physics* **72** 167
[2] Sinars D B, Gregorian L, Hammer D A and Maron Y 2004 *Proc. IEEE* **92** 1110
[3] Aleksandrov V V; Gasilov V A; Grabovski E V, et al. *Plasma Physics Reports* 2014 **40(12)** 939-954
[4] Labetsky A Yu, Baksht R B, Oreshkin V I, Rousskikh A G, Shishlov A V *IEEE Transactions on Plasma Science* 2002 **30(2)** 524-531
[5] Chaikovsky S A, Labetsky A Yu, Oreshkin V I, Shishlov A V, Baksht R B, Fedunin A V, Rousskikh A G *Laser and Particle Beams* 2003 **21(2)** 255-264
[6] Zaitsev V I, Volkov G S, Barykov I A, et al. *JETP Letters* 2008 **88(9)** 582-585
[7] Baksht R B, Oreshkin V I and Rousskikh A G *Phys. Plasmas* 2013 **20** 082701
[8] Russkikh A G, Oreshkin V I, Zhigalin A S, Chaikovsky S A and Baksht R B 2011 *Phys. Plasmas* **18** 092707
[9] Rousskikh A G, Shishlov A V, Zhigalin A S, Oreshkin V I, Chaikovasky S A and Baksht R B *Plasma Sources Sci. Technol.* 2011 **20** 035011
[10] Rousskikh A G, Zhigalin A S, Oreshkin V I, Labetskaya N A, Chaikovsky S A, Batrakov A V, Yushkov G Yu, Baksht R B *Physics of Plasmas* 2014 **21(5)** 052701
[11] Oreshkin V I 1995 *Rus. Phys. J.* **38** 1203
[12] Mosher D, Krishnan M and Qi N 1998 *IEEE Trans. Plasma Sci.* **26(3)** 1052

Modernization of cathode assemblies of electron sources based on low pressure arc discharge

V N Devyatkov[1,2], M.S. Vorobyov[1], N.N. Koval[1,2], V.V. Shugurov[1]

[1]Institute of High Current Electronics SB RAS, 2/3 Akademichesky Ave., Tomsk, 634055, Russian Federation, E-mail: vlad@opee.hcei.tsc.ru
[2]National Research Tomsk State University, 36 Lenin Ave., Tomsk, 634050, Russian Federation

E-mail: vlad@opee.hcei.tsc.ru

Abstract. The paper presents modified cathode assemblies, including their main characteristics, for plasma electron sources based on a constricted low-pressure arc initiated by a discharge in crossed electric and magnetic fields. One of the modified cathode assemblies allows the electrode system of the auxiliary discharge to operate at a pressure of $\approx$ 0.3 Pa, as against $\approx$ 10 Pa before the modification, provides a larger diameter constriction of the pulsed (up to $\tau = 250$ μs) arc, and thus extends the range of operating currents for the plasma-cathode discharge system up to $I_d = 300$ A. The other modified assembly operates at widely varying discharge currents $I_d = 5$–100 A, provides lesser amounts of cathode microdroplets at the discharge electrodes, and allows the arc to operate at comparatively low voltages. The use of cathode assemblies adapted for specific discharge systems extends the capabilities of plasma-emitter electron sources as well as the range of their applications in scientific and technological fields.

1. Introduction

Low-pressure arc discharges are widely used in plasma-cathode electron sources [1,2]. This type of discharge allows generation of emission plasma in a wide pressure range and provides discharge currents of hundred amperes with a single cathode and of several kiloamperes with multicathode discharge systems. Due to the low arc operating voltage (tens of volts), the discharge system requires low power to operate. In a discharge system of this type, depending on the method of cathode spot initiation, an open arc can be used or the discharge current can take the path through a constriction channel. This channel provides a pressure difference between the cathode and anode regions of the discharge system and hence assists the initiation of an arc. It is known that during the operation of a cathode spot at the cathode surface, the cathode is severely sputtered [3]. Separation of the cathode and anode regions of an arc discharge impedes the penetration of the sputtered cathode material into the anode region. Also, the positioning of cathode assemblies crosswise the electron extraction axis in a plasma emitter allows an increase in the dielectric strength of the high-voltage acceleration gap [4]. Likely, the increase is also due to lesser amounts of cathode microdroplets formed at the emission grid and in the region of the high-voltage acceleration gap during the cathode spot operation.
The paper presents modified cathode assemblies for two types of plasma-cathode electron sources with a grid-stabilized plasma emission boundary. The characteristics of the discharge systems before and after modification of its cathode assemblies are compared.

2. Cathode assembly for a wide-aperture pulsed electron source

The cathode assembly was developed for a modified DUET wide-aperture ($\geq 10^3$ cm^2) pulsed electron source with a grid plasma cathode [5]. The proposed cathode assembly (figure 1), which replaces the unit of cathode spot initiation by a dielectric flashover [6,7], is such that the hollow anode and the emission grid are screened from the cathode spot (from its line of sight) and the penetration of the cathode material into the anode region of the discharge is thus decreased. The cathode spot in the hollow cathode is initiated by a low-pressure gas breakdown at a point on the left branch of the Paschen curve close to its minimum. The use of hollow electrodes (cathode and trigger electrode) at fixed gas flow rates makes it possible to decrease the voltage for a discharge to arise and transform into an arc. At an applied voltage of amplitude $\approx$ 10 kV, a glow gas discharge is ignited between hollow cathode 4 and trigger electrode 1 with subsequent initiation of a cathode spot and glow-to-arc discharge transition. The gas breakdown obviates the necessity of using a dielectric which, being contaminated by the cathode material, decreases the lifetime of the cathode assembly [6,7].

For easier ignition of the discharge and its stabilization at low gas pressures, when the discharge operates mostly in vapors of cathode material 4 (figure 1), the cathode is made of magnesium whose ionization potential and melting temperature are rather low [8].

The use of permanent ring magnet 3 (figure 1) provides a magnetic field distribution with a maximum field strength in the expanded region of trigger electrode 1, and this lengthens the paths of electrons and increases their interactions with the gas supplied to the discharge system. At the same time, due to the ability of a cathode spot to shift toward the acute angle the magnetic field lines make with the cathode, the spot is confined at the cathode surface by the magnetic field [9]. Therefore, in this magnetic field geometry, the cathode spot is held at the inner face of hollow cathode 4 facing trigger electrode 1 and provides stable operation of the discharge with decreasing minimum arc current. If the cathode spot leaves the working face of the cathode for its lateral surface, the discharge continues to operate because the magnetic field returns the cathode spot back to the cathode face.

It should be noted that in cathode assemblies based on a Penning cell [10], the arc discharge is constricted in the hole of one of the electrodes which, as a rule, is placed at floating potential, and this increases the discharge operating voltage. A plasma bunch surrounded by the electric double layer of the space charge, which provides electron acceleration and focusing, is normally formed from the cathode side of the constriction channel [11]. The typical voltage across the layer is 50–80 V and this provides a high ionizing capacity of electrons accelerated in the layer. The constriction channel produces an additional pressure difference between the plasma generation region and the high-voltage acceleration gap, thus allowing an increase in the dielectric strength of the latter. However, a decrease in gas flow rate or an increase in discharge current results in arc current cutoffs in the constriction channel, making possible the initiation and stable operation of a cascade arc at a comparatively high voltage but widely varying discharge current amplitudes and durations [10].

The constriction hole in the new cathode assembly is 4 mm in diameter which provides comparatively high discharge current densities at fixed durations with no arc current cutoffs, as opposed to cathode assemblies based on a Penning cell [10,12]. If we compare the current-voltage characteristics (figure 2) for low-pressure arc discharges with cathode assemblies based on a Penning cell [13] (curve 1), open arc [16] (curves 3,4), and arc produced in the proposed design (curves 2,5), it can be seen that the discharge operating voltage in the latter case is lower. The lower discharge operating voltage is due to the absence of external magnetic field effects on the main discharge, which actually decrease the anode area for the discharge current to take path through (compared to curve 1), and to the processes occurring in the cathode region of the discharge (compared to curves 3,4). In the new cathode assembly, the effect of the external magnetic field on the current of the main discharge is eliminated by using ferromagnetic insert 6 (figure 1) which precludes the magnetic flux from the permanent magnet in the hollow anode region of the main discharge.

With an external magnetic field, the ionization processes occurring in the cathode assembly, namely the enhanced interaction of primary magnetized electrons with atoms of the working and residual gas at a constant electron free path, lead to a decrease in the voltage across the near-cathode layer compared to the cathode assembly based on an open arc [6,7]. With no external magnetic field in the new cathode assembly, the discharge operating voltage increases (figure 2, curve 2) because the number of interaction events between the primary electrons and the working gas at the same electron free path length decreases. The chosen hole diameter (4 mm) in the proposed cathode assembly allows a decrease in the amount of microdroplets from the cathode to the plasma emitter and acceleration gap with no constriction-induced cutoffs of the discharge current at a low discharge operating voltage.

Life tests of the new cathode assembly show that compared to the cathode assembly with cathode spot initiation by a dielectric flashover, the increased mass of the cathode with its re-sputtering due to the hollow design and the absence of a dielectric insert on which the cathode material is deposited have made it possible to decrease the wear of the cathode and lengthen the time of stable operation of the cathode assembly to $\approx 10^7$ pulses.

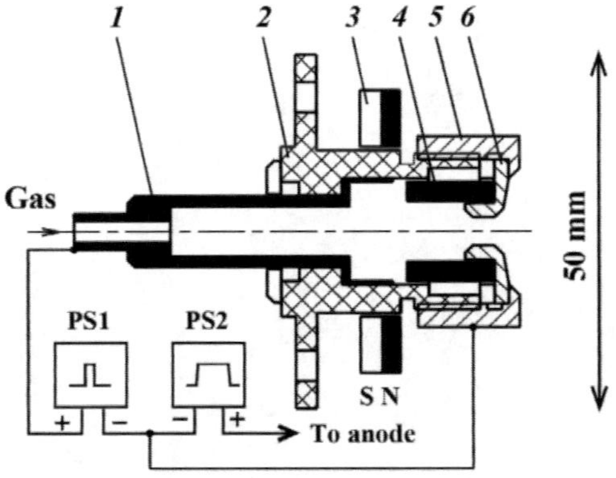

Figure 1. Cathode assembly: 1 – trigger electrode with gas supply, 2 – insulator, 3 – ring magnet, 4 – hollow magnesium cathode, 5 – stainless steel nut, 6 – ferromagnetic insert; PS1 – auxiliary discharge power supply, PS2 – arc discharge power supply.

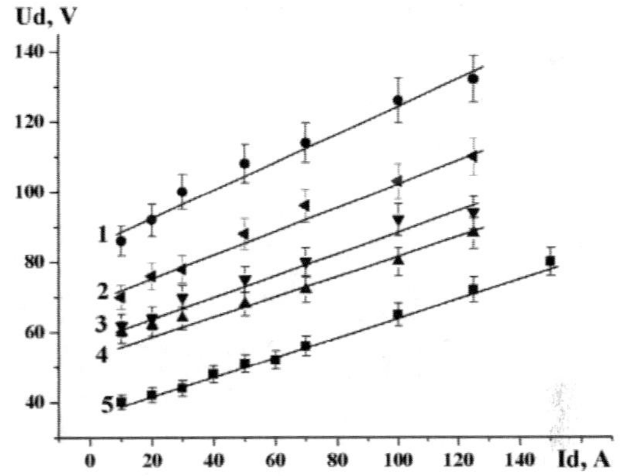

Figure 2. Current-voltage characteristics of arc discharges at p = 50 mPa (Ar) for different cathode assemblies: 1 – cathode assembly based on a Penning cell [13]; 3,4 – cathode assembly based on an open arc with a tungsten and magnesium cathode, respectively [7]; 2,5 – new cathode assembly with no and with a ring magnet, respectively.

3. Cathode assembly for an intense pulsed electron source

The cathode assembly was developed for a SOLO electron source [14–16]. The source is based on an arc discharge with a grid plasma cathode; the discharge current is I_d=20–300 A, the pulse duration is up to $\tau = 200$ μs, the initial beam diameter is 40–60 mm. The cathode assembly is similar in its basic design and principle of operation to that shown in figure 1. Its modification brings the auxiliary discharge system closer to a reflex discharge (figure 3). Ring electrode 2 serves as the anode of the reflex discharge. Resistor Rc provides the formation of a cathode spot at main magnesium cathode 1 without its formation at auxiliary cathode 5. After the initiation of the cathode spot, the main arc discharge takes the path from cathode 1 through constriction channel 4 to an emission grid anode (not shown in the figure).

The change in the electrode configuration allowed us to decrease the pressure required for auxiliary discharge initiation from 5–10 to 0.2–0.3 Pa. The decreased requirements on the pressure difference in the discharge system made it possible to increase the diameter of 12-mm constriction channel 4 from 6 to 10–11 mm with a simultaneous decrease in the auxiliary discharge initiating voltage (figure 4). Till the instant at which the auxiliary discharge was ignited, the rate of voltage rise was 0.8 kV/μs. The discharge initiating voltage was fixed at the point of a steep voltage drop. Noteworthy is the effect of the anode geometry in the reflex discharge on its initiating voltage. In experiments, the anode width was 6 and 23 mm. The anode of larger size made it possible to decrease the auxiliary discharge initiating voltage at low gas (Ar) pressures but somewhat impaired the conditions for discharge ignition in the operating pressure range of the electron source (p = 25–100 mPa).

Journal of Physics: Conference Series **652** (2015) 012066 doi:10.1088/1742-6596/652/1/012066

The increase in the diameter of channel 4 allowed us to stabilize the arc discharge at low pressures and to increase its maximum current, which was limited before the modification (in part of the pulses) by cutoffs of the discharge current at low gas flow rates and pressures in the discharge system (figure 5). On cutoffs of the discharge current, the rate of current rise is defined by the inductance of the isolation transformer in the discharge power supply; the transformer is also responsible for the voltage polarity reversal in the discharge gas on completion of the discharge current pulse. The amplitude of the discharge current with no cutoffs for the channel of diameter 11 mm is no less than $I_d = 300$ A (figure 6). Figure 7 presents current-voltage characteristics of the main arc discharge for the cathode assembly before and after the modification. The increase in the discharge operating voltage U_d with decreasing discharge current can be due to a decrease in the no-load voltage of the arc power supply and associated impairment of the discharge operating conditions. The arc discharge operating voltage U_d depending on the pressure p is shown in figure 8. It is seen that the pressure dependence of the voltage is strong.

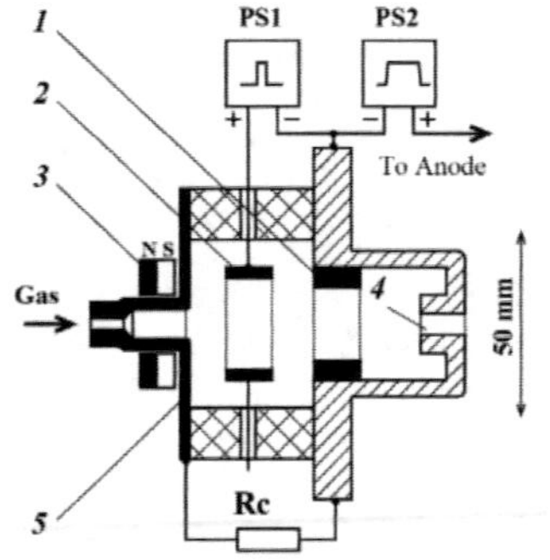

Figure 3. Cathode assembly with arc initiation by a reflex discharge. 1 – Mg cathode, 2 – anode, 3 – permanent magnet, 4 – constriction channel, 5 – auxiliary cathode. PS1 – auxiliary discharge power supply, PS2 – arc discharge power supply.

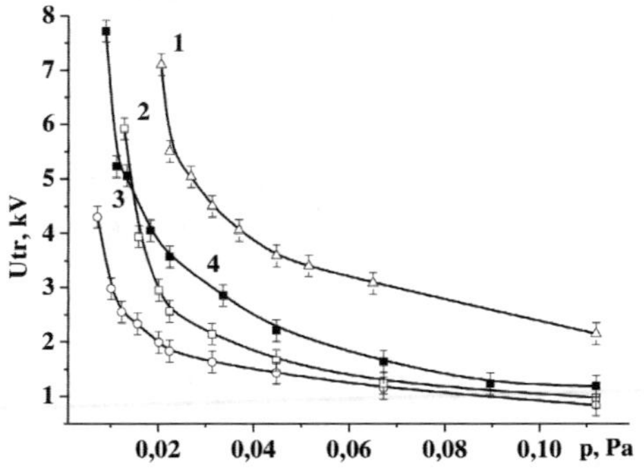

Figure 4. Auxiliary discharge initiating voltage Utr vs the pressure p at a constriction channel diameter of 6 (1), 11 (2), 10 (3), and 11 mm (4); the width of the reflex discharge ring anode is increased from 6 to 23 mm.

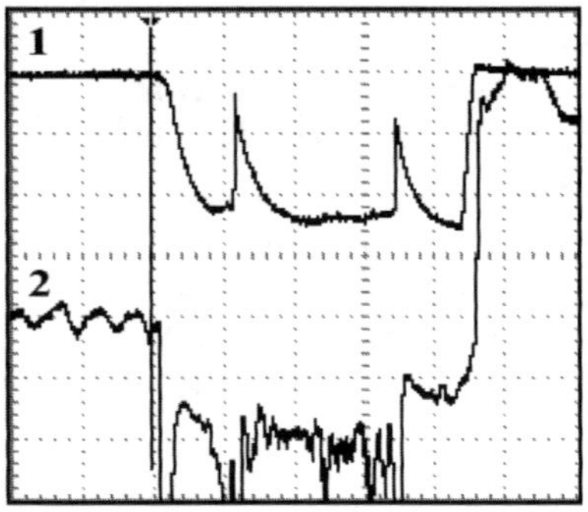

Figure 5. Typical waveforms of the discharge current I_d (1) and voltage U_d (2) for the non-modifed cathode assembly at p = 45 mPa. Scale: 100 A/div, 50 μs/div, 40 V/div.

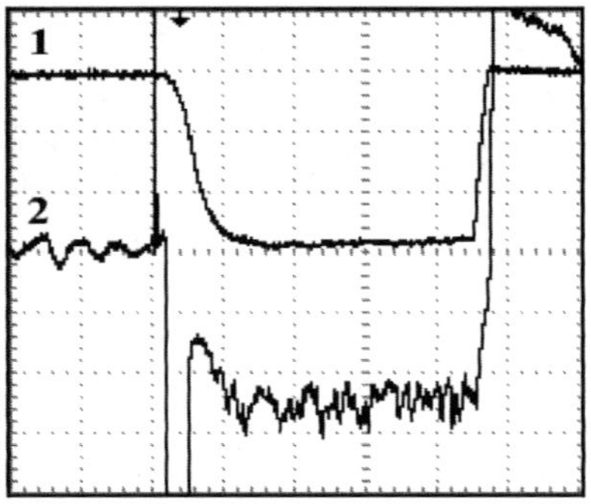

Figure 6. Typical waveforms of the discharge current I_d (1) and voltage U_d (2) for the modified cathode assembly at p = 45 mPa. Scale: 100 A/div, 50 μs/div, 40 V/div.

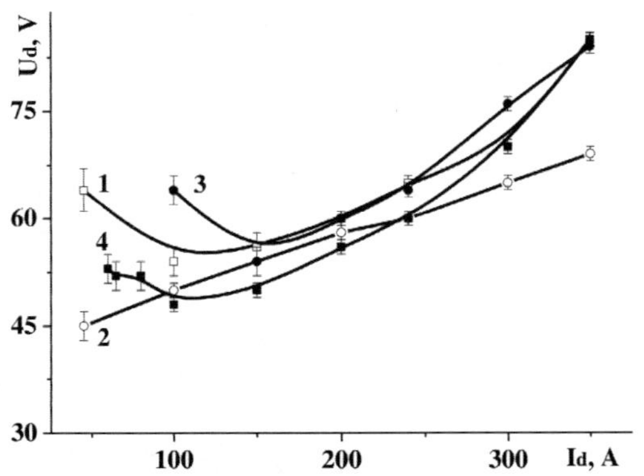

Figure 7. Current-voltage characteristics of the arc discharge for the cathode assemblies with a constriction channel diameter of 6 (1,2) and 11 mm (3,4) at p=0.067 Pa (1,3) and p=0.11 Pa (2,4).

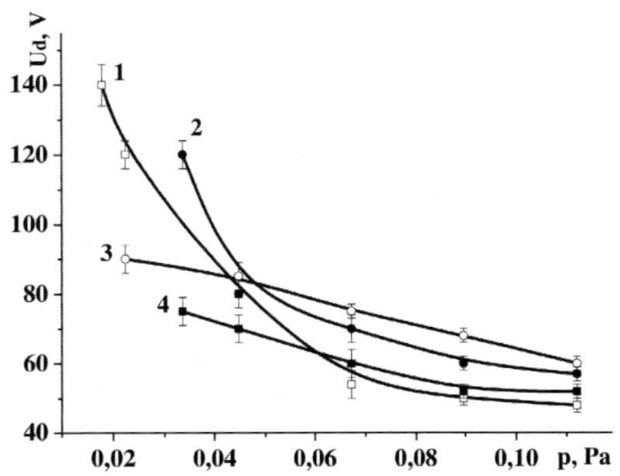

Figure 8. Arc discharge operating voltage Ud vs the pressure p at a discharge current of 150 (1,4) and 250 A (2,3) for the cathode assemblies with a constriction channel diameter of 6 (2,4) and 11 mm (1,3).

4. Conclusion

Thus, the modification of the cathode assemblies of electron sources through changing the mechanism of cathode spot initiation (the use of a gas discharge and a reflex discharge, instead of a dielectric flashover) and through optimizing the constriction channel for a specific operating range of the plasma cathode discharge systems made it possible to greatly improve the operation of the devices. The operating current range of the discharge systems was extended and the time of stable operation of the electron sources in the mode of electron beam generation was increased.

Acknowledgement

This work was supported by the Russian Science Foundation (project # 14-29-00091).

References

[1] Kreindel Yu E 1977 Plasma Electron Sources (Moscow, Atomizdat) (in Russian)

[2] Koval N N, Oks E M, Protasov Yu S, Semashko N N 2009 *Emission electronics* (Moscow, MGTU) (in Russian)

[3] Mesyats G A 2011 *Explosive Electron Emission* (Moscow: Izd. Fiz. Mat. Lit.) (in Russian) 280 pp.

[4] Gavrilov N V, Kovalchuk B M, Kreindel Yu E, Tolkachev V S, Schanin P M 1981 Plasma-emitter high-voltage electron source for generation of large-cross-section beams *J Prib. Tekh. Eksp* **24** Issue **3(2)** 727

[5] Vorob'ev M S, Denisov V V, Koval' N N, Shugurov V V, Yakovlev V V, Uemura K and Raharjo P 2015 Radiation Processing of Natural Latex Using a Wide Aperture Electron Accelerator with a Plasma Emitter *J High Energy Chemistry* **49**, Issue 3 143

[6] Koval N N, Kreindel Yu E, Месяц G A, Tolkachev V S, Schanin P M 1983 Efficient use of a low-pressure arc in a grid plasma electron emitter *J Pis'ma Zh. Tekh. Fiz.* **9** Issue 9 568

[7] Koval N N, Kreindel Yu E, Tolkachev V S, Schanin P M 1985 The effect of gas on the development of a vacuum arc with a hollow anode. *IEEE Transactions on electrical insulation V.EI-20* **4** 735

[8] Gulyaev A P. 1986 *Metal Science. Textbook for higher educational institutions, 6rd ed., rev. ad.* (Moscow, Metallurgia) (in Russian) 544 p.

[9] Andreev A A, Sablev L P, Grigoryev S N 2010 *Vacuum arc coatings* (Kharkov, NSC KIPT) (in Russian) 318 p.

[10] Kreindel Yu E, Oks E M, Schanin P M 1984 Plasma-emitter pulsed electron source based on a cascade arc. *J Prib. Tekh. Eksp.* **4** 127

[11] Zlobina A F, Kazmin G S, Koval N N, Kreindel Yu E 1980 Plasma parameters in the expander of an electron emitter with a constricted arc discharge *J Zh. Tekh. Fiz.* **50** Issue 6 1203

[12] Gavrilov N V, Kreindel Yu E, Oks E M, Schanin P M 1983 Transition of a low-pressure arc discharge from constricted to cascade mode *J Zh. Tekh. Fiz.* **53** Issue 10 1947

[13] Gavrilov N V, Kreindel Yu E, Schanin P M 1984 Pulsed electron source based on a non-self-sustained high-voltage glow discharge *J Prib. Tekh. Eksp.* **2** 143

[14] Koval' N N, Shchanin P M, Devyatkov V N, Tolkachev V S, and Vintizenko L G 2005 A Facility for Metal Surface Treatment with an Electron Beam *J Instruments and Experimental Techniques*, **48** Issue **1** 117

[15] Koval N N, Devyatkov V N, Grigoryev S V, Sochugov N S 2006 'SOLO' Plasma electron source *Proc.II Intern. Kreindel Workshop «Plasma emission electronics»* (Ulan-Ude: Russia) 79

[16] Grigoriev S V, Devjatkov V N, Koval N N and Teresov A D. 2008 The automated installation for surface modification of metal and ceramic-metal materials and products by intensive pulse submillisecond electron beam *Proc. 9th Intern. Conf. on Modification of Materials with Particle Beams and Plasma Flows* (Tomsk: Russia) 19

Journal of Physics: Conference Series **652** (2015) 012067 doi:10.1088/1742-6596/652/1/012067

Electron source with a multi-apertured plasma emitter

M.S. Vorobyov[1,2], **N.N. Koval**[1,2], **S.A. Sulakshin**[1]

[1]Institute of High Current Electronics SB RAS, 2/3 Akademichesky Ave., Tomsk, 634055, Russian Federation, E-mail: vlad@opee.hcei.tsc.ru
[2]National Research Tomsk State University, 36 Lenin Ave., Tomsk, 634050, Russian Federation

E-mail: vorobyovms@yandex.ru

Abstract. In the present study, we investigated the energy efficiency of an electron source with a multi-aperture plasma emitter where the generated beam is extracted into the atmosphere through a thin metal foil. The boundary of the plasma produced in this type of emitter is stabilized with a fine metal grid. To prevent the loss of electrons at the circle-holed support grid of the extraction foil window, a metal mask with holes of smaller diameter arranged coaxially to the support grid holes is put on the emission grid. Thus, the electron beam is a superposition of beamlets formed by individual electron emitting units with the plasma boundary stabilized by the fine metal grid. The efficiency of current extraction from the acceleration gap into the atmosphere reached 75% with respect to the gap current, making possible to increase the average power of the extracted electron beam. With a 200-kV accelerating voltage, a 16-A current in the acceleration gap, and 40 μs FWHM pulse duration, 4 kW of the average beam power was extracted into the atmosphere from the acceleration gap. With the geometric transparency of the support grid of the extraction foil window equal to 56%, this made 65% of the beam power in the gap. Further increasing the beam power was limited by the power of the high-voltage power supply.

1. Introduction

The electron sources using cathodes of any type to produce large-cross-section beams extracted into the atmosphere through a foil window are widely used in industry, medicine, environmental protection, processing of agricultural products, etc. [1–4]. With the relatively high average beam power provided by these electron sources, special attention is given to minimization of the loss of electrons both at the support grid of the extraction foil window and in the foil. The foil material is chosen so that, on the one hand, it would have a low density for more current could be extracted at a given beam energy and, on the other hand, it should be as mechanically strong as possible at the smallest thickness.

As the support grid of an extraction foil window typically has a geometric transparency less than 90% and part of the electron beam is dissipated as heat at the grid edges, the large-cross-section electron beam extracted into the atmosphere consists of a plurality of beamlets whose cross-section shape and area are determined by the design of the support grid. Therefore, to reduce the beam current losses at the support grid edges, it is necessary to eliminate the electron flow toward the edges. To do this requires that the large-cross-section beam consist of a plurality of small-cross-section beamlets even upon generation and during acceleration [5–7].

2. Experimental procedure

The wide-aperture electron source used in the experimental study under consideration has a plasma emitter operating based on a low-pressure arc with a grid-stabilized plasma emission boundary. The electron source is capable of generating electron beams of specified structure. Electron sources of this type are rather indifferent to vacuum conditions and feature a long service life, mutually independent basic parameters of the beam, and a reasonably uniform current density distribution over the beam cross section.

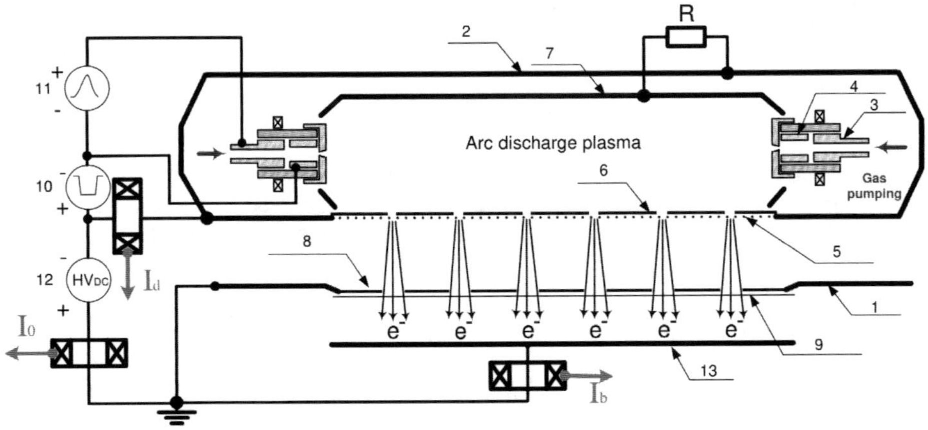

Figure 1. Schematic of the wide-aperture electron source: 1 – vacuum chamber, 2 – plasma emitter, 3 – igniter, 4 – cathode, 5 – emission grid, 6 – mask, 7 – hollow anode, 8 – support grid, 9 – extraction foil, 10 – discharge power supply, 11 – igniter power supply, 12 – high-voltage power supply, and 13 – collector.

Measurements of the efficiency of electron extraction from the plasma emitter into the acceleration gap, α, equal to the ratio of the current flowing through the gap, I_0, to the discharge current I_d ($\alpha = I_0/I_d$) and the efficiency of current extraction from the acceleration gap into the atmosphere, β, equal to the ratio of the extracted beam current I_b to the gap current I_0 ($\beta = I_b/I_0$), were carried out on DUET, a modernized wide-aperture pulsed electron source with a grid plasma cathode (Fig. 1) [8, 9].

The plasma emitter (2) of the source is a hollow stainless-steel half-cylinder of dimensions 750 mm × 150 mm with two low-pressure-arc cathode units (3, 4) fixed on the ends [10]. A stainless-steel mask (6) of thickness 200 μm is put on the emission grid (5), which serves as the anode for the arc. There are 344 round holes of diameter 8–12 mm in the mask, which are individual emission units of the plasma emitter. The hollow anode (7) is electrically connected via a resistor to the emission grid (5), which is necessary to facilitate the discharge ignition, switching, and operation in the region of the emission grid. Electrons are extracted from the plasma surface of the emission units under the action of a dc accelerating voltage of up to 200 kV applied to the 120-mm gap between the emitter (2) and the extraction foil window (8, 9). The aligned holes in support grid 8, having a total geometric transparency of 56%, are of the same number as those in mask 6, but of larger diameter (15 mm). Alignment of the holes in mask 6 and in support grid 8, and thereby attainment of a coaxial plane-parallel geometry of the acceleration gap, minimizes the beam losses in the support grid of the extraction foil window. Thus, the electron beam produced is a superposition of beamlets formed by individual emission units with the plasma boundary stabilized by the fine metal grid. Support grid 8 is covered with 30-μm-thick foil 9 made of AlMg–2n alloy. The beam current Ib extracted from the acceleration gap into the atmosphere was measured in the circuit of a collector (13) of dimensions 800 mm × 200 mm using a Rogowski loop.

3. Results and discussion

When electrons were extracted from the plasma emitter into the acceleration gap using a 0.4 mm × 0.4 mm mesh grid of total geometric transparency 44% in the absence of mask 6, the maximum electron extraction efficiency, $\alpha \approx 0.6$, was achieved with the resistance of the resistor connected in the hollow anode circuit $R \geq 10\ \Omega$ (Fig. 2a). The electron extraction efficiency α also increased with accelerating voltage and discharge current. In the first case, this was due to the penetration of the electric field in the emission grid meshes and to the switching of the discharge current electronic component to the meshes. In the second case, an increase in discharge current, and, hence, plasma density, reduced the thickness of the near-electrode ion layer, which in turn increased the area of the open plasma emission surface and, hence, the extraction efficiency α [10]. As the geometric transparency of the emission grid was lower than the extraction efficiency α, the mechanism of the discharge current switching into the emission grid meshes becomes apparent. This method of current switching was demonstrated to be very efficient, as in this case, the discharge voltage varied slightly (Fig. 2b) and the extraction efficiency increased more than twice (Fig. 2a). Additionally, when a resistor of resistance $R \geq 10\ \Omega$ was connected in the circuit of hollow anode 7, the emission current pulse had a sharper leading edge and a more pronounced flat top.

Figure 3 (curve 1) shows the efficiency of extraction of the beam current from the accelerating gap into the atmosphere, β, as a function of accelerating voltage for the case of no mask in the emitter. It can be seen that the beam extraction efficiency β increases with accelerating voltage. It can be stated that in the absence of the mask, the increase in β with accelerating voltage is only due to the increase in transparency of the foil [11]. Thus, in the investigated range of beam electron energies, the percentage of the current losses at the edges of support grid 8 weakly depends on accelerating voltage. This can be accounted for by the small angular divergence of the generated electron beam associated with the relatively low temperature of the electrons extracted from the plasma emitter. The weak dependence of β on the time point within the pulse duration for which it was determined (see Fig. 3, curve 3) is evidence of the source stable operation throughout the range of beam current pulse durations with a time-invariable efficiency of electron beam extraction into the atmosphere, even supposing that the parameters of the emission plasma varied in time.

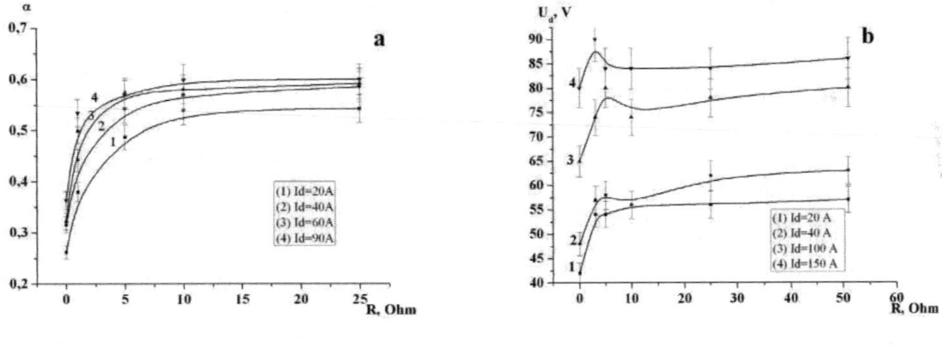

Figure 2. The extraction efficiency α in the absence of a mask at U0 = 160 kV (a) and the discharge voltage Ud (b) as functions of the resistance R of the resistor connected in the hollow anode circuit. An emission grid of mesh size 0.4 mm × 0.4 mm was used.

The data of an experiment on minimization of the electron losses in the support grid of the extraction foil window covered with a mask having 12-mm diameter holes are also plotted in Fig. 3 (curves 2 and 4). In this case, at an accelerating voltage of 200 kV, the current losses in the support grid of the extraction foil window were reduced to 30%. According to the data given in Ref. 12, about 12% of the electron current was lost in 30-μm thick aluminum foil at an accelerating voltage of 200 kV. It can be supposed that the remaining 18% included the losses due to high-energy electrons reflected from the foil and ions produced by the interaction of the reflected electrons with the desorbed gas and the beam losses at the support grid of the extraction foil window due to nonideal alignment of the holes in mask 6 and in support grid 8.

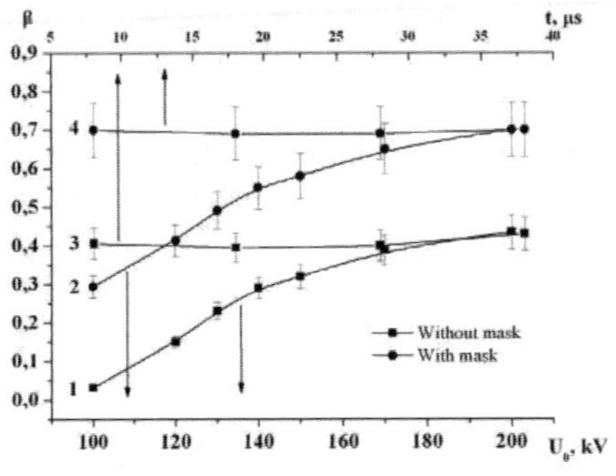

Figure 3. Beam extraction factor β as a function of accelerating voltage and of the time of β determination within the pulse duration. The data were obtained without a mask on the support grid (curves 1, 3) and with a mask having 12-mm dia holes aligned with the support grid 15-mm dia holes (curves 2, 4).

When a metal mask with 12-mm diameter holes and ~30% total geometric transparency was put on the emission grid of mesh size 0.4 mm × 0.4 mm, the extraction efficiency α decreased in direct proportion to the mask geometric transparency to $\alpha \approx 0.6 \times 0.3 \approx 0.2$. Using such an emission grid at discharge currents of 10–100 A, a discharge current pulse duration of 40 μs, an acceleration gap spacing of 70–200 mm, and a working gas pressure of 20–60 mPa, we could not obtained the extraction efficiency α greater than the geometric transparency of the mask.

To facilitate the alignment of the holes in the mask and in the plasma emitter support grid, the mask holes were made 8 mm in diameter. Obviously, using a grid with meshes 0.4 mm × 0.4 mm in size and decreasing the mask hole diameter to 8 mm, with the mask geometric transparency equal to about 13%, should reduce the extraction efficiency to $\alpha \approx 0.6 \times 0.13 \approx 0.08$ if additional measures are not taken. Therefore, to facilitate the switching of the discharge current to the emission grid meshes, and thereby increase α, the emission grid was replaced by a grid of mesh size 0.6 mm × 0.6 mm.

With the grid of mesh size 0.6 mm × 0.6 mm used in the emitter, but in the absence of a mask, the electron source operation was extremely unstable. When the accelerating voltage was high and a discharge was ignited, a high-frequency (1–2 MHz) modulation occurred on the pulse waveforms of both the discharge and the emission current (Fig. 5a). It was found that the high voltage affected the discharge current waveform and amplitude, the electric strength of the acceleration gap decreased, and the operation of the plasma emitter power supply became unstable to the extent that some components of its electric circuits could fail. Nevertheless, a pronounced increase in extraction efficiency α was observed: at a discharge current $I_d = 20$ A, an accelerating voltage of 160 kV, and an acceleration gap spacing of about 120 mm, we had $\alpha = 0.75$ (see Fig. 4a). This can be accounted for by the increased area of the open plasma emission surface at the same wall layer as with the grid of 0.4 mm × 0.4 mm mesh size, as the measurements were carried out at the same discharge current and the same working gas pressure in the chamber.

When a mask with 8-mm diameter holes and 13% geometric transparency was put on the emission grid of 0.6 mm × 0.6 mm mesh size, all high-frequency modulations on the discharge current pulses were ceased and the accelerating voltage ceased to affect the shape and amplitude of the discharge current pulse (Fig. 5b). Thus, one of the main advantages of using a plasma emitter was realized, namely the possibility of independent control of the beam current parameters (pulse amplitude, duration, and repetition frequency) by varying the discharge current.

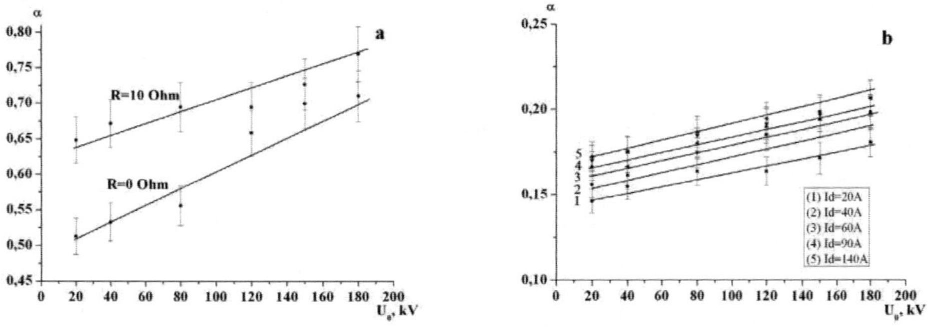

Figure 4. Extraction efficiency αas a function of accelerating voltage U0 in the absence of a mask (a) and with a 8-mm diameter hole mask of geometric transparency 13% (b). The emission grid mesh size was 0.6 mm × 0.6 mm.

For this case, the extraction efficiency α was estimated and plotted as a function of accelerating voltage U_0 (Fig. 4b). For the discharge current I_d = 100 A, U_0 = 160 kV, and 13% geometric transparency of the mask, α = 0.21 was obtained. The emission current pulse waveform had a slightly rising flat top at an invariable discharge current. This is well consistent with the model of the discharge current switching to the emission grid meshes under the action of the applied accelerating voltage [11], and the current rise may indicate an increase in plasma density during the discharge current pulse and a gradual decrease in thickness of the wall layer in the grid meshes.

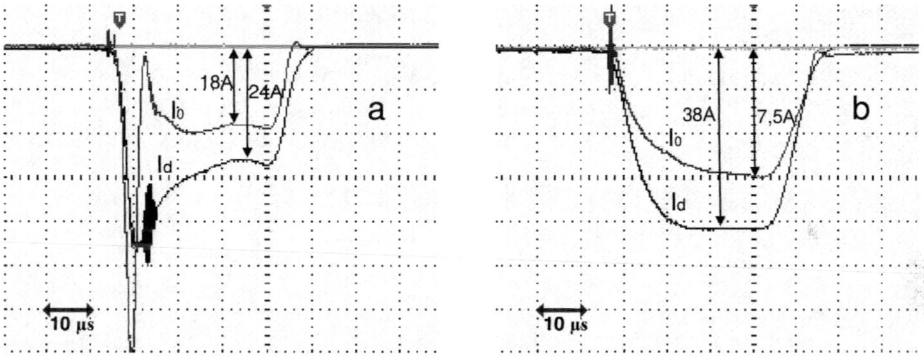

Figure 5. Typical waveforms of the discharge and emission currents at U_0 = 150 kV in the absence of a mask (a) and with a 8-mm diameter hole mask of geometric transparency 13% (b). The emission grid mesh size was 0.6 mm × 0.6 mm.

Using a multi-aperture system of electron extraction from the plasma emitter and a plane-parallel geometry of the mask aligned with the support grid of the extraction foil window, about 75% of the emission current and more than 60% of the beam power were extracted into the atmosphere from the acceleration gap. Thus, at an accelerating voltage U_0 = 200 kV, acceleration gap current I_0 = 16 A, pulse duration t = 40 µs, pulse repetition rate f = 50 s^{-1}, and an about 6.4 kW average beam power in the gap, about 4 kW of the average power were extracted from the acceleration gap through the foil window. Further increasing the beam power was limited by the power of the high-voltage power supply.

4. Conclusion

The study performed has shown that the proposed and investigated simple method of minimization or even, with precision alignment of the equipment, almost complete elimination of the beam current losses at the support grid of the extraction foil window makes it possible to substantially increase the efficiency of the electron source. In addition, its operation can be made more reliable while retaining all of its key advantages, such as the mutual independence of the main beam parameters (energy, current, duration, pulse repetition frequency) and the possibility of varying them over wide limits, the energy homogeneity of the beam, the high pulse repetition frequency of the beam current, and the weak dependence of the beam parameters on vacuum conditions. This opens new prospects for both scientific and technological use of electron sources of this type.

Acknowledgement

This work was supported by the Russian Science Foundation (project №14-29-00091).

References

[1] Bugaev S P, Kreindel Yu E, Schanin P M 1984 *Large section electron beams* (Moscow, EAI) (in Russian)
[2] Hentse M 2004 *Sewage treatment* (Moscow, Mir)
[3] Rostov V V, Alekseenko P I, Vyhodtsev P V, Shteinle A V, Mazin V I etc 2012 *Siberian medical journal* **27**, 1 141
[4] Sokovnin S Yu 2007 Nanosecond electron accelerators and radiation technologies based on them (Yekaterinburg, UD RAS) 224 pp. ISBN 5-7691-1840-7
[5] Abroyan M A, Uspenskiy N A, Fedyakov V P 1984 *J Prib. Tekh. Eksp* Issue 4 24
[6] Gavrilov N V, Kreindel Yu E, Schanin P M 1984 Pulsed electron source based on a non-self-sustained high-voltage glow discharge *J Prib. Tekh. Eksp*. 2 143
[7] Hegeler F, Myers M C, Friedman M, Giuliani J L, Sethian JD, Swanekamp S B, Wolford M F. The electra KrF Laser. IEEE 25 International power modulator symposium and high voltage workshop, Hollywood, California, 2002, 121–125.
[8] Koval N N, Kreindel Yu E, Tolkachev V S, Schanin P M 1985. *IEEE Trans. Electr. Insul.* **El-20** 735-737.
[9] Gielkens S W A, Peters P J M, Witteman W J, Borovikov P V, Stepanov A V, Tskhai V M, Zaivjalov M A, Guschenets V I, Koval N N. 1996 *Rev. Sci. Instr*. **67** 7 2449-2452.
[10] Koval N N, Oks E M, Protasov Yu S, Semashko N N 2009 Emission electronics (Moscow, MGTU) (in Russian)
[11] Seltser S M, Berger M J 1974 *Nucl. Instrum. and Methods*. **119** 157-179.

Journal of Physics: Conference Series **652** (2015) 012068 doi:10.1088/1742-6596/652/1/012068

Modification of the cylindrical products outer surface influenced by radial beam of argon ions at automatic mode

R A Valikov[1], A S Yashin[1], T V Yakutkina[1], B A Kalin[1], N V Volkov[1], V P Krivobokov[2], S N Yanin[2], O Kh Asainov[2] and Yu N Yurev[2]

[1] National Research Nuclear University MEPhI (Moscow Engineering Physics Institute), 115409, Russia, Moscow, Kashirskoe highway, 31

[2] National research Tomsk polytechnic university, 634050, Russia Tomsk, Lenina avenue, 30

E-mail: Romansbl@bk.ru

Abstract. Obtaining surface with high purity and good roughness is important for increasing the corrosion resistance and wear resistance of products working in corrosion-active environment. Installation ILUR-03 with the coaxial ion beam wide energy spectrum source for cleaning, polishing and surface doping of long cylindrical items has been developed. Upgraded installation ILUR-03 provides effective technological defects cleaning (abrasives after mechanical polishing, acid residues after chemical etching, adsorbed gases), surface polishing, film deposition by using magnetrons and surface doping by ion mixing method in one technological cycle.

1. Introduction

Such surface`s characteristics as topography, elemental composition, grain size of the surface layer have a significant effect on the wear resistance and corrosion resistance of the final products. Surface impurities and bad roughness parameter reduces resistance of the material to nodular corrosion. In this regard the development of cleaning and the polishing surface technology by ion beams is promising because during mechanical or chemical polishing product`s surface is contaminated by abrasive particles or acid.

2. Materials and tools

Installation ILUR-03 (fig. 1) has been developed in our university. It allows cleaning and polishing cylindrical items outer surface by ion beam with wide energy spectrum [1]. The ion source with wide energy spectrum increases speed of the process and also increases thickness of the modified layer (fig. 2). The chamber with magnetron source allows sputtering the metal films and ion source allows doping surface by ion mixing method after that.

Fragments of zirconium alloy E110 tubes (diameter - 9.15 mm, wall thickness - 0.65 mm, length - 30 mm and 500 mm) were used as samples.

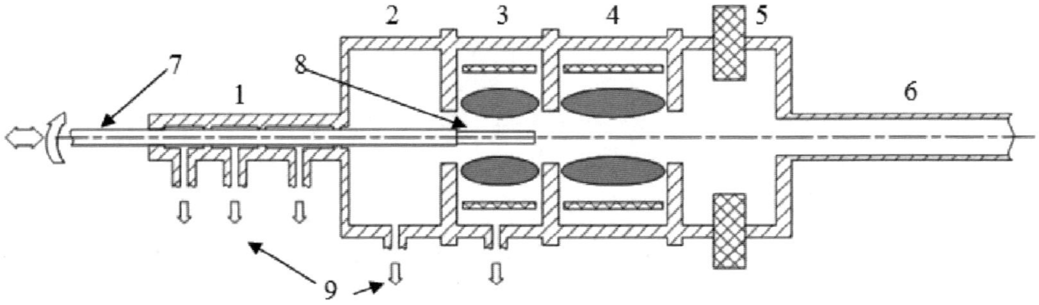

Figure 1. Scheme of the installation ILUR-03: 1 – gateway, 2 – transition chamber, 3 – ion cleaning chamber, 4 – ion polishing chamber, 5 – magnetron chamber, 6 – exit items chamber, 7 – manipulator with sample holder, 8 – sample, 9 – vacuum pump connection

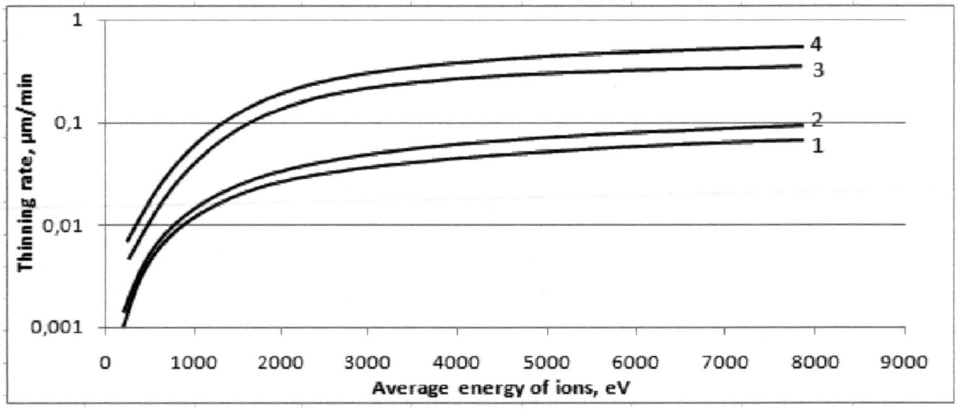

Figure 2. Calculated sputtering rate of zirconium: 1 – monoenergetic ion beam, the ion current 8 mA, 2 – monoenergetic ion beam, the ion current 16 mA, 3 – ion beam with the wide energy spectrum, the ion current 8 mA, 4 – ion beam with the wide energy spectrum, the ion current 16 mA

To determine the limit processing mode the thermal distribution along the length of the tube during process and the maximum of the temperature for different thermal load were simulated (Fig. 3). Experimental measurements of sample heating in the processing zone at different power beam were conducted (Fig. 4). Ion cleaning and polishing modes were determined by using these data and with calculations of zirconium sputtering rate (the average ion energy in the beam 3-4 kV, the ion current of 0.4-0.7 mA/cm^2, the radiation dose to 5×10^{18} ions/cm^2). The maximum speed of the surface cleaning is 4 m/min. Polishing rate depends on the initial state of the surface and can reach 0.2-0.3 m/min.

Journal of Physics: Conference Series **652** (2015) 012068 doi:10.1088/1742-6596/652/1/012068

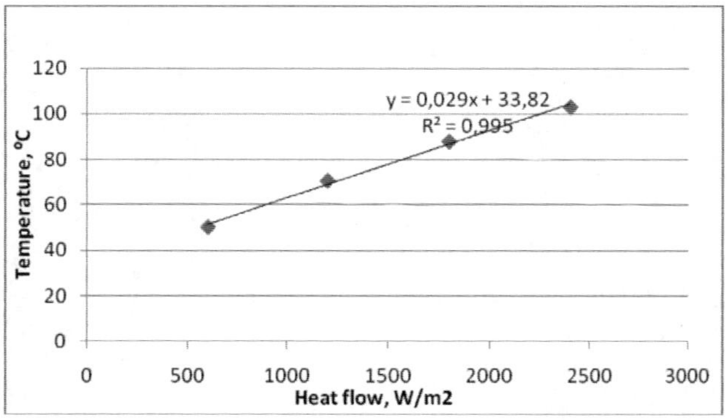

Figure 3. Calculated graph of processing item maximum temperature

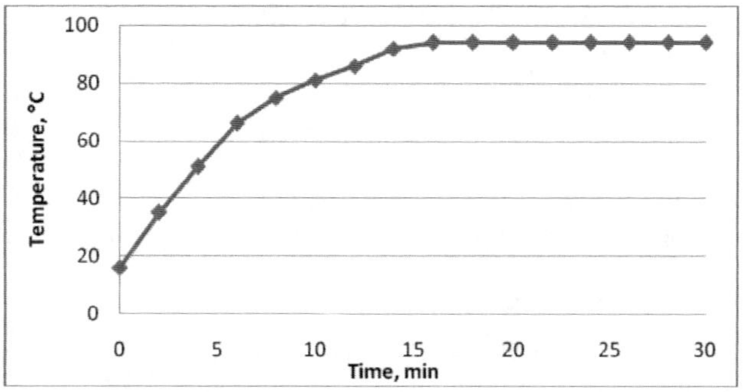

Figure 4. Graph of the sample temperature rise in the treatment zone

3. Experimental results and discussion

After the treatment samples microhardness and roughness were measured. Some partы of the samples were tested in autoclave in condition simulating a nuclear reactor operating parameters (environment - water, temperature –350°C, pressure - 17 MPa). The condition of the surface and oxide films was studied using scanning ion microscopy (SIM).

Microhardness measurements along the length of the sample before and after treatment showed that the ion polishing doesn't change microhardness value. At the same time the variation of values after treatment is considerably reduces [2]. This changes indicates great homogeneity of the surface layer commensurate with the depth of indenter penetration which is equal 3-4 microns at selected load conditions. This increase of the homogeneity is probably happened due to removal of the material surface layer enriched by different type of technological defects [3].

Studying the surface roughness before and after ion treatment was carried out on profilometer TR-200 (f. A&D, Japan). The initial roughness R_a of the samples (after mechanical polishing) was 1.0 microns. Ion polishing provides great scratch smoothing. It reduces average surface roughness of up to R_a = 0,3-0,5 μm (fig. 5, 6).

Journal of Physics: Conference Series **652** (2015) 012068 doi:10.1088/1742-6596/652/1/012068

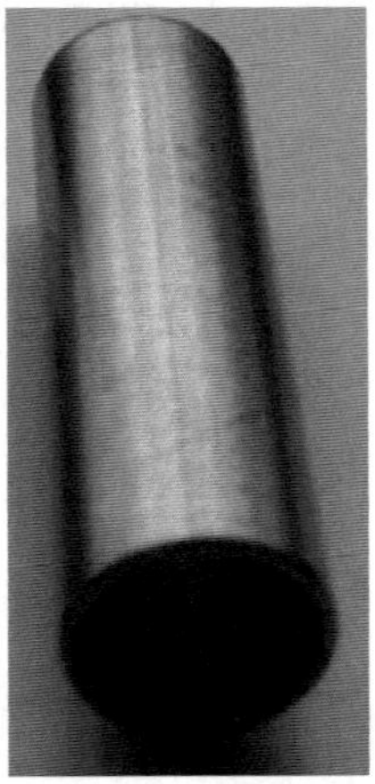

Figure 5. Samples before ion polishing (Ø9.15 mm)

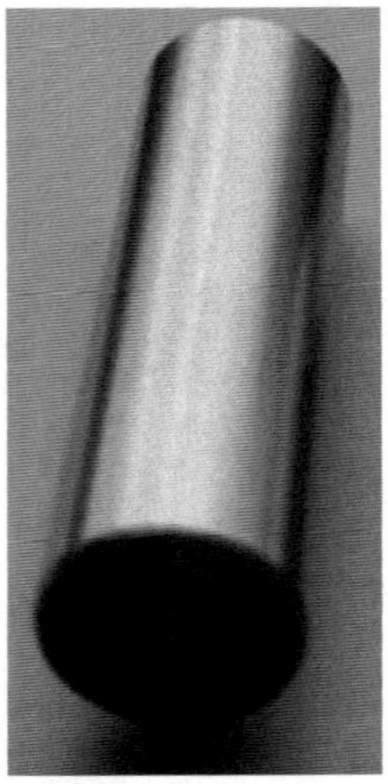

Figure 6. Samples after ion polishing (Ø9.15 mm)

Estimation of the surface layer condition by the SIM has showed that ion treatment removes deformed layer that occurs at the mechanical polishing**.** The amorphous material layer after ion polishing is forms by the accumulation of defects on the surface. This layer helps to reduce stress in the oxide film at the first stages of oxidation and also reduces the oxygen penetration rate through the 'metal-oxide' border.

To examine the oxide films after the autoclave test ion scanning microscopy was used. To study the internal structure of the oxide (estimated thickness, the presence of pores and cracks) the holes were cut by using gallium ion beam. Surface alignment and amorphous layer formation on the surface by ion treatment promotes formation of the oxide film with a monoclinic crystal lattice. This lattice has protective properties and strongly adheres to the substrate. It also reduces the oxide growth rate. The slow down growth rate tendency of the oxide after ion polishing occurs after 500 hours of the oxidation (fig. 7, 8). After 3000 hours autoclave tests in the initial samples occurs a significant number of pores and cracks near the boundary 'metal-oxide'. That increases the migration of the oxygen to the metal and as a result increases oxide film growth rate (fig. 9, 10). At the same time concentration and the pore`s size in the bulk of the sample`s oxide after ion polishing and further corrosion test for 3000 hours is much lower (the difference is 30-40%).

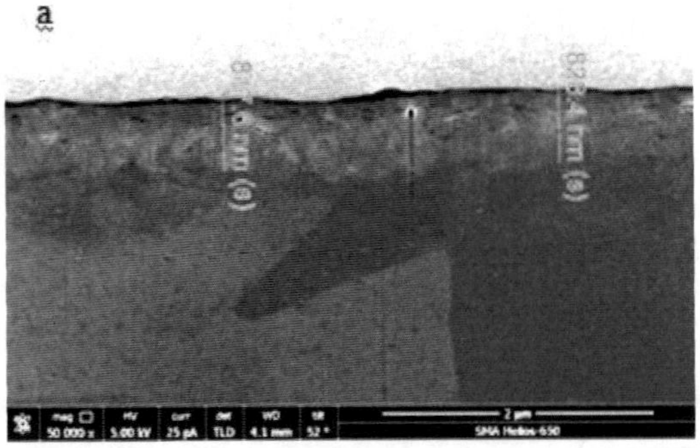

Figure 7. Cross structure photo of the oxide film after 500 hours autoclave test obtained by SIM (the samples in the initial state)

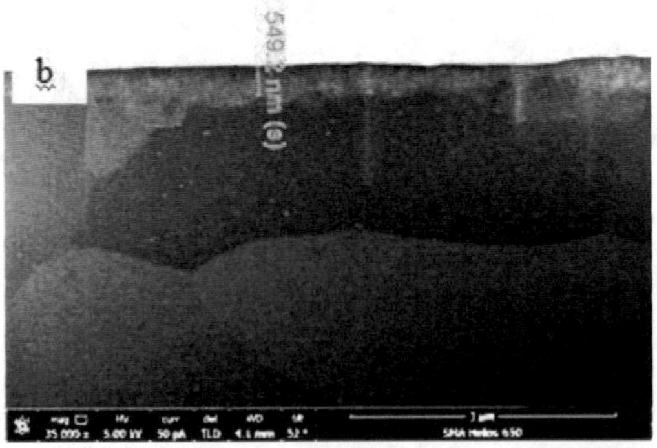

Figure 8. Cross structure photo of the oxide film after 500 hours autoclave test obtained by SIM (the samples after ion polishing)

Figure 9. Cross structure photo of the oxide film after 3000 hours autoclave test obtained by SIM (the samples in the initial state)

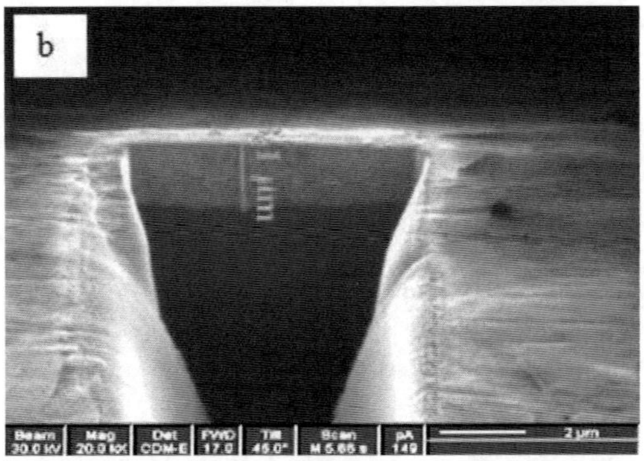

Figure 10. Cross structure photo of the oxide film after 3000 hours autoclave test obtained by SIM (the samples after ion polishing)

4. Conclusions

- The possibility of using ion polishing as a finish operation of manufacturing long cylindrical items that requires high surface quality (surface roughness achieved 0.3 ± 0.1 mm) is shown.
- The experimental automatic ion cleaning and polishing installation ILUR-03 was created. The installation provides cleaning and polishing of the long products and doping the surface layers by the ion-mixing method.
- The experimental modes that allow cleaning of the surface at a speed up to 4 meters per minute were received.
- An experimental result shows that ion treatment improves corrosion resistance by 30-40%.

References

[1] Volkov N V, Kalin B A, Valikov R A, Yashin A S and Yalcev V N 2012 *Russian Phys. J.* **55 12-3** 186-190

[2] Kalin B A, Volkov N V, Valikov R A, Yashin A S 2014 *Bulletin of the Russian Academy of Sciences: Physics* **78(6)** 553-557

[3] Volkov N V, Kalin B A, Valikov R A 2014 *Journal of Surface Investigation* **8(6)** 1161-1164

Journal of Physics: Conference Series **652** (2015) 012069 doi:10.1088/1742-6596/652/1/012069

Study of the evolution of the metal surfaces topography under ion beam impact using statistic methods

A S Yashin, R A Valikov, T V Yakutkina and N V Volkov
National Research Nuclear University MEPhI (Moscow Engineering Physics Institute) 115409, Russia, Moscow, Kashirskoe highway, 31

E-mail: yashin_itf@mail.ru

Abstract. Evolution of the state of cladding tubes outer surface from zirconium E110 alloy during ion-beam treatment was investigated using statistic methods. The samples have been irradiated on specialized installation ILUR-03 be radial Ar^+ beam with wide energetic spectrum 0.5-5.0 keV up to doses (5-10)×10^{18} ion cm^{-2}. It is shown, that profile of the ion-modified surface may be descripted by means of autocorrelation function. Average relief roughnesses size increase after ion beam treatment is found. It is revealed that there are regular as well as random roughnesses at the surface investigated. It is established, that random component of relief, caused by mechanical abrasion, decreases during ion-treatment under given conditions.

1. Introduction

One of the most effective methods to improve quality of the different products surfaces is ion beam treatment. It is well-known, that surface of material and its near surface layer may contain a lot of technological impurities, which desorbed and sprayed during ion bombardment. Under certain parameters of ion beam, the ion polishing effect occurs. The surface roughness decreases, protrusions smooths and scratches heal [1].

It is well-known, that a lot of materials properties, such as friction, corrosion- and wear-resistance, are greatly affected by topography of its surfaces. In terms of that effect, parametric description of profile roughnesses is not universe because of strong dependence of physical processes, taking place at the surface, on relief features at equal values of R_a, R_q, R_z etc. Furthermore, using standardized criteria in precise surfaces analysis is not informative because of metrological causes [2-3].

In this way using statistic methods in analysis of ion-modified surfaces seems to be promising and actual goal.

2. Materials and tools

The experiments were performed on fragments of the cladding tubes from E110 alloy (Zr-1%Nb, length – 500 mm, outer diameter – 9.15 mm). Outer surface of the specimens was previously polished mechanically and then treated on installation ILUR-03 [4] by radial beam of Ar^+ ions with wide energetic spectrum 0.5-5.0 keV up to (5-10)×10^{18} ion cm^{-2}. Dose value has been chosen to provide sputtering of 1-2 μm layer of near-surface material. Sputtered layer thickness Δr has been determined from the results of microweighting on the basis of calculation 1 mg = 0.17 μm. Microweighting was carried out on analytic weighing-machine GH-252 (A&D, Japan). Samples temperature during irradiation controlled by means of IR-pyrometers array and did not exceed 150-200 °C.

Journal of Physics: Conference Series **652** (2015) 012069 doi:10.1088/1742-6596/652/1/012069

Profilograms of the samples outer surface *h(x)* were filmed along a generator line on mechanical profilograph-profilometer TR-200 (Time Group Inc.) followed by automatic data transfer to PC. Traversing length was 1 mm, the measure step – 0.5 μm.
Summary data of the samples state are presented in table 1.

Table 1. Parameters of the samples for investigation.

#	Surface state	Δr (μm)	R_a (μm)
1	Staff (mechanical abrasion)	-	0.23
2	Ion treatment, D=5×10^{18} ion cm^{-2}	1	0.19
3	Ion treatment, D=10×10^{18} ion cm^{-2}	2	0.17

3. Profilograms processing technique

Profilograms of outer surface of the samples were processed in Microsoft Excel 2010. The straight line was used as a middle line of profile, coefficients of which were determined by the least squares method. For convenience of the surface state analysis all the features of the profile could be divided into systematic and random. Systematic component is a periodically arranged roughness of certain size that can occur as a result of regular technological factors operating. Random component is a result of the combination of a large number of irregular effects on the surface of the material during the formation of the relief. In first approximation, the random component of roughnesses can be regarded as a random stationary process. Thus, the actual surface profile can be represented as superposition of two selected components [3]:

$$h(x) = h_{syst}(x) + h_{rand}(x)\,, \tag{1}$$

where $h_{syst}(x)$ – is certain periodic function, $h_{rand}(x)$ – is stationary random function with zero mean value.
For detailed analysis of the surface state autocorrelation function $K(\tau)$ for each of profilograms has been built. Autocorrelation function is known to be one of the most informative methods to describe stationary random processes and characterizes the dependence of process values at given point x on the value of the same process at a different point $x+\tau$ with shift τ. For a limited number of discrete values, the normalized autocorrelation function can be estimated by the formula [3]:

$$K(\tau) = \frac{\sum_{i=1}^{n-1}(h(x_i) - \langle h \rangle)(h(x_i+\tau) - \langle h \rangle)}{\sqrt{\sum_{i=1}^{n-1}(h(x_i) - \langle h \rangle)^2 \sum_{i=1}^{n-1}(h(x_i+\tau) - \langle h \rangle)^2}}, \tag{2}$$

where n – is number of measurements, x_i – is coordinate of the i- measured point, $\langle h \rangle$ - is arithmetic mean of the values $h(x_i)$ measured. The $K(\tau)$ error is calculated as follows:

$$\Delta K = \sqrt{\frac{1-K^2}{n-2}}\,. \tag{3}$$

In view of the equation (1), the autocorrelation function of the profile can be represented by a superposition of individually calculated autocorrelation functions for the systematic and random components of roughnesses:

$$K(\tau) = K_{syst}(\tau) + K_{rand}(\tau)\,, \tag{4}$$

where $K_{syst}(\tau)$, $K_{rand}(\tau)$ – are autocorrelation functions of systematic and random profile components $h_{syst}(x)$, $h_{rand}(x)$ respectively. In first approximation profile systematic component can be represented by limited set of harmonic function in the form:

$$h_{syst}(x) = A\cos(2\pi x/T + \varphi), \tag{5}$$

where A – is roughness height, T – period (wavelength), φ– is random value on the interval $[0, 2\pi]$, connected with the starting point of the profilogram selection. For the random component of roughnesses description autocorrelation function corresponding to homogeneous isotropic field can be used:

$$K_{rand}(\tau) = B^2 \exp(-\tau/a), \tag{6}$$

where B^2 – is certain constant, a – is coefficient characterizing frequency content of the random roughnesses.
Thus autocorrelation functions may be approximated by the formula:

$$K(\tau) = \sum_{j=1}^{m} A_j^2 \cos(2\pi\tau/T_j) + B^2 \exp(-\tau/a), \tag{7}$$

where m – is a number of harmonic components with a wavelength T_j and amplitude A_j. Approximation parameters were determined using the Curve Fitting Tool package of MatLab R2012a program based on analysis of spectral density function of profile roughnesses, which estimated by the formula:

$$I(T) = 1/2\pi L \left| \int_0^L \exp(i\,2\pi x/T) h(x) dx \right|^2, \tag{8}$$

where L – is profilogram length, i – is imaginary unit.
Fraction of the profiles random component was estimated by the formula: [3]:

$$v = B^2 \Big/ (B^2 + \sum_{j=1}^{m} A_j^2) \times 100\%. \tag{9}$$

4. Results and discussion

There are typical profilograms from outer surface of the samples in different states and corresponding normalized autocorrelation functions in fig.1. As can be seen, the ion treatment leads to visually noticeable increase in the average size of roughnesses with a slight decrease of peaks height and dents depth (fig. 1. a, c, e). It is observed, that regular component of the autocorrelation function allocates, and profile of the surface smooths while irradiation dose increases (fig. 1. b, d, f).

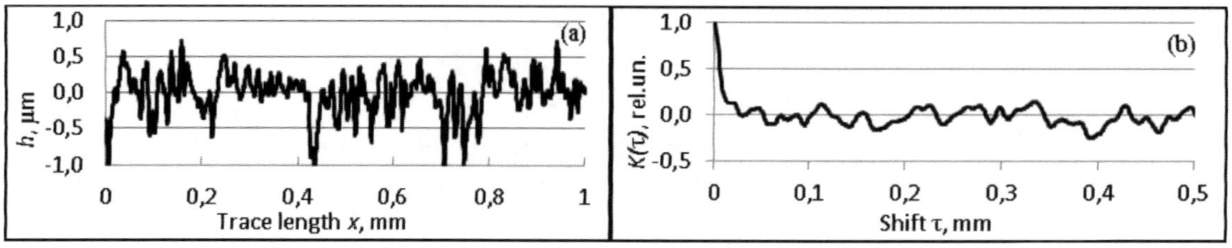

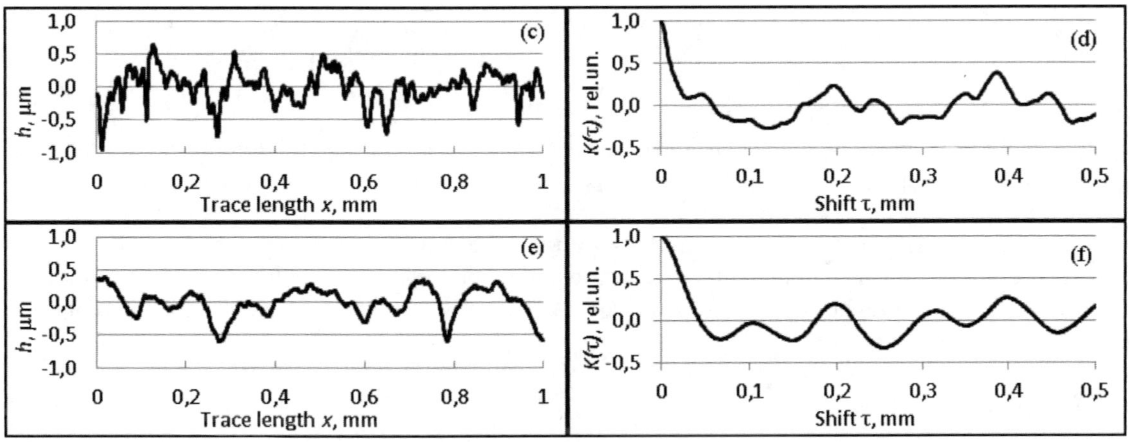

Figure 1. Profilograms (a, c, e) and autocorrelation functions (b, d, f) from the outer surface of the samples after mechanical polishing (a, b) and ion-beam treatment to doses 5×10^{18} ion cm^{-2} (c, d), 10×10^{18} ion cm^{-2} (e, f).

Functions obtained have been approximated by the form (7) for different number of harmonic components m and taking into account error in assessment of the autocorrelation functions calculated by formula (3). It was observed, the accuracy of the approximation increases from R^2=0.86-0.94 for m=0 and R^2=0.97-0.99 for m=3. Parameters of the approximation functions for selected number of harmonic components are shown in table 2. As can be seen, with increasing irradiation dose, the proportion of random component of the profile roughnesses ν decreases from 85% to 55%, and its characteristic size grows indicating increase of uniformity of the surface relief.

Table 2. Autocorrelation functions approximation parameters.

	#1	#2	#3
Surface state	Staff	5×10^{18} ion cm^{-2}	10×10^{18} ion cm^{-2}
B^2 (rel.un.)	0.94	0.70	0.57
A_1^2 (rel.un.)	0.08	0.08	0.19
A_2^2 (rel.un.)	0.06	0.16	0.16
A_3^2 (rel.un.)	0.03	0.12	0.12
ν (%)	85	66	55
a (µm)	6	10	28
T_1 (µm)	110	49	102
T_2 (µm)	278	197	174
T_3 (µm)	726	361	377
R^2 (rel.un.)	0.97	0.97	0.99

It is well-known, that mechanical polishing process causes a great number of irregular impacts because of great difference in sizes of abrasive particles, their location and pressure on the surface treated. That is why there is a big part of random roughnesses of surface of the samples in initial state. Mechanical polishing trails smooth with increase of ion beam treatment duration, because their characteristic size does not exceed the boundaries of the ion polishing maximum efficiency for the specified ion beam parameters.

Journal of Physics: Conference Series **652** (2015) 012069 doi:10.1088/1742-6596/652/1/012069

Formation of the relief regular component may be caused by some features of the tubes production technological process, such parameters as shape, size, speed, and vibration of the tool.

5. Conclusion

Evolution of the state of cladding tubes outer surface from zirconium E110 alloy during ion treatment by radial Ar^+ beam up to $(5\text{-}10)\times 10^{18}$ ion cm^{-2} has been investigated using statistic methods.
It is shown, that profile of the ion-modified surface may be descripted by means of autocorrelation function. Regular as well as random roughnesses at the surface investigated are found.
It is revealed, that ion treatment at given beam parameters leads to smoothing of the surface relief while average relief roughnesses size grows and random component of the profile, concerning with mechanical abrasion, decreases.

References

[1] Kalin B A, Krivobokov V P, Volkov N V, Yanin S N, Asainov O Kh, Valikov R A and Yashin A S 2012 *Russian Phys. J.* **55** 136-40
[2] Tabenkin A N, Tarasov S B and Stepanov S N 2007 *Roughness, waviness profile. International experience* (SPb: Polytech. Univers. Publ.) p 136
[3] Khusu A P, Vitenberg Yu R and Palmov V A 1975 *Surfaces roughness: theoretical-probability method* (Moscow: Science) p 344
[4] Volkov N V, Kalin B A, Valikov R A, Yashin A S and Yalcev V N 2012 *Russian Phys. J.* **55** 186-90

Journal of Physics: Conference Series **652** (2015) 012070 doi:10.1088/1742-6596/652/1/012070

Thermal-spikes temperature measurement in pure metals under argon ion irradiation (E = 5-15 keV)

V V Ovchinnikov[1,2], F F Makhin'ko[1] and V I Solomonov[1,2]

[1] Institute of Electrophysics, Ural Branch of Russian Academy of Sciences, Amundsena street 106, Yekaterinburg, 620016, Russia
[2] Ural Federal Technical University named after the First President of Russia B.E. Yeltsin, Mira street 19, Yekaterinburg, 620002, Russia

E-mail: viae05@rambler.ru

Abstract. A method for determining the parameters of energy release in dense cascades of atomic displacements and for estimating the temperature and the level of pressures in thermalized cascade regions (thermal spikes) is proposed. It is based on the measurement and analysis of the spectral composition of the surface glow of condensed media in the course of irradiation with Ar^+ 5-15 keV accelerated ions, under the assumption of the presence of a thermal component of glow defined by the presence of thermalized cascade regions.

1. Introduction

Many authors observed the surface glow of various substances in the visual range under the action of accelerated ion beams. The emission spectra displayed both discrete lines and continuous bands. When irradiating metals, the presence of oxides on their surface should be considered. Fortunately, these oxides are sputtered by continuous ion beams (P_s > 1 W/cm^2) for a few minutes of exposure.

None of the theoretical models proposed so far, which can be conditionally subdivided into thermodynamic, molecular, detachment, and collisional ones [2, 3], is able to explain all the peculiarities of the luminescence provoked by ion irradiation and describe its spectrum. It was found that the bands of continuous emission does not exhibits some features of luminescence.

In connection with this, to explain the presence of the continuous spectrum (continuum), it can be assumed that we are dealing with some equilibrium (in other words, thermal) or at least quasiequilibrium emission. This can be, e.g., the emission from so-called "thermal spikes" that are formed in the immediate vicinity of the surfaces of solids [4, 5] bombarded by ions of low and medium energy (1-100 keV). These spikes appear as a result of the evolution of dense (nonbranched) atomic collision cascades. This hypothesis is illustrated by Figure 1. Figure 1 also shows the possible effects of nanoscale shock wave generation by heated rapidly expanding regions of thermal spikes. This possibility is confirmed by the data below.

2. Experimental

Irradiation was performed with Ar^+ 5-15 keV ions using an ILM-1 ion beam implanter equipped with a PULSAR-1M ion source based on a low-pressure glow discharge with a hollow cold cathode [1].

The emission spectra of the targets in the course of irradiation were measured with a multichannel photodetector based on OC-12 diffraction spectrograph and a CCD array in the range from 360 to 850

Journal of Physics: Conference Series **652** (2015) 012070 doi:10.1088/1742-6596/652/1/012070

nm. We used a quartz optical fiber, the receiving end of which was set at a distance of 1 cm from the edge of the sample and was directed onto the sample surface at an angle of 60°.

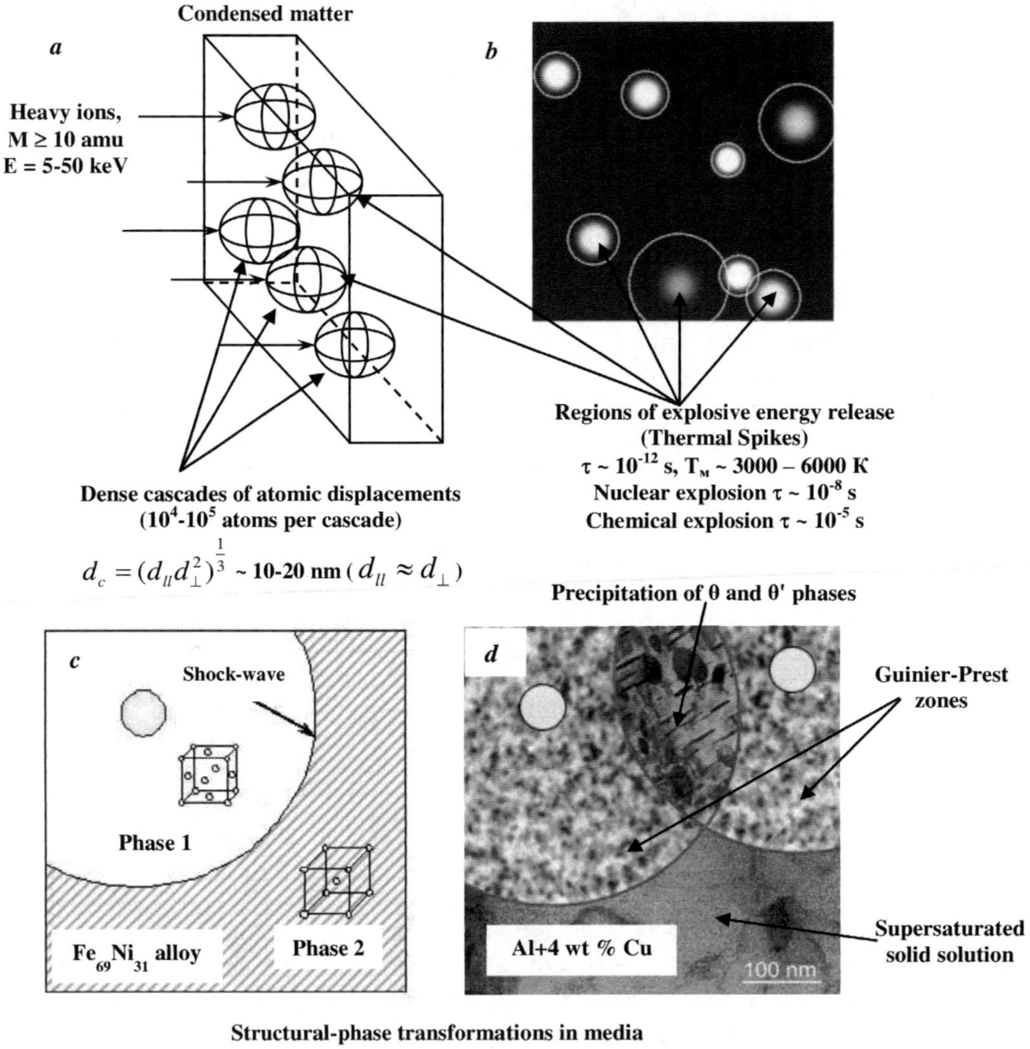

Figure 1. Radiation-dynamic effects under ion irradiation (c, d illustrate the data of [6] and [7]).

It should be noted that the depth of visible radiation penetration into metals is on the order of $\lambda/2$ (where λ is the wave length of radiation), which greatly exceeds the depth of the formation and the linear dimensions of thermal spikes.

3. Results and discussion

In order to verify the aforementioned hypothesis, we have analyzed the spectral composition of optical emission from high-purity tungsten (99.99 W), iron (99.96 Fe), and commercial aluminum (99.5 Al) bombarded by Ar^+ ions with energies E = 5-15 keV. It was of interest to compare the experimental temperatures and those calculated using the theory and computer simulations in the field of thermal spikes (as average energy per cascade atom), depending on the energy of the accelerated ions.

The emission spectra (Figure 2) displayed both discrete lines and continuous bands. The discrete emission can be generated by a fraction of atoms, ions, and molecules dissipated from the surface layers of the targets being in excited states. Any nonequilibrium emission outside a solid body must be discrete because of the lack of quasi-continuous energy bands. For this reason, none of the proposed so far theoretical models (all of them belong to the different types of luminescence) can explain all the features of this kind of emission, in particular, to describe the continuous component of the spectrum.

In accordance with the above assumption we considered that the spectra of all the targets have two broad bands of Planck thermal radiation (Figure 2), which can be described by the relation:

$$\varepsilon_{\lambda,T_1,T_2} = \varepsilon_0 \cdot \frac{2\pi hc^2}{\lambda^5}\left[\frac{a_1}{\exp(\frac{hc}{k\lambda T_1})-1} + \frac{a_2}{\exp(\frac{hc}{k\lambda T_2})-1}\right], \qquad (1)$$

where h is the Planck's constant and; c is the speed of light in vacuum; k is the Boltzmann constant; $T_1 \equiv T$ is the temperature of nano-sized areas of explosive energy release (thermal spikes); $T_2 \equiv T'$ is the temperature of integrally heated matrix (target) measured by a thermocouple, ε_0, a_1, a_2 are the constants taking into account the degree of blackness and integral sections (area) of radiating regions (thermal spikes and matrix) as well as the absorption of the surface layer and instrumental factors.

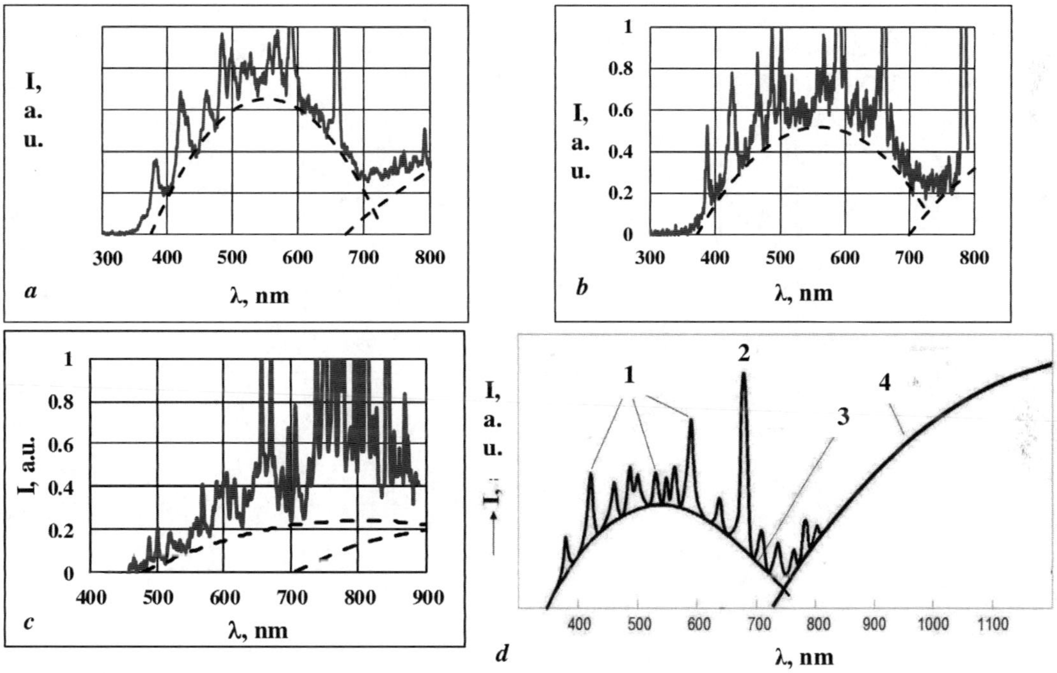

Figure 2. Emission spectra of targets bombarded by Ar^+ ions: (*a*) α-Fe, E = 15 keV, T = 5270 K, T′ = 533 K; (*b*) W, E = 15 keV, T = 5080 K, T′ = 520 K; (*c*) Al, E = 10 keV, T = 3600 K, T′ = 490 K; (d) schematic diagram illustrating formation of the emission spectra of the irradiating target.

The first band has a maximum at wavelength λ_{ml}, which varies within 500-805 nm for investigated metals, according to the hypothesis under consideration, corresponds to equilibrium Planck emission from strongly heated near-surface regions (thermal spikes) of ion-bombarded targets. The typical radius of these regions is ~3-5 nm[1]. Perhaps it can also refer to glowing of dense surface plasma

[1] The time of cascade thermalization (i.e., thermal spike formation) is ~ 10^{-12} s, while the time of its cooling is ~ 10^{-11} s. There simultaneously exist no more than 10^4 of such nano-sized "hot spots" on an area of ~ 1 cm^2 at an ion beam current density within 50-150 $\mu A/cm^2$. The average distance between them is on the order of several hundred microns.

formed during "splashing" of thermal spikes [2, 3], in which the pressure, as shown below can be up to several GPa. This may explain a glowing halo experimentally observed near the target surfaces [2].

The maximum of the second emission band at wavelength λ_{m2} falls in the IR spectral range. This band is only manifested by its short-wavelength wing that is due to thermal radiation from a target integrally heated by ion beam (λ_{m2} was determined based on thermocouple readings).

On the background of these broad bands, all spectra displayed much narrower bands of nonequilibrium emission. Some of these bands corresponded well to the emission from atoms of the bombarded metals and from argon atoms. Figure 2e shows a diagram of the formation of the observed spectral composition of the target emission under bombardment with accelerated ion beams.

In the assumption that the system attains *quasiequilibrium* state in the region of thermal spikes formed as a result of thermalization of the dense cascades of atomic displacements, using Wien's displacement law, we obtain estimations of temperature in the region of thermal spikes as a function of parameters of ion bombardment (Table 1). Estimated errors are about 50-200 K.

A decrease in the temperature of a thermal peak with an increase in ion energy (Table 1) is due to the advanced growth in the cascade volume with respect to the ion energy generating a cascade. Figure 3 shows a comparison of the calculated (using Monte-Carlo TRIM [8] and kinetic Boltzmann equation [9] methods) and experimental values of the temperature of the thermal peaks.

Table 1. Results of determination of the temperature and pressure in the region of thermal spikes in metals during their irradiation with accelerated argon ions (E = 5-15 keV)

Energy Ar^+, keV	Wave length of peak position, nm	Target tempera-ture, K	Temperature of the cascade region, K and respective estimated pressure, GPa				Experimental yield strength, GPa	Theoretical yield strength, GPa
E	λ_{m_1}	T' (exper.)	$T = b/\lambda_{m_1}$ (exper.)	P	T^* (TRIM)	P^*	σ_T^{real}	$\sigma_T^{teor} = \left(\frac{\mu}{6} - \frac{\mu}{30}\right)$
			Al					
5	620	348	4674	3.67	6007	4.73	0.05-0.120	3.46
10	805	490	3600	2.59	3526	2.54		3.46
			Fe					
5	500	343	5800	6.56	13897	15.7	0.1-0.25	9.33
10	535	388	5420	5.97	7569	8.34		9.33
15	550	533	5270	5.50	5250	5.48		9.33
			W					
5	505	328	5740	4.71	6566	5.39	0.7-1.4	21.46
10	525	413	5520	4.26	3894	3.01		21.46
15	570	520	5080	3.61	2928	2.08		21.46

As the Table 1 shows, the wavelength of the iron target that corresponds to the maximum of the Planck radiation at an energy of Ar^+ ions of 5 keV is 500 nm. In accordance with Wien's displacement law, this corresponds to T = 5800 K (b = 0.002898 mK is the Wien's constant).

It is easy to calculate that the rate of energy release in dense cascade of atomic displacements with a radius of about 3-5 nm (a form of cascade in the case of heavy ions is an ellipsoid of revolution with similar semi-axes size) during its thermalization ($\sim 10^{-12}$ s) is nearly the same as in a nuclear explosion (with a nuclear plasma at 10^8 K and characteristic time $\sim 10^{-8}$ s).

For the target made of commercially pure aluminum, the temperature in region of thermalized cascades (Figure 3, Table 1) is lower than that in heavy metals, that is consistent with the results obtained by the numerical methods [8, 9]. A large excess of the calculated data over the experimental ones for the Fe target at low argon ion energy (and, consequently, extremely small sizes of the cascade) may be due to the decisive role of the heat removal. At high energies, the opposite effect may be associated with cascade fragmentation, which is not taken into account due to averaging in simulation methods.

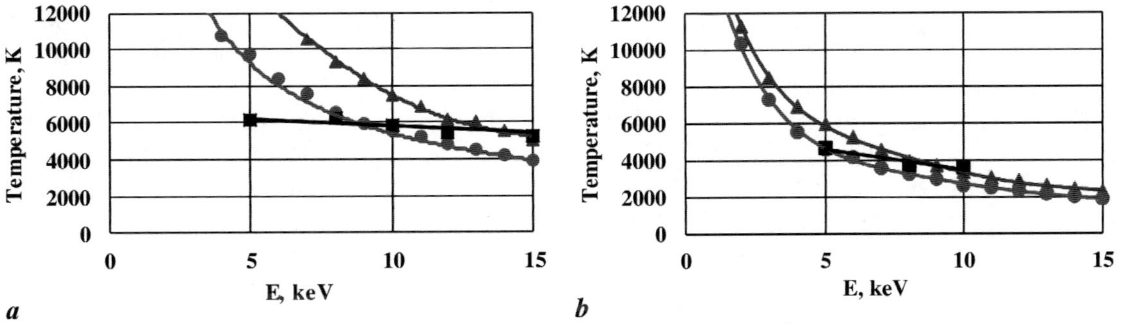

Figure 3. Theoretical (▲) [8], (●) [9] and experimental (■– from the analysis of the emission spectrum of targets) temperature of thermal spikes in pure iron (*a*) and aluminum (*b*), depending on the Ar^+ ion energy. For the calculated curves T′=300 K.

Calculated estimates of the temperature T^* in regions of thermalized cascades presented in Figure 3 and Table 1 were obtained, considering the following reasons.

In the thermalized dense cascade generated by an ion with energy E, the temperature increment (average energy per atom) is determined by:

$$E/N = 3/2 \cdot k \cdot \Delta T \quad \left(\Delta T = T^* - T'\right), \tag{2}$$

where $N = \nu \cdot N_A$ is the number of atoms in a cascade ($\nu = M/A$ is the number of moles of substance; M is the mass of the cascade, A is the atomic mass of a target); and N_A is the Avogadro's number.

Linear dimensions of a cascade ΔR_{II} and $\Delta R_\perp$ and its volume $V(E) = 4/3 \cdot \pi \cdot \Delta R_{II} \cdot \Delta R_\perp^2$ were calculated using [8] (Table 1) and [8, 9] (Figure 1) algorithms.

As a result, the following relationship can be written for calculating thermal spike temperature T^*:

$$T^* = T' + \Delta T = T' + \frac{1}{2\pi} \cdot \left(\frac{E \cdot A}{k\rho \cdot \Delta R_{II} \cdot \Delta R_\perp^2 \cdot N_A} \right), \tag{3}$$

where T′ is the initial temperature of the matrix (integrally heated target).

Estimates of the lower pressure limit in the regions of thermalized cascades (Table 1) were obtained on the basis of both experimental *T* and calculated T^* spike temperatures using the ratio [10, 11]:

$$p = \frac{E}{V}(c_P / c_V - 1). \tag{4}$$

Eexperimental *V(E)* and *P* estimates were obtained using the values of *T* (Table 1) and equations (2), (3), by substituting T^* for *T*.

Table 1 lists data on the theoretical [8] and experimental yield strengths for W, Fe, and Al. As the table suggests the pressure in the area of thermal spikes can reach values of several GPa, which is the confirmation of the probable emission of post-cascade solitary waves with the stresses at their fronts exceeding the yield strength of the material (including iron, aluminum and their alloys). Such waves may be responsible for the rearrangement of condensed matters under ion bombardment.

In [5, 6, 7], the experimental data on the rearrangement (including structural, phase, and interphase processes) of metastable media proceeding at an abnormally large depth much greater than the projected range of ions are given. As an explanation, hypotheses and models of processes self-propagating in metastable media are proposed, the rough analogy of which can serve the processes of explosion crystallization of supercooled liquids and decomposition of supersaturated aqueous solutions during impact or shaking.

Figure 1 schematically shows the processes of diffusion-free fcc → bcc phase transition in the $Fe_{69}Ni_{31}$ alloy [6] and Al-4 wt. % Cu supersaturated solid solution decomposition [7] resulting from post-cascade shock waves effects in the volume of condensed matter. In [5], such processes are referred to the dynamic long-range effects or radiation-dynamic effects. The results of this study are proof of the principal possibility of considered effects.

The probable degree of the deviation of the state of thermalized regions from equilibrium one depends on various factors. It is determined by the energy of bombarding ions, the elemental composition of targets, thermalized region sizes arisen in them and the features of their evolution. It is necessary to consider separately the electronic and lattice processes [5].

The maximum contribution to the formation of emission spectra ($\sim T^4$) is due to initial stages of the lifetime of thermal spikes.

The growth in the intensity of the IR wing of radiation correlates with an increase in the ion energy, ion current density, the irradiation fluence, and, as already noted, is associated with an integral heating of targets with ion beam (monitored via the thermocouple).

The data obtained for pure iron are in a good agreement with those of [3, 12]. The work [12] showed the presence of scattered atoms with thermal energies under irradiation of pure iron with 10-keV argon ions and using the time-of-flight techniques. The most probable velocity in the measured rate distribution, which appeared similar to the Maxwell distribution, corresponded to a thermal spike temperature of T = 5800 K. It is shown that the majority atoms with thermal energies are neutral, since it does not feel any impact from the outside imposed electric field perpendicular to the surface with a field density up to 4 kV/cm.

Important evidence in favor of a special role of thermal spikes is the fact that the continuous emission (continuum) was not observed for the other methods of surface excitation.

Gas discharge, arc, spark, laser irradiation, and gas phase collisions of accelerated particles – all these factors generate only well resolved narrow lines corresponding to the chemical composition of targets [2], whereas only ion bombardment of solid bodies induces reliably observed continuum.

4. Summary

It is shown that the continuous spectral bands in the range 400-700 nm, as part of the surface emission spectra of metal targets (Al, Fe, and W), in the course of their bombardment by Ar^+ 5-20 keV can be approximated by the Planck curves of thermal radiation. The temperatures derived from such approximation are in acceptable agreement with those simulated by Monte Carlo and Boltzmann-equation-based methods for the respective ion energies. Numerical models of the evolution of atomic displacement cascades to the stage of thermal spikes can explain the tendency to a decrease in the temperature of these spikes with increasing ion energy. In common it was shown the possibility for the direct experimental determination of the spike temperature (and, therefore, the density of energy release in cascade regions) by measuring the position of the maximum of the spectral density in the visible range. The pressures in the thermalized cascade regions in the studied metals are estimated for different ion energies.

Acknowledgments

The research was carried out within the state assignment of FASO of Russia (No. 0389-2014-0002), supported in part by RFBR (project No. 15-08-06744-A).

References

[1] Gavrilov N V and Emlin D R 2003 *Tech. Phys.* **48** 1186

[2] Pleshivtsev N V and Bazhin A I 1998 *Physic of Ion Beam Action on Materials* (Moscow: Vuz. Kniga)

[3] Ovchinnikov V V, Makhin'ko F F, Solomonov V I, Gushchina N V, Kaigorodova O A 2012 *Tech. Phys. Lett.* **38** 42
[4] Nastasi M and Mayer J W 2006 *Ion Implantation and Synthesis of Materials* (Springer Series in Materials Science)
[5] Ovchinnikov V V 2008 *Phys. Usp.* **51** 955
[6] Ovchinnikov V V 1994 Self-propagating phases transformations in metastable media induced by ion bombardment *Proc. XVI International Symposium on Discharges and Electrical Insulation in Vacuum* **2259** ed. Mesyats G A (Moscow-St.Petersburg. SPIE) pp 605-608
[7] Gushchina N V, Ovchinnikov V V, Mücklich A, Wieser E 2011 *Izv. Vyssh. Uchebn. Zaved., Fiz.* **54** 1/2 204
[8] Biersack J P and Haggmark L G 1980 *Nucl. Instr. & Meth.* **174** 257
[9] Burenkov A F, Komarov F F, Kumachov M A and Temkin M M 1985 The spatial distribution of the energy released in a cascade of atomic collisions in solids (Moscow: Energoizdat)
[10] Zeldovich Ya B and Raizer Yu P 1966 Physics of Shock Waves and High-Temperature Hydrodynamic Phenomena. Moscow: Nauka. 1966.
[11] Didenko A N, Ligachev A E, Kurakin I B 1987 *The impact of beams of charged particles on the surface of metals and alloys* (Moscow: Energoatomizdat)
[12] Schweer B and Bay H L 1982 *Appl. Phys. A* **29** 53

Journal of Physics: Conference Series **652** (2015) 012071 doi:10.1088/1742-6596/652/1/012071

Experimental and Theoretical Studies of Distribution of Vacancy Clusters in Depth of Material Irradiated by High-Energy Particles

A.A. Kupchishin[1], A.I. Kupchishin[1,2,*], N.A. Voronova[1,2], V.I. Kirdiashkin[1,2], AM Bogdan[3]
[1]Abay Kazakh National Pedagogical University, Almaty, Kazakhstan
[2] al-Farabi Kazakh National University, Almaty, Kazakhstan
[3] Tomsk Polytechnic University, Tomsk, Russia

E-mail: ankupchishin@mail.ru

Abstract. Non-destructive method of positron depth sensing of the radiation defects in solids was proposed and implemented. The depth distribution of vacancy clusters in molybdenum irradiated by protons with energy of 30 MeV was found based on the experimental data. Theoretical calculations of this value were made and sufficient agreement between calculations and experiment was obtained.

Keywords: vacancy cluster, material, radiation, particle, proton, primary knock-onatom, energy spectrum, concentration, positron, annihilation, method, energy.

1. Introduction

Due to the intensive development of nuclear power, various studies of the physical processes taking place in near-Earth space, including in radiation belts around Earth, study of the influence of electron, gamma, and ion irradiation on the properties of materials is currently one of the most important trends in radiation physics and solid state chemistry. As a result of bombardment of solids by charged particles, such as ions, not only light particles such as secondary electrons are generated (when power is lost, mainly through ionization of the atoms and molecules), but also heavy atoms and ions of the medium, which are the progenitor of the atomic nuclear cascades, can be generated along the path of their movement. Neutrons interacting with matter also form primary knock-on atoms (PKA). However, due to their great penetration capability they create radiation damage over much greater distances unlike charged particles. The energy spectrum of PKA at different depths in the material depends on the energy, the mass of the incident particles and the type of target (atomic mass, density), integral and differential cross sections of particle-atom system and atom-atom collisions, and the energy losses due to ionization and excitation. Propagating through the medium, PKAs generate secondary knock-on atoms, and then tertiary ones, etc. Ultimately, after the first stage of the process in the solid, there is a formation of radiation-induced defects such as vacancy-interstitial atom, divacancy, two knock-on atoms and so on. As a result of the passage of particles and development of diffusion processes defects interacting with each other transform into vacancy complex, clusters of interstitials. Types of defects and their distribution by size and depth of the irradiated sample depend not only on the type and energy of particle, material type, but also on the impurity concentrations, dose, temperature, uniformity and time of the irradiation, and are responsible for a change of almost all properties of the irradiated material [1]. The complexity of the processes occurring in solids under irradiation by different particles leads to considerable difficulties in the interpretation of the experimental results, especially because at the moment there is not enough complete theory of radiation damage of materials, particularly metals that form the basis of nuclear power plants.

2. The method of positron depth sensing of radiation defects

In [2], it was shown that in many experiments on the annihilation of positrons, the distribution of the concentration of radiation-induced defects in depth can be determined in defective materials by varying thickness of the absorber involving selection of the irradiated material in the form of a stack of foils. The number of foils is desirable to choose as equal to $N \geq N_0$ (N_0 ~10-15), and their thickness – from the condition of $d = x_{max}/N$, where x_{max} is a maximum depth of material where there are defects, annihilating in which positrons are still able to contribute to the spectra of photons. Since actual experiments often have to deal with whole materials (rather than with a stack of foils), there is a need for a method to determine the profiles of radiation-induced defects in materials having a greater thickness. To this end, we propose a method of defects depth sensing by high-energy positrons. The maximum depth of x_{max}, where the defects can be explored, is determined by the energy of positrons, which is desirable to be changed. The method of deep sensing of radiation defects can be carried out using both units of angular correlation (2γ, 3γ, etc. matches) and the unit of the lifetime or the Doppler effect with small changes in the patterns of these experiments. Schematic diagram of the experimental depth sensing is shown in Fig.1. It is based on the principle of gradual displacement of the annealed sample the first and then the defective sample perpendicular to the line D1 - D2 (connecting the detectors) to a relatively small distance Δx. The principle of moving the samples had previously been used by researchers to study the process of passage of positrons through materials [3]. However, our experiment has entirely different purpose - to establish the spatial distribution of defects in the depth of the irradiated sample. However, ultimately distribution (2γ-matches) of J(x) for both annealed and for defective samples is measured.

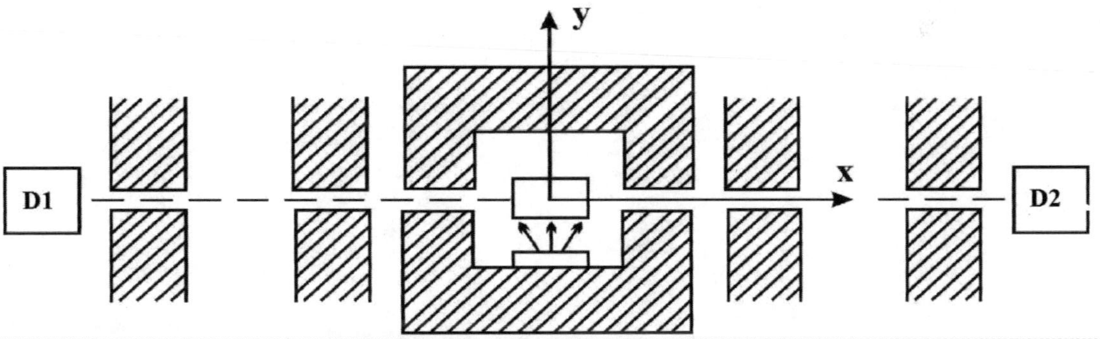

Figure 1. Schematic diagram of the defects deep sensing method

We investigated samples representing monolithic tablets of polycrystalline molybdenum with diameter of 20 mm, annealed in a vacuum of $1.3 \cdot 10^{-3}$ Pa at 1100 °C for 1 hour. The thickness of the sample corresponds to the full path of the protons with an initial energy of 30 MeV in molybdenum. Irradiation was carried out on a cyclic accelerator by the flow of $9 \cdot 10^{12}$ h/cm^2 to the integral dose of $6 \cdot 10^{17}$ particles/cm^2. The sample temperature during irradiation was maintained at 80 °C. The measurement of the angular distributions of annihilation photons was carried out using UKAF automated instrumentation with program software [3]. During this experiment, the angular resolution of the instrumentation was 1 mrad, and the pitch of the movable detector was 0.5 mrad.

Journal of Physics: Conference Series **652** (2015) 012071 doi:10.1088/1742-6596/652/1/012071

3. Experimentals

Fig. 2 shows the dependences of $J_1(x)$, $J_2(x)$ on x annealed Mo samples and samples irradiated by protons with an energy of 30 MeV to the fluence of $6{\bullet}10^{17}$ r/cm^2. It can be seen that the irradiation leads to a shift of the curve to the left, and to a substantial increase in the intensity of annihilation photons at the same values of displacement x. Analysing the change in the intensity of J(x) for annealed and irradiated states in different layers of Δx, a change in the concentration of these defects can be also studied. In addition to moving the sample, the method can be based on moving the slits with width Δx. There are other options, but the essence will not change.

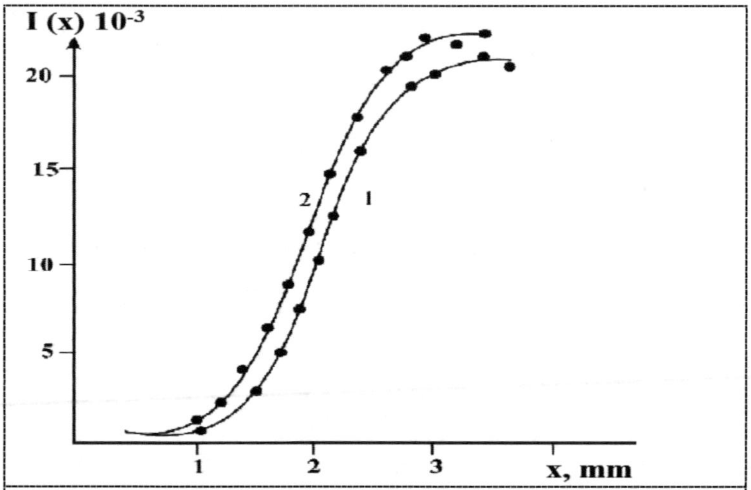

Figure 2. Dependence of annihilation photons intensity on the displacement of annealed (1) and the defect (2) samples

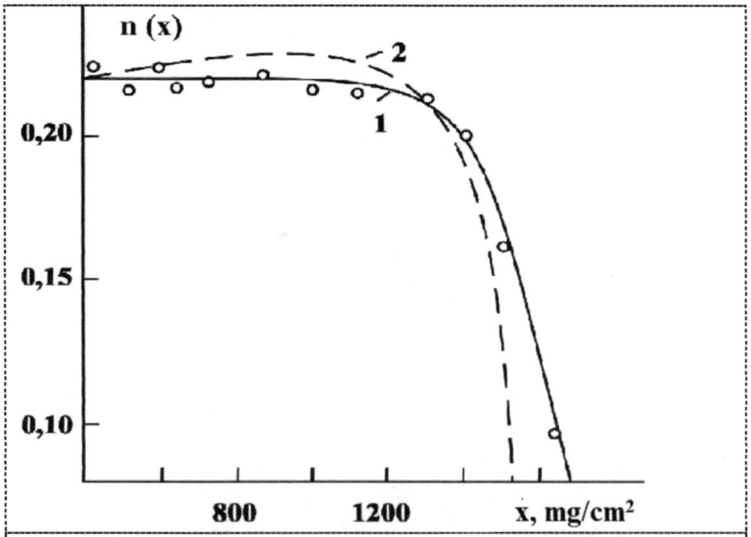

Figure 3. Dependence of the concentration on the depth of molybdenum irradiated by protons with energy of 30 MeV; 1 – experiment; 2 – calculation by CP-method at T = 100 keV

In the case of defects in the material of only one type (or by the separation of the photon flow to i-component, for example during the annealing) the following formula can be obtained [3]:

$$n(h) = \frac{(J_{ir} - J_{an})}{\Delta x} n_0, \tag{1}$$

where n (h) is the concentration of defects at a depth *h*; J_{ir} and J_{an} are normalized values of the intensity of photons in the maximum of the curve for irradiated and annealed states; n0 is distribution constant.

Dependence of the defect concentration on the depth has been calculated according to the formula (1) (Fig. 3). Fig. 3 shows that n(h) is constant up to a certain depth $h = h_{cr}$ and then sharply decreases to zero. These results are also confirmed by electron-microscopic studies. Then, using the data on positron annihilation (Fig. 4), the energy dependence of the total concentration of defects produced by proton along the track in Mo has been calculated.

Further, CP-method was used for calculation of the spectra of PKA in molybdenum irradiated by protons with energy of 30 MeV. It was assumed that between the elastic collisions, leading to the displacement of PKA, the proton loses its energy continuously (described by the formula such as the Bethe-Bloch as in the case of electrons) and after each collision it retains its direction of motion. Introducing the concept of minimum energy of the formation of cluster T_k, in the first approximation the concentration of clusters C_k (h) at the depth of h can be found from the expression [1]:

$$C_k(E_1, h) \approx \frac{1}{\bar{\lambda}(E_1)} \frac{E_d}{T_k} \left(1 - \frac{T_k}{E_d(E_1)} \right), \tag{2}$$

where $\bar{\lambda}(E_1)$ is the average value of the displacement path; p is the average value of atomic collisions along the track of the proton up to the depth of h; Ed (E_1) is damage energy.

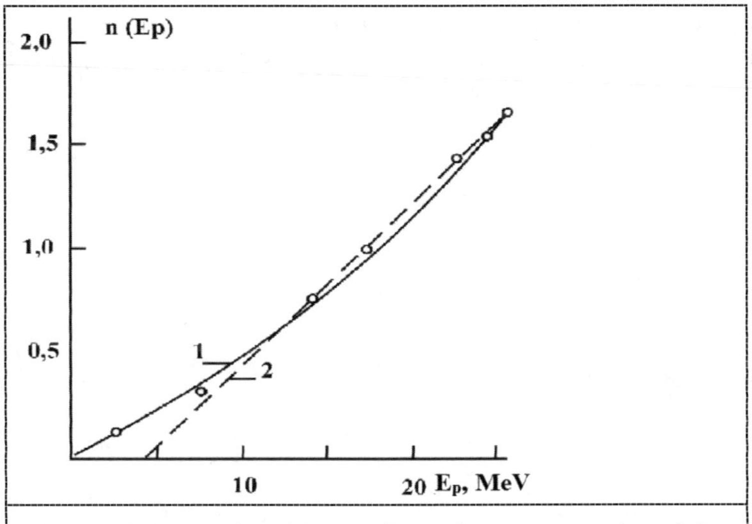

Figure 4. Power dependence of complete concentration of the defects created by protons along all the track in Mo
1 – experiment; 2 – calculation by CP-method at T = 100 keV

According to this formula at T = 100 keV, spatial distribution of vacancy clusters and energy dependence of the total concentration of defects produced by protons along the track in Mo (Fig.4, dashed line 2) were calculated. It is clear that C_K (h, E_1) is essentially constant and then changes

Journal of Physics: Conference Series **652** (2015) 012071 doi:10.1088/1742-6596/652/1/012071

abruptly at the end of the track. Fig. 4 shows the energy dependence of C_K (E_1) for Mo calculated by the formula (2). Comparison of calculations with experimental data on positron annihilation shows that there is a satisfactory agreement for both the spatial distribution of C_K (h, E_1) (Fig.3) and the energy dependence of C_K (E_1) (Figure 4, dashed curve 2).
Thus, the expression for C_K (h, E_1) found by CP-method in the first approximation explains the experimental data on positron annihilation and can be used by researcher in their work.

4. Conclusion

1. Non-destructive method of positron depth sensing of radiation defects in solids was developed. Distribution of intensities of gamma rays directly in the annihilation zone was experimentally measured for irradiated (by the protons with energy of 30 MeV) and non-irradiated samples of molybdenum. These data and the proposed method were used for studying the distribution of vacancy clusters in the depth of the material.

2. Cascade-probabilistic method was used for calculation of C_K (h), and a satisfactory agreement with experiment was obtained. The value of the concentration of vacancy clusters initially hardly changes with depth, and then abruptly falls to zero at the end of the path.

Acknowledgement
This study was financially supported by grants of the Ministry of Education and Science of Kazakhstan Republic (a. N 32, 349).

References

[1] Boos E G, Kupchishin A I, Kupchishin A A 2015 *Cascade-probabilistic method. Solution of radiation-physical problems, Boltzmann equations. Connection with Markov chains.* (Monograph, Almaty, Abay KazNPU, al-Farabi KazNU) [In Russian)]

[2] Ibragimov S S, Mukashev K M, Kupchishin A I et al. 1981 Study of the energy dependence of radiation damage of molybdenum irradiated by protons by the method of positron annihilation. *Issues of atomic science and technology. Series: Physics of radiation damages and radiation materials*. Kharkov. Publ. house KhPTI, Is. **3** (17) pp. 3-4. [In Russian]

[3] West R H 1973 Positron studies of condensed matter, *Advances in Physics,*. **23** **3** 263-283

Investigation of Mechanical Properties of Polyimide-Polyethyl-eneterephthalate System

A.I. Kupchishin[1,2]*, B.G. Taipova[1], V.M. Lisitsyn[3], A.A. Kupchishin[1], N.A. Voronova[1,2], V.I. Kirdiashkin[1,2]

[1]Abay Kazakh National Pedagogical University, Almaty, Kazakhstan
[2] al-Farabi Kazakh National University, Almaty, Kazakhstan
[3] Tomsk Polytechnic University, Tomsk, Russia

*E-mail: ankupchishin@mail.ru

Abstract. The study of the physico-mechanical properties of new composite materials based on polyimide (PI) and polyethyleneterephthalate (PET) was carried out. Dependences of σ from ε for different concentrations of the second component and irradiation doses were measured. The experimental data are described satisfactorily by the exponential cascade-probabilistic model.

Keywords: mechanical properties, composite material, polyimide, polyethyleneterephthalate, strain, irradiation, plasma particle flux, strength.

1. Introduction

Currently, much attention is paid to the development of new thin-film materials [1]. Examples of their use are elements of integrated circuits in modern micro- and nanoelectronics, coatings (protective, hardening, semi-transparent, dielectric, magnetic, and so on). Depending on the function of material its thickness varies widely – from a few angstroms to several tens of micrometers [2].

Due to the high requirements of technology, modification of polymers has become predominated, because it allows creating new compositions for special purposes, which form a system with the specified properties: specific electrical conductivity, strength, ductility, resistance to radiation, high and low temperatures [3]. The multifunctional material which combines a number of useful consumer properties is obtained by introducing different active compounds into the polymer matrix [4].

Studying the influence of a certain type of filler on physico-mechanical properties of polymeric composite materials (PCM) is not only of scientific but also of practical interest [5]. Introduction of additives alters the structure and properties of the system. The interaction of particles at the polymer-filler interface is the basis of the mechanism of modification of PCM characteristics. Strength and plastic properties of polymer composites are significantly affected by the fillers used and their concentrations. Irradiation of different systems by a stream of plasma particles, accelerated ions and electrons leads to changing their characteristics.

Polymer composites consist of two main parts: a binder polymer and a reinforcing filler. In polymer mixtures hard-chain polymers stimulate the orientation of macromolecules with a flexible chain or amorphous flexible chain polymers lead to disordering of crystalline polymers and create new possibilities for controlling their properties. Due to the heat resistance, polyimide composite materials are used as structural materials in main branches of industry. As it is known, polyimides (PI) are cyclochain heterocyclic polymers containing macromolecules of cyclic imide groups in the main or side chains. PI have high thermal stability, extremely high resistance to effect of plasma flows and radiation, as well as high strength and flexibility. Polyethylene terephthalate was selected as the filler.

Polyethylene terephthalate is a polycondensate of terephthalic acid (OH)-(CO)-C6H4-(CO)-(OH) and monoethyleneglycol (OH)-C2H4-(OH). During polycondensation, linear molecule of polyethylene terephthalate [-O-(CH2)2-O-(CO)-C6H4-(CO)-]n and water are formed. Polyethylene terephthalate is durable, hard and light material.

2. Experimental

Dependence of stress σ on deformation ε of new composite materials (based on PI and PET) (irradiated and non-irradiated) under uniaxial stress for several concentrations of the second component (polyethylene terephthalate - PET) was studied.

Composite films were prepared according to the scheme shown in Fig.1. Current technologies [1] were used for preparation of composites as a mixture of PI+PET with the following concentra-tions of filler: 0.5; 2 and 5 wt.%.

Mechanical testing of the materials was carried out on a tensile testing machine of RMU-0.05-1 type with the rate of clamps moving of 36.09 ± 0.05 mm/min. Capture moving associated with measuring does not exceed 0.1 mm. Tests were conducted as uniaxial stretching with a special reverser at a constant temperature (20 ± 2) °C and relative humidity (45 ± 5)%. Tensile testing machine was equipped with appropriate software – standard Windows application.

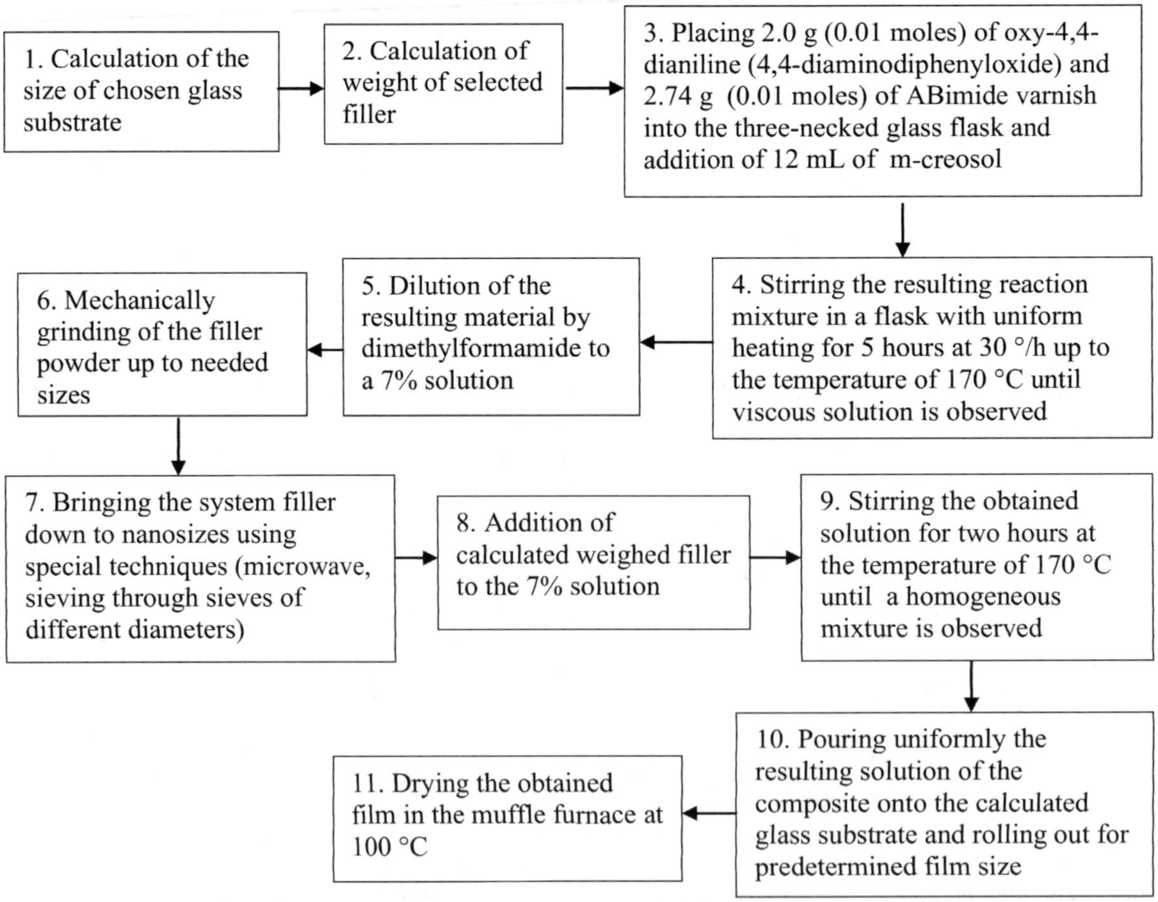

Figure 1. Technological scheme of the composite based on polyimide

Samples of the films were fixed with special clamps. Workspace was 50 mm (length), 5 mm (width), and thickness - 35 microns for polyimide and (40 ÷ 75) mm for composite materials.
Irradiation of samples was carried out in the air on ELU-6 electronic linear accelerator with the energy of the incident particle of 2 MeV to the absorbed dose (D) = 250 kGy.

3. Results and discussions

Introduction of 0.5 wt.% of PET increases the strength of the material by 37 % and elongation by 100% compared with the pure polyimide. Improvement of the mechanical properties of composite materials with a filler of PET is due to the fact that they were obtained by stirring the polyethylene terephthalate and polyimide, and filler particles fill the pores of the matrix fixing the molecules of lavsan and polyimide i.e. structuring of polyimide chains by plastic skeleton of polyethylene terephthalate occurs. Simultaneous and uniform heating to 170 °C promotes cross-linking of the radicals of polyethylene terephthalate with polyimide macromolecules. Increase in concentration of filler C and stress σ increases the deformation ε. So for C = 5 wt.% and σ = 70 MPa, ε was increased 2 times as compared with C = 0.5 wt.% (Table 1).

Table 1. Dependence of the relative lengthening on the concentration of the second component

σ =10 [MPa]		σ =30 [MPa]	σ =50 [MPa]	σ =60 [MPa]	σ =70 [MPa]
C [wt.%]	ε [%]	ε [%]	ε [%]	ε [%]	ε [%]
0	6	17.9	34.5	-	-
0.5	7.5	20.2	28.1	47.2	75.1
2	6.8	18.7	27	51.6	80.8
5	8	22.8	32.6	80.8	204.4

Effect of electron irradiation of 250 kGy on the composite (filled with 0.5 wt.% PET) increases strength by 17 % and relative extension by 23 % in relation to the original value of the pure polyimide. In comparison with non-irradiated polyimide, the impact of these factors leads to improvement of the mechanical properties of composite materials. This is explained by additional polymerization of the polymer matrix and cross linking of macromolecules that occurs during irradiation. Filler as lavsan forms a homogeneous composite with polyimide and enhances its mechanical properties.

Exponential model was used to describe the physico-mechanical properties based on the balance equation within the cascade-probabilistic method under uniaxial stress of materials. In particular, the dependence of the deformation (ε) of the stress (σ) for the case of the exponential model is represented as [6]:

$$\varepsilon = exp\left(\frac{\sigma}{\sigma_0^*}\right) - 1,$$

or, after conversion:

$\sigma = \sigma_0 * \ln(\varepsilon+1)$,

where σ_0 is generalized strength module.

Fig. 2 shows the stress-elongation dependence for non-irradiated (a) and irradiated (b) composite samples for C = 0.5 wt.% and D = 0 and 250 kGy.

Comparing the theoretical (calculated by the exponential model, curve 3) and experimental (curve 2) curves, a good agreement can be observed. The same figure shows also the theoretical curve calculated by Hooke's law (curve 1). It can be seen that at maximum elongation Hooke curve odds with experiment by 50 %. For irradiated composite, the difference reaches 68 %.

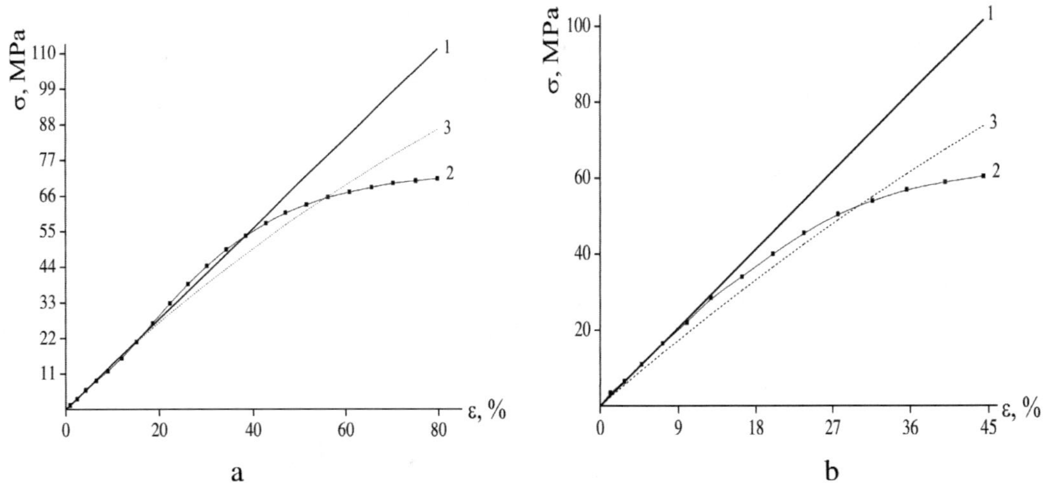

Figure 2. Dependence of the stress on the relative elongation for the composite material of PIAB + 0.5 wt.% PET: a) D = 0; b) D = 250 kHr
1 – Hooke's law, 2 – experiment, 3 – calculation by cascade-probabilistic model

4. Conclusion

1. It was found that the mechanical properties of composite film materials based on polyimide and polyethylene terephthalate before and after irradiation depend on the concentration of the filler and are due to the features of internal structure.
2. It was shown that the exponential model gives the best agreement with experiment. At the same time, calculation according to Hooke's law at a maximum ε overstates σ by 50-70 %.

Acknowledgement

This study was financially supported by grants of the Ministry of Education and Science of Kazakhstan Republic (a. N 32, 349).

References

[1] Kudaikulova S K, Iskakov R M, Kravtsova V D, Umerzakova M B, Abadie M, Zhubanov B A 2006 Polymers of special purposes (Almaty) [In Russian]
[2] Matrenin S V, Ovechkin B B 2008 Composite materials and coatings based on polymers: Textbook (Tomsk: TPU) [In Russian]
[3] Kerber M L, Vinogradov V M, Golovkin G S, Gorbatkina Y A, Kryzhanovsky V K, Kuperman A N, Simonov-Yemelyanov I D, Khaliulin V I, Bunakov V A 2008 Polymeric composites: structure, properties, technology: Textbook (Saint Petersburg, SPb: Professiya) [In Russian]
[4] Smirnova V E, Gofman I V, Ivan'kova Y M, Didenko A L, Krestinin A V, Zvereva G I, Svetlichnyi V M, Yudin V Y 2013 Polymers 55 5 427-437 [In Russian]
[5] Panova L G 2010 Fillers for polymeric composite materials: Textbook (Saratov: Saratov state technical university) [In Russian]
[6] Kupchishin A I, Muradov A D, Taipova B G, Voronova N A 2009 Eurasian Physical Tehnical Journal 6 1(11) 16-22